BRICC

丝路国际产能合作促进中心

The Silk Road Promotion Center for International Production Capacity Cooperation

国家重点研发计划资助（编号：2016YFC0700200）

National Key Research and Development Program Funding

“一带一路”建筑设计系列丛书

A“Belt and Road” Series for Architecture Design

医疗保健设施工程咨询与设计手册

Manual for Consultation and Design of Health Care Facilities

（中英文对照）

(Chinese and English)

朱希　主编

Xi Zhu　Editor in Chief

深圳市联合创艺建筑设计有限公司　组织编写

Shenzhen United Architecture Design Co., Ltd. Organize to Write

中国建材工业出版社

图书在版编目（CIP）数据

医疗保健设施工程咨询与设计手册：汉英对照/朱希主编；深圳市联合创艺建筑设计有限公司组织编写. --北京：中国建材工业出版社，2020. 10

（“一带一路”建筑设计系列丛书）

ISBN 978-7-5160-3069-1

Ⅰ. ①医…　Ⅱ. ①朱…　②深…　Ⅲ. ①医疗保健—民用建筑—建筑设计—咨询服务—手册—汉、英　Ⅳ. ①TU246-62

中国版本图书馆 CIP 数据核字（2020）第 181149 号

内容简介

本书对医疗保健设施进行了系统分类，除介绍其基本功能构成、建设特点、建设标准与规范外，还介绍了此类设施建设项目在策划、可行性研究、建筑设计、医用专项系统设计、设计管理与设计总承包方面的核心知识。全书内容言简意赅，可供读者进行专业法规查询、咨询与设计参考、工作方法检索等。

本书适合从事医疗保健设施的投资、咨询、设计、管理人员阅读，也可作为高等院校建筑及其相关学科的辅助教学资料。

医疗保健设施工程咨询与设计手册

Yiliao Baojian Sheshi Gongcheng Zixun yu Sheji Shouce

朱 希　主编

深圳市联合创艺建筑设计有限公司　组织编写

出版发行：中国建材工业出版社

地　　址：北京市海淀区三里河路 1 号

邮　　编：100044

经　　销：全国各地新华书店

印　　刷：北京鑫正大印刷有限公司

开　　本：787mm×1092mm　1/16

印　　张：29

字　　数：650 千字

版　　次：2020 年 10 月第 1 版

印　　次：2020 年 10 月第 1 次

定　　价：198.00 元

本社网址：www. jccbs. com，微信公众号：zgjcgycbs

本 书 编 委 会
Editorial Committee

出品人： 廖志斌

Producer: Zhibin Liao

出品助理： 丘志强

Production Assietant: Zhiqiang Qiu

主编： 朱　希

Editor in Chief: Xi Zhu

审稿： 张伯林

Reviewer: Bolin Zhang

编委人员：

Editorial Committee:

廖志斌　朱　希　张伯林　徐小田　孟庆林　赵立华　丘志强

Zhibin Liao, Xi Zhu, Bolin Zhang, Xiaotian Xu, Qinglin Meng, Lihua Zhao, Zhiqiang Qiu

徐怀平　范美态　谭晚芳　罗　红　骆　艳　张振华　袁　清

Huaiping Xu, Meitai Fan, Wanfang Tan, Hong Luo, Yan Luo, Zhenhua Zhang, Qing Yuan

汤　琼　叶少玲　许伟忠　潘秋兰　张　伟　张旭明　张颖慧

Qiong Tang, Shaoling Ye, Weizhong Xu, Qiulan Pan, Wei Zhang, Xuming Zhang, Yinghui Zhang

韩艳红　林建刚　南伟力　罗新松　袁永久　黄学勋　韩广智

Yanhong Han, Jiangang Lin, Weili Nan, Xinsong Luo, Yongjiu Yuan, Xuexun Huang, Guangzhi Han

高德华　胡　健　徐保鹏　陈业军　段留栓　柳旭良　代　路

Dehua Gao, Jian Hu, Baopeng Xu, Yejun Chen, Liushuan Duan, Xuliang Liu, Lu Dai

参编单位：

Coeditors:

深圳市联合创艺建筑设计有限公司

Shenzhen United Architecture Design Co. , Ltd.

深圳市柏鹏建筑设计事务所有限公司

Shenzhen Baipeng Architects &Asociates Co. , Ltd.

华南理工大学亚热带建筑科学国家重点实验室

South China University of Technology State Key Laboratory of Subtropical Building Science

广东省建筑节能协会

Guangdong Building Energy Conservation Association

中机十院国际工程有限公司

Sinomach TDI International Engineering Co. , Ltd.

深圳协康实业发展有限公司

Shenzhen Xiekang Industry Development Co. , Ltd.

深圳市辰普森信息科技有限公司

Shenzhen CPSON Information Consulting Co. , Ltd.

深圳市得益节能科技股份有限公司

Shenzhen DEE Energy Saving Technology Co. , Ltd.

深圳市尚筑建筑设计有限公司

Shenzhen Sunture Architectural Design Co. , Ltd.

深圳市森磊镒铭设计顾问有限公司

Shenzhen Sldic International Design Consultant Co. , Ltd.

诺尔国际设计（深圳）有限公司

Noel International Design (Shenzhen) Co. , Ltd.

广东宏拓项目管理有限公司

Guangdong HongTuo Project Management Co. , Ltd.

序　一
Preface 1

助力医疗保健设施建设，携手共筑健康丝绸之路
Support the construction of health care facilities and join hands to build a healthy Silk Road

自 2013 年国家主席习近平提出“一带一路”倡议至今已有近七年时间，共建“一带一路”从总体布局的“大写意”，转入精谨细腻的“工笔画”，全面推进政策沟通、设施联通、贸易畅通、资金融通、民心相通，同联合国、东盟、非盟、欧盟、欧亚经济联盟等国际和地区组织的发展和合作规划对接，同各国发展战略对接。其共商、共建、共享的和平发展、共同发展理念获得许多国家的共鸣与认可，为世界经济增长注入了新动力，为国际贸易和投资搭建了新平台，为完善全球经济治理拓展了实践渠道，并致力于为增进各国民生福祉作出贡献，成就共同的机遇之路、繁荣之路。

It has been nearly seven years since President Xi Jinping put forward the “Belt and Road Initiative (BRI)” in 2013. The past years have seen the transformation of BRI construction from “freehand brushwork” in the overall layout to “fine brushwork” in meticulous detail. Now comprehensive efforts are being made to promote policy communication, infrastructure connectivity, free trade, financial integration and friendly exchanges between the people, aiming at integrating with the development and cooperation plans of international and regional organizations such as the United Nations (UN), the Association of Southeast Asian Nations (ASEAN), the African Union (AU), the European Union (EU) and the Eurasian Economic Union (EEU), as well as the development strategies of different countries. Advocating peaceful and common development based on the principles of extensive consultation, joint contribution and shared benefits, BRI has aroused resonance and recognition from many countries, and has injected new impetus into the global economic growth, set up a new platform for international trade and investment, and expanded the practice channels to improve global economic governance. Moreover, it is also committed to contributing to the well-being of the people of all countries and to building a common path to opportunity and prosperity.

2020 年是不同寻常的一年，突如其来的新冠肺炎疫情给中国和世界经济发展带来

前所未有的冲击，多国卫生系统面临不能承受之重，人际交流被阻隔，社会治理面临挑战……新冠肺炎疫情防控局势促进了加强全球公共卫生合作的紧迫性，也再次彰显了推进“一带一路”合作、构建人类命运共同体的重要意义。中国在抗击疫情过程中采取的果断举措、取得的积极成效、展现的必胜信念，得到国际社会普遍赞誉。同时，中国应急医疗保障设施建设的能力举世瞩目，成就斐然，创出了一条中国特色的建设之路。为了弘扬国际人道主义精神，中国的医疗建筑设计师们也在积极为其他疫情国家提供力所能及的支持和帮助，迅速编撰用于指导医疗保健设施建设的应用技术手册，助力“一带一路”上的卫生事业发展。

The year 2020 is an unusual year. The sudden outbreak of COVID-19 has brought unprecedented impact on the global economy, including China, resulting in a series of problems such as overburden on the health systems of many countries, obstruction in interpersonal communication, challenge in social governance...The prevention and control of COVID-19 has highlighted the urgency of strengthening global public health cooperation, and has once again manifested the significance of promoting the BRI cooperation and building a community with a shared future for mankind. The decisive measures taken by China to fight against the epidemic, the positive results achieved and the triumphalism exhibited by Chinese people have been widely praised by the international community. At the same time, China has drawn worldwide attention by virtue of its ability to build emergency medical security facilities and the great achievements made, which have opened up a road of construction with Chinese characteristics. In order to carry forward the international humanitarian spirit, Chinese medical architects are also actively providing support and assistance to other countries attacked by the epidemic, including compilation of applied technical manuals to guide the construction of health care facilities and assist the development of health services of BRI countries.

危中有机，不破不立。在抗击疫情的艰苦斗争中，各国人民对医疗健康的意识显著提高，而“一带一路”沿线国家医疗保健设施无论是在数量上还是质量内涵上都蕴含着巨大的市场需求。中国医疗保健设施工程设计与建设企业应抓住机遇，帮助沿线国家建立高质量的医疗保健设施，并把来自中国的设计技术、设计管理与品牌、建设标准带进更多“一带一路”沿线国家，提高中国在地区性、全球性卫生多边治理机制中的影响力。

Opportunity always comes with challenge. In the arduous struggle against the epidemic, people' s awareness of health care has enhanced significantly, while BRI countries also contain huge market demand for health care facilities in terms of both quantity and quality. Chinese enterprises engaged in the design and construction of health care facilities should seize this opportunity to help BRI countries establish high-quality health care facilities and bring the design technologies, design management and brands and construction standards from China into more

BRI countries, thus enhancing China' s influence in regional and global multilateral health governance mechanisms.

医疗保健设施建设水平是一个国家民生质量的重要反映。随着基础医学研究的不断深入、医疗技术手段的日新月异以及医疗诊断和治疗设备的更新换代，作为医疗卫生服务的载体——医疗保健设施的设计建造也在不断完善。而医疗保健设施在满足医疗功能和卫生学要求的前提下，还需满足人性化、智能化、环保化的要求，可以说是所有民用建筑中最为复杂和特殊的建筑类型之一，也正是这些复杂性和特殊性使医疗建筑区别于其他建筑，形成了独特的专业设计领域。

The construction level of health care facilities is an important manifestation of the quality of people' s livelihood in a country. With the intensification of basic medical research, the rapid development of medical technology and the upgrading of medical diagnosis and treatment equipment, the design and construction of health care facilities, as the carrier of health care services, are also under constant perfection. Health care facilities are regarded as one of the most complex and special building types among all civil buildings, because they need to satisfy the requirements of humanization, intelligentization and environmental protection on the premise of meeting medical purposes and hygiene requirements. Such complexities and particularities distinguish health care facilities from other buildings and make them a unique field of professional design.

作为本书的出品人，廖志斌先生长期致力于建筑工程设计领域的相关研究工作，其创立的深圳市联合创艺建筑设计有限公司更是多年从事医疗建筑设计与建设专业化发展，不断探索行业发展需求，努力提升设计品质和服务质量。本书的出版，将有助于行业相关人士在融会贯通医疗功能、科室体系、就诊流程等知识的基础上，设计功能齐全、安全可靠、节能降耗、环境友好的医疗保健设施，并对医疗保健设施的咨询与设计水平提升和标准设立起到积极的推动作用。

As the producer of this book, Mr Liao Zhibin has long devoted himself to the research on construction engineering design. The Shenzhen United Architecture Design Company he founded has been engaged in professional medical building design and construction for many years, and it is still constantly exploring the development requirements of the industry to improve the design quality and service quality. This book is expected to help industrial professionals design fully functional, safe, reliable, energy-saving and environment-friendly health care facilities on the basis of comprehensive knowledge of medical functions, department systems and treatment process. It will also play a positive role in promoting the level of consultation and design and the establishment of standards related to health care facilities.

大道同行！“一带一路”倡议不仅仅关注传统能源和基建领域的合作，在医疗健康领域的发展也大有可为。愿医疗保健设施工程设计企业不断提升自身本领，同时本着命运与共的情怀，做健康丝绸之路的建设者和合作发展的推动者，与沿线国家在医疗卫生领域深化合作，携手应对公共卫生安全挑战，破解各种世纪性难题，朝着构建人类命运共同体目标向前迈进。

A promising road should be open to all! BRI focuses not only on cooperation in traditional energy and infrastructure, but also on the development of health care. I hope that enterprises for engineering design of health care facilities will continue to improve their own abilities, and at the same time, act as builders of a healthy Silk Road and agents of cooperative development in the spirit of shared destiny to deepen cooperation with BRI countries in health care, work together to deal with the challenges to public health security, solve all kinds of long-standing problems, and move forward towards the goal of building a community with a shared future for mankind.

谢阳军

Yangjun Xie

丝路国际产能合作促进中心主任

Director of the Silk Road Promotion Center for International Production Capacity Cooperation

2020 年 5 月于北京

Beijing, May, 2020

序　二
Preface 2

欣闻《医疗保健设施工程咨询与设计手册》即将付梓，向所有参与编辑和付出努力的同仁们致以崇高的敬意和热烈的祝贺！

I am glad to hear that Manual for Consultation and Design of Health Care Facilities is about to be delivered for publication. I would like to express the sincere respect and warm congratulations to all editors and colleagues who worked hard.

在快节奏的社会背景下，中国的健康产业发展潜力巨大，建筑行业的技术进步对推动医疗保健行业的发展更是发挥了不可替代的作用，就医疗建筑而言，它设计的好坏直接影响到为病患提供医疗保健服务的质量，关系到人们的健康与安全。也正因如此，它在现代化、专业化、精细化设计方面有着极高的要求。尤其进入21世纪以来，随着生活质量的普遍提升、医疗科技的突飞猛进、医院管理模式的日新月异，人们对社会医疗保健的需求越来越高，越来越多元化，由此引发了现代医院体系和医疗制度的一系列变革，进而带来了医疗建筑从设计理念到功能布局、空间设计以及工程建造、项目管理等的全方位改变。

Under the fast-paced social background, China's health industry has great potential for development, and the technological progress of the construction industry has played an irreplaceable role in promoting the development of the health care industry. As far as medical buildings are concerned, the quality of their design directly affects the quality of health care services provided to patients and is related to people's health and safety. For this reason, there are extremely high requirements for medical buildings in terms of modernization, specialization and detailed design. Especially since the beginning of the 21st century, with the general improvement of the quality of life, the rapid development of medical technology, and the never-ending changes and improvements of hospital management mode, people's demand for social medical care has become higher and higher and more diversified, which has led to a series of changes in the modern hospital system and medical system, thus bringing about all-round changes in medical buildings from design concept to functional layout, space design, engineering construction, project management, etc.

医疗建筑设计怎样才能满足持续不断变化的医疗技术的需要？怎样在保障医疗效果的同时，拓展智慧化、多元化服务？怎样更智能、更人性、更安全，是摆在建筑设计师面前的问题。

How can medical buildings meet the needs of constantly changing medical technologies? How to expand intelligent and diversified services while ensuring medical effects? How to be more intelligent, more humane and safer? These are problems facing architectural designers.

20 世纪 90 年代至今，我国经历了一轮又一轮的医疗设施建设高潮，新建、改扩建项目众多，可持续设计、一体化建设等新的理念与模式也随之涌现。对此，一批专家、学者，基于实际项目的经验总结，就医疗保健设施建造技术展开了研究。但是，就目前现状而言，研究的广度和深度仍然有限，比如，针对医疗保健设施建设的前期可行性研究、设计管理模式及环境工程等方面的研究就比较缺乏，而且，关于医疗建筑设计的专项性分析偏多，系统性分析研究偏少。更为深远的课题还包括医患共居空间的物理环境质量对免疫力的影响等。鉴于医疗建筑的专业复杂性，关于医疗保健设施建造技术的研究未来还有很长一段路要走，也需要更多理论型、实践型人才的共同投入。

Since the 1990s, China has experienced rounds of medical facility construction upsurge, with numerous new construction, renovation and expansion projects, and new concepts and modes such as sustainable design and integrated construction have emerged. In response, a group of experts and scholars, based on the experience of practical projects, have studied the construction technology of health care facilities. However, as far as the current situation is concerned, the breadth and depth of the research are still limited. For example, the research on the preliminary feasibility research, design management mode and environmental engineering for the construction of health care facilities is relatively lacking. Moreover, there are more special analysis and less systematic analysis and research on the design of medical buildings. More far-reaching topics include the impact of the physical environment quality of the doctor-patient shared space on immunity and other aspects. In view of the professional complexity of medical buildings, there is still a long way to go in the research on the construction technology of health care facilities, and more talents with theoretical knowledge and practical experience are also needed for joint research.

《医疗保健设施工程咨询与设计手册》一书便是在这样的社会背景和行业需求下出版问世。本书融合了业内权威及医疗建筑设计与运营管理资深专业人士的经验成果，通过编者对医疗保健设施在策划与管理、设计与营造、技术与创新等多个方面的思考与研究，在适当注重理论介绍与探讨的同时，更注重实践和经验的总结，内容丰富、观点鲜明、文字简洁，非常值得一读。

Manual for Consultation and Design of Health Care Facilities is published under such social background and industry demands. This book includes the experience and achievements of the industry authorities and senior professionals in medical building design and operation management. Through the editors' thinking and research on the planning and management, design and creating, technology and innovation of health care facilities, while paying due attention to the introduction and discussion of theories, it also lays more stress on the summary of practice and experience. It is rich in content, clear in viewpoint and concise in writing, and is worth reading.

相信本书的出版，将为今后医疗保健设施建设的理论研究与实践操作提供很好的参考并发挥启示作用。

I believe the publication of this book will provide great reference and enlightenment for the theoretical research and practices of the construction of health care facilities in the future.

孟庆林

Qinglin Meng

华南理工大学教授

Professor at South China University of Technology

亚热带建筑科学国家重点实验室副主任

Deputy Director of State Key Laboratory of Subtropical Building Science

2020 年 6 月

June, 2020

前　言

Foreword

2020年新冠肺炎疫情爆发期间，暴露出我国现有公共卫生系统和医疗设施存在不充足、不合理、不平衡问题：大城市综合医院发热门诊及感染病房无法应对接诊，传染病医院床位数量不足，设施配置不到位；中小城市医疗资源配置不齐全、不合理；分级诊疗难以实施；既有医疗保健设施水平参差不齐，难以满足保障人民健康的基本需求。大城市、大医院面对患者剧增而应接不暇，中小城市和县级医疗机构设施差别太大，小医院各种医疗设施不健全，故而百姓不得不往大医院跑。分析发达国家的医疗体系可以看到其分级诊疗制度完备，家庭医生、社区医疗、诊所、专科医院、综合医院等的设施配备均好，便于实行分散就医，人群不聚集。因此，我国还需相当长一段时期不断改进医疗保健设施之不足，提升各级、各类医疗保健设施建设水平。

During the outbreak of COVID-19 in 2020, it has been revealed that the existing public health system and medical facilities in China are inadequate, unreasonable and unbalanced: fever clinics and infection wards in general hospitals in large cities are unable to cope with reception, the number of beds in hospitals for infectious disease is insufficient, and the facilities are not allocated properly; the allocation of medical resources in small and medium-sized cities is incomplete and unreasonable; tiered diagnosis and treatment is difficult to implement; the levels of the existing health care facilities vary greatly and can hardly meet the basic requirements for ensuring people's health. Large cities and hospitals are overwhelmed by a dramatically increasing number of patients. There is a wide gap in the facilities of small and medium-sized cities and county-level medical institutions. Medical facilities in small hospitals are inadequate, so people have to visit large hospitals. It can be seen from the medical systems of developed countries that their tiered diagnosis and treatment systems are complete, and family doctors, community health care centers, clinics, specialized hospitals, general hospitals are all well-equipped, so that decentralized medical treatment can be implemented and patients do not need to flock to large hospitals. Therefore, there is still a long way for China to go towards filling the gap in its health care facilities and constantly improving its various levels and types of health care facilities.

全球医疗保健设施建造技术随着工程科学的发展不断升级，每当经历各种疫病检验后也都不断调整和持续改进，以确立更科学的设计理论、建立完善的建造机制。医疗保健设施工程建设按照项目管理模式开展，项目依据类型、规模、功能和运营方式的不同而体现多样性，其工程设计因此各具特点。值此背景下，策划与工程咨询价值渐居顶层，工艺设计必须为建筑创作提供支持，建筑工程技术之规律性和通用性将提高设计效率，有效的设计管理已被作为重要的生产力。

With the development of engineering science, the construction technology of global health care facilities has been continuously upgraded, and has been constantly adjusted and continuously improved after undergoing various disease tests, so as to establish a more scientific design theory and a perfect construction mechanism. The construction of health care facilities is carried out in accordance with the mode of project management, and such projects are diversified in terms of types, scales, functions and operation modes. Therefore, their engineering designs have their own characteristics. Under this background, the value of planning and engineering consultation is gradually at the top level. Process design must support the architectural creation. The regularity and universality of construction engineering technology will improve design efficiency. Effective design management has been important productivity.

医疗保健设施工程咨询与设计由众多工种协同作业、共同完成，包括：工艺、建筑、结构、机电、物流、景观、经济等。医疗工艺是医疗保健设施建筑设计的依据，它通过各个专业在空间、负荷、原理、系统、末端、材质等方面分别实现，在建筑专业统筹下协调各工种确立主次分工、互为条件、逐级落实。医疗保健设施作为公共建筑应实现其为公众服务的社会价值，并具有符合大众审美的艺术性，而非仅仅实现业主或建筑师个人的好恶。作为具有工艺特性的建筑，医疗保健设施设计应力求实现人类的身心需求，包括急救、医疗、康复、预防、保健、安养、科研、教学、医药制剂等方面的功能，促进医学科学进步，符合城市公共卫生配套、健康服务、建筑运维、生物安全、舒适、卫生、人文、环保、绿色等各方面的要求，平等关爱置身于其中的患者、家属、医护、保障等工作人员。

The consultation and design of health care facilities projects are jointly completed by different types of work, including technology, architecture, structure, electromechanical, logistics, landscape, economy, etc. Medical process is the basis for the architectural design of health care facilities. It is realized by several disciplines in terms of space, load, principle, system, terminal, material and others. Under the overall planning of the architecture discipline, various types of work must be coordinated to establish primary and secondary division of labor, and these works are mutually conditional and implemented step by step. As public buildings, health care facilities should realize their social value of serving the public and have artistic quality conforming to the public aesthetics, rather than merely the personal likes and dislikes of owners or architects. As buildings with technological characteristics, the design of health care facilities should strive to meet the physical and mental needs of human beings, including the functions of first-aid, medical treatment, rehabilitation, prevention, health care, nursing, scientific research, teaching, pharmaceutical preparations, etc., promote the progress of medical science, meet the requirements of urban public health facilities, health services, building operation and maintenance, bio-safety, comfort, hygiene, humanities, environmental protection, green, etc., and equally care for patients, their family members, medical care personnel, security personnel and other staff involved.

医疗保健设施工程建设全过程具有特定的阶段和时序，有别于其他类型的建筑，要从项目的选址、可行性研究起步，完成医疗工艺设计、建筑策划，到拟订出详细的设计任务书之后，才能开展建筑设计。在建筑方案创作环节，要把医疗工艺方案设计落实到建筑空间中，既满足建筑学和其相关规范要求，又符合医疗工艺和公共卫生学要求；在建筑初步设计和施工图阶段，医疗工艺条件设计要满足所有工种末端设计输入要求；工程设计不仅要达到一般公共建筑的共性要求，更关键的是要实现医疗工艺设计要求。

The whole process of the construction of health care facilities has specific stages and sequence, which is different from other types of buildings. The architectural design can only be performed after the project site selection and feasibility study have been launched, the design for medical process and architectural planning have been completed, and the detailed requirements specification has been prepared. In the creation of architecture proposal, the design of medical process SD should be implemented into the architectural space, which not only meets the requirements of architecture and its related specifications, but also meets the requirements of medical process and public health science. In the PD and CD stages, the design of medical process conditions shall meet the terminal design input of all types of work. The engineering design should meet the common requirements of common public buildings, and more importantly realize the requirements of design for medical process.

随着建筑行业的专业分工越来越细，从咨询中分化出医疗工艺咨询；从建筑中分化出室内设计；从室内设计中又细分出一般公共建筑空间和特种医用空间，如手术室、实验室、放射防护与屏蔽等；从结构中分离出钢结构、装配式；从空调中分离出净化空调；从给排水中分离出医用水；从动力中分离出医用气体；从弱电中分离出智能化、信息化；从物流中分离出气动、轨道、箱式；从机电中分离出管线综合；从管线综合中衍生出 BIM；从停车设计中分化出机械车库、全自动仓储车库。但不论怎么细分，建筑学永远是工程设计所有专业的牵头者，有责任担负起所有专业的定位、主次、品质、工程质量之总责任，这正是国际上通行建筑师负责制的缘由所在。

As the division of disciplines in the construction industry is becoming more and more detailed, medical process consultation is subdivided from consultation; and interior design is subdivided from construction and then subdivided into general public building space and special medical space, such as operating room, laboratory, and radiation protection and shielding; the steel structure and assembly are subdivided from the structure; the purified air conditioners are subdivided from the air conditioners; medical water is subdivided from water supply and drainage; medical gas is subdivided from power; intelligence and informatization are subdivided from low voltage; pneumatic, rail and box-type are subdivided from logistics; pipeline synthesis is subdivided from electromechanical system; BIM is derived from pipeline synthesis; mechanical garages and fully automatic storage garages are subdivided from parking design. However, no matter how subdivided, architecture is always the leader of all disciplines in engineering design, and has the responsibility to assume the overall responsibility for the positioning, priority, quality and engineering quality of all disciplines. This is the reason why architect responsibility system is internationally popular.

我国许多医疗保健设施的设计做不到位或难以做到投资方和业主所想象的那样，其原因：一是业主没有提出具体的设计要求，由设计师按自己假想的要求生造，没有明确设计任务书或只有简单设计要求的项目比比皆是；二是许多投资方不了解咨询和设计全过程的程序和工作内容，既没有前期研究也没有策划；三是长期习惯的“脑体倒挂”观念，把前期研究创造性技术工作及相应费用延后积压到施工阶段，导致施工后期无图可用，迫使施工机构用二次设计的形式为前期设计买单，甚至胁迫设计变更；违背了建筑工程也是科学创造的规律。

Many health care facilities in China are not designed in place or are difficult to achieve what investors and owners except. One of the reasons is that the owners have not put forward specific design requirements, and designers have created according to their own imagination. There are many projects with no clear requirements specification or only simple design requirements; second, many investors do not understand the whole procedures and work content of consultation and design, neither preliminary research nor planning; third, the long-standing concept of “wage disparity between white-collar and blue-collar workers” postpones the early research on creative technology and corresponding costs to the construction stage, resulting in no drawings available at the later stage of construction, forcing the construction company to pay for the early design in the form of secondary design, and even forcing design changes, which violates the law that construction engineering is also scientific creation.

尽管现在与医院设计有关的规范、标准和著述很多，但不是零星分散在不同的专业资料里就是长篇巨著，阅读不够方便。设计师在实际设计工作中没有全面系统的研究就开始设计，甚至整个设计过程中顾不上去全面查询资料的情况很多。为此，本手册将各类知识进行整理、汇总、罗列、中英文双语，以简短语句作为提示，使投资人、业主、咨询、设计、管理、建造方可以借此迅速查询到相关知识体系。本手册意在覆盖面广，不涉深度和来由，也不展开具体应用的陈述，以作导引型轻便工具书，更易随身携带。

Although there are many specifications, standards and writings related to hospital design, they are either scattered in materials of different disciplines or voluminous works, making reading inconvenient. Actually, designers start design without comprehensive and systematic study, and even there are many cases in which they fail to search for information in the whole design process. For this reason, this Manual arranges, summarizes and lists all kinds of knowledge in both Chinese and English, with short sentences as hints, so that investors, owners, consultants, designers, management and builders can quickly search the knowledge system. This Manual is intended to cover a wide range of areas rather than depth and causes or statements of specific applications, so as to serve as a portable handbook, which is easier to take along.

本书适合从事医疗保健设施的投资、咨询、设计、管理人员阅读，也可作为高等院校建筑及其相关学科的辅助教学资料。

This Manual is suitable for personnel engaged in investment, consultation, design and management of health care facilities, or used as auxiliary teaching materials for buildings and related disciplines in institutions of higher learning.

朱　希

Xi Zhu

研究员级高级工程师

Senior Engineer (Professor Level)

享受国务院政府特殊津贴专家

Expert Obtaining the special allowance of State Council

2020 年 6 月

June, 2020

目　录

Table of Contents

第1章

医疗保健设施前期咨询

Chapter Ⅰ

Preliminary Consultation for Health Care Facilities

徐怀平 朱 希

Huaiping Xu Xi Zhu

1.1 初步可行性研究（项目建议书）与选址

1.1 Preliminary feasibility study (project proposal) and site selection

1.1.1 初步可行性研究目的

1.1.1 Purpose of preliminary feasibility study

1.1.1.1 医疗保健设施项目按投资来源主要分为社会投资项目和政府投资项目两类，政府投资项目的初步可行性研究也称项目建议书。

1.1.1.1 Health care facilities projects are mainly divided into society-invested projects and government-invested projects from the sources of investment. The preliminary feasibility study of government-invested projects is also called project proposal.

1.1.1.2 社会投资项目初步可行性研究目的，是在投资方进行医疗保健设施项目投资机会研究或企业发展规划的基础上，对项目投资的必要性和初步可行性进行研究，初步判断项目方案设想是否具有生命力，提出是否需要进行下一步可行性研究。

1.1.1.2 The purpose of the preliminary feasibility study of society-invested projects is to study the necessity and preliminary feasibility of project investment on the basis of investment opportunity study or enterprise development plan for health care facilities project of investors, to preliminarily judge whether the project proposal assumption is viable and to propose whether the feasibility study in the next step is needed.

1.1.1.3 项目建议书是按照国家及地方相关政策、发展规划或其他建设依据，对医疗保健设施项目建设的必要性和初步可行性进行研究，初步判断项目是否必要和可行，作为政府项目投资决策和立项的依据。

1.1.1.3 The project proposal is to study the necessity and preliminary feasibility of the construction of health care facilities in accordance with relevant national and local policies, development plans or other construction basis, to preliminarily judge whether the project is necessary and feasible, and to serve as the basis for government investment decisions and project initiation.

1.1.2 初步可行性研究各相关方

1.1.2 Stakeholders in the preliminary feasibility study

参加项目初步可行性研究的相关方有：社会投资方、政府建设部门（或代建方）、咨询方、医疗工艺设计方、建设方案设计方、项目评估方、政府审批部门等，通过各方协作、互审才能完成研究工作。

Stakeholders participating in the preliminary feasibility study of the project include social investors, government construction departments (or construction agents), consultants, design institution of medical process, construction SD institution, project evaluators, government approval departments, etc. Only through cooperation and mutual review can the study be completed.

1.1.2.1 社会投资方

1.1.2.1 Social investors

1. 遴选初步可行性研究咨询、医疗工艺设计、建设方案设计、评估等机构；

1. Select institutions for preliminary feasibility study consulting, design for medical process, construction SD, and evaluation;

2. 为各相关方提供研究所需的依据和资料；

2. Provide the basis and data required for all stakeholders;

3. 组织项目调研及考察；

3. Organize project survey and investigation;

4. 审核可行性研究各相关方提交的成果，并提出意见；

4. Review the results submitted by all stakeholders in the feasibility study and put forward opinions;

5. 组织开展项目沟通会、论证会等；

5. Organize and carry out project communication meetings, demonstration meetings, etc. ;

6. 做出投资决策。

6. Make investment decisions.

1.1.2.2 政府建设部门

1.1.2.2 Government construction departments

1. 遴选项目建议书咨询、医疗工艺设计、建设方案设计、评估等机构；

1. Select institutions for project proposal consulting, design for medical process, construction SD, and evaluation;

2. 为各相关方提供工作必需的依据和资料；

2. Provide necessary basis and data for all stakeholders;

3. 组织项目调研及考察；

3. Organize project survey and investigation;

4. 配合业主或医疗工艺咨询方的研究工作；

4. Cooperate with the owner or medical process consultant for research;

5. 审核咨询单位提交的成果，并提出意见；

5. Review the results submitted by consultant and put forward opinions;

6. 组织开展项目沟通会、论证会等；

6. Organize and carry out project communication meetings, demonstration meetings, etc. ;

7. 项目建议书上报政府审批部门。

7. The project proposal shall be submitted to the government approval departments.

1.1.2.3 咨询方

1.1.2.3 Consultants

1. 项目调研；

1. Project survey;

2. 选址及建设方案研究；

2. Research site selection and construction proposal;

3. 编制初步可行性研究报告或项目建议书；

3. Prepare the preliminary feasibility study report or project proposal;

4. 配合投资方和政府建设部门的项目论证；

4. Cooperate with investors and government construction departments for project demonstration;

5. 配合评估方的项目评审；

5. Cooperate with the evaluator for project review;

6. 向投资方和政府建设部门提出合理化建议。

6. Put forward reasonable suggestions to investors and government construction departments.

1.1.2.4 医疗工艺设计方

1.1.2.4 Medical process design institutions

1. 市场调研与策划；

1. Market survey and planning;

2. 提出项目功能需求、医疗规划、医疗设备规划、信息化规划等；

2. Propose project functional requirements, medical sources planning, medical device planning, information planning, etc.

3. 完成医疗工艺前期设计，满足初步可行性研究的需要；

3. Complete the PD of medical process to meet the needs of preliminary feasibility study;

4. 配合投资方和政府建设部门的工艺技术论证；

4. Cooperate with investors and government construction departments for process technology demonstration;

5. 配合评估方的项目评审；

5. Cooperate with the evaluator for project review;

6. 为投资方和政府建设部门提出医疗工艺合理化建议。

6. Put forward reasonable medical process suggestions to investors and government construction departments.

1.1.2.5 建设方案设计方

1.1.2.5 Construction SD institutions

1. 参加项目调研与选址研究；

1. Participate in project survey and site selection research;

2. 研究规划设计条件，进行建筑策划；

2. Study the planning and design conditions and carry out architectural planning;

3. 完成概念性规划设计，为初步可行性研究提供建设方案；

3. Complete conceptual planning and design, and provide construction proposal for preliminary feasibility study;

4. 配合投资方和政府建设部门的项目论证；

4. Cooperate with investors and government construction departments for project demonstration;

5. 配合评估方的项目评审；

5. Cooperate with the evaluator for project review;

6. 为投资方和政府建设部门提出规划设计合理化建议。

6. Put forward reasonable planning and design suggestions to investors and government construction departments.

1.1.2.6 项目评估方

1.1.2.6 Project evaluators

1. 评估初步可行性研究报告或建议书；

1. Evaluate the preliminary feasibility study report or proposal;

2. 组织开展项目评审会；

2. Organize the project review meeting;

3. 出具初步可行性研究报告或建议书评估报告。

3. Issue a preliminary feasibility study report or proposal evaluation report.

1.1.2.7 政府审批部门

1.1.2.7 Government approval departments

1. 委托项目评估单位；

1. Entrust a project evaluators;

2. 预审项目建议书、可行性研究报告，并提出修改意见；

2. Preliminarily review the project proposal, feasibility study report, and put forward amendments;

3. 受理上报的项目建议书、可行性研究报告；

3. Accept the project proposal and feasibility study report;

4. 对项目评估单位工作进行指导和监督；

4. Guide and supervise the work of project evaluators;

5. 下达项目立项批复文件。

5. Issue the approval document for the project initiation.

1.1.3 初步可行性研究工作程序

1.1.3 Procedures of preliminary feasibility study

1.1.3.1 委托签约

1.1.3.1 Entrustment and contracting

社会投资方、政府建设部门或代建方，通过招标、委托等方式把初步可行性研究或项目建议书咨询工作发包给有咨询资质和咨询专业的机构开展初步可行性研究。

Social investors, government construction departments or construction agents will contract the preliminary feasibility study or project proposal consulting to institutions with consulting qualifications and consulting discipline to carry out preliminary feasibility study through bidding, entrustment, etc.

1.1.3.2 组建项目组

1.1.3.2 Establishment of a project team

1. 项目组包括社会投资方项目组、政府建设部门项目筹建组、咨询方项目组。

1. A project team includes the social investor project team, the government construction department project planning team and the consultant project team.

2. 社会投资方或政府建设部门项目组由总负责人、医疗保健业务专业人员、医疗保健工程专业人员、前期工作协调人员、后勤保障人员等组成。

2. The project team of social investors or government construction departments is composed of a general leader, health care professionals, health care engineering professionals, pre-work coordinators, logistics support personnel, etc.

3. 咨询方项目组由项目总负责人、技术负责人、各专业负责人、技术人员、编写人员、后勤保障人员组成，如有需要可外聘医疗保健的管理和建设专家做技术顾问。

3. Consultant project team is composed of a project chief, technical principal, principal of each discipline, technician, writers and logistics support staff, if necessary, health care management and construction experts can be hired as technical consultants.

1.1.3.3 项目启动

1.1.3.3 Project initiation

咨询方和投资建设方对项目的前期研究和开展情况，以及项目文件依据进行交底；如已预选址，进行现场勘察。

The consultant and the investor and undertaker shall disclose the preliminary study and development of the project and the basis of the project documents, and carry out site investigation if the site has been pre-selected.

1.1.3.4 调研

1.1.3.4 Survey

调研包括一手资料调研和二手资料调研（或称文案调研）。一手资料调研：由投资方或建设方组织参与可行性研究的相关各方，对预选国内外同类医疗保健机构和项目进行调研，咨询方出具调研报告。

Survey includes primary data survey and secondary data survey (or document survey). Primary data survey: The investor or the undertaker shall organize the stakeholders involved in the feasibility study to carry out survey on pre-selected domestic and foreign similar health care institutions and projects, and the consultant shall issue a survey report.

1.1.3.5 建设方案研究

1.1.3.5 Research on construction proposal

在调研成果的基础上，对项目定位及需求、建设规模、选址及建设条件、规划建设方案设想等进行专题研究。

Specially study the project orientation and demand, construction scale, site selection and construction conditions, planning and construction plan assumption on the basis of research results.

1.1.3.6 编制报告

1.1.3.6 Report preparation

在项目调研、专题研究、医疗工艺方案研究成果的基础上，按照初步可行性研究文件编制的要求编制初步可行性报告初稿。

On the basis of results of project survey, monographic study and medical process proposal research, prepare the first draft of the preliminary feasibility report in accordance with the requirements for the preparation of preliminary feasibility study documents.

1.1.3.7 报告的论证和修改

1.1.3.7 Demonstration and revision of the report

由投资建设方组织对报告初稿审核，并组织召开专家论证会论证，咨询方按专家及多方意见和建议进行修改。

The investor and undertaker shall organize the review of the first draft of the report, and hold an expert meeting to demonstrate it. The consultant shall revise the report according to the opinions and suggestions of experts and various parties.

1.1.3.8 报告的上报审核或审批

1.1.3.8 Report submission for review or approval

报告提交投资建设方决策审核或政府审批部门审批。

Submit the report to the investor and undertaker for decision-making and review or the government approval departments for approval.

1.1.3.9 后期服务

1.1.3.9 Later services

在投资方决策或政府审批部门审核或审批中，咨询方需做好配合评审工作，并在下一步的工作中做好交底对接工作。

In the investor's decision-making or the review or approval by the government approval departments, the consultant shall cooperate with the review, and make disclosure and coordination in the next step.

1.1.4 初步可行性研究主要内容

1.1.4 Main contents of preliminary feasibility study

1.1.4.1 社会投资项目

1.1.4.1 Society-invested project

1. 背景、依据、必要性

1. Background, basis and necessity

研究医疗保健项目的政策背景、产业发展背景、企业发展战略，分析项目建设的必要性。

Research the policy background, industrial development background and enterprise development strategy of health care projects, and analyze the necessity of project construction.

2. 项目目标及功能定位

2. Objectives and functional orientation of the project

通过项目背景和企业发展规划研究、同类项目调研，确定项目的目标及功能定位。

Determine the objectives and functional orientation of the project through the research on project background, enterprise development planning and similar projects.

3. 项目需求、建设内容及规模

3. Project requirements, construction contents and scale

参考医疗保健相关建设标准、规范，结合项目的目标及功能定位，测算项目需求，确定建设内容及规模。

Refer to construction standards and specifications for health care. Estimate the project requirements according to the objectives and functional orientation of the project, and then determine the content and scale of the construction.

4. 选址及建设条件

4. Site selection and construction conditions

按照项目建设内容及规模，参考医疗保健相关建设标准、规范，测算项目用地需求，确定用地规模，并提出符合医疗保健项目的选址方案，如在初步可行性研究阶段已选址，需进一步研究选址的合理性。

Estimate the project land and determine the land scale according to the content and scale of the project construction and referring to the construction standards and specifications of health care, and put forward the site selection proposal that accords with the health care project. Further study the rationality of the site selection if the site has been selected during the preliminary feasibility study.

5. 建设方案

5. Construction proposal

按照项目建设内容及规模，提出规划设计、公用工程、交通等初步建设方案设想。

Propose the preliminary construction proposals such as planning and design, public works and transportation according to the content and scale of project construction.

6. 实施方案

6. Implementation proposal

对项目建设实施条件、实施进度、招投标方案等进行筹划，扩建或改造项目还需研究其腾挪和周转方案。

Plan the implementation conditions, progress and bidding proposal of the project construction, and study its transfer and turnover proposal in case of expanded or reconstructed project.

7. 绿色建筑、节能、海绵城市方案

7. Proposal for green buildings, energy saving and sponge city

按照医疗保健项目特点，提出绿色建筑、节能、海绵城市建设方案基本要求。

Put forward the basic requirements for construction proposal of green buildings, energy saving and sponge city according to the characteristics of health care projects.

8. 环境保护方案

8. Environmental protection proposal

按照项目所在地执行的环境质量标准，对项目建设期、运营期的废气、医疗废水、医疗固体废物、噪声、辐射、感染等环境污染因素进行分析，提出环境保护初步方案。

Analyze the environmental pollution factors such as waste gas, medical wastewater, medical solid waste, noise, radiation, and infection during the construction and operation of the project according to the environmental quality standards implemented at the project site, and put forward preliminary environmental protection proposal.

9. 劳动安全、卫生与消防方案

9. Proposal for labor safety, health and fire protection

为保障医疗保健项目建设、运营安全，对项目的劳动安全、卫生、消防提出初步方案。

Propose a preliminary proposal for labor safety, health and fire protection of the project in order to ensure the safety of the construction and operation of the health care project.

10. 投资估算

10. Investment estimation

按照医疗保健工程的特点、复杂程度，参考同地区、同类项目近期竣工价、结算价，以及工程造价指数，以单位指标估算法，进行项目投资估算。社会投资项目总投资范围包括建设投资、建设期利息、流动资金三部分，建设投资由工程费用（建筑工程、医疗设备购置、安装工程、医疗信息化、智能化）、工程其他费（用地拆迁、建设运营）和预备费组成。政府投资项目总投资范围只含建设投资（不含用地拆迁和运营其他费），如果另有规定，医疗设备购置和医疗信息化、智能化可作为专项单独立项。

Estimate the project investment by unit index estimation method according to the characteristics and complexity of health care project and referring to the recent completion price, settlement price and construction cost indexes of similar projects in the same region. The total investment of a society-invested project includes three parts, namely construction investment, interest during the construction period and working capital. The construction investment consists of engineering costs (construction engineering, purchase of medical device, installation works, medical informatization and intelligence), other project costs (land acquisition, construction and operation), and reserve funds. The total investment of a government-invested project only includes construction investment (excluding land acquisition and other operating expenses). If otherwise specified, the cost for purchase of medical device and medical informatization and intelligence can be calculated separately.

11. 融资方案

11. Financing proposal

在项目投资确定的基础上，分析确定项目的融资主体、资金来源渠道和方式，优化项目融资方案。

On the basis of the determination of project investment, analyze and determine the financing subject, funding sources and methods of the project, and optimize the project financing proposal.

12. 财务评价

12. Financial evaluation

结合医疗保健项目的特点，进行项目财务现金流量分析，评价项目的投资盈利能力、偿债能力、财务生存能力，判断财务可行性。

Analyze the project's cash flow according to the characteristics of health care projects, so as to evaluate the project's investment profitability, solvency and financial viability, and to judge the financial feasibility.

13. 不确定性和风险分析

13. Uncertainty and risk analysis

按照项目的相关财务指标，进行敏感性分析、临界点分析、盈亏平衡分析；并对项目风险识别、估计和评估，提出风险预警、风险规避的对策，为决策提供依据。

Conduct sensitivity analysis, critical point analysis and break even analysis according to the financial indicators of the project; identify, estimate and evaluate project risks, and put forward countermeasures of risk early warning and risk avoidance to provide basis for decision.

14. 经济影响及社会效益分析

14. Analysis of economic impact and social benefit

分析项目对医疗保健行业、区域经济、宏观经济的影响，并评价项目的社会效益和当地社会环境对项目的适应性和可接受程度。

Analyze the impact of the project on the health care industry, regional economy and macro economy, and evaluate the social benefits of the project and the adaptability and acceptability of the local social environment to the project.

15. 研究结论和建议

15. Research conclusions and recommendations

通过对医疗保健项目的初步可行性研究，对项目必要性及依据，项目定位目标、需求及规模，初步建设方案，社会和经济效益的可行性提出结论，提出是否可行的建议，并提出下一步工作和项目实施中需要解决的问题和建议。

Based on the preliminary feasibility study of health care, put forward conclusions on the necessity and basis of the project, the project's orientation and objectives, requirements and scale, the feasibility of the preliminary construction proposal, social and economic benefits, make suggestions about whether the project is feasible or not, and put forward the problems to be solved and suggestions in the next step and project implementation.

1.1.4.2 政府投资项目

1.1.4.2 Government-invested project

政府投资医疗保健设施项目和社会投资医疗保健设施项目相比，前者是为社会提供公共产品，侧重于公共产品的服务能力和社会效益。后者是市场投资行为，更侧重于经济效益和投资回报。政府投资医疗保健设施项目在研究的主要内容上和社会投资医疗保健项目对比有以下不同：

Compared with society-invested health care facilities, government-invested health care facilities are to provide public products for the society, and focus on the service capability and social benefits of public products. The social investment is a market investment behavior, with more emphasis on economic benefits and investment returns. There are the following differences in the main contents of the research between the government-invested health care facilities and society-invested health care facilities:

1. 无须融资方案；

1. No financing proposal is required;

2. 只需经济影响及社会效益分析，无须财务评价；

2. Only economic impact and social benefit analysis are required, and no financial evaluation is required;

3. 只需风险分析，无须不确定性分析；

3. Only risk analysis is required, and no uncertainty analysis is required;

4. 需新增对项目代建方式的分析；

4. An analysis of the construction agent of the project is added;

5. 如有征地，需研究征地拆迁方案；

5. If there is any land acquisition, it is necessary to study the land acquisition proposal;

6. 需新增对后期运营费用的研究。

6. Research on later operating expenses is added.

1.1.5 初步可行性研究成果及标准

1.1.5 Results and standards of preliminary feasibility study

初步可行性研究的成果是初步可行性研究报告或项目建议书。成果以《投资项目可行性研究指南》《建设项目经济评价方法与参数》为参考，符合国家和地方发布的医疗保健相关建设标准和规范，以及项目本地发布的项目前期研究的相关标准，达到项目初期决策和审批立项的深度和要求。

The results of the preliminary feasibility study are the preliminary feasibility study report or the project proposal. The results are based on Guideline of Investment Project Feasibility Study and Methods and Parameters for Economic Evaluation of Construction Projects and conform to the national and local health care construction standards and norms, as well as the standards for preliminary study of the project at the project location, reaching the depth and requirements for initial decision-making and project approval.

1.1.6 选址

1.1.6 Site selection

在初步可行性研究阶段，需提出选址需求及选址条件，以便为下一步规划部门确定选址和下达选址意见书提供依据和参考。为了加快项目前期工作，初步可行性研究阶段对项目的意向地块，提前征求规划部门意见。医疗保健项目选址需满足如下基本条件：

In the preliminary feasibility study, put forward site selection requirements and site selection conditions, so as to provide basis and reference for the planning department to determine the site selection and issue site selection opinions in the next step. Consult with the planning department in advance on the intended plot for the project in the preliminary feasibility study, so as to speed up the preliminary work of the project. The following basic conditions must be met for the location of health care project:

1.1.6.1 符合城市规划、用地规划等要求。

1.1.6.1 Meet the requirements of city planning, land use planning, etc.

1.1.6.2 统筹医疗资源布局，选择适合的建设地点和地址。

1.1.6.2 Coordinate the distribution of medical resources and select suitable construction sites and addresses.

1.1.6.3 考虑自然环境和地貌特征，选择自然环境较好，地质结构稳定，无暗河、箱涵等建设条件较好的区域。

1.1.6.3 Consider the natural environment and geomorphic features, and select areas with good natural environment, stable geological structure, no hidden rivers, no box culverts and good construction conditions.

1.1.6.4 考虑服务人群和辐射半径，诊所和医院选择在居住人口较集中、公共交通便利、有地铁站或轨道交通的区域。精神病院、传染病院、疗养院选择在离市区较远的区域。

1.1.6.4 Consider the service population and radiation radius, and select an area with relatively concentrated residential population, convenient public transportation, subway stations or rail transit. Health care projects can be located from areas far away from the urban area.

1.1.6.5 保证环保和周边人群卫生安全，选择在和居住小区、商业区、学校、公园有一定安全距离，相关环保条件符合环保要求的区域。

1.1.6.5 Ensure environmental protection and the health and safety of the surrounding population, and select an area with a certain safe distance from residential areas, commercial areas, schools and parks with the environmental protection conditions meeting the requirements of environmental protection.

1.1.6.6 考虑地下水污染及周边生态保护、水源保护，选择不影响地下水污染及周边生态，离水源保护区较远的区域。

1.1.6.6 Consider groundwater pollution, surrounding ecological protection and water source protection, and select the area far away from the water source protection area without affecting groundwater and surrounding ecology.

1.1.6.7 选址尽量避免涉及占用林地、交通、水务、海域等问题的地块。

1.1.6.7 The site selection should avoid plots involving the occupation of forest land, transportation, water supplies, sea areas and other issues.

1.2　可行性研究

1.2　Feasibility study

1.2.1　可行性研究目的

1.2.1　Purpose of feasibility study

医疗保健设施项目可行性研究的目的，是在初步可行性研究成果的基础上，对项目的必要性、需求、项目建设方案、实施方案、投融资方案、财务效益、社会效益等更全面、深层次、科学性的研究，提出可行性方案，作为投资方或政府最终投资决策和审批的依据。

The feasibility study of the health care facilities project is to make a more comprehensive, in-depth and scientific research on the necessity, requirements, project construction proposal, implementation proposal, investment and financing proposal, financial benefits and social benefits of the project on the basis of the preliminary feasibility study results, and to put forward the feasibility plan as the basis for the final investment decision and approval of the investor or the government.

1.2.2　可行性研究流程

1.2.2　Process of feasibility study

不是所有的项目都必须按照初步可行性研究或项目建议书、可行性研究报告流程进行，对投资额较小、不太复杂的项目，按照投资方的要求可直接编制可行性研究报告。政府投资项目如地方有相关规定，对投资额较小、不太复杂的项目也可只编制项目建议书，立项后直接进入概算阶段，对政府有文件依据的加急或特殊项目可直接进入可行性研究或概算阶段。

Not all projects must be carried out according to the preliminary feasibility study or project proposal and feasibility study report. For project with small investment and less complexity, feasibility study report can be directly prepared according to the requirements of the investor. If there are local regulations for a government-invested project, project proposal can be prepared only for project with small investment and less complexity. After the project is approved, the budget estimation can be directly performed. For urgent or special project with government documents, the feasibility study or budget estimation can be directly performed.

1.2.3　可行性研究工作程序

1.2.3　Procedures of feasibility study

可行性研究工作程序和初步可行性研究基本相同，但在委托环节做如下补充：在条件容许时，初步可行性研究或项目建议书、可行性研究、医疗工艺方案、设计方案、工程设计等不同阶段的前期咨询工作可组合打包招标、委托、咨询方或工程设计方作为总包方，医疗工艺方案、概念设计方案等工作可以分包给有资质或经验的专业公司，可减少多次招标、委托，加快前期工作进度。

The procedure of the feasibility study is basically the same as that in preliminary feasibility study. However, the supplement in the entrustment is as follows: If the conditions permit, the preliminary consultation in different stages involving preliminary feasibility study or project proposal, feasibility study, medical process proposal, design proposal, engineering design institution can be bidden, entrusted and consulted as a contract package. As the general contractor, the consultant or engineering design institution can subcontract the medical process proposal and conceptual design proposal to qualified or experienced professional companies, which can reduce the number of times of bidding and entrustment, and accelerate the preliminary works.

1.2.4 可行性研究主要内容

1.2.4 Main contents of feasibility study

可行性研究的主要内容和初步可行性研究基本相同，但在研究内容的深度上有更高的要求，交通规划及评价、环境保护与环境影响、绿色节能等内容需做专篇研究。

The main content of the feasibility study is basically the same as that of the preliminary feasibility study, but there are higher requirements for the depth of the study. Special studies are required for traffic planning and evaluation, environmental protection and environmental impact and green energy.

1.2.5 可行性研究成果与深度

1.2.5 Results and depth of feasibility study

可行性研究的成果为可行性研究报告，可行性研究报告是在初步可行性研究的基础上进行，研究深度达到项目最后决策和审批的要求。深度如下：

The results of feasibility study are stated in the feasibility study report, which is prepared based on the preliminary feasibility study, and the depth of the report meets the requirements of final decision-making and approval of the project. The depth is as follows:

1.2.5.1 内容更齐全，数据更准确可靠，论据更充分，结论更明确，满足决策者定项目、定方案的要求；

1.2.5.1 The content is more complete; the data is more accurate and reliable; the arguments are more sufficient, and the conclusion is more definite, and meet requirements of decision-maker to determine projects and proposals;

1.2.5.2 确定的重大技术方案、财务方案，应有两个或以上的比选；

1.2.5.2 There are two or more identified major technical proposals and financial proposals for option;

1.2.5.3 提供医疗工艺报告书，即医疗工艺前期设计，包括医院项目策划、功能规划及医疗流程设计；

1.2.5.3 Provide medical process report, namely PD for medical process, including hospital project planning, functional planning and design for medical flows;

1.2.5.4 确定技术数据、工程技术参数、设备选型等，指导下一步初步设计和概算工作；

1.2.5.4 There is determined technical data, engineering technical parameters and device selection, which can guide the next PD and budget estimate;

1.2.5.5 对建设投资或运营成本应进行详细分项估算，其误差达到 ±10% 以内，精度高于初步可行性研究报告或建议书的 ±20%。

1.2.5.5 Estimate the construction investment or operation cost in items with an error within 10%, and the accuracy is ± 20% higher than that of the preliminary feasibility study report or proposal.

1.2.6 对外投资及国际总承包项目可行性研究

1.2.6 Feasibility study on foreign investment and international EPC project

1.2.6.1 分类研究

1.2.6.1 Study on classification

对外投资及国际总承包保健设施项目类型主要有：国内企业对外投资类、政府捐建类、国内企业代建类等，国内企业对外投资更关注国际市场及投资的可行性，注重投入产出和经济效益；政府捐建类更注重国际合作、建设地的社会经济背景、当地实际需求、国际影响力及社会效益；国内企业代建类要把国内前期研究和建设经验推广出去，还要符合当地政府或企业的需求和要求。

The types of foreign investment and international general contracting health care facilities projects mainly include: foreign projects invested by domestic enterprises, projects donated by the government and projects agented by domestic enterprises. Foreign projects invested by domestic enterprises pay more attention to the international market and feasibility, focusing on input-output and economic benefits; projects donated by the government pay more attention to international cooperation, social and economic backgrounds of the construction location, local actual demands, international influence and social benefits; projects agented by domestic enterprises mainly to promote the domestic preliminary study and construction experience according to the requirements of local government or enterprises.

1.2.6.2 现场调研

1.2.6.2 Site survey

没有调研就没有发言权，基于“一带一路”国家的国情不同，医疗需求及建设标准和国内不同，可行性研究在确定项目定位、功能及需求、规划建设等核心内容方面，首先要在当地深入调研，掌握“一带一路”国家和地区的实际情况，为研究打好基础。

No survey, no right to speak. In terms of the national conditions of “the Belt and Road”, the medical demands and construction standards in any other country are different from those in China. When determining the core contents of the projects in the feasibility study, such as project orientation, functions and demands, planning and construction, conduct in-depth local survey first, so as to understand the actual situation of the countries and regions along “the Belt and Road” and lay a good foundation for the study.

1.2.6.3 医疗工艺策划

1.2.6.3 Medical process planning

“一带一路”国家和地区医疗保健设施项目的运营及管理与我国有差别，医疗工艺方案中的业务发展规划、医疗专科规划、医疗工艺流程、医疗设备方案等要聘请医疗工艺咨询经验丰富、了解当地实际的咨询单位研究，为可行性研究做支撑。

The operation and management of health care facilities projects in the countries and regions along “the Belt and Road” are different from those in China. In terms of business development planning, medical discipline planning, medical process flow and medical device proposal in the medical process proposal, a consultant with rich experience regarding medical process consultation and understanding of local conditions shall be employed, so as to support the feasibility study.

1.2.6.4 选址论证

1.2.6.4 Site selection demonstration

“一带一路”国家和地区的气候、环境、地质、水文、环保，以及用地政策、规划建设标准、规范都有当地的特殊性，选址不但要考虑医疗保健项目在公共卫生的感染、医用设备辐射、医疗废水废物排放、医用气体安全等方面对选址的影响，还要考虑当地的环境要求、地质特点、建设标准、规范等，在选址中要做适合当地的多方案比选和可行性论证。

The climate, environment, geology, hydrology and environmental protection, land use policies, planning and construction standards and specifications of the countries and regions along “the Belt and Road” are significantly different from each other. During the site selection for health care facilities projects, not only the impact of public health infection, medical device radiation, discharge of medical waste water and wastes and medical gas safety, but also local environmental requirements, geological characteristics, construction standards and specifications and others shall be given into consideration. It is necessary to make the demonstration on comparison and selection of proposals required and feasibility that conform to the local situation.

1.3 医疗保健设施前期策划

1.3 Preliminary planning of health care facilities

1.3.1 运营策划

1.3.1 Operation planning

运营模式决定医疗保健设施的规划布局、建筑规模、形式、内容、标准，建设一项设施前，首要工作就是对运营模式进行策划，在此列出国际、国内现有的各类运营模式及其对应的建筑形式，以作为策划依据。

The operation mode determines the planning layout, construction scale, form, contents and standards of health care facilities. Before building a facility, plan the operation mode first. The following is various existing international and domestic operation modes and their corresponding construction forms as the planning basis.

1.3.1.1 国际典型医疗保健设施运营模式

1.3.1.1 Operation mode of typical international health care facilities

1. 市场主导型（美国）

1. Market-oriented operation (America)

在医院管理上实行自由经济、自主经营、自由竞争，政府干预有限。美国医院分为政府医院、非政府非营利性医院和营利性医院。政府医院有联邦政府医院、州及地方政府医院，如退伍军人医院、伤残医院、精神病医院、印第安人医院等，占医院总数的27%左右，占住院治疗人数的18%和门诊治疗人数的23%。非政府非营利性医院有医学院附属医院、教学医院、教会医院以及由慈善机构开办的社区医院。营利性医院大多数是为了获取利润的综合医院。通常由一家大医院联合不同等级医院构成健康系统，体系内每个医院各有特色，医院之间有互补性，内部有转诊制度，医疗资源及信息共享，对外提供医技服务和护理服务。除一些政府医院外，医生通常不是医院的雇员。建筑设施只有医技和住院两部分，部分综合医院设急诊，门诊由诊所承担。

Free economy, independent management and free competition are implemented in hospital management, with a limited intervention of the government. The hospitals in America are divided into government-based hospitals, non-government and non-profit hospitals and for-profit hospitals. Government-based hospitals include federal hospitals, state and local government-based hospitals, such as veterans hospitals, disabled hospitals, psychiatric hospitals, Indian hospitals, accounting for about 27% of the total number of hospitals, 18% of the number of in-patients and 23% of the number of out-patients. Non-government and non-profit hospitals include affiliated hospitals of medical schools, teaching hospitals, mission hospitals and community hospitals founded by charitable organizations, and most of for-profit hospitals are the general hospitals to make profits. In general, a health system consists of a large hospital and many hospitals at different levels. Each hospital in the system has its own characteristics, and all hospitals complement each other. In addition, there is a referral system among such hospitals, and medical resources and information can be shared. These hospitals mainly provide medical technology service and nursing service to the external organizations. Except for some government-based hospitals, the doctors are usually not the employees of hospitals. The construction facilities only include medical technology and hospitalization. Some general hospitals have emergency treatment departments, and the outpatient service is undertaken by the clinics.

美国医院建筑依照“医疗保健设施设计与建设指南”进行设计，由医院空间委员会出具空间规划清单和服务流程要求。建筑通常都有紧凑的平面布局，呈综合体形式，公共空间主要用作交通枢纽和商业餐饮休闲服务，建筑空间具有适应未来发展的灵活性，设计师运用建筑信息模型管控复杂的情形。建筑用材是医院装修设计中很重要的一个方面，运用极具现代感的玻璃材料，大幅提升医院内部的自然采光，降低能耗，并且从视觉上增加医疗建筑的吸引力，创造舒适的护理环境、明亮的光线、漂亮的颜色，促进患者的身心健康。病房区域采用大量降噪设计，包括天花隔声板、橡胶地板、移动玻璃门等。在确保实用性能的条件下，尽可能地建立起医疗建筑与天然采光、花园等自然环境的联系，创造一个安静并且充满生机的护理环境。普遍运用装饰壁画，色彩丰富的自然环境图像与高科技医疗环境相互融合，令空间富有艺术性。医疗安全和护理质量，是美国医院评选等级非常重要的两个指标，医院装修设计要求降低人员跌倒和受伤的概率，减少医疗差错和感染，提高医院员工的生产力和沟通能力，提升患者及其家属的舒适度和满意度，最终达到促进医疗质量管理和护理环境建设的目的。

The hospital buildings in America are designed in accordance with the Guidelines for Design and Construction of Health Care Facilities. The spatial planning list and service flow requirements are issued by the hospital space committee. The buildings are usually in compact layout in the form of complex. Public spaces are mainly used for transportation hubs and commercial catering and leisure services. Building space is flexible for development in the future, and designers control complicated situations through building information models. Building materials are an important aspect during hospital decoration design. The use of highly modern glass materials can greatly improve the natural lighting inside the hospital, reduce energy consumption, visually increase the attractiveness of healthcare buildings, create comfortable nursing environment with bright light and artistic colors, so as to promote the physical and mental health of patients. The wards feature significantly noise reduction design, including sound-proof ceilings, rubber floors, removable glass doors and others. Under the condition of ensuring practicability, the connection between healthcare buildings and natural environment, such as natural lighting and gardens is established as much as possible to create quiet and vibrant nursing environment. Decorative murals are widely used, making colorful natural environment images and high-tech medical environment integrated, so that the space is artistic. Medical safety and nursing quality are two important indicators in hospital grade evaluation in American. The hospital decoration is required to reduce the probability of falling and injuries, medical errors and infections, and improve the productivity and communication of hospital employees as well as enhance the comfort level and satisfaction of patients and their families, so as to promote medical quality management and nursing environment construction.

2. 国家福利型

2. National welfare operation

（1）英国

（1）UK

医疗体制特点是政府高投入，国民高税收，全民强制参保，全民免费医疗。英国实行的国民卫生服务体系 NHS 被认为是世界上最有效、最公平的医疗服务体系之一。卫生设施主要为国有，医务人员为国家工作人员，医疗服务由国家税收购买。私营医疗保健服务仅占 10%。医疗服务分为初级卫生保健（全科医师服务）、二级保健（医院）、三级医疗服务（专科服务），呈金字塔形，底层是初级保健，中间是二级医疗，塔尖是三级专科服务。患者必须通过初级保健方能转诊至二级医疗服务，然后才能享受三级医疗服务。

The medical system is characterized by high government investment, high national taxes, compulsory universal insurance and free universal medical care. The national health service system (NHS) implemented in UK is considered as one of the most effective and fair medical service systems in the world. Health facilities are mainly state-owned; medical personnel are state functionaries, and the medical services are paid by national tax revenues. Private health care services only account for 10%. Medical services are divided into first-class health care (general practitioner services), second-class health care (hospital services) and third-class health care (specialized services), which are in a pyramid shape. The first-class health care services are at the bottom, the second-class health care services are at the middle, and the third-class specialized services are at the top. Patients only can be referred from first-class health care service organizations to second-class health care service organizations before they can enjoy third-class health care services.

大部分健康问题由全科医师识别、分流，并通过健康教育等预防手段予以控制，医疗资源利用高效合理。二级医院通常都是一座综合体，带有急诊室，位于城区或居民区，方便患者步行或乘坐公交系统抵达，服务便利。三级医院是大型建筑组群，位于郊外，用地面积较大，必须乘车前往，与市政公共交通直接接驳，呈开放式，通常采用综合体附带多个独立功能的建筑形式，主体部分是综合医院、急诊部和医技部分，附带部分是专科，如心血管、骨科、妇产、儿科，采取中央医技平台共享的方式，围绕医疗综合体建筑周边是科研楼、停车楼、教学楼及学术交流中心；只在医疗综合体中存在较复杂的医疗功能关系，其他的专科部分均为底部门诊、上部住院，在底部与综合体相连，空间关系比较简单，公共空间里提供餐饮和便利店服务，并提供舒适的休息、会客设施。

Most health care is identified by general practitioners for diversion, and controlled by preventive means, such as health education. The medical resources can be used efficiently and reasonably. A second-class hospital is usually complex with emergency rooms, which is located in an urban or residential area and convenient for patients to visit the hospital on foot or by bus. A third-class hospital is a large building group and located in the suburbs with a large piece of land. Patients can visit such hospital by car. Such hospital is directly connected with municipal public transportation in an open form. In general, it consists of a complex and multiple buildings with independent functions. The main part is the general hospital, emergency department and medical technology department; the attached part is the specialized departments, such as cardiovascular department, orthopedics department, obstetrics and gynecology department and pediatric department, which adopt a central medical technology platform sharing mode. There are scientific research buildings, parking buildings, teaching buildings and academic exchange centers around the medical complex. Complicated medical functional relations only exist in the medical complex. For other specialized departments, outpatient departments are on the lower floors, and hospitalization departments are on the upper floors. These departments are connected to the complex at the bottom with a simple spatial relation. Catering services and convenience stores are provided in public spaces where comfortable rest and visitor facilities are provided.

（2）澳大利亚

（2）Australia

卫生管理体系分联邦、州和地方三级。医疗机构分为三类：私人诊所，诊治一般病症；专科诊所，如牙科、眼科等；公立医院和私立医院：诊治需要住院的病人，不设门诊。公立医院占72%，包括退伍军人医院、公共精神病医院。宗教机构或慈善机构举办的教会医院属于公立医院，其人员一种是全职职工（管理人员、实习生、住院医生以及教学医院的教授岗位），是固定全日制工作，另一种是自由职业者（护士和绝大多数高年资医生、专科医生）与医院签订合同，同时在多家医院执业。

The health management system in Australia is divided into federal, state and local levels. Medical institutions are divided into private clinics for common diseases, specialized clinics such as dental department and ophthalmology department as well as public and private hospitals for the patients who require hospitalization without outpatient services. Public hospitals account for 72%, including veterans hospitals and public psychiatric hospitals. The mission hospitals founded by religious or charitable organizations are public hospitals. In public hospitals, staff is divided into full-time employees (management, interns, resident doctors and professors in teaching hospitals) who work on a regular full-time basis and freelancers (nurses and most of senior doctors and specialist doctors) who sign contracts with the hospital and practice in several hospitals at the same time.

澳大利亚医院建筑形式多样，有大型综合体附带专科医院的综合体形式，有多个不同类型的独立专科医院综合体建筑分布在相邻街区，相互靠近或以连廊连接。建筑多为集中式小高层建筑，带地下停车场，专科医院大多靠近医学院或直接设在融入市政街区的开放式医学院校内，医院建筑与学校的教学建筑共享校区的绿地、停车场、运动场、培训设施。

The healthcare buildings in Australia are diversified, including large complexes with specialized hospitals. Many different types of independent specialized hospitals of the building complexes are located in adjacent blocks, which are close to each other or connected by corridors. Most of the buildings are middle-height buildings with underground parking lots, and most of the specialized hospitals are close to medical schools or directly located in open medical schools in municipal blocks. Hospital buildings and teaching buildings of the schools share green land, parking lots, playground and training facilities of the campus.

3. 公共合同型

3. Public contracting operation

（1）日本

（1）Japan

实行政府导向性市场经济，推行医疗保险全覆盖政策，分为一级至三级诊疗。一级诊疗任务由诊所承担，二级诊疗由中小型医院承担，三级诊疗由特定医院承担。但由于不同等级医院医保报销比例相同，所以分级诊疗形同虚设，病人集中在设施条件好的医院，造成门诊等候时间较长。综合医院科室设置包括：管理部、临床部、急救部、药剂部、医技部、护理部，采取以病人为中心、以就诊流线为主线的运行体制。管理部负责挂号、收费、出入院、病案管理、信息管理，把非医疗性业务全部委托给企业，以企业管理方法将设备、仪器中心化，以提高使用效率、减少开支、减低成本。

The government-oriented market economy mode is implemented in Japan, and the full coverage policy of medical insurance is promoted, with three levels of diagnosis and treatment. The first-class diagnosis and treatment are undertaken by clinics, the second-class diagnosis and treatment are undertaken by small and medium-sized hospitals, and the third-class diagnosis and treatment are undertaken by specific hospitals. However, as the medical insurance reimbursement ratio of the hospitals in different levels is the same, the hierarchical diagnosis and treatment system is useless. Patients are concentrated in the hospitals with good facilities, resulting in a long waiting time for outpatient service. The departments of a general hospital include management department, clinical department, emergency department, pharmacy department, medical technology department and nursing department. Under the patient-centered operation system with the medical treatment streamline as the main line, the management department is responsible for registration, charging, admission and discharging, case management and information management, and entrusts all non-medical businesses to external enterprises, centralizing device and instruments by means of enterprise management, so as to improve use efficiency and reduce expenses and costs.

建筑形式是高度集约化的多层或中高层建筑，常以综合医院为主体，附带专科如儿科医院，紧密结合、资源共享，但出入口和门诊独立，配套设立体车库和学生公寓。通常建筑地下为机电、药剂、放射、消毒供应室、病案科室，下部为急诊、门诊、医技科室，上部为住院。护理单元分为ICU、HCU、一般单元三种护理等级。医疗模式分为高度急性期医院、急性期医院、恢复期医院、慢性期医院，以各种各样的医疗福利设施协同合作，为患者提供多元化医疗服务。医院建筑追求能够自由改变结构的功能类型，根据医学、医疗技术以及医疗器械的不断跨越式发展，医院建筑具有一定的灵活性，能通过内部改建或扩建来适应使用要求。

The buildings are highly intensive multi-storey or mid and high-rise buildings, which take the general hospital as the main body, with specialized hospitals such as pediatric hospital. The buildings are closely combined and share resources. However, the entrance and exit are independent of the outpatient department with mechanical parking systems and student apartments. Generally, the electromechanical device room, drug room, radiology room, disinfection supply room and medical record department are located underground; emergency department, outpatient department and medical technology department are located in the lower floors, and inpatient department is located in the upper floors. The nursing units are divided into ICU, HCU and general units. The hospitals are divided into highly acute hospitals, acute hospitals, convalescent hospitals and chronic hospitals by medical model, which cooperate with various medical welfare facilities to provide diversified medical services for patients. The structure of hospital buildings can be freely changed. With the continuous leap-forward development of medicine, medical technology and medical devices, hospital buildings can be flexibly reconstructed and expanded to adapt to the use requirements.

日本医院病房楼每个护士站所要照料的病床数比欧美医院的平均35床要多，一般为40～45床。普遍采取门诊化疗、扩大日间手术、多学科诊疗形式，按病种/器官/系统将多个科室集中为一个中心，按患者属性分围产中心、高龄中心。将医院、研究所、医疗相关企业、大学等组织联合起来，促进医学的发展，使医院成为诊疗、再生医疗的临床应用和实践的场所，推进医疗器械的研究开发，支撑医药品临床研究，构建医疗产业都市。

The number of hospital beds to be cared for is about 40 to 45 in each nurse station in the inpatient building in Japan, which is more than the average of 35 in the hospitals of Europe and America. Outpatient chemotherapy, extended day surgery and multidisciplinary diagnosis and treatment are generally adopted. Multiple departments are centralized in the same center according to disease species / organs / systems. The center is divided into perinatal center and elderly center according to patient attributes. Hospitals, institutes, medical enterprises, universities and other organizations are united to promote the development of medicine, making the hospitals as the places for clinical application and practice of diagnosis and treatment and regenerative medicine, so as to promote the research and development of medical devices, support clinical research on pharmaceuticals and build a medical industry city.

（2）德国

（2） Germany

实行社会市场经济模式（市场经济+总体协调+社会保障），公立医院占75%，其余是私立医院和教会医院。医疗体系实行分级管理，以大中城市为中心划分成数百个区域性医疗服务体系，在每一个区域内按标准配置四级医疗机构：社区服务（门诊）、跨区级服务级（专科医院）、中心服务级（综合医院）、最高服务级（大型医院）。每一个区域类医疗服务体系由一所最高服务级医院、两三所中心服务级医院、数十所跨社区服务级医院和数百家社区服务级诊所组成。

Social market economy mode (market economy + overall coordination + social security) is implemented in Germany. Public hospitals account for 75%, and the rest are private hospitals and mission hospitals. The level-to-level management is implemented in the medical system, which is divided into hundreds of regional medical service systems with large and medium-sized cities as the centers. There are four levels of medical institutions allocated according to the standards in each region: community service (outpatient departments), cross-region service (specialized hospitals), central service (general hospitals) and the highest service (large hospitals). Each regional medical service system consists of a hospital of the highest service level, two or three central service hospitals, dozens of cross-community service hospitals and hundreds of community service clinics.

私人诊所广泛分布在城乡各个角落，每个德国家庭都有自己的“家庭医生”，负责健康检查、保存病历、快速诊治，遇到急病重症立刻转到医院，一个私人诊所服务几十到数百人。小型区级医院包含内科、五官科、妇产科、儿科等常规科室，提供门诊和急诊服务，有少量的病房和小手术室，服务人数在一万以上。大型综合性医院包含急诊、门诊、医技、手术和大规模住院部，一般百万人以上配备一到两个大型综合性医院。大学医学院在教学科研的同时也给市民提供相当于综合性医院的服务。另外还有一定数量的专科医院，只收治特定的病症患者或者特定年龄段的病人。这样的机制使人员分流在院外，医院内人少，流程简单，环境舒适，建筑形式模块化、标准化，用材精致，建造技术工业化程度高。

Private clinics are widely distributed in urban and rural areas. Every family has its own “family doctor” in Germany, who is responsible for physical examination, medical history preservation and rapid diagnosis and treatment, and such doctor could immediately transfer the patients with acute and severe diseases to the hospital. A private clinic serves dozens to hundreds of patients. A small district hospital includes several routine departments, such as internal medical department, ophthalmology and otorhinolaryngology department, obstetrics and gynecology department and pediatric department, and provides outpatient and emergency services. There are a few wards and small operating rooms with a service population of over 10000. There are emergency department, outpatient department, medical technology department, surgery department and large inpatient department in a large general hospital. Generally, one or two large general hospitals are provided for the place with millions of people. The medical schools of universities provide the citizens with the services equivalent to those in general hospitals in addition to teaching and research. There are also a certain number of specialized hospitals that only treat the patients with specific diseases or of specific ages. Such mechanism keeps a majority of patients diverted outside the hospital. Therefore, there are few patients in the hospitals, with simple flow and comfortable environment. The buildings are modular, standardized with exquisite materials and highly industrialized construction technology.

4. 公私功能互补型（新加坡）

4. Operation with public and private function complementation (Singapore)

新加坡的医疗保健集团（包括公立医院和诊所）是国立公有制、法人治理、自主经营，受政府与公司法双重管控，注重精细化、标准化管理，执行国际职业卫生与安全管理体系认证，以提高服务水平和服务效率、控制医院服务成本。医院提供不同档次的床位供病人选择，床位区别在于能否自主选择住院条件和主治医师，不同类型的床位享受不同的政府津贴。A 类为私人性质的床位，政府不给津贴，B 类为半私人性质的床位，分 B1, B2 两层，政府津贴为总费用的 20% ~65%，C 类床位政府津贴达 80%。

Health care organizations in Singapore (including public hospitals and clinics) are state-owned, governed by the legal person and operated independently, which are controlled by both the government and corporate laws. These organizations pay attention to refine and standardized management and implement the certification of international occupational health and safety management system to improve service level and efficiency and control the service cost of the hospitals. The hospitals provide various beds of different levels for patients. Different beds determine whether the patients can choose their own hospitalization conditions and attending doctors. In addition, the patients with different types of beds enjoy different government subsidies. Class A beds are private, enjoying no government subsidies. Class B beds of two layers (B1 and B2) are semi-private, enjoying government subsidies of 20% to 65% of the total cost, and class C beds enjoy government subsidies of 80% of the total cost.

医院建筑强调在亚热带环境中的自然通风和遮阳防晒，广泛融入空中绿化，创造花园医院；将公共设施如公交车、出租车、救护车引入院内的架空层及平台上，立体交通、风雨无阻；将餐饮、书店、保健品商店、药店、健康管理中心、医学教育、专科私人诊所等相关业态的服务性商铺融入综合医院的医疗街内，形成消费力强、富有活力的商业内街，而诊疗单元内则安静、有序，达到动、静分区。

Hospital buildings emphasize natural ventilation, sun shading and sun screening in subtropical environment, widely integrate into skyrise greenery, and create garden hospitals; introduce public facilities such as buses, taxis, ambulances into the hospital's overhead floor and platform, realizing vertical transportation regardless of the weather; integrate catering stores, book stores, health products stores, pharmacies, health management centers, medical education institutions, specialized private clinics, and other service shops of relevant business types into the medical street of the general hospital, so as to form a commercial street with strong consumption and vitality, and quiet and orderly diagnosis and treatment unit, thus achieving dynamic and static division.

5. 混合市场经济型（法国）

5. Mixed market economy (France)

法国卫生部通过控制财权、人权和业务发展权等，控制全国所有医院的建设和发展，实行政府指导的混合市场经济模式。医院、诊所分公立和私立两类，公立医院为政府开办的福利事业单位，私立医院由慈善组织或其他机构开办。公立医院分为地区大学医院（医学中心）、省级中心医院、地方医院、专科医院和急诊医院五类，面向低收入居民。私立医院面向高收入家庭，主要接纳外科、妇产科病人，慈善机构医院负责恶性肿瘤等疾病的诊疗。私立医院提供超过 1/3 的住院治疗，较少提供急诊、康复、长期护理及精神疾病服务。私立医院数量多于公立医院，但公立医院在医疗服务中起主导作用。

The French Ministry of Health controls the construction and development of all hospitals in the country by controlling the right of property, human rights and the right of business development, and implements the mixed market economy mode guided by the government. Hospitals and clinics are public and private. Public hospitals are welfare institutions run by the government, while private hospitals are run by charitable organizations or other institutions. Public hospitals are divided into five categories: regional university hospitals (medical centers), provincial central hospitals, local hospitals, specialized hospitals and emergency hospitals, for low-income residents. Private hospitals are for high-income families, mainly admitting patients from surgical department, and obstetrics and gynecology department, and hospitals of charitable organizations undertake the diagnosis and treatment of cancer and other diseases. Private hospitals provide more than 1/3 of hospitalization, and fewer services in emergency, rehabilitation, long-term care and mental disease. The number of private hospitals is more than that of public hospitals, but public hospitals play a leading role in medical services.

综合医院在院区内分散设置专科门诊和私人门诊，门诊不设药房，病房分为普通病房、特护病房、私人病房、日间病房等。院内建筑按病种或专业分散楼座，即使合并一座大楼，也独立分区以内庭院隔开不同专科，分成有独立出入口、单独小门厅、有分散便利店服务的组团，不设共享大厅，将外来人群分散，避免人群聚集，只在后勤和医技手术部分设内部联通。建筑之间和内部没有特别复杂的分区和流程，多设内部采光庭院，少见黑房间，造型活泼。救护车均直接开进室内，不使外部看见急救患者，保护隐私，急诊部附近设直升机停机场。

In general hospitals, specialized and private outpatient departments (excluding pharmacy) are scattered in the hospital. The ward is divided into general ward, special care ward, private ward, daytime ward, etc. The buildings in the hospital are scattered by diseases or specialties. Even if they are combined into a building, the building is also divided into independent partitions by an inner courtyard for different specialties, forming units with independent entrances and exits, separate small lobbies, and decentralized convenience store services. No sharing hall is set, so as to decentralize the outside population and avoid people from flocking. Only the logistics department is internally linked with medical and technical as well as operating rooms. There is no particularly complex division and flow between and inside the buildings. More internal lighting courtyards are set, making dark rooms fewer, with vivid modeling. The ambulances are directly driven into the room to prevent the emergency patients from being seen from the outside and to protect the privacy. The emergency department is close to the helicopter airport.

1.3.1.2 中国医疗保健设施运营模式

1.3.1.2 Operation mode of health care facilities in China

1. 大陆医疗机构模式

1. Mode of medical institutions in mainland China

实行全民医保制度，有公立、中外合资合作、股份制、股份合作制、私有制等多种所有制形式。公立医院实行管办分开、政事分开、医药分开、营利性与非营利性分开。医院管理实行院长负责制或医院法人治理，实行绩效考核、岗位管理、竞争机制，医院和员工之间双向选择采用聘任制和聘用制。医院追求提高管理效率、提升运营绩效、提高医疗质量、提高医护人员待遇。

They adopt the national medical insurance system, in public system, Sino-foreign joint venture system, joint-stock system, joint-stock cooperative system, private system and other various forms of ownership. Public hospitals separate the management authorities from the founding authorities, the administrative organs from the institutional organizations, the medical services from the medicine supply, and the for-profit departments from the non-profit departments. Hospitals are managed through president responsibility system or subject to hospital corporate governance. Performance appraisal, post management and competition mechanism are adopted, and appointment system and employment system are used for two-way selection between hospitals and employees. Hospitals pursue to improve management efficiency, operation performance, medical treatment quality and remuneration of medical care staff.

政策要求医疗服务层级化，中央、省级医院设置少量承担医学科研、教学功能的医学中心或区域医疗中心，以及承担全国和区域性疑难病症诊治的专科医院等医疗机构，县市负责开办县级医院、乡村卫生院和社区卫生服务机构，其余公立医院由市负责开办。公立三级医院床位数量和建筑体积庞大，门诊规模大，护理模式单一，人流聚集度高，建筑空间流线复杂；普遍因病种不分、轻重不分，致使诊疗流程复杂，院感控制难度高。分级诊疗制度难以落实，民众的多样化需求得不到满足。

The policy requires that medical services should be tiered. Central and provincial hospitals should set up a small number of medical centers or regional medical centers for medical treatment, research and teaching, as well as specialized hospitals and other medical institutions for diagnosis and treatment of national and regional difficult and complicated diseases. Counties and cities should organize county hospitals, rural hospitals and community health service institutions, and cities should set up the rest of public hospitals. Public third-class hospitals are characterized by a large number of beds and huge building volume, large service scale of outpatient department, single nursing mode, high people aggregation, complex streamline of the building space; generally, the diagnosis and treatment flow is complex and the hospital sensing control is difficult due to no division of disease species and severity. It is difficult to implement the tiered diagnosis and treatment system, and impossible to meet the diverse needs of the public.

2. 中国香港医院管理模式

2. Management mode of hospitals in Hong Kong, China

香港医疗沿袭类似 NHS 的英国体制，公立医院病床虽已达到 4 张/千人，但数量仍在不断增多。政府出资、管办分离，其管理主体是医院管理局，负责整合公共医疗资源，提高公共资源使用效率。公立医院采取社区首诊制度、分流制度、轮候制度，看病需要预约、排队。未经首诊的患者不能进入上一级医院就诊，急诊除外。急诊病人先要经评估分为危殆、危急、紧急、次紧急、非紧急五类，再根据病人的不同类别确定治疗顺序后加以分流。轮候制度对象是有专科医疗服务需求的病人。

Hong Kong's medical institutions follow the British system similar to the NHS. Although the number of beds in public hospitals has reached 4/1000 people, the number is still increasing. With contribution of capital from the government and separation of management authorities from the founding authorities, hospitals are managed by the hospital administration that integrates public medical resources and improves the service efficiency of public resources. Public hospitals adopt community first diagnosis system, diversion system and waiting system, which requires appointment and queuing. Patients without first diagnosis are not allowed to seek medical services in upper class hospitals, except for emergency treatment. Emergency patients should be first divided into five categories upon evaluation: more critical, critical, urgent, sub-urgent and non-urgent, and then the treatment sequence should be determined according to different categories of patients, for diversion. The waiting system is applicable to patients in need of specialized medical services.

大型医院规划常按治疗方式或病种分不同的独立建筑，通过空中连廊连接，方便工作人员通行和转运病人，如癌症、妇产、康复。慈善机构建设的私立综合医疗大楼多为高层建筑，设有日间治疗、住院，每个病区内设置隔离或感染病房，提供学术交流、教堂、职工俱乐部、餐饮等配套服务。

During planning, large-scale hospitals are usually divided into different independent buildings according to treatment methods or disease species, which are connected by air connecting corridors to facilitate access of the staff and transfer of patients, involving cancer, maternity and rehabilitation. The private comprehensive healthcare buildings built by charitable organizations are mostly high-rise buildings, with functions of daytime treatment and hospitalization. Each ward is equipped with isolation or infection ward, providing academic exchange, church, staff club, catering and other supporting services.

3. 中国台湾医院管理模式

3. Management mode of hospitals in Taiwan, China

台湾卫生行政管理部门是行政院卫生署下辖的中央健康保健局、疾病管理局、国民健康局、中医药委员会、管制药品管理局等的分支机构，实行全民健康保险政策和健保基金统筹支付。疾病管理局负责疾病防疫，国民健康局负责健康教育、疾病干预。台湾医疗服务体系由医学中心、区域医院、地区医院、精神专科医院和卫生所组成，医院分为公立医院、财团法人医院和私人医院三类。医疗服务管理以法治和医院自治为主。各医疗机构的经营管理百花齐放、各取所需、各有所长，呈现多元化、多样性特征，多数医院不设门诊部。台湾私立医院设计中有较多中国特色的保健服务，并采取医疗养老相结合、医疗服务与商业服务相结合、健康管理与疾病治疗相结合的模式。

Taiwan's health administration is a branch of the Central Health Care Bureau, the Disease Administration Bureau, the National Health Bureau, the traditional Chinese Medicine Committee, and the Controlled Drugs Administration under the Health Department of the Executive. It implements the national health insurance policy and the overall payment of health insurance fund. The Disease Administration is responsible for disease prevention, and the National Health Bureau is in charge of health education and disease intervention. Taiwan's medical service system consists of medical centers, regional hospitals, district hospitals, psychiatric specialized hospitals and clinics. Hospitals are divided into three categories: public hospitals, consortium corporate hospitals and private hospitals. The management of medical services is based on the rule of law and hospital autonomy. The operation and management of various medical institutions are flourishing. Each takes what it needs and has its own strengths. Diversified and various characteristics are shown. Most hospitals do not have outpatient departments. There are many health care services with Chinese characteristics in the design of Taiwan's private hospitals, and the mode of combining medical care with pension, medical services with commercial services, and health management with disease treatment is adopted.

分析各国及地区医疗体制及相应的医院建筑得知，由于分级分类诊疗，各国及地区医疗机构建筑规模都不大，功能也不太复杂，无复杂的工艺流程。只要家庭医生和诊所可以解决，人们就尽量不去医院。建筑师力求把医院设计得像“天堂”或“艺术馆”那么轻松，致力于分散、分流人群，让医院内显得人越少越好，努力掩盖医疗行为，让公共空间只看到像家庭、商场、酒店、儿童乐园那样的美好舒适状态，让拥有高技术知识和能力的医学工作者在最美好的环境里工作、研发，使他们充满优越感。凡是有患者和医生停留的空间，普遍使用家庭化建筑装饰材料，包括仿制木材、人造地毯、陶瓷、雕塑、装置、壁画、舒适的沙发、背景音乐、吸声材料；只在技师操作的物理空间如实验室、消毒室内使用易洁材料。

Upon analysis on the medical system and corresponding hospital buildings in various countries and regions, the scale of medical institutions in various countries and regions is not large, the function is not too complex, and there is no complex process flow. As long as family doctors and clinics can work out, people try not to go to hospitals. Architects strive to design hospitals to be as easy as "paradise" or "art gallery", and devote to dispersing and diverting people, making the people in the hospital the fewer the better, try to cover up medical behaviors, making public space beautiful and comfortable like family, shopping malls, hotels and children's paradise, so that medical care staff with high-tech knowledge and ability can be engaged in work, research and development in the most beautiful environment, and be full of superiority. For the space where patients and doctors stay, household building decoration materials are generally used, including imitation wood, artificial carpet, ceramics, sculpture, installation, mural, comfortable sofa, background music and sound-absorbing materials; materials that are easy to clean are used only in the physical space operated by technicians, such as laboratory and disinfection room.

1.3.2 投资与管理策划

1.3.2 Investment and management planning

1.3.2.1 市场调查

1.3.2.1 Market survey

1. 城市所在地区的医疗资源总量，包括医疗机构数量、专业分类、医疗床位总量、千人床位占比率。

1. It covers the total amount of medical resources in the region where the city is located, including the number of medical institutions, specialty classification, total number of medical beds, and the ratio of beds/1000 people.

2. 确定医疗、医技、护理人员总量，区分公立与民营医院在区域内总门诊量、住院人数所占比例、床位使用率、周转次数、平均住院日。

2. Determine the total number of medical, medical and technical and nursing personnel, and distinguish between public and private hospitals in the total outpatient visits, proportion of inpatients, bed utilization rate, turnover times and average length of hospitalization in the region.

1.3.2.2 投资分析

1.3.2.2 Investment analysis

计算区域内医疗床位的增量需求，统计两个人均费用比：人均门诊费用比，其中药费所占比例；人均住院费用，其中药费所占比例。

Calculate the incremental demand of medical beds in the region, and make statistics of two per capita cost ratios: per capita outpatient cost ratio, including the proportion of expenses for medicine; per capita expenses for hospitalization, including the proportion of expenses for medicine.

1.3.2.3 市场分析

1.3.2.3 Market analysis

进行两个测算：规模预测、收入预测。

Make estimation on the following two items: scale and income.

1.3.2.4 医院规模测算

1.3.2.4 Hospital scale estimation

确定医院规模的前提条件：区域规划要求达到相应的常住人口数，医疗市场有刚性需求，当地优质医疗资源较少；在此基础上确定医院的规模和定位，如果超前扩大，建成后要长期负债经营。

The prerequisite for determining the hospital scale: According to regional planning requirements, the corresponding number of permanent residents should be reached, the medical market has rigid demand, and the local high-quality medical resources are fewer; on this basis, the scale and positioning of the hospital should be determined. If it is expanded ahead of time, it will be in operation on borrowings for a long time after completion.

1. 综合医院可以500床规模的医院为一个基本单位，对于一个300万人口的城市，按每千人4张床位计算，应设置医疗床位12000张，现有医疗床位9000张，则得知剩余市场空间。

1. As for general hospitals, hospitals with 500 beds are used as a basic unit. For a city with a population of 3 million, 12000 medical beds should be set up according to 4 beds/1000 people. Now, there are 9000 medical beds, so the remaining market space is clear.

2. 以一个日门诊量为1200人次的综合医院测算，住院人员为就诊人数的4%，则日均有48人需要住院治疗，以平均住院日为10天/（人/日），则医院的床位数是500张左右。

2. Upon estimation of a general hospital with 1200 daily outpatient visits, if the number of inpatients is 4% of that of patients, average 48 people need to be hospitalized every day. If the average number of hospitalization is 10 days/(person/day), the number of beds in the hospital is about 500.

3. 若为教学科研性综合医院，则宜采用800床医院为一个基本单位测算。

3. If it is a general hospital for teaching and scientific research, hospitals with 800 beds should be used as a basic unit for estimation.

1.3.2.5 市场定位

1.3.2.5 Market positioning

医疗保健应以当地的常见病种与已有医疗保健市场需求为依据。若公立机构是以当地常见病、多发病的医疗为主导，并已达到区域内一流的专业诊疗水平，民营机构选择投资的主营方向宜为：

Health care should target at the local common diseases and meet the existing market demand. If public institutions take the lead in the medical treatment of locally common and frequently-occurring diseases, and have reached the first-class professional diagnosis and treatment level in the region, the main investment direction of the private institutions should be as follows:

1. 弥补医院所在区域内的整体医疗保健需求的不足。

1. Make up for the deficiency of overall health care needs in the region where the hospital is located.

2. 满足未来周边城市与乡村的医疗保健需求。

2. Meet health care needs of surrounding cities and villages in the future.

3. 满足当地特殊人群的高端医疗保健需求。

3. Meet the high-end health care needs of local special groups.

1.3.2.6 发展步骤

1.3.2.6 Development steps

一所医院的发展步骤与建设周期和投资密切相关，一次性投资总资金量大，但单价低；分期投资总造价少，但单价高。若是分为两期，则设备也应分两期，并根据专业学科设置重点进行投资估算。当一所医院确定为三级规模时，必须按照卫生部医院等级设定标准的相关规定，满足科室设置及设备配置要求，否则医院就无法开展经营。

The development steps of a hospital are closely related to the construction period and investment. The total financing volume of one-time investment is large, but the unit price is low; as for phased investment, the total cost is small but the unit price is high. If it is divided into two phases, devices shall also be divided into two phases. In addition, the investment estimation shall be made based on the key points set in specialty. When it is determined as a three-class hospital, it must meet the requirements of department setting and device configuration in accordance with the relevant provisions of the hospital level setting standards of the Ministry of Health; otherwise, the hospital will not be in operation.

1.3.2.7 风险与效益分析

1.3.2.7 Analysis on risks and benefits

即评估医院经营风险，梳理医院经营思路；风险分析是确定医院经营策略的重要步骤。

It refers to evaluation of operation risks of the hospital and sorting of operation ideas of the hospital; risk analysis is an important step to determine operation strategies of the hospital.

1.3.2.8 经济效益分析

1.3.2.8 Analysis on economic benefits

1. 参照当地医院为基础的现行成本要素的平均值计算。

1. Conduct calculation with reference to the average of current cost elements based on local hospitals.

2. 按照国内医院管理最优的指标结构为基础计算，按医疗成本要素结构进行预测。

2. Carry out calculation based on the optimal index structure of hospital management in China, and make prediction according to the structure of medical cost elements.

1.3.2.9 投资管理

1.3.2.9 Investment management

1. 科室设置

1. Department setting

综合性医院编制要按照等级规模设定科室，并配置必要的装备，申请收费标准；专科医院要按照专科医院的标准设定规模，并配置必要的装备；要符合当地市场客观情况。

General hospitals should set up departments according to the level scale, equip necessary devices, and apply for charging standards; specialized hospitals should set up the scale according to the standards of specialized hospitals, and equip necessary devices; they should conform to the objective situation of the local market.

2. 设备配置

2. Device configuration

一所医院如果定位为二级医院或三级医院时，必须遵循中华人民共和国国家卫生健康委员会颁发的医院等级规定中床位设备与装备配置的要求。

If it is located as a second-class hospital or a third-class hospital, it must comply with the requirements of bed devices and device configuration in the hospital level regulations issued by the National Health Commission of the People's Republic of China.

1.3.2.10 组织结构与运营管理

1.3.2.10 Organizational structure and operation management

1. 正确的方向定位

1. Correct orientation

以市场需求为导向，服务对象明确，专科特点鲜明。

Guided by the market demand, service objects are clear and specialty characteristics are distinct.

2. 良好的组织体系

2. Good organization system

机构扁平化、简单化、分权管理，各部门独立核算；组织结构达到：后勤围绕医疗，医技围绕临床，医生、护士围绕客户；管理者由技术型专家转型为管理型专家。

The organization is flat, simple and decentralized, with independent accounting of each department; the organizational structure meets the following requirements: Logistics is around medical treatment, medical and technical treatment is around clinical treatment, doctors and nurses are around patients; managers are transformed from technical experts to management experts.

3. 服务以人为本

3. People-oriented services

从医疗流程服务内容、环境、便捷等方面考虑患者的最佳服务路径，求得最佳的就医者时间付出、精力消耗及费用性价比；以人为本的主体为客户和员工，要实现双方均满意。

Consider the best service path of patients from medical flow service content, environment, convenience and other aspects, to obtain the best time consumption, energy consumption and cost performance ratio of the patients; people-oriented services, should achieve the satisfaction of patients and employees.

4. 人力资源发挥作用

4. Role of human resources

统筹安排，合理规划，注重医护人员的教育培训，合理使用人才。

Make overall arrangement and reasonable planning, pay attention to the education and training of medical care staff, and intelligently use talents.

5. 经营管理的策略

5. Operation and management strategies

临床一线、辅助科室及保障科室运营管理均以价值为导向，建立全方位的经营管理体系；无论公立还是私立医院都应以品牌为导向，应进行业务流程设计、业务组织设计并持续改进或再造。

The operation and management of clinical front-line, auxiliary departments and guarantee departments are value-oriented, and an all-round operation and management system is established; both public and private hospitals should be brand-oriented, and should carry out business flow design, business organization design, and continuous reconstruction.

1.3.2.11 医院经营目标

1.3.2.11 Hospital operation objectives

应明确社会责任/医疗人数、患者满意度/医疗质量、发展后劲/医院收入、员工发展/薪酬水准；建立医疗安全的三大防线，事先防范、事中规范、事后补救。

Make clear social responsibilities/number of patients, patient satisfaction/medical quality, development potential/hospital income, employee development/salary level; establish three defense lines of medical safety, namely, prevention, standardization and remediation respectively before, during and after the process.

1. 科室建设目标

1. Department construction objectives

以市场为导向的学科建设，以协作为基础的梯队建设，以效率为指标的内部管理，以人性化为目标的过程管理。

They cover market-oriented discipline construction, cooperation-based echelon construction, internal management with efficiency as index and people-oriented process management.

2. 个人成长目标

2. Personal growth objectives

以名医为方向的专业技能，以互动为过程的医患关系，自我培育的道德素养，人生价值追求的幸福感。

They cover professional skills with famous doctors as the direction, doctor-patient relationship with interaction as the process, self-cultivation of moral quality, and happiness in pursuit of life value.

1.3.2.12 投资风险性与可行性

1.3.2.12 Investment risk and feasibility

1. 医疗保健设施建设投资包括：建筑、学科设置、设备配备、人力资本、市场前景、政策支持。

1. The investment in the construction of health care facilities includes: construction, discipline setting, device configuration, human capital, market prospect and policy support.

2. 投资决策因素：政策风险、经营风险、存在的商机。

2. Investment decision factors: policy risks, operational risks and existing business opportunities.

(1) 地区常住人口数量及其增长趋势是医院经营的前提，名医、名科的影响力与辐射力是医院发展的动力。

(1) The permanent resident population and its growth trend are the premise of hospital operation in a new region, and the influence and radiation of famous doctors and departments are the driving force of hospital development.

(2) 医疗行业成长周期长、技术要求高、投资巨大，民营资本注入存在风险，但若品牌与专业选择恰当，仍存在机会。

(2) The medical industry is characterized by a long growth cycle, high technical requirements and huge investment. There are risks in private capital injection, but there are still opportunities if the brand and professional selection are appropriate.

(3) 我国医疗资源总量不足，大型医院的发展表现为床位规模与医疗技术水平、专科水平成正比，与门诊量成正比，中心医院医疗技术影响力的辐射性促使中心医院在一定时期内要走外延扩张的道路。

(3) The total medical resources in China are insufficient. The development of large-scale hospitals shows that the size of beds is in direct proportion to the level of medical technology and specialty, as well as the amount of outpatient visits. The influence of medical technology used in central hospitals led to the extension and expansion in a certain period of time.

(4) 只有当投资者有能力立足长远进行规划经营，医疗保健行业的投资才是可行的。

(4) Only when the investor is able to perform long-term planning and operation, can the

investment in the health care industry be feasible.

(5) 投资医疗行业必须坚持高起点原则，坚持医教研结合的方向，才能确保医院长远发展。要避免同质化，要通过差异化的经营、优质化服务、质量化管理、良好的医疗环境，在民众中建立信誉，才具有发展优势。

(5) To ensure the long-term development of hospitals, we must adhere to the principle of high starting point and the direction of combination of medical treatment, teaching and research. To avoid homogenization, it is necessary to establish credibility among the public through differentiated operation, quality services, quality management and good medical environment, so as to have development advantages.

(6) 公立医院的发展受宏观政策的约束，基本任务是多发病、常见病，基本方向是适宜性；民营医疗保健设施的发展方向是弥补公立医疗之不足，两者之间可以优势互补，达到医疗保健市场多元化。

(6) The development of public hospitals is constrained by macro policies. The basic tasks are frequently-occurring and common diseases, and the basic direction is suitability. The development direction of private health care facilities is to make up for the deficiency of public health care, and the two can complement each other's advantages to achieve the diversification of health care market.

1.3.2.13 我国医院管理发展趋势

1.3.2.13 Development trend of hospital management in China

1. 管理制度现代化

1. Modernization of management system

产权清晰、责权明确、政事分开、管理科学。

Property rights are clear, responsibilities and rights are definite, administrative organs are separated from institutional organizations, and management is scientific.

2. 设置规划层级化

2. Tiering of setting planning

卫生部门实行统一规划、统一准入、统一监管，中央、省级可以设置少量承担医学、科研、教学功能的医学中心或区域医疗中心，以及承担全国或区域性疑难病症诊治的专科医院等医疗机构。县市主要负责开办县级医院，乡村卫生和社区卫生服务机构，其余公立医院由市负责开办。强化区域卫生规划，充分利用和优化配置现有医疗卫生资源，对不符合规划要求的医疗机构要逐步进行整合，提高医疗卫生资源利用效率。

The health department organizes unified planning, admission and supervision. Central and provincial hospitals may set up a small number of medical centers or regional medical centers for medical treatment, research and teaching, as well as specialized hospitals and other medical institutions for diagnosis and treatment of national and regional difficult and complicated diseases. Counties and cities should organize county hospitals, rural hospitals and community health service institutions, and cities should set up the rest of public hospitals. We should strengthen regional health planning, fully use and optimize the configuration of existing medical and health resources, and gradually integrate medical institutions failing to meet the planning requirements, so as to improve the utilization efficiency of medical and health resources.

3. 管理决策科学化

3. Scientization of management decision

循证医院管理：基于信息技术的精细化医院管理。

Evidence-based hospital management: refined hospital management based on information technology.

4. 管理人员职业化

4. Professionalization of management personnel

医院管理要有管理理论作为指导，要由经过医院管理教育和培训的人担任医院管理者。

Hospitals should be managed by the person who has received hospital management education and training under the guidance of management theories.

5. 管理手段信息化

5. Informatization of management means

电子病历系统，移动医疗及远程医疗，大数据利用，普及基层医疗机构信息化。

The electronic medical record system, mobile medical treatment and telemedicine, big data utilization, popularization of informatization of basic medical institutions.

6. 后勤管理社会化

6. Socialization of logistics management

把非医疗卫生技术服务工作从医院中分离出去，由社会承担，包括：基建、维修、洗涤、绿化、保洁、停车等后勤服务。

The non-medical technical services are separated from the hospital and undertaken by the society, including: infrastructure construction, maintenance, washing, greening, cleaning, parking and other logistics services.

1.3.3 医疗服务流程策划

1.3.3 Planning of medical service flow

1.3.3.1 医疗服务流程

1.3.3.1 Medical service flow

1. 医疗服务流程是满足病人健康需求而进行的一系列活动。医疗服务流程策划目的是在设施内部减少浪费，以最小的投入获得最大的产出，以最优质量和最低成本的产品对用户需求做出迅速响应；以合理而又有效的空间提供最好的服务，并能够降低空间成本；应持续地、规范地对流程进行策划和改进，实现精益管理和精益医疗。

1. Medical service flow is a series of activities to meet the health needs of patients. The planning of medical service flow is to reduce waste within the facility, obtain the maximum output with the minimum input, respond to the user's demand with the products with the best quali-

ty and the lowest cost; provide the best services with reasonable and effective space, and reduce the space cost; plan and improve the flow continuously and normatively for lean management and medical treatment.

2. 通过医疗服务流程策划，可使病人日就诊率增加，CT、MRI 日扫描量增加，癌症治疗中心的日间化疗和抗菌药物灌注人次提高，急诊室病人等待就医的平均时长降低，手术室数量减少，从而可以减少相应建筑空间的面积，这对于医院建设指标的制定意义重大。

2. Through the planning of medical service flow, the daily treatment rate of patients can increase, the daily scanning volume of CT and MRI increases, the number of patients receiving daily chemotherapy and antibiotics infusion in the cancer treatment center increases, and the average waiting time of patients in the emergency room decreases, so that the area of corresponding building space can be reduced. This is of great significance to the formulation of hospital construction indicators.

3. 通过医疗服务流程改善，减少病人不必要的等待时间，不仅能够缓解医院的拥挤现象，还可以提高工作效率。

3. The improvement of medical service flow reduces the unnecessary waiting time of patients, which can not only alleviate the crowding of hospitals but also improve the work efficiency.

1.3.3.2 临床路径

1.3.3.2 Clinical pathway

1. 按流程制定多学科综合服务规范，不仅局限于外科手术病人，从急性病向慢性病、从外科向内科扩展。

1. The formulation of multidisciplinary comprehensive service standards according to the flow is not only limited to surgical patients, but also extends from acute diseases to chronic diseases and from surgical department to internal medical department.

2. 应策划好医疗流程的临床路径，这涉及临床多学科和职能管理部门，要根据流程设定原则，基于循证医学、临床诊疗指南进行，把诊疗线路图转化成表单，配以相关规定和制度结合实际实施，并根据路径变异持续改善优化路径。

2. The clinical pathway of medical flow should be planned properly, which involves clinical multi-disciplinary and functional management departments. It shall be conducted according to the principle of flow setting, evidence-based medicine and clinical diagnosis and treatment guidelines, the diagnosis and treatment roadmap should be transformed into a form, with relevant regulations and systems combined with actual implementation, and the pathway should be continuously improved and optimized according to path variation.

3. 利用计算机技术，结合护理路径和病人反馈，设计防错监控、自动提醒、发布自动统计、信息流告知的视屏，让策划出的医疗服务流程处于低风险之中。

3. Use computer technology and combined with nursing pathway and patient feedback, video screens for error prevention monitoring, automatic reminder, automatic statistics and information flow notification, so that the medical service flow planned is at low risk.

1.3.3.3 基于病人需求的医疗服务流程

1.3.3.3 Medical service flow based on patient demand

1. 科学处理好战略流程、核心流程、支持流程三者之间的关系，确保支持流程是战略流程和核心流程的基础。

1. Scientifically handle the relationship among strategic flow, core flow and support flow to ensure that support flow is the basis of strategic flow and core flow.

2. 平衡不同的客户需求、定义差距、审核团队及信息的资源需求。

2. Balance different patient needs, define gaps, and review team and information resource needs.

3. 从检验医疗服务结果的设计转向检查医疗服务措施和行为的设计。

3. Design from medical service results of examination to medical service measures and behaviors of examination.

确保医疗服务流程与病人的需求相一致，理解病人需求与医疗服务需求的关系，按照病人需求的先后顺序确定服务需求的重要程度；与竞争者相比较，确定本机构在技术功能竞争性方面的差距，将各种需求做进一步分析，确定改善与设计的重点；病人需求重要度是医疗流程设计质量功能展开中极其重要的数量指标。

Ensure that the medical service flow is consistent with patient needs, understand the relationship between patient needs and medical service needs, and determine the importance of service needs in accordance with the order of patient needs; compare with competitors, to determine the gap in the competitiveness of technical functions of the organization, further analyze various needs, and determine the focus of improvement and design; the importance of patient needs is an extremely important quantity index in the deployment of design quality functions of medical flow.

4. 对医疗服务流程设计的结果要从患者满意度来评价，患者有三种类型的期望值：基本型、期望型和兴奋型。在流程设计时，要尊重流程使用者的建议和意见，激发他们的潜能，让他们参与到流程的设计中，这样可提高其对流程的依从性。

4. The design results of medical service flow should be evaluated from patient satisfaction. Patients have three types of expectation: basic type, expectation type and excitement type. In the flow design, we should respect the suggestions and opinions of process users, stimulate their potential, and ask them to participate in the flow design, so as to improve their compliance with the process.

5. 任何流程都有优化和改进的空间，要尽可能清除和减少不必要的非增值的环节，利用表格、程序、沟通、技术指导、自动物流，简化所有过于复杂的环节；通过活动整合、组织整合、顾客整合、供应商整合，理顺流程过程；对脏活、累活、险活、乏味的工作、数据采集、数据传输、数据分析，尽可能利用外部资源和自动化设备仪器、计算机程序来解决；应促进医院高层领导、中层管理者和一线医务人员寻找新的医疗服务流程管理方法，助推医院可持续发展。

5. There is room for optimization and improvement in any flow. Unnecessary and non value-

added links should be eliminated and reduced as much as possible. Forms, procedures, communication, technical guidance and automatic logistics should be used to simplify all overly complex links. Sort out the flow process by integration of activity, organization, patient and supplier; try to complete dirty, tiring, dangerous and tedious work, data collection, data transmission and data analysis, with external resources, automatic devices and instruments, computer programs as much as possible; promotes senior leaders, middle managers and front-line medical care staff of the hospital to look for new medical service flow management methods, so as to promote the sustainable development operation of hospitals.

1.3.3.4 消除医疗机构存在的八种浪费

1.3.3.4 Elimination of waste of medical institutions in the following eight aspects

1. 八种浪费

1. Eight aspects

(1) 质量缺陷——是为患者用药时药品或剂量有误、手术并发症。

(1) Quality defect——It refers to the wrong medicine or dosage and the complications of operation during medication for patients.

(2) 等待——是患者等待检查结果、等待预约、等待床位等流程。

(2) Waiting——It refers to the flow that patients wait for examination results, appointment, beds, etc.

(3) 不必要的走动——建筑布局不合理，导致非必要的走动。

(3) Unnecessary walking——It refers to the unnecessary walking for unreasonable building layout.

(4) 不必要的运输——把患者送到距离远的诊断室，去进行不必要的诊断，护士转移患者到新的病房或科室。

(4) Unnecessary transfer——It refers to the process of sending patients to a remote diagnosis room for unnecessary diagnosis, and nurses transfer patients to a new ward or department.

(5) 过量负荷——对患者多余的、过度的、不必要的诊断或治疗。

(5) Overload——It refers to extra, excessive and unnecessary diagnosis or treatment of patient.

(6) 重复作业——患者需要多次回答相同的问题，填写不必要的表格，护士需要记录不必要的信息。

(6) Repetition——It refers to the condition that patients need to answer the same question for many times, fill in unnecessary forms, and nurses need to record unnecessary information.

(7) 库存——柜子里存放多余或过期的药物，药物供应不足，导致患者延长等待治疗的时间。

(7) Inventory——It refers to storage of redundant or expired medicine, and insufficient medicine supply, resulting in that patients prolong waiting time for treatment.

(8) 未被利用的员工能力——不听取员工关于改善的建议，不支持员工发展学习新的医疗诊断和技术。

(8) Unused employee competence——Do not listen to employees' suggestions for improvement or support employees to develop and learn new medical diagnosis and technology.

2. 在八种浪费中，等待对患者治疗效果、医务人员的参与机制及治疗成本起到关键作用，要给患者带来更多的价值，必须尽可能缩短治疗时间，特别是对于心肌梗死和中风这类突发疾病快速治疗，意味着恢复得更好更快，甚至挽救生命，必须把缩短治疗时间作为流程设计的标准。

2. As for the waste among the above eight aspects, waiting plays a key role in treatment effect of patients, participation mechanism of medical care staff, and treatment cost. In order to bring more value to patients, we must shorten the treatment time as much as possible. Especially for sudden diseases such as myocardial infarction and stroke, rapid treatment means better and faster recovery, or even saving of lives. We must take shortening the treatment time as the flow design standard.

3. 医疗保健服务业在当日诊疗过程中，病人数量不断波动，诊疗量难以事先预测，要充分考虑现有服务能力的调整，合理及时安排。精益管理是在质量安全、效益成本、服务能力与速度三方面取得平衡的行之有效的医疗机构管理方法。

3. As for the health care service industry, in the process of diagnosis and treatment on that day, the number of patients is constantly fluctuating, and it is difficult to predict diagnosis and treatment volume in advance. We should fully consider the adjustment of existing service capacity, and make reasonable and timely arrangements. Lean management is an effective way for medical institutions to balance quality and safety, benefit and cost, service ability and speed.

1.3.3.5 在业务流程中应用信息化技术

1.3.3.5 Application of information-based technology in business flow

1. 预约挂号系统

1. Appointment registration system

预约挂号系统流程设计——是基于就医患者凭本人就诊卡或医保卡、新农合卡、市民卡就医，医院在预约、就诊和费用结算等过程中核对患者身份的就医模式，可以实现整个就诊流程的集约化、统一化、系统化管理。

The flow design of appointment registration system is based on the medical treatment mode that patients use their own medical card or medical insurance card, new rural cooperative medical card, citizen card, and the hospital checks the patient's identity during appointment, medical treatment and settlement, so as to realize the intensive medical treatment flow, and unified and systematic management.

2. 缴费结算系统

2. Payment and settlement system

(1) 患者对就诊卡进行预充值，医院将住院收费和门诊收费合二为一，并提供多种形式的结算渠道，如微信、支付宝，方便患者。

(1) Patients conduct pre-charge for medical cards, and hospitals combine inpatient expen-

ses with outpatient expenses, and provide settlement channels in various forms, such as WeChat and Alipay, for convenience.

（2）采用一站式服务的医生工作站，以实名制预存款代替挂号、收费、预约等环节，还可提供辅助检查预约、复诊预约、会诊预约、缴费结算等一站式服务。

（2） The doctor workstation with one-stop services can replace registration, charge, appointment and other links with pre-deposit of real-name system, and can also provide one-stop services such as auxiliary examination appointment, follow-up appointment, consultation appointment and payment settlement.

（3）在住院区的自助服务机，可以实现预存转住院预交款、银联卡转账、住院预交款、住院费用查询、一日清单查询和打印、住院费用汇总查询和打印、病历打印、出院结算余额转就诊卡、住院患者满意度调查等功能，因此产生门诊大厅和住院大厅没有具体功能的情况，可大大减少建筑公共服务空间的面积，或可将这些厅堂空间用于员工和访客的餐厅以及商业服务空间。

（3） The self-service machine in the inpatient area can enable functions of transfer of pre-deposit to inpatient prepayment, transfer of Unionpay card to inpatient prepayment, inquiry of inpatient expenses and daily list, printing of inquiry results of inpatient expense summary, printing of medical record and discharge record, transfer of the settlement balance to the medical card, inpatient satisfaction survey, etc. As a result, the outpatient hall and inpatient hall have no specific functions, which can greatly reduce the area of public service space in the building or use such space for restaurants and commercial service spaces for employees and visitors.

3. 辅助检查系统

3. Auxiliary examination system

辅助检查系统优化的原则是人—机—环境达到最佳匹配，充分利用现有资源，最大程度减少阻塞环节，提高工作人员的效率，提高患者的满意度，患者在医院停留时间即为流程各个串联服务环节的停留时间之和。因此，流程设计必须把串联服务环节变成并联服务环节，减少非增值环节，取消无效环节，减少患者在医院停留时间。

The optimization principle of auxiliary examination system is to achieve the best matching of man-machine-environment, fully use existing resources, minimize blocking links, and improve staff efficiency and patient satisfaction. The stay time of patients in the hospital is the sum of the stay time of all series service links in the flow. Therefore, the flow design must turn series service links into parallel service links, reduce non value-added links, cancel invalid links, and reduce the stay time of patients in the hospital.

4. 药品分发流程

4. Medicine distribution flow

（1）在信息系统的支持下，医院可以在中心药房的基础上设置众多分药房，患者不再需要携带打印出来的处方单，也不再需要将处方单递交给药房工作人员，而是直接凭就诊卡到就近的药房领取相应药品，整个过程无纸化操作。

（1） With the support of the information system, the hospital can set up many branches on the basis of the central pharmacy. Patients no longer need to carry the prescription printed or submit the prescription to the staff of the pharmacy. Instead, they can go to the nearest pharmacy to get the corresponding medicine directly with the medical card. The whole process is paperless.

（2）中心药房采用自动发药机设备，节省了人力成本，为患者取药提供了更为安全精准的方式。医生诊断开药后，药房自动发药机自动调配药品向患者发放，避免患者在诊室和药房之间往返。需要输液打针的患者无须自行提着药品，从药房走到输液注射室，患者只需要去输液注射室旁的分药房刷一下就诊卡，相关药品的配送就能够在药房的后台由护士操作完毕，减少输液患者往返缴费奔波。

(2) The central pharmacy is equipped with automatic dispensing machines, which saves labor cost and provides a more secure and accurate way for patients to take medicine. After doctors give diagnosis and medication, automatic dispensing machines in the pharmacy automatically dispenses medicine to patients, avoiding patients from going back and forth between the consulting room and the pharmacy. Patients who need infusion and injection do not need to carry the medicine by themselves. When walking from the pharmacy to the infusion and injection room, they only need to go to the pharmacy branch next to the infusion and injection room to swipe the medical card. The distribution of related medicine can be completed by the nurse in the backstage of the pharmacy, reducing rushing of infusion patients for round-trip payment.

5. 良好的业务流程策划可以缩短就诊时间、节约人力成本、改善就诊体验。业务流程设计或再造是永无止境的，是一个持续变化、不断改进的方法。对于医疗保健设施的建筑空间来说，应该保持有一定的弹性能够适应这种变化。

5. Appropriate business flow planning can shorten the visit time, save the labor cost and improve the medical treatment experience. The design or rebuilding of business flow is endless. It is a method of continuous change and constant improvement. For the building space of health care facilities, it should be certainly flexible to adapt to this change.

1.3.4 设备及设施策划

1.3.4 Planning of devices and facilities

1.3.4.1 医疗设备设施配置

1.3.4.1 Configuration of medical devices and facilities

1. 医疗设备设施配置及分期投入必须符合项目医疗保健业务规划、服务规划，所需仪器设备根据分期运营进行规划，除运营初期开业需购置的设备外，其余设备将依据实际运营目标分期、逐步购置，并备注拟投入周期，所投入设备设施在投资分析中均以预期财务费用支出时间为准，列入财务模型。

1. The configuration and phased investment of medical devices and facilities must conform to health care business planning and service planning of the project. The instruments and devices required shall be planned according to phased operation. Except for the devices to be purchased at opening in the initial operation, the rest of the devices will be purchased in phases and steps according to the actual operation objectives, and the proposed investment period shall be noted.

2. 医疗保健设备投入以学科战略规划为依据，来开展各项技术服务。财务测算应在实际开展该项技术并取得经营收入前至少六个月投资到位，为设备验收、人员培训上岗、设备申报资质及计量做好时间准备。

2. The devices and facilities invested shall be subject to the spending time of expected financial expenses in the investment analysis and included in the financial model. The investment of health care devices is based on the disciplinary strategic planning, to carry out various technical services. The financial estimation shall be invested at least six months before the actual implementation of the technology and the business income, so as to make time preparations for device acceptance, personnel training, device application qualification and measurement.

3. 策划表：见表 1-1。

3. Format of planning table: see Table 1-1.

表 1-1 策划表

Table 1-1 Format of planning table

序号 No.	设备名称 Device name	科室 Department	单价 Unit price	数量 Quantity	预算金额 Budget amount	备注 Remarks
1	X 光机 X-ray machine	医学影像科 Medical imaging department	…	1	…	开业配置 Opening configuration
2	移动式 X 光机 Mobile X-ray machine	医学影像科 Medical imaging department	…	1	…	开业配置 Opening configuration
3	数字 DSA 造影机 Digital subtraction angiography(DSA) machine	介入中心 Intervention center	…	1	…	运营第三年配置 Configuration in the third year of operation
4	心电生理三维标测系统 ECG physiological 3D mapping system	介入中心 Intervention center	…	1	…	运营第三年配置 Configuration in the third year of operation
…	…	…	…	…	…	…
合计 Total		…	…	…	…	…

1.3.4.2 信息系统

1.3.4.2 Information system

1. 信息系统投入包含三项内容：硬件设备、软件设备、系统运营维护费用。

1. The investment of information system includes expenses for the three parts: hardware, software, and system operation and maintenance.

2. 信息化建设是一个持续投入的过程，与医疗规划、科研、教学密切相关。

2. Information-based construction is a process of continuous investment, which is closely re-

lated to medical sources planning, scientific research and teaching.

3. 应考虑信息系统每年运营维护、系统开发、硬件设备周期性更新等费用的持续性投入。

3. The continuous investment of annual operation and maintenance, system development, periodic update of hardware and other costs of information system shall be considered.

1.3.4.3 运营基础设施投入

1.3.4.3 Investment of operation infrastructure

运营基础设施包括：办公家具、培训教学、营养食堂、员工餐厅等。

Operation infrastructure includes: office furniture, training and teaching room, nutrition canteen, staff canteen, etc.

1.3.4.4 开办费

1.3.4.4 Organization expense

1. 指设施在筹备期间所发生的各项费用，包括设施筹建期人力成本、办公费、培训费、差旅费、印刷费、申请医疗保健机构注册登记费用以及不计入固定资产和无形资产购建成本的汇兑损益和利息支出。

1. It refers to all expenses for facilities incurred during preparations, including labor cost, office expense, training expense, travel expense, printing expense, expense for application for registration of health care institutions, exchange gains and losses and interest expense excluded in the acquisition and construction cost of fixed assets and intangible assets.

2. 筹建期是指从机构获得上级行政主管部门“同意设置”批准之日起至正式取得医疗保健机构执业许可证（或正式开业运营）的期间时段。

2. The preparation period refers to the period from the date when the institution obtains “setting approval” from the superior administrative department to the time when the practice license (or formal opening for operation) of the health care institution is formally obtained.

3. 开办费一次性进入当期损益，在管理费用中核算，不在长期待摊费用中核算。

3. The organization expense shall be included in the current profits and losses all at once, and accounted in the management expense, instead of long-term unamortized expense.

1.3.5 以医院为导向的开发（HOD）策划

1.3.5 Hospital Oriented Development (HOD) planning

1.3.5.1 HOD模式是以医院为社区发展的核心驱动力，基于“服务设施-土地利用”相互关系进行土地开发、设施规划及社区管理。

1.3.5.1 HOD mode is to take the hospital as the core driving force of community development, which is based on the relationship of “service facilities-land use” to carry out land development, facility planning and community management.

1.3.5.2 将医院置于城市或社区的大环境之中，将医院内的部分功能和服务社会化，把能让社会承担的部分分离出来，利用市场机制，获取全面、低耗、优质的服务，细化社会分工，提高专业化程度，使医院管理和运营的重心更加集中于医、教、研方面。

1.3.5.2 Put the hospital in the environment of the city or community, socialize some functions and services in the hospital, separate the parts that can be undertaken by the society, use the market mechanism to obtain comprehensive, low-cost and high-quality services, refine the social division of labor, improve specialization, and focus more on medical treatment, teaching and research during hospital management and operation.

1.3.5.3 发掘公共交通、城市公共配套设施与医院的相关作用。

1.3.5.3 Explore the role of public transport and urban public auxiliary facilities related to hospitals.

1.3.5.4 实现可持续发展，既是医院自身的可持续发展，更是所在社区和城市的可持续发展。

1.3.5.4 Sustainable development is for both the hospital and the community and the city.

1.3.5.5 医院要以发展的眼光考虑规划设施结构，以永续经营的思维和眼光看待医院建设工程，坚持动态发展的主导思想。新建医院运营初期对城市周边影响不大，一旦医院满负荷甚至超负荷运转，对周边就会产生很大的影响，如交通压力，商业要求，旅业要求，停车要求等，因此城市规划和卫生管理部门应对此有远近结合的城市功能协调性部署。

1.3.5.5 The hospital should consider planning the structure of facilities from the perspective of development, treat the hospital construction project from the perspective of sustainable operation, and adhere to the leading idea of dynamic development. During initial operation, the new hospital has a little impact on the surrounding areas of the city. Once the hospital is fully loaded or even overloaded, it will have a great impact on the surrounding areas, such as traffic pressure, business requirements, tourism requirements and parking requirements. Therefore, the urban planning and health management departments should have a coordinated deployment of urban functions combining long-term and short-term considerations.

1.3.5.6 重视医院的核心驱动作用和医疗服务产业的发展，使之除了为居民提供医疗卫生服务、自身创造价值以外，而且还成为一个地区重要的人群聚集地和发展驱动力，凭借自身需求带来的大量人流、物流、信息流，拉动其周边产业、物业、商业的持续发展。

1.3.5.6 Pay attention to the core driving role of the hospital and the development of the medical service industry, to not only provide medical and health services for residents and create value for itself, but also become an important crowd gathering place and development driving force in a region, thus promoting the sustainable development of its surrounding industries, properties and businesses by virtue of a large number of people flow, logistics, information brought by its own needs.

1.3.5.7 以全局的思维、用发展的眼光，重视医院对商业的驱动作用，对城市公共建筑风貌的促进作用；充分利用社会资源，促进医院和城市、社区的和谐互动。

1.3.5.7 Focus on the driving role of the hospital for business and the promotion role of the urban public building style from the perspective of the overall thinking and development;

make full use of social resources to promote the harmonious interaction between the hospital, the city and community.

1.4 设计任务书

1.4 Requirements specifications

1.4.1 编制设计任务书的时机

1.4.1 Time for preparation of requirements specifications

1.4.1.1 当新建或改扩建医院建设项目被批准立项，完成了前期策划，承建方获得正式批复的建设规模和投资规模之后，建设方需要进入设计文件制定阶段时，应编制医院建设的设计任务书，向设计机构下达设计任务。

1.4.1.1 After the construction projects for construction or reconstruction and expansion of hospitals are approved, the preliminary planning is completed, and the construction scale and investment scale are officially approved by the project owner, the construction owner needs to formulate design documents, prepare the requirements specifications for the hospital construction, and assign the design tasks to the design institution.

1.4.1.2 当医院需要对已有建筑或场地进行改造、扩建、修缮时，应当布置设计任务，交由专业人员进行施工图纸设计工作。

1.4.1.2 When the existing buildings or sites of a hospital need to be reconstructed, expanded and repaired, the design tasks shall be assigned and the CD stage shall be handed over to professionals.

1.4.1.3 设计任务书是工程建设大纲，制约着工程建设的全过程和各个方面，建设项目的决策得当与设计任务书的正确编制关系十分密切；设计任务书是项目建设方案的决策性文件，是编制设计文件的主要依据。

1.4.1.3 Requirements specifications, the outline of project construction, restrict the whole process and all aspects of the project construction. Proper decision-making of the construction project is closely related to the correct preparation of the requirements specifications. Requirements specifications are the decision-making documents of the project construction proposal and a main basis for preparing the design documents.

1.4.2 设计任务书类型

1.4.2 Type of requirements specifications

1.4.2.1 简单的设计任务可以用委托函的形式提出，或以传真邀约、电子邮件告知，也可以直接在设计合同中陈述设计要求，只要承揽任务一方能明白所要承担的任务即可，而不必专门编制设计任务。例如：

1.4.2.1 Simple design tasks can be put forward in the form of letter of entrustment, invited by fax, informed by e-mail, or stated directly in the design contract. It is not necessary to specially prepare the design tasks as long as the project owner can understand the tasks to be undertaken. For example:

1. 委托编制用于申办规划设计条件的总平面指标图;

1. General layout plan prepared by entrustment for bidding planning and design conditions;

2. 帮助业主在现状中找出可以改扩建或拆除的旧建筑或可清空的场地，画出改扩建用地范围或室内改造空间范围;

2. Help the owner to find out the old buildings or empty sites that can be reconstructed, expanded or demolished in the current situations, and draw out the scope of land for reconstruction and expansion or the scope of indoor reconstruction space;

3. 标准化设计任务，要求按某个规定标准进行设计，该标准已经描述了完整的建设要求，而建设方除此之外没有其他个性化要求;

3. Standardized design tasks require the design to be carried out according to a specified standard, which has already described the complete construction requirements, beyond that the project owner has no other personalized requirements;

4. 某医院复苏室装修设计;

4. Decoration design of recovery room in a hospital;

5. 某医院配电房改造设计;

5. Reconstruction design of power distribution room in a hospital;

6. 大门、围墙设计。

6. Design of gate and enclosure.

对于规模大、有个性化要求的医院建设项目，必须编制专门的设计任务书。设计任务书根据建设方交办任务不同，主要有以下类型:

Special requirements specifications must be prepared for large-scale hospital construction projects with personalized requirements. According to different tasks assigned by the project owner, the requirements specifications mainly include the following types:

1.4.2.2 总体设计任务书

1.4.2.2 Overall requirements specifications

1. 概念性规划设计任务书

1. Conceptual planning and requirements specifications

2. 总体（发展）规划设计任务书

2. Overall (development) planning and requirements specifications

3. 医院改扩建规划设计任务书

3. Hospital reconstruction and expansion planning and requirements specifications

4. 医疗工艺设计任务书

4. Specifications of design for medical process

5. 医院信息系统设计任务书

5. Hospital information system requirements specifications

1.4.2.3 单体工程设计任务书

1.4.2.3 Individual works requirements specifications

1. 建筑设计任务书（包含医院全部建筑单体设计要求）

1. Architectural requirements specifications (including the design requirements for all individual buildings of the hospital)

2. 门诊楼、住院楼等单体建筑设计任务书（个别建筑）

2. Requirements specifications for outpatient building, inpatient building and other individual buildings (single building)

1.4.2.4 单项工程设计任务书

1.4.2.4 Individual works requirements specifications

1. 医院室内装修设计任务书

1. Hospital interior decoration requirements specifications

2. 管线系统改造设计任务书

2. Pipeline system reconstruction requirements specifications

3. 中心手术部设计任务书

3. Central operation department requirements specifications

1.4.3 设计任务书内容

1.4.3 Content of requirements specifications

1.4.3.1 设计任务书应符合项目实际需要，结合项目自身特点，整合各医疗分项功能的策划成果、结合医院管理和职能部门的实际情况以及医院未来业务发展规划，应介绍项目概况，提供设计条件、依据、标准，提出设计内容、要求、范围、深度。

1.4.3.1 The requirements specifications shall meet the actual requirements of the project, integrate the planning results of various medical sub-functions, and combine the characteristics of the project itself as well as the actual situations of hospital management and functional departments and the future business development planning of the hospital. The general overview of the project shall be introduced, the design conditions, basis and standards shall be provided, and the design contents, requirements, scope and depth shall be proposed.

1.4.3.2 建筑设计任务书内容

1.4.3.2 Content of architectural requirements specifications

1. 建设目的、依据和设计指导思想；

1. Construction purpose, basis and design guiding ideology;

2. 建设地点、建设内容与规模、规划设计条件（包括容积率、覆盖率、绿化率、建筑退线、道路机动车开口限制、古建筑保护要求、建筑限高、外形限制条件、日照要求等）、技术经济指标；

2. Construction site, construction contents and scale, planning and design conditions (including plot ratio, coverage rate, greening rate, building setback, road vehicle opening limit, ancient building protection requirements, building height limit, shape limit conditions, sunshine requirements, etc.), technical and economic indicators;

3. 场地建设条件、基础设施、四周建筑物、交通管制、供排水、供电、通信、能源、动力供应等情况；

3. Site construction conditions, infrastructure, surrounding buildings, traffic control, water supply and drainage, power supply, communication, energy, supply of power, etc. ;

4. 建设项目功能要求：应包括医院编制规模、医疗规划、工艺要求、医疗设备规划与场地空间要求、各科各部门感染控制要求、医疗设施（含水、电、通风、空调、气体、物流站点、信息点等）要求、工作人员岗位配备数量、班次等；

4. Functional requirements of construction project: It shall include the hospital preparation scale, medical planning, process requirements, medical device planning and site space requirements, infection control requirements of all departments, medical facility requirements (including water, electricity, ventilation, air conditioner, gas, logistics stations, information stations, etc.), number and shift of staff posts, etc. ;

5. 防空、防震、环保、绿色、节能与循环经济等要求；

5. Requirements for air defense, earthquake protection, environmental protection, green, energy conservation and circular economy, etc. ;

6. 设计周期及建设工期要求；

6. Requirements for design period and construction period;

7. 投资造价控制额；

7. Investment cost control amount;

8. 图纸及文件要求。

8. Requirements for drawings and documents.

1.4.3.3 改扩建项目设计任务书

1.4.3.3 Requirements specifications of reconstruction and expansion projects

1. 改扩建建筑选址情况描述；

1. Description of site selection for reconstruction and expansion buildings;

2. 既有建筑、构筑物与新建筑或改扩建建筑之间的连接要求；

2. Connection requirements between existing buildings, structures and new buildings or reconstruction and expansion buildings;

3. 因局部扩建而引起新旧连接体共同装修改造的要求；

3. Requirements of joint decoration and renovation for new and old connected bodies caused by local expansion;

4. 运营当中的既有建筑在施工期间的使用防护要求；

4. Protection requirements for use of existing buildings in operation during construction period;

5. 分步骤拆除及分步建设的要求；

5. Requirements for step-by-step demolition and construction;

6. 新建筑与既有建筑的风格（统一）、材料要求；

6. Requirements for (unified) style and materials of new buildings and existing buildings;

7. 原有室外管线因改扩建而需要增容改造的要求；

7. Requirements for capacity expansion and reconstruction of original outdoor pipeline due to reconstruction and expansion;

8. 原有污水处理站因改扩建而需要增容改造的要求；

8. Requirements for capacity expansion and reconstruction of the original sewage treatment station due to reconstruction and expansion;

9. 为改扩建创作而提供的现状条件图，弹性规划的发挥余地。

9. Current condition diagram provided for reconstruction and expansion, and scope of flexible planning.

1.4.4 设计任务书作用

1.4.4 Role of requirements specifications

设计任务书能使各专业的设计师从方案、初步设计到施工图全过程得到明确的指示，使设计能满足各项功能要求。如果医院设计任务书深度不足、内容欠缺、数据不完善、数据错误或缺乏依据，将导致设计过程的不断变更，甚至误导而产生错误的设计方案。任务书不合格产生以下后果：

Requirements specifications can provide the designers of all kinds of specialty with clear instructions from the whole process of SD, PD and CD, and ensure the design meets the functional requirements. If the hospital's requirements specifications lack depth, contents, data and basis or error in data, it shall lead to the continuous changes of the design process, and even mislead designers to make a wrong design proposal. The following consequences shall result from the unqualified specifications:

1.4.4.1 第一种情况：由招标代理制定设计任务书，医院方面没有参与决策。在招标产生建筑方案后，医院方面抛开该设计任务书，随意提出设计要求，导致设计方无所适从。

1.4.4.1 The first situation is that the bidding agency prepares the requirements specifications, and the hospital does not participate in the decision-making. After the construction proposal has been produced by bidding, the hospital puts aside the requirements specifications and puts forward the design requirements at will, resulting in the design institution's confusion.

1.4.4.2 第二种情况：医院口头下达任务，重外观，不重功能，由设计师自由发挥，导致随意性大的盲目设计。

1.4.4.2 The second situation is that the hospital gives tasks orally, emphasizing appearance rather than function, and the designers are free to give full play to them, resulting in blind design with great randomness.

1.4.4.3 第三种情况：任务书只规定了一般建筑功能，没有明确医疗运营模式、医疗空间规划要求、工艺原则、医疗设备布置要求和医疗设施要求，不能指导详细设计。

1.4.4.3 The third situation is that the specifications only specify the general building functions, do not specify the medical operation mode, medical space planning requirements, process principles, medical equipment layout requirements and medical facility requirements, and cannot guide the detailed design.

1.4.4.4 第四种情况：任务书所规定的功能指标错误，数据自相矛盾，如：在规划限制条件下无法实现的指标，不可行的设计要求，各分项科室或房间面积指标之和大于总面积指标。

1.4.4.4 The fourth situation is that the functional indicators specified in the specifications are wrong, and the data is contradictory. For example, the indicators that cannot be achieved and the infeasible design requirements under the planning restrictions, and the sum of the area indicators of each department is greater than the total area indicators.

1.4.4.5 第五种情况：强调美观，忽视可行性和造价，违背投资控制原则，导致工程投资失控。

1.4.4.5 The fifth situation is emphasizing beauty instead of feasibility and cost, violating the principle of investment control, thus resulting in out-of-control project investment.

1.5 设计咨询与设计监理

1.5 Design consultation and supervision

1.5.1 设计咨询

1.5.1 Design consultation

顾问受聘于医院、建设方（投资方或代建方）、设计方、施工方，为之提供有关设计的咨询服务，属于设计咨询。有下列几种情况：

The consultants are employed by the hospital, the project owner (investor or construction agent), the design institution, and the construction company, to provide consulting services related to the design, namely design consultation. Design consultation includes the following situations:

1.5.1.1 医院或建设方聘请顾问对建筑设计院的设计工作进行优化。

1.5.1.1 The hospital or the project owner employs the medical consultants to optimize the design of the architectural design institute.

1.5.1.2 建筑设计院聘请医疗顾问机构为其提供医疗咨询服务。

1.5.1.2 The architectural design institute employs the medical consultant organization to provide medical consulting services.

设计机构的咨询顾问应站在先进合理的专业角度向设计师提供医疗工艺要求和咨询服务，做出医疗工艺设计指导，使设计师能够在医院不提供设计任务书或任务书不详细的情况下完成设计，帮助设计师深化医院设计的流程、合理配置医疗设施、解答有关医疗的专业问题。同时也可以代表设计方向医院或建设方讲解医疗专业流程，帮助其理解图纸、做出科学的决策。

The consultants of design agencies shall provide medical process requirements and consulting services for the designers from an advanced and reasonable professional perspective, to guide the medical process design, enable the designers to complete the design in the case that the design specification is not provided by the hospital or not detailed, help the designers deepen the hospital design process, rationally allocate the health care facilities, and answer the professional questions on medical treatment. The consultants can also explain the medical professional process to the hospital or the project owner on behalf of the design institutions, and help them understand the drawings and make scientific decisions.

1.5.1.3　医院工程总承包方聘请医疗顾问机构为其提供医疗咨询服务。

1.5.1.3　The general contractor of hospital project employs the medical consultant organization to provide medical consulting services.

医院工程的施工机构通常需要聘请精通医疗工艺的咨询顾问或专家，为其所做的室内装修工程提供医疗功能深化设计要求。工程总承包机构的咨询顾问应提供医疗工艺流程和医疗设施的施工指导，使总承包机构能够在承建方不能提供技术要求的情况下完成施工，并解答有关医疗的专业问题。

The construction organizations of hospital projects need to employ the consultants or experts who are proficient in medical process to provide the deepening design requirements of medical function for their interior decoration projects. The consultants of the general contractor shall provide the medical process flow and the construction guidance of healthcare facilities, so that the general contractor can complete the construction in the case that the specifications are not provided by the project owner, and answer the professional questions on medical treatment.

1.5.2　设计监理

1.5.2　Design supervision

是指医院或建设方聘请医疗顾问监督建筑设计院的设计工作，落实医院的医疗要求。医疗顾问的行为属于设计监理。

Design supervision refers to that the hospital or the project owner employs the medical consultants to supervise the design of the architectural design institute, and implement the medical requirements of the hospital. The function of medical consultants is design supervision.

作为医院或建设方的咨询顾问，应充分了解医院或建设方的需求，编写设计任务书，协助建设方进行设计全过程的医疗功能把关，根据建设期间医院或建设方可能提出的建设修改意见及时调整或修改设计任务书，并负责检查设计工作的完成情况。

The consultant of the hospital or project owner shall fully understand the demands of the hospital or project owner, prepare the design specifications, assist the project owner to check the medical function in the whole design process, timely adjust or modify the design specifications according to the construction modification suggestions that the hospital or project owner may put forward during the construction period, and check the completion of the design work.

1.5.3 装备设施与系统优化咨询

1.5.3 Consultation of equipment facilities and system optimization

1.5.3.1 医疗保健设施建设中涉及的建筑设备、医疗装备、保健设施、系统、产品种类繁多，应根据不同的医疗规模、管理特点等进行选择，确定适宜的数量，配置适合的品种，获得合理的性价比和良好的使用效果。

1.5.3.1 Hospital construction involves various types of devices, facilities, systems and products related to architecture and medicine, which shall be selected according to different medical scale and management characteristics, to determine the appropriate quantity and variety and obtain reasonable cost performance and good use effect.

1.5.3.2 这类咨询有很好的经济价值，能节省投资和运营费用。通常由经验丰富的设计师、医疗规划师、咨询师等进行调查、分析研究、计算、数据比较、技术访问和交流，提出建议，或提供咨询报告。

1.5.3.2 This type of consultation features good economic value, saving investment and operating costs. The consultation is usually conducted by experienced designers, medical planners, consultants and others through research, analysis and study, calculation, data comparison, technical access and communication in the forms of providing suggestions and consulting reports.

1.6 建设前期管理与法规审办

1.6 Pre-construction management and audit of regulation conformity

1.6.1 法规管理审办部门与职责

1.6.1 Regulation administration and audit authorities and the duties

1.6.1.1 发展与改革部门：负责项目审批、核准、备案。

1.6.1.1 Development and reform authority: Perform project approval, check, and record.

1.6.1.2 工程咨询机构：负责项目全程咨询、项目建议（含划拨用地选址）、可行性研究、医疗工艺设计咨询、建筑策划、造价咨询、交通影响评价、环境影响评价。

1.6.1.2 Engineering consultation organization: Provide consultation throughout the project, provide project proposal (including site selection of allocated land), feasibility study, medical process design consultation, architectural planning, cost consultation, traffic impact assessment, and environmental impact assessment.

1.6.1.3 规划勘察设计机构：负责医疗卫生专项规划、控制性详细规划、修建性详细规划、建筑设计、地质勘察、岩土工程设计（基坑支护）、专项工程设计、园林和市政工程设计。

1.6.1.3 Planning, investigation and design institution: Carry out special planning for medical and health care, detailed control planning, detailed construction planning, architectural design, geological survey, geotechnical design (foundation pit support), special engineering design, and garden and municipal engineering design.

1.6.1.4 住房和城乡建设部门：负责工程开工审批，核发划拨用地的建设项目选址意见书，核发工程设计、施工、监理、勘察、招标代理资质，办理建设用地划拨决定书、国有土地使用证、用地规划许可证、建设工程规划许可证、建筑工程施工许可证。

1.6.1.4 Ministry of housing and urban-rural development: Approve the project commencement; approve and issue the proposal for project site selection for the allocated land; verify and issue the qualifications of engineering design, construction, supervision, survey and bidding agency; handle the decision of allocation of construction land, state-owned land use certificate, land planning permit, construction project planning permit, and construction engineering construction permit.

1.6.1.5 卫生健康委员会：负责编制和实施医疗机构设置规划，办理医疗机构执业许可证，对放射诊疗建设项目职业病危害及放射防护做预评价。

1.6.1.5 Health commission: Prepare and implement the program of the establishment of healthcare institutions; handle the permit for practicing of healthcare institutions, and make the pre-evaluation of occupational hazards and radiation protection for the construction project of radiological diagnosis and radiotherapy.

1.6.1.6 环境保护部门：负责审批项目环境影响评价文件，负责环境监理。

1.6.1.6 Environmental protection department: Approve the environmental impact assessment documents of the project and be responsible for environmental supervision.

1.6.1.7 水利部门：负责水资源论证报告及水土保持方案审批，审批医院沿河、湖取水许可及入河排污口设置论证报告，进行施工监理和竣工验收。凡征占土地面积在一公顷以上或者挖填土石方总量在 10000m^3 以上的建设项目，应当编报水土保持方案报告。

1.6.1.7 Ministry of water resources: Approve the water resources demonstration report and water and soil conservation proposal; approve the demonstration report of the hospital's water permit along the river and lake and the setting of sewage outlet to the river; conduct construction supervision and completion acceptance. For the construction projects of which the land acquisition area is more than a hectare or the total amount of earth and stone excavation and filling is more than 10000m^3, a water and soil conservation proposal report shall be prepared.

1.6.1.8 安全监察部门：负责项目安全评价验收及职业病评价验收工作，核发安全、职业病及检测机构资质。医院工程项目涉及危险化学品的生产、储存、有危险化学品产生的生产装置和设施，建设单位应委托设计单位进行项目安全设施设计并交付审查。

1.6.1.8 Safety supervision department: Carry out the safety assessment and acceptance and the occupational disease assessment and acceptance of the project; approve and issue the qualification of safety, occupational disease and testing organizations. If the hospital engineering project involves the production and storage of hazardous chemicals or has the production devices and facilities that produce hazardous chemicals, the project owner shall entrust the design institute to design the project safety facilities and submit the design for review.

1.6.1.9 国土资源部门：负责建设用地的审批，我国土地用途包括农用地、建设用地和未利用地，严格限制农用地转为建设用地，对耕地实行特殊保护，医疗保健设施用地必须属于建设用地。我国颁发三种土地证书：集体土地所有权证、集体土地建设用地使用权证、国有土地使用权证。我国的土地所有权归国家或集体经济组织所有，卫生（医院）用地使用年限统一为 50 年（一般公共建筑设计耐久年限亦为 50 年），土地使用类型只有土地划拨和土地出让两种形式。政府全额投资的公立医疗机构经批准获得划拨土地使用权，非政府全额投资的医疗保健设施需通过购买，有偿获得有规定使用期限的出让土地使用物权。

1.6.1.9 Ministry of land and resources: Approve the construction land. The land use in China includes agricultural land, construction land and unused land. The conversion of agricultural land into construction land is strictly restricted, and agricultural land is under special protection. The land for medical and health care facilities must be construction land. Three kinds of land use certificates are issued in China: collective land ownership certificate, collective land construction land use right certificate, and state-owned land use right certificate. In China, the land ownership belongs to the state or collective economic organizations. The land usage term for health (hospital) land is 50 years uniformly, (the designed life of durability for general public buildings is also 50 years) and the land use types are only land allocation and land grant. Public healthcare institutions that are fully invested by the government can obtain the allocated land use right after approval, while health care facilities that are not fully invested by the government need to obtain the real right to use the granted land in a specified period of use through purchase.

1.6.1.10 气象部门：负责重大工程气候可行性论证，对项目防雷设计进行检测。

1.6.1.10 Meteorological department: Carry out climate feasibility study of major projects; test the lightning protection design of the projects.

1.6.1.11 地震部门：负责项目地震安全性评价，为工程规划设计提供一定概率水准下的地震动参数和相应资料。

1.6.1.11 Earthquake department: Carry out the seismic safety evaluation of the project; provide the ground motion parameters and corresponding data in certain probability level for the project planning and design.

1.6.1.12 消防部门：负责项目消防设计审核及验收，工程项目的总平面、建筑各专业消防设计均须经消防部门审核。

1.6.1.12 Fire department: Review and accept the fire control design of the project; review and accept the fire control design in general layout, building and other disciplines of the project.

1.6.2 “一书”“四证”申办

1.6.2 Application for “one Approval” and “four Permits”

包括《医疗设置批准书》《建设用地规划许可证》《国有土地使用证》《建设工程规划许可证》和《建筑工程施工许可证》。

It covers Approval of Establishment of Healthcare Institutions, Construction Land Planning Permit, State-owned Land Use Certificate, Construction Project Planning Permit, and Construction Permit for Construction Engineering.

1.6.2.1 医疗设置批准书

1.6.2.1 Approval of Establishment of Healthcare Institutions

申请设置医疗机构应当提交下列文件：

An applicant who applies for establishment of healthcare institution shall submit the following documents:

1. 设置申请书；

1. Application for establishment;

2. 设置可行性研究报告；

2. Feasibility study report on establishment;

3. 选址报告和建筑设计平面图；

3. Site selection report and architectural design plan;

4. 单位或者个人设置医疗机构，应当按照以下规定提出设置申请：不设床位或者床位不满100张的医疗机构，向所在地的县级人民政府卫生行政部门申请；床位在100张以上的医疗机构和专科医院按照省级人民政府卫生行政部门的规定申请；

4. Any organization or individual that establishes a healthcare institution shall submit an application for establishment in accordance with the following provisions: Healthcare institutions that do not have beds or have fewer than 100 beds shall apply to the health administrative authority of people's government at the county level; healthcare institutions and specialized hospitals with more than 100 beds shall apply in accordance with the provisions of the health administrative authority of the people's government at the provincial level;

5. 县级以上地方人民政府卫生行政部门应当自受理设置申请之日起30日内，做出批准或者不批准的书面答复；批准设置的，下发设置医疗机构批准书。

5. The health administrative authority of people's government at or above the county level shall, within 30 days from the date of accepting the application for establishment, give a written reply of approval or disapproval; if an approval is given, an Approval of Establishment of Healthcare Institution shall be issued.

1.6.2.2 建设用地规划许可证

1.6.2.2 Construction Land Planning Permit

在城市范围内开发建设项目，建设项目的用地单位应按规定程序申请办理建设用地规划审批手续，取得建设用地规划许可证后，以出让方式取得的国有土地使用权、以划拨方式取得的国有土地使用权或征用集体土地的各项审批工作，办理程序如下：

For the construction projects developed within the scope of city, the land user of the construction project shall handle the approval procedures for the construction land planning in accordance with the provisions; after the Construction Land Planning Permit is obtained, the approval to the use right of state-owned land acquired by granting, the use right of state-owned land acquired by allocation, or collective land requisition, are as the following procedures:

1. 许可证办理的申请

1. Application for permit

用地单位需要办理建设用地规划许可证的，应向审批行政主管部门提交规定的申请文件和资料，填报建设项目审批申报表，提出审批申请。

For the land user which needs to apply for a Construction Land Planning Permit, it shall submit the application documents and data to the competent administrative authority of approval as required, fill in and submit the application form for approval of construction projects, and submit an application for approval.

2. 许可证申请的受理

2. Acceptance of an application for permit

审批行政主管部门接受用地单位提交的申请文件和资料，对于符合要求的即时予以受理，制作“建设用地规划许可证立案表”。行政主管部门接受用地单位提交的申请文件和资料，对于不符合要求的不予受理，将所需补齐补正的全部内容等相关情况告知用地单位。

When the competent administrative authority of approval receives the application documents and data submitted by the land user, for the projects meeting the requirements, it timely accepts the application and prepares the Construction Land Planning Permit Registration Form, while for the projects failing to meet the requirements, it does not accept the application, but notifies the land user of all the necessary contents to be made up and other relevant information.

3. 许可证申请的审核

3. Review of an application for permit

审批行政主管部门对用地单位提交的文件和资料，按照初审的标准进行初步审查。审批行政主管部门根据初审意见及审核标准对用地单位提出的申请进行审核，对符合要求的制定《建设用地规划许可证》文稿，对不符合要求的终结审批。

For the documents and data submitted by the land user, the competent administrative authority of approval carries out preliminary examination according to the standards of preliminary examination, reviews the application submitted by the land user according to the preliminary review opinions and standards of review, and formulates the manuscripts of Construction Land Planning Permit for the projects meeting the requirements, and ends the approval for the projects failing to meet the requirements.

4. 许可证申请的审定

4. Determination of an application for permit

审批行政主管部门的相关主管人员根据审核意见及复审标准对用地单位提出的申请进行复审，做出同意或不同意的复审意见。审批行政主管部门的相关主管人员根据复审意见及审定标准对用地单位提出的申请进行审定，出具同意或不同意的审定意见。

The relevant supervisors of the competent administrative authorities of approval recheck the application submitted by the land user according to the review opinions and standards, and give an opinion of approval or disapproval; then determine the application submitted by the land user according to the determination opinions and standards of determination, and issue an opinion of approval or disapproval.

5. 许可证申请的批准

5. Approval of application for permit

审批行政主管部门根据审定结论对用地单位核发《建设用地规划许可证》或《退件通知书》。

The competent administrative authority of approval issues the Construction Land Planning Permit or the Notice of Return to the land user according to the conclusion of determination.

1.6.2.3 国有土地使用证

1.6.2.3 State-owned Land Use Certificate

《国有土地使用证》是证明土地使用者向国家支付土地使用权出让金，获得了在一定年限内某块国有土地使用权的法律凭证，办理程序如下：

The State-owned Land Use Certificate is a legal certificate proving that the land user paid the land use right grant fee to the state and obtained the use right of certain state-owned land within a certain number of years, as the following procedures:

1. 土地登记申请；

1. Application for land registration;

2. 地籍调查；

2. Cadastral investigation;

3. 土地权属审核；

3. Review of land ownership;

4. 颁发国有土地使用权证书。

4. Issue the State-Owned Land Use Certificate.

1.6.2.4 建设工程规划许可证

1.6.2.4 Construction Project Planning Permit

《建设工程规划许可证》是有关建设工程符合城市规划要求的法律凭证。

The Construction Project Planning Permit is a legal certificate proving that the construction project conforms to the requirements of city planning.

在城市规划区内新建、扩建和改建建筑物、构筑物、道路、管线和其他工程设施，必须持有关批准文件向城市规划行政主管部门提出申请，由城市规划行政主管部门根据城市规划提出的规划设计要求，核发建设工程规划许可证件，办理程序如下：

For the construction, expansion and reconstruction of buildings, structures, roads, pipelines and other engineering facilities within a city planning area, application shall be submitted to the administrative authority in charge of city planning together with the related documents of approval. The administrative authority in charge of city planning issues a Construction Project Planning

Permit according to the planning and design requirements for city planning, as the following procedures:

1. 建设单位提出建设申请；

1. The project owner puts forward a construction application;

2. 城市规划行政主管部门根据城市规划审定设计方案；

2. The administrative authority in charge of city planning determines the design proposal according to city planning;

3. 城市规划行政主管部门对各项技术指标进行审定；

3. The administrative authority in charge of city planning determines the technical indicators;

4. 建设单位须完成开发规模的审批、节能审批、文物勘探、人防报批、缴费等程序；

4. The project owner shall complete the procedures such as approval of the development scale, energy conservation, and exploration of cultural relics, application for approval of civil air defense, and payment of fees;

5. 跟进最后审定结果；

5. Follow up the final results;

6. 对建设单位的建设工程规划许可证予以公示；

6. Publicize the Construction Project Planning Permit of the project owner;

7. 没有异议的核发证书，有异议的退回。

7. Issue the Permit if there is no objection, and withdraw the Permit if there is any objection.

1.6.2.5 建设工程施工许可证

1.6.2.5 Construction Permit for Construction Project

《建设工程施工许可证》是建筑施工单位符合各种施工条件、允许开工的批准文件，是建设单位进行工程施工的法律凭证，也是房屋权属登记的主要依据之一，办理程序如下：

The Construction Permit for Construction Project is an approval document proving that the construction company meets various construction conditions and is allowed to start work; it is the legal certificate for the project owner to carry out the project construction; it is also one of the main bases of housing ownership registration, as the following procedures:

1. 建设单位向发证机关领取《建设工程施工许可证申请表》；

1. The project owner gets the Application for Construction Permit for Construction Project from the issuing authority.

2. 建设单位持加盖单位及法定代表人印鉴的《建设工程施工许可证申请表》和规定的证明文件，向发证机关提出申请；

2. The project owner files an application to the issuing authority, holding the Application for Construction Permit for Construction Project affixed with the seals of the project owner and the legal representatives and the required supporting documents.

3. 发证机关在收到建设单位报送的《建设工程施工许可证申请表》和所附证明文

件后，对于符合条件的，应当自收到申请之日起规定时间内颁发《建设工程施工许可证》；对于证明文件不齐全或者失效的，应当限期要求建设单位补正，审批时间可以自证明文件补齐后作相应顺延，对于不符合条件的，应当自收到申请之日起规定时间内书面通知建设单位，并说明理由。

3. After receiving the Application for Construction Permit for Construction Project and the attached supporting documents that are submitted by the project owner, the issuing authority shall issue the Construction Permit for Construction Project as for the project that meets the requirements within the prescribed time from the date of receiving the application; for the project with incomplete or invalid supporting documents, the issuing authority shall require the project owner to supplement and correct all the contents within a limited time, and the approval time may be postponed correspondingly after the documents are completed; as for the project failing to meet the requirements, the issuing authority shall notify the project owner in writing and explain the reasons within the prescribed time from the date of receiving the application.

第2章

医疗保健设施设计管理

Chapter Ⅱ

Design and Management of Health Care Facilities

朱 希

Xi Zhu

2.1 医疗保健设施设计管理

2.1 Design and management of health care facilities

2.1.1 设计项目管理

2.1.1 Design project management

2.1.1.1 设计项目管理特征

2.1.1.1 Design of project management characteristics

1. 医疗保健设施尤其是大型医院非常复杂，其设计很难由一个设计机构独立完成，需要通过设计管理来组织实施全部的设计工作，通常采取项目管理的方式来进行设计管理。

1. The health care facilities especially those in large hospitals, are very complex, so it is difficult for a design institution to independently complete the design. It is necessary to organize and implement all the design through design management, which is usually carried out in the way of project management.

2. 设计管理是医疗保健设施项目管理中的一个管理环节，是通过组织来实施完成的工作，这个组织可以是一个设计机构或多个设计机构的组合，在多个机构组合的情况下，由一个总牵头或总承包机构，通过多个合同或一个总包合同的下辖分包来执行。

2. Design management is a management step of health care facility project management, which is completed through the implementation of organization. The organization can be a design institution or a combination of various design institutions. In the case of combination of multiple organizations, the design is led by a general leader or general contracting organization, and implemented through multiple contracts or subcontracts under a general contract.

3. 设计机构的工作分为两种基本类型：一种是连续不断、周而复始，具有具体操作的管理活动，即日常运作；另一种是有时间周期限制的一次性活动，称之为项目；它们的共同点是需要人力资源、物力资源限制、计划、执行、控制。

3. The work of design institution includes two basic types: one type is the management activities that are continuous, cyclical, with specific operations, that is, daily operations; the other type is the one-time activity within a time period, called a project; both the two require human resources and physical resource limitation, planning, execution, and control.

2.1.1.2 设计项目的时限性

2.1.1.2 Time limitation of design project

每个医疗保健设施设计项目都有确定的开始和结束时间。一般是从项目委托或设计招投标开始，经过前期策划、方案设计、初步设计、施工图设计到施工配合，最后形成竣工验收报告为止。

Each health care facility design project has a defined start and end time. The design project generally starts from project entrustment or design bidding, and ends as the final completion acceptance report is formed through preliminary planning, SD, PD, CD and construction cooperation.

2.1.1.3　设计项目的独特性

2.1.1.3　Uniqueness of design project

每个医疗保健设施设计项目均是个性化的，不可通用，表现在项目建设目标、规模、环境、组织、资源、过程等多方面。

Each health care facility design project is personalized but not universal, which is reflected in the objective, scale, environment, organization, resources, process and other aspects of project construction.

2.1.1.4　设计项目的多目标性

2.1.1.4　Multiple objectives of the project

1. 体现在经营方面：要考虑经济效益，对设计的成本、人力和时间有所管控。

1. In terms of operation: consider the economic benefits, and control the design cost, labor and time.

2. 体现在作品创优方面：要投入更多的创造性资源，包括高层级的建筑师、资深顾问、外部咨询服务、有经验的施工图设计人员。

2. In terms of creating excellent works: invest more creative resources, including high-level architects, senior consultants, external consulting services and experienced construction drawing designers.

3. 体现在人才培养方面：有梯队的人员安排，包括实习生、助理设计师、主任设计师及主管、项目经理之间的协同作业。

3. In terms of talent training: echelon the personnel, including the collaborative operation among interns, assistant designers, chief designers, supervisors and project managers.

4. 体现在技术升级方面：需要研发，研究改进设计流程，研究标准化设计，包括标准化构件、标准化空间、标准房型、通用流程、标准工艺、标准化装饰材料等。这种内部的设计项目没有具体的项目目标，但研究的目标是适用于一部分通用项目，也需要按内部立项进行设计项目管理。

4. In terms of technology upgrading: research, develop and improve the design process and study standardized design, including standardized components, standardized space, standardized room type, general flow, standard technology, standardized decorative materials, and others. This kind of internal design project has no specific project objectives, but the research objectives are applicable to some general projects, so the design project shall also be managed according to the internal initiation.

5. 体现在客户经营方面：制定更多的服务类型，包括前期策划、设计、咨询、顾问等服务，广泛应用于各种项目。

5. In terms of customer operation: develop more service types, including preliminary planning, design, consulting, consulting and other services, which are widely used in various projects.

2.1.2 设计的一次性与结果不可挽回性

2.1.2 One-time nature and result irretrievability of design

2.1.2.1 设施的项目设计结果是一次性的，只能单方向推进，不可逆。当发生设计事故后，或许能有所弥补，但不可挽回，如设计造成房屋倒塌；当设计方向选择错误，项目可能被终止，如投标设计方案落标。

2.1.2.1 The project design result of the facility is one-time, which can only be developed in one direction and is irreversible. When a design accident occurs, it may be remedied, but it is irreparable, for example, the collapse of the house caused by the design; when an incorrect design direction is selected, the project may be terminated, such as losing the bid of design proposal.

2.1.2.2 任何设计变更必然导致工程成本发生或多或少的改变；由于设计错误、返工，导致追加投资、延后时间甚至降低目标要求，则项目的实质目标已发生了改变。过多的设计变更产生设计变更的设计管理，由此产生更多的设计过程资料管理、成果档案管理，为此，增加设计成本、增加更长的时间和更多的人力。

2.1.2.2 Any design change will inevitably lead to more or less changes in the project costs; if the design errors and rework result in additional investment, delay or even reducing of target requirements, the substantive objectives of the project have changed. Too many design changes lead to the design management of design changes, which leads to more design process information management and results archive management, then the design costs, longer time and more manpower increase.

2.1.2.3 若设计变更是设计方自身造成的，设计方应承担这部分的成本，包括超时工作、增加人力；若设计变更是甲方提出的，当变更量超出合同规定的范围或设计任务书的内容较多时，甲方应支付变更费用。

2.1.2.3 If the design change is caused by the design institution itself, the design institution shall bear the corresponding cost, including overtime work and increase of manpower; if the design change is proposed by Party A, Party A shall pay the change fee when the amount of changes exceeds the scope specified in the contract or the contents of the design specifications.

2.1.3 设计团队的临时性和开放性

2.1.3 The temporality and openness of the design team

2.1.3.1 每一个医疗保健设施项目的设计团队都不同于一般的组织，它是为实现项目设计目标专门且临时设立的，是一种按照团队模式开展项目设计工作的组织，是项目人力资源的聚集体。

2.1.3.1 The design team of each health care facility project is different from general organization. It is set up specially and temporarily to achieve the project design objectives. It is an organization carrying out project design in team mode, and an aggregation of human resources of projects.

2.1.3.2 项目设计团队在项目开始时建立，在项目结束时解散。项目团队的项目负责人、设计人员或秘书、文员都是临时工作人员，使得项目中人力资源的管理相对于一般的人力资源管理有着明显的特殊性，需要项目负责人处理好本项目可能出现的问题，否则将极大地威胁到项目的成功。

2.1.3.2 The project design team is established at the beginning of the project and disbanded at the end of the project. The project principal, designers, secretaries and clerks of the project team are all temporary, staff which makes the human resource management of the project have obvious particularity compared with the general human resource management. The project principal is required to deal with the possible problems that may occur in the project; otherwise it will greatly threaten the success of the project.

2.1.4 设计项目的三重约束性

2.1.4 Three constraints of design project

2.1.4.1 任何设计项目总是受到有限的时间、有限的人力和物力资源制约。

2.1.4.1 Any design project is always restricted by limited time, human power and material resources.

2.1.4.2 设计团队各自不同的目标之间，常常互相影响、掣肘，某一专业的强化，往往需要另一专业的目标做出让步，如建筑艺术可能受制于医疗工艺，机电设备管线可能受制于结构。所以项目负责人或项目总建筑师要在不同目标之间进行判断、权衡，以最后获得总体设计性价比最高、不突破投资控制或收益最佳为总目标。

2.1.4.2 The different objectives of the design team affect and restrict each other. The strengthening of one discipline often requires concessions of the objective of another discipline. For example, the architectural art may be restricted by the medical process, and the mechanical and electrical equipment pipelines may be restricted by the structure. Therefore, the project principal or the project chief architect shall judge and weigh between different objectives, so as to achieve the general objective of the highest design cost performance, no breaking through of investment control and the best return.

2.1.4.3 设计项目的三重约束性是指进度、成本和质量。设计行业最高的成本是人力资源成本，越优秀的设计师其薪酬、提成或股份也越高。设计项目是否优秀也受到时间的约束，很好的设计师如果只有很短的时间，也难以产生好的设计作品。一个优秀的创作概念，如果没有卓越的表现技术、成图技术来支撑，也难以体现好的设计，包括效果图、动画、多媒体制作，CAD 辅助技术、BIM 技术、模型制作技术、打印技术，这些大都需要购买服务、增加成本，在成本受到严格控制的情况下，这些表现技术难以使设计成果达到最佳状态。

2.1.4.3 The three constraints of a design project refer to progress, cost, and quality. The highest cost in the design industry is the cost of human resources. The better the designer, the higher the salary, commission and share. Whether a design project is excellent or not is also constrained by time. If a good designer has only a short time, it is difficult to produce a good design work. An excellent creative concept, if not supported by distinguished expression technology and drawing making technology, is difficult to embody good design, including design sketch, animation, multimedia making, CAD auxiliary technology, BIM technology, model making technology and printing technology. These technologies mostly require purchase of services and increase of costs. In the case of strict cost control, these expression technologies are difficult to present the best design results.

2.1.4.4 项目设计管理的控制过程是保证项目各方面因素从整体上能够互相协调。选择优秀的、有经验的项目负责人或项目经理，能够较好地实现项目设计管理，满足多方面的需求和约束。

2.1.4.4 The control process of project design management is to ensure that all factors of the project can coordinate with each other as a whole. Selecting excellent and experienced project principals or project managers can better achieve project design management and meet various needs and constraints.

2.1.5 设计项目审查

2.1.5 Design project audit

2.1.5.1 每一个医疗保健设施项目设计的全过程中，要经历各种项目评审和审查。每一个环节的设计成果通过评审之后，才能进入下一个环节的设计。

2.1.5.1 Each health care facility project is subject to various project reviews and audits during the whole design process. Only after the design results of each step have passed the review, can the next step of design continues.

2.1.5.2 在设计机构的外部审查方面，对于重大项目、政府投资项目，有相关部门规定的审查环节、审查成果要求，包括但不限于：规划核查、医疗工艺审查、建筑方案评审、初步设计评审（含消防审查、人防审查、概算审查等）、施工图审查。

2.1.5.2 In terms of external audit of design institution, for major projects and government-invested projects, the steps and audit results requirements stipulated by relevant departments include but are not limited to: planning for verification, medical process audit, building proposal review, PD review (including audits of fire protection, civil air defense, budget and others), and CD audit.

2.1.5.3 在设计机构内部有各自不同的质量管理体系，包括总师办审查、ISO 质量体系作业文件、程序文件等内审。对拟定创优的重要或大型项目制定保障质量的特殊项目管理规定，通常要比一般项目更为严格。一般项目通过 ISO 质量管理体系、三校两审、外部专家评审，以保障有较好的设计质量。

2.1.5.3 There are different quality management systems within the design institution, including internal audit of chief architect office, ISO quality system operation documents, procedure documents and others. For the important or large-scale projects that are proposed to achieve excellence, the special project management regulations are formulated to guarantee that the quality is stricter than that for general projects. General projects are subject to the ISO quality management system, three inspections and two reviews (the system of self-inspection, mutual-inspection, inspection by a third-party, and review and approval), and external expert reviews to ensure good design quality.

2.1.5.4 设计突破行业规范和标准需进行超限审查，包括建筑高度或长度超限、结构超限和消防超限等。

2.1.5.4 The design breaking through industry norms and standards shall be subject to out-of-code evaluation, including out-of-code in building height or length, structure and fire fighting.

2.2　设计总承包项目管理

2.2　Management of EPC design project

2.2.1　总承包设计管理特点

2.2.1　Characteristics of general contracting design management

2.2.1.1　项目高度复杂性、专业化程度高、有工艺要求、系统配置复杂、建筑技术要求高。

2.2.1.1　The project is highly complex and specialized, with technology requirements, complex system configuration and high construction technology requirements.

2.2.1.2　专业分工细、分包项目多、工期长、任务繁重、交叉作业面多、相互影响。

2.2.1.2　The project has detailed division of labor and specialization, more subcontracted projects, with long construction period, heavy tasks, more cross operation surfaces and mutual effects.

2.2.1.3　特种设备设施多，对建筑防护、屏蔽、荷载、水电气暖空调、电梯、节能、智能化、信息化、停车、绿色要求高，造价高。

2.2.1.3　The project has many special devices and facilities, and has high requirements and costs for building protection, shielding, load, water, power, gas, heating and air conditioning, elevator, energy conservation, intellectualization, informatization, parking and greening.

2.2.1.4　使用方需求变化多、设计变更多，对工期、成本影响较大。

2.2.1.4　The project features more user demand changes and design changes, which have a greater impact on the construction period and costs.

2.2.2　总承包项目管理重点

2.2.2　Focus of EPC project management

2.2.2.1　前期策划

2.2.2.1　Preliminary planning

设计前期阶段应获得详细的设计任务书，应进行前期策划，切实做好项目可行性研究，以完善设计输入条件。除了常规公共建筑应获得的规划设计条件之外，还应获得医疗工艺前期设计成果。

In the early stage of design, the detailed design specifications shall be obtained, the preliminary planning and the project feasibility study shall be carried out to complete the design input conditions. In addition to the planning and design conditions of conventional public buildings, the PD results of medical process shall also be obtained.

2.2.2.2 医疗工艺设计

2.2.2.2 Design for medical process

医疗工艺设计应由设计委托方提供，若委托方没有能力做出医疗工艺设计，则需要第三方提供医疗工艺设计咨询服务。医疗工艺设计应该能够带动并协助建筑工程所有专业和设计的全过程，包括设计变更。应要求项目业主和使用方重视医疗工艺设计以减少设计返工，绝大多数医疗设施的设计变更都是由于医疗工艺设计发生改变而导致的。

The medical process design shall be provided by the design client. If the client is unable to make the medical process design, the third party is required to provide consulting services for medical process design. Medical process design shall be able to drive and assist the whole design process of all disciplines of construction engineering, including design changes. It is necessary to require the project Employer and users to pay attention to the medical process design, to reduce the design rework. Most of the design changes of medical facilities are caused by the changes of medical process design.

2.2.2.3 专项分包设计

2.2.2.3 Special subcontract design

医疗设施的专项设计包括：医用气体、空气净化、放射防护与屏蔽、医疗用水、自动物流、信息系统、标识系统、医用家具等。这些专项设计不属于一般公共建筑设计范畴，必须结合产品、厂家产能、造价来进行，设计机构内不具备这种设计能力。

The special design for medical facilities includes: medical gas, air purification, radiation protection and shielding, medical water, automatic logistics, information system, identification system, medical furniture and others. These special designs, out of the scope of general public building design, must be carried out in combination with products, manufacturers' production capacity and costs. The design institution does not have such design capability.

设计总包方需要对这些专项设计进行总的设计管理和控制，使其服从于整体建筑的要求以及受到各项规范的制约。由任何产品厂家进行的专项深化设计，不可避免有扩大工程量、推介专利技术的因素，或可导致设计不完全符合使用需要。总包单位项目负责人应对专项设计进行合理控制，重视产品的可实现度以及相关规范，通过分包合同实施设计管理。

The design general contractor shall carry out overall design management and control on these special designs, so that the special designs meet the requirements of the whole building and are in the constraints of various specifications. The special in-depth design conducted by any product manufacturer inevitably has the factors of expanding the quantities and promoting the patented technologies, or may cause the design not fully meet the needs of use. The project principal of the general contractor shall reasonably control the special designs, pay attention to the product realizability and relevant specifications, and implement the design management through subcontracts.

2.2.3 设计总承包过程管理

2.2.3 Management of design general contracting process

2.2.3.1 医疗保健设施设计过程分为概念设计、方案设计、初步设计、施工图设计、专项设计、深化加工设计、现场服务（隐蔽工程验收、装修收口、设备材料样品选择、变更）等环节。

2.2.3.1 The design process of health care facilities includes conceptual design, SD, PD, CD, special design, in-depth processing design, on-site service (concealed works acceptance, decoration closing, equipment and material sample selection, changes) and others.

2.2.3.2 设计管理的过程是输入条件管理的过程，对各专业之间的互提资料深度、正确性、准确度的管理是最重要的环节，应采用事先指导、中间检查、事后审核的方式，关注每个阶段对医疗工艺的落实，要求业主方分阶段签署确认。制定设计计划周期的时候，要植入管理的过程和时间，预留项目管理、审核、成果检查、档案归存的时间以及修改的时间。

2.2.3.2 The process of design management is the process of input condition management. The most important step is the management of the depth, correctness and accuracy of mutual information between various disciplines. The way of prior guidance, intermediate inspection and post review shall be used to focus on the implementation of medical process in each stage, which shall be signed and confirmed in stages by the Employer. When developing the design plan cycle, the management process and time shall be included, and the time for project management, review, result inspection, archive storage and modification shall be reserved.

2.2.3.3 在施工图环节，汇签是很重要的设计管理工作，必须通过这个环节，检查所有各专业之间相互的适合性、满足度以及交叉作业的结果是否合理，不能流于形式，“闭着眼睛”签字。

2.2.3.3 In the link of construction drawing, counter sign is a very important design management task. It is necessary to check whether the mutual suitability, satisfaction and cross-operation results among all the disciplines are reasonable through this link, rather than being formalized or signing with “eyes closed”.

2.2.3.4 可以利用 BIM 技术解决管线综合及汇签不到位的问题，直观表达所有专业设计之间的关系以及最后的空间效果，但是 BIM 只能发现问题，不能解决问题，解决 BIM 发现的问题还要靠各专业自身的设计来完成。

2.2.3.4 BIM technology can be used to solve the problems of pipeline integration and inadequate counter sign, and intuitively express the relationship between all specialty designs and the final spatial effect. But BIM can only find problems, not solve problems. To solve the problems found by BIM, the design of each discipline is relied on.

2.2.4 设计总承包分包管理

2.2.4 Subcontracting management of design general contracting

2.2.4.1 应事先制订分包计划，让各个分包的子项目能够与整体设计在各个阶段同步进行，应确保分包界面无遗漏。每一个阶段的设计成果都应包含总包范围内所有主体以及分包的设计成果。

2.2.4.1 The subcontracting plan shall be prepared in advance so that the subprojects of each subcontract can be carried out synchronously with the overall design at all stages, to ensure that the subcontract interfaces are free of omission. The design results of each stage shall include the design results of all the subjects within the scope of the general contract and subcontract.

2.2.4.2 分包设计的成果应纳入统一的设计档案管理，分包设计的机构与其设计人员应可追溯。分包的设计质量、责任与风险由分包合同加以制约，实现责任和风险共担。对于甲方规定分包必须报备的情况，应向甲方告知分包方的相关资质、人员以及联系方式。

2.2.4.2 The results of subcontract design shall be included in the unified design archive management, and the mechanism of subcontract design and its designers shall be traceable. The design quality, responsibilities and risks of the subcontracting shall be restricted by the subcontracts to realize the sharing of responsibilities and risks. For the situations of subcontracting that must be reported as stipulated by Party A, Party A shall be informed of the relevant qualifications, personnel and contact information of the subcontractor.

2.2.4.3 分包设计成果在文件和图纸中应完整表达总包方和分包方相关的信息，包括设计机构名称、资质、联系方式、项目及子项目编号、设计阶段、专业、设计与校对人员姓名、设计时间、版次。

2.2.4.3 The subcontract design results shall fully express the relevant information of the general contractor and subcontractors in the documents and drawings, including the name, qualification, contact information of design institution, project and subproject No. , design stage, discipline, name of designer and proofreading personnel, design time and version.

2.2.4.4 总包方应对分包设计制定统一技术措施和指导，对完成的分包设计图纸进行审核。

2.2.4.4 The general contractor shall formulate unified technical measures and guidance for the subcontract design and review the completed subcontract design drawings.

2.2.5 设计变更管理

2.2.5 Design change management

2.2.5.1 设计变更的文件、图纸、数据是设计文件的一部分。

2.2.5.1 The documents, drawings and data of design changes constitute part of the design documents.

2.2.5.2 分包设计变更应获得总包设计方的认可。

2.2.5.2 The subcontract design changes shall be approved by the design institution of general contractor.

2.2.5.3 发生设计变更应知会甲方、监理、施工单位以及其他相关方，在取得主管单位的认可后，变更才能执行。

2.2.5.3 In the case of design change, Party A, the supervisor, the construction company and other stakeholders shall be informed, and the changes can only be implemented after being approved by the competent organization.

2.2.5.4 在项目竣工验收之前，所有设计变更应进行汇总；设计变更的图纸、文件应可纳入竣工图，成为竣工结算的依据。

2.2.5.4 Before the completion acceptance of the project, all design changes shall be summarized; the drawings and documents of design changes shall be included in the as-built drawings as the basis for completion settlement.

2.3　建筑师负责制

2.3　Architect responsibility system

2.3.1　工作模式

2.3.1　Operating mode

2.3.1.1 建筑师负责制是以担任建筑工程项目的设计主持人或注册建筑师（以下称为建筑师）为核心的设计团队，依托所在的设计企业为实施主体，依据合同约定，对民用建筑工程全过程或部分阶段提供全寿命周期设计咨询管理服务，最终将符合建设单位要求的建筑产品和服务交付给建设单位的一种工作模式。

2.3.1.1 The architect responsibility system is an operating mode, it takes the design directors or registered architects (hereinafter referred to as architects) undertaking the construction engineering project as the core design team, and the design enterprise they work in as the implementation subject, to provide the full life cycle design consulting management service for the whole process or part of the civil construction engineering projects according to the contract, and finally deliver the building products and services meeting the project owner' s requirements to the project owner.

2.3.1.2 建筑师负责制是国际通行的工作模式，中国香港、中国台湾、中国澳门均实行建筑师负责制。

2.3.1.2 The architect responsibility system is an international operating mode, and is implemented in China's Hong Kong, Taiwan, and Macao.

2.3.1.3 中国大陆未实行建筑师负责制，而是由建设单位（如政府公务属、卫生局、民政局，医院后勤部、开发商、投资方等）来执行这项工作，通过组建基建办、项目部，委任临时项目主管、分管后勤院长、项目管家，临时聘任工程师、顾问，来承担建筑师工作。即建筑师负责制的乙方工作、责任、义务及权力由甲方自理。

2.3.1.3 The mainland China has not implemented the architect responsibility system, the corresponding works are undertaken by the project owners (such as government official, Health

Bureau, Civil Affairs Bureau, hospital logistics department, developers, and investors) through establishing Infrastructure Office and Project Department, appointing temporary project directors, logistics directors in charge, project managers, and temporary engineers and consultants. That is to say, the tasks, responsibilities, obligations and rights of Party B under the architect responsibility system shall be undertaken by Party A.

2.3.2 建筑师的服务内容

2.3.2 Architect services

2.3.2.1 建筑师可参与规划：参与城市修建性详细规划和城市设计，统筹建筑设计和城市设计协调统一。

2.3.2.1 Participation in planning: Architects participate in the detailed urban construction planning and urban design, and coordinate and harmonize the building design and urban design.

2.3.2.2 提出策划：参与项目建议书、可行性研究报告与开发计划的制定，确认环境与规划条件，提出建筑总体要求，提供项目策划咨询报告、概念性设计方案及设计要求任务书，代理建设单位完成前期报批手续。

2.3.2.2 Proposing plans: Architects participate in the formulation of project proposal, feasibility study report and development plan, confirm the environment and planning conditions, put forward the general building requirements, provide the project planning and consultation report, conceptual design proposal and design specifications, and complete the pre-approval procedures as an agent of the project owner.

2.3.2.3 完成设计：完成方案设计、初步设计、施工图技术设计和施工现场设计服务；综合协调把控幕墙、装饰、景观、照明等专项设计，审核承包商完成的施工图深化设计；建筑师负责的施工图技术设计重点解决建筑使用功能、品质价值与投资控制；承包商负责的施工图深化设计重点解决设计施工一体化，准确控制施工节点大样详图，促进建筑精细化。

2.3.2.3 Completion of design: Architects complete the SD, PD, CD and design on construction site; comprehensively coordinate and control the special designs for curtain walls, decoration, landscape and lighting, and review the detailed CD completed by the contractor. The technical CD undertaken by the architects focuses on the use function, quality value and investment control of the building; the in-depth CD undertaken by the contractors focus on the integration of design and construction, accurate control of detailed drawings of construction nodes, and promotion of building refinement.

2.3.2.4 监督施工：代理建设单位进行施工招投标管理和施工合同管理服务，对总承包商、分包商、供应商和指定服务商履行监管职责，监督工程建设项目按照设计文件要求进行施工，协助组织工程验收服务。

2.3.2.4 Supervision of construction: Architects, as the agent of the project owner, provide construction bidding and construction contract management services, perform supervision duties to the general contractor, subcontractors, suppliers and designated service providers, to supervise that the construction project is constructed in accordance with the design documents, and assist in organization of project acceptance.

2.3.2.5 指导运维：组织编制建筑使用说明书，督促、核查承包商编制房屋维修手册，指导编制使用后维护计划。

2.3.2.5 Guide in operation and maintenance: Architects organize the preparation of building operation instructions, urge and verify the contractor on preparation of building maintenance manual, and guide the preparation of post-use maintenance plan.

2.3.2.6 更新改造：参与制定建筑更新改造、扩建与翻新计划，为实施城市修补、城市更新和生态修复提供设计咨询管理服务。

2.3.2.6 Renewal and renovation: Architects participate in the formulation of building renovation and reconstruction, expansion and renewal plan, and provide design consultation and management services for urban repair, urban renewal and ecological restoration.

2.3.2.7 辅助拆除：提供建筑全寿命期提示制度，协助专业拆除公司制定建筑安全绿色拆除方案等。

2.3.2.7 Auxiliary demolition: Architects provide the life-cycle prompt system for the building, and assist the professional demolition company in the formulation of safe and green demolition plan for the building.

2.3.3 建筑师负责制合同

2.3.3 Contract of the architect responsibility system

2.3.3.1 实行建筑师负责制的项目，建设单位应在与设计企业、总承包商、分包商、供应商和指定服务商的合同中明确建筑师的权力，并保障建筑师权力的有效实施。建筑师应自觉遵守国家法律法规，诚信执业，公正处理社会公众利益和建设单位利益，维护社会公共利益，及时向建设单位汇报所有与其利益密切相关的重要信息，保证专业品质和建设单位利益。

2.3.3.1 For projects using architect responsibility system, the project owner shall clarify the architects' rights in the contracts with design enterprises, general contractors, subcontractors, suppliers and designated service providers, and ensure the effective implementation of the architects' rights. Architects shall consciously abide by national laws and regulations, practice in good faith, fairly handle the interests of the general public and the project owners, safeguard the public interests, timely report all important information closely related to their interests to the project owners, and ensure professional quality and the interests of the project owner.

2.3.3.2 实行建筑师负责制的建设单位应当根据设计企业和建筑师承担的服务业务内容和周期，结合项目的规模和复杂程度等要素合理确定建筑师服务报酬，在合同中明确约定并及时支付，或者采用“人工时”的计价模式取费。建筑师负责制服务收费，应纳入工程概算。

2.3.3.2 The project owners implementing architect responsibility system shall reasonably determine the architect service remuneration according to the service business content and cycle undertaken by the design enterprises and the architects, in combination with the scale and complexity of the projects and other factors, and clearly stipulate and pay in time in the contract, or adopt the pricing mode of "man-hour" to collect fees. The service fees of the architect responsibility system shall be included in the project estimate.

2.3.3.3 推行建筑师负责制职业责任保险，建立企业、团队与个人保险相互补充机制。

2.3.3.3 The professional liability insurance for architect responsibility system shall be implemented, to establish a complementary mechanism among enterprises, teams and individuals.

2.3.3.4 建筑师在提供建筑师负责制的项目中，应承担相应法定责任和合同义务，因设计质量造成的经济损失，由设计企业承担赔偿责任，并有权向签章的建筑师进行追偿。建筑师负责制不能免除总承包商、分包商、供应商和指定服务商的法律责任和合同义务。

2.3.3.4 In the projects using architect responsibility system, the architects shall bear the corresponding statutory responsibilities and contractual obligations. The design enterprises shall bear the compensation responsibilities for the economic losses caused by the design quality, and are entitled to claim against the signed architects. The architect responsibilities system cannot exempt the general contractor, subcontractors, suppliers and designated service providers from their legal responsibilities and contractual obligations.

2.3.4 国际通行建筑师负责制

2.3.4 International general architect responsibility system

2.3.4.1 建筑师负责制是全球建筑业的通行制度，是保障建筑品质的基础，中国企业在境外承接设计和工程总承包任务，必须遵循国际惯例实行建筑师负责制。

2.3.4.1 The architect responsibility system is a general system in the global construction industry and the basis for guaranteeing the building quality. The Chinese enterprises undertaking design and general contracting tasks abroad must follow the international practice to implement the architect responsibility system.

2.3.4.2 建筑师负责制下的建筑师是一个抽象概念，不是指一个单独的个人、一个单独的建筑专业，甚至不限于一个设计单位，而是指一个设计咨询团队，是以注册建筑师的责任建筑师为核心的社会咨询团队。在这个团队中，责任建筑师发挥主导、协调、监督的作用，咨询团队或个人对其成果的专业性、正确性及完整性负相关的责任。

2.3.4.2 The architect under the architect responsibility system is an abstract concept, not a single person, a single architectural specialty, or even not limited to a design unit, but a design consulting team, that is, a social consulting team with the responsible architect of the registered architect as the core. In this team, the responsible architect plays a leading, coordinating and supervising role, and the consulting team or individual is responsible for the professionalism, correctness and integrity of its results.

2.3.4.3 国际通行的"建筑师负责制"是工程项目全过程服务，建筑师在工程中拥有灵魂地位和指挥作用，承担总体协调和总设计师的角色，对项目最终价值与品质起着关键作用。建筑师要具备广泛的专业技术能力和组织管理能力，具备极强的专业服务精神和责任心，还要有成本概念与投资控制能力，只有能力和素质全面的注册建筑师才能胜任。

2.3.4.3 The international "architect responsibility system" is a whole project process service. The architect is in the core status and plays the command role in the project, undertakes the role of overall coordination and chief designer, and plays a key role in the final value and quality of the project. Architects shall have extensive professional technical capability and organization and management ability, strong professional service spirit and sense of responsibilities, as well as the cost concept and investment control capacity. Only a registered architect with comprehensive ability and quality can be competent.

2.3.5 中国建筑师负责制

2.3.5 Architect responsibility system in China

2.3.5.1 2018年7月4日住房城乡建设部同意在上海市、深圳市开展建筑师负责制试点。

2.3.5.1 On July 4, 2018, the Ministry of Housing and Urban-Rural Development agreed to carry out the pilot projects of architect responsibility system in Shanghai and Shenzhen.

2.3.5.2 2019年2月13日发布《深圳市建筑师负责制试点工作实施方案》《福田区政府投资项目工程设计计费参考体系》，明确建筑师负责制的工作内容和计费方式，设计服务参考收费 = 基础服务收费 ×（1 + 调整系数），为工程总造价的6% ~8%，实现政府投资建设模式与国际工程建设模式接轨。

2.3.5.2 On February 13, 2019, Scheme for Pilot Implementation of Architect Responsibility System in Shenzhen and Charging Reference System for Engineering Design of Futian District Government Invested Projects were issued, which clearly defined the work content and charging method of the architect responsibility system. The reference charge for design service = basic service charge × (1 + adjustment coefficient), accounting for 6% ~ 8% of the total cost, realizing the connection between the government investment construction mode and the international engineering construction mode.

2.3.5.3 采用建筑师负责制的项目，责任建筑师除了提供基础设计服务外，还提供全过程报批报建管理、设计管理（含施工图审查）、运维指导、拆除辅助、工程监理等专业技术服务，服务内容覆盖从项目立项到竣工验收的建设全过程。

2.3.5.3 For projects using architect responsibility system, the responsible architect provides not only basic design services but also professional technical services such as the whole process of application for approval and construction management, design management (including construction drawing audit), operation and maintenance guidance, demolition assistance, and project supervision, covering the whole process of construction from project approval to completion acceptance.

2.4 EPC 总承包项目设计管理

2.4 EPC general contract project design management

2.4.1 EPC 模式

2.4.1 EPC mode

EPC 是国际工程建设中广泛应用的总承包模式，是工程企业按合同约定承担项目的设计、采购、施工及试运行工作，并对项目实施期间的质量、安全、工期、造价全面负责，又称“交钥匙”工程。根据业主具体的需求，可以在合同中约定更多的管理服务内容，如建设前期报批配合服务、施工期间管理服务等。

EPC is a general contracting mode widely used in international engineering construction, in which the engineering enterprises undertake the design, procurement, construction and commissioning of the project according to the contract, and are fully responsible for the quality, safety, construction period and cost during the implementation of the project, also known as “turnkey” project. According to the specific demand of the employer, more management services can be agreed in the contract, such as coordination in approval in the early stage of construction, and management services during construction.

2.4.2 EPC 设计管理特点

2.4.2 Characteristics of EPC design management

2.4.2.1 国际工程实行全程建筑师负责制，建筑师主导整个工程建设，其 EPC 由建筑师牵头。中国没有实行全程建筑师负责制，其 EPC 均由施工企业牵头，目前的 EPC 有从建设前期开始和从初步设计完成后开始两种模式，设计被分为两个阶段。

2.4.2.1 The international project implements the whole-process architect responsibility system, the architect leads the whole project construction, and the EPC is led by the architect. China has not implemented the whole-process architect responsibility system, and the EPC is led by the construction enterprise. Currently there are two forms of EPC in China, namely starting from the early stage of construction and starting from the completion of the PD, totally two design stages.

2.4.2.2 EPC 设计管理目标是克服设计、采购、施工相互制约和相互脱节的矛盾，将设计、采购、施工各阶段工作合理衔接，实现建设项目进度、成本和质量控制符合建设工程承包合同约定，确保获得较好的投资效益。

2.4.2.2 The objectives of EPC design management are to overcome the contradiction of mutual restriction and disconnection between design, procurement and construction, reasonably connect the design, procurement and construction project in each stage, realize the construction project schedule, cost and quality control in line with the construction contract, and ensure better investment benefits.

2.4.2.3 实行工程总造价限额设计。

2.4.2.3 Implement the design of total cost quota.

2.4.3 EPC 设计阶段划分

2.4.3 Division of EPC design stage

2.4.3.1 从立项阶段开始 EPC 总承包，则前期咨询、设计全过程、审批申报都包含在总承包合同内，咨询方和设计方服从总承包方管理。业主负责完成医疗工艺设计，提出建筑方案需求，参加方案设计评审。

2.4.3.1 For the EPC general contracting starting from the project approval stage, the preliminary consultation, whole design process, approval and application are included in the general contract, and the consultants and design institution are subject to the management of the general contractor. The employer completes the technological design for hospital, proposing the architectural proposal requirements, and participating in the SD review.

2.4.3.2 初步设计有时是流程需要，某些地区政府投资立项需要达到初步设计深度才予以评审；若无政府部门外审要求，则初步设计是设计项目内部各专业方案论证、自我验证可行性和设计是否超投资的环节，对于复杂、大型、限额项目是必要的。

2.4.3.2 PD is sometimes required in process, since some local governments only review the investment and project approval on the premise that the PD depth is reached; if the government authority has no external audit requirement, the PD is a link of proposal demonstration, and self-verification of feasibility and whether the design is beyond the investment for various disciplines within the design project, which is necessary for complex, large and quota projects.

2.4.3.3 从施工图设计阶段开始 EPC 总承包，业主对前期工作的决策力加大，施工招标前置到初步设计概算完成时；设计被分为初步设计前总承包和 ECP 施工图设计，前期工作应完成至能用初步设计文件和概算进行施工招标的程度，初步设计范围要遍及投资所控各个子项目，要加大设计深度直至全部工程量可被准确计算。

2.4.3.3 For the EPC general contracting starting from the CD stage, the employer's decision-making power on the preliminary work is enhanced, and the construction bidding is being forward to the completion of PD budgetary estimate. The design is divided into general contracting before PD and EPC construction drawing design. The preliminary work shall be completed to the extent that the PD documents and budgetary estimates can be used for construction bidding. The scope of PD shall cover all sub-projects under the control of investment, and the design depth shall be increased until the total quantities can be accurately calculated.

2.4.3.4 在初步设计完成后进行 EPC 总承包发包，初步设计应向施工图设计方或直接向施工单位交底；总承包方边做土建施工图边进行主体施工，同时展开机电设备、医用专项工程系统及设备采购，由产品生产厂家和设备供货方根据现场进行安装工程施工图设计。

2.4.3.4 The bidding of EPC general contracting is carried out after the completion of the PD, and the PD shall be disclosed to the construction drawing design institution or directly to the construction company; the general contractor shall carry out the main construction while making the civil construction drawings, and purchase the electromechanical devices, medical special engineering system and devices at the same time, and the product manufacturers and device suppliers shall design the construction drawings of installation works according to the site situation.

2.4.4 EPC 初步设计文件深度

2.4.4 Depth of EPC PD documents

2.4.4.1 EPC 模式下的初步设计广度和深度是项目成败的关键，业主方和设计方对初步设计的管理应从了解初步设计深度、标准开始，对各专业初步设计图纸内容进行把控。

2.4.4.1 The breadth and depth of PD under EPC mode is the key to the success or failure of the project. The owner and the designer should understand the depth and standard of PD and control the contents of PD drawings of various disciplines.

2.4.4.2 初步设计文件内容：效果图、设计说明、设计图纸、概算书，其中设计图纸应包含以下内容：

2.4.4.2 Contents of PD documents: renderings, design specification, design drawings and bill of estimates, of which the design drawings shall include the following contents:

1. 总体设计：初步设计总说明书、主要数据和技术经济指标、消防专篇、节水专篇、节能专篇、环境保护专篇、卫生防疫专篇、人防专篇、绿色建筑专篇、初步设计总概算。

1. Overall design: PD general specifications, main data and technical and economic indicators, special section on fire protection, special section on water saving and energy saving, special section on environmental protection, special section on health and epidemic prevention, special section on civil air defense, special section on green building and PD general budget.

2. 建筑专业：设计说明、总体及单体建筑效果图、夜景亮化效果图，总平面图、竖向设计图、各层平面图、立面图、剖面图，防火分区图，特殊空间和外墙控制性节点做法，室、内外装修做法与材料表，电梯、扶梯参数表，门窗表、门窗及幕墙分格开启图。

2. Architecture discipline: design specification, general and individual building renderings, night view lighting renderings, general layout, vertical design, plans, elevations, sections of each floor, fire compartment diagram, special space and exterior wall control node practices, interior and exterior decoration practices and material list, elevator and escalator parameter table, door and window list, door and window and curtain wall compartment opening diagram.

3. 工艺专业：设计说明，医疗工艺一级流程图、二级流程图，医疗工艺条件图。

3. Process discipline: design specification, first-class flow chart and second-class flow chart of medical process, medical process conditions chart.

4. 结构专业：设计说明、基础图、地下室结构图，剪力墙、柱平面布置图，裙房楼面梁板平面布置图，标准层梁板平面布置图，转换层梁板平面布置图，屋面结构平面布置图，所有图纸均应有构件尺寸及沿竖向变化尺寸（含钢筋混凝土结构及钢结构、幕墙结构）。

4. Structural discipline: design specification, foundation drawing, basement structure drawing, shear wall and column plane layout, podium floor beam and slab plane layout, standard floor beam and slab plane layout, transfer floor beam and slab plane layout, roof structure plane layout. All drawings shall have component size and vertical variation size (including reinforced concrete structure, steel structure and curtain wall structure).

5. 暖通空调专业：设计说明、设备材料表、系统流程图、总平面管线，各层平面图（含采暖、通风、排风、空调、空气净化、防排烟、冷热源机房），空调、通风及制冷机房放大平面图，水系统原理图。

5. HVAC discipline: design specification, devices and materials list, system flow chart, pipeline general layout, plane layout of each floor (including heating, ventilation, exhaust, air conditioning, air purification, smoke control and exhaust, cooling and heat source room), enlarged plan of air conditioning, ventilation and refrigeration room, and schematic diagram of water system.

6. 给排水专业：设计说明、设备材料表，卫生器具清单，总平面外网管线综合图；给排水、热水、污水、废水、雨水、消火栓给水、自动喷水灭火及泡沫灭火系统原理图；直饮水、医用纯水、太阳能热水系统图；给排水及消防、消防水箱间给水排水、雨水排水及雨水收集平面图；设备布置图（含泵房、水池、水箱，冷却塔、水处理间、污水处理站、核废物处理衰变池）。

6. Water supply and drainage discipline: design specification, devices and materials list, sanitary fixtures list, and comprehensive drawing of external network pipeline general layout; schematic diagram of water supply and drainage, hot water, sewage, wastewater, rainwater, hydrant water supply, automatic sprinkling and foam fire extinguishing systems; direct drinking water, medical pure water, solar hot water system diagram; plane layout of water supply and drainage, firefighting, water supply and drainage between firefighting water tanks, rainwater drainage and rainwater collection; devices layout (including pump room, water pool, water tank, cooling tower, water treatment room, sewage treatment station and nuclear waste treatment and decay tank).

7. 电气及智能化专业：设计说明、设备材料表，总平面管线、总等电位联结图；高低压供配电系统图，综合布线、电视、电话、对讲、呼叫、广播、数字教学系统图，火灾自动报警及联动、安防及防盗、视频监控、车库管理、建筑设备监控系统图，室内照明、室外照明、夜景亮化系统图；高低压变配电室、发电机房、弱电设备平面布置图，各层主要干线、火灾自动报警、防雷布置平面图，纵向配电平面图。

7. Electrification and intellectualization discipline: design specification, devices and material list, pipeline general layout, general equipotential connection diagram; high and low voltage power supply and distribution system diagram, comprehensive wiring, television, telephone, intercom, call, broadcast, digital teaching system diagram, automatic fire alarm and linkage, security and theft prevention, video monitoring, garage management, building automation system diagram,

indoor lighting, outdoor lighting, night view lighting system diagram; plane layout of high and low voltage transformation and distribution rooms, generator room and low voltage equipment, plane layout of main lines, automatic fire alarm and lightning protection of each floor, and longitudinal power distribution plan.

8. 动力专业：设计说明，设备材料表，室外供热管道，室内供热管道；锅炉房工艺设备布置平面图，热力系统图，蒸汽系统图；燃油燃气锅炉房应有燃料系统图，锅炉房为独立工程或者在总平面布置图上表示不清楚时，应绘制区域布置图。

8. Power discipline: design specification, devices and materials list, outdoor heat supply pipeline, indoor heat supply pipeline; plane layout, thermal system diagram, steam system diagram of process device in boiler room. The fuel-fired and gas-fired boiler room shall have a fuel system diagram. If the boiler room is an independent works or is not clearly indicated on the general layout, an area layout shall be drawn.

9. 气体专业：设计说明、系统流程图、设备材料表、总平面管线、气源或供气站布置图、平面图。

9. Gas discipline: design specification, system flow chart, devices and materials list, pipeline general layout, gas source or gas supply station layout and plan.

10. 物流专业：设计说明、站点表、图例、系统流程图、设备材料表、平面图、轴测/竖向图。

10. Logistics discipline: design specification, stations list, legend, system flow chart, devices and materials list, plan, axonometric drawing/vertical diagram.

11. 人防专业：设计说明、人防门表、设备材料表、平时平面图、战时平面图。

11. Civil air defense discipline: design specification, civil air defense door list, devices and materials list, peacetime plan and wartime plan.

12. 室内专业：设计说明、重点空间室内效果图，固定家具与部品选型表，室内固定设备设施与固定家具平面布置图，顶棚平面图，固定隔断平面定位及构造图。

12. Interior discipline: design specification, key space interior renderings, list of fixed furniture and decorations, plane layout of indoor fixed devices and facilities and fixed furniture, ceiling plan, fixed partition planar positioning and structure diagram.

13. 景观专业：效果图、设计说明、图例、苗木表、室外家具表、室外场地装饰材料表、平面图、竖向图。

13. Landscape discipline: renderings, design specification, legend, seedling list, outdoor furniture list, outdoor site decoration material list, plan and vertical diagram.

14. 放射防护与屏蔽专业：设计说明、防护与屏蔽材料表、平面图、剖面图。

14. Radiation protection and shielding discipline: design specification, radiation protection and shielding material list, plans and sections.

15. 标识专业：效果图、设计说明、图例、布点表、室内外平面布置图。

15. Identification discipline: renderings, design specification, legend, layout table, indoor and outdoor plane layout.

16. 经济专业：初步设计概算编制说明、综合概算，应分成：单位工程概算、单项工程概算和建筑项目总概算。概算编制方法可采用：概算定额法、概算指标法、类似工程预算法。

16. Economics discipline: PD budget estimates preparation description and comprehensive budget estimates, including unit project budget estimates, single project budget estimates and construction project general budget estimates. Preparation methods adopted for the budget estimates include: estimate rating method, estimate index method and similar project budget method.

（1）初步设计达到一定的深度，建筑结构较为明确，基本能按初步设计图纸计算出楼面、地面、墙体、门窗和屋面等分部工程的工程量时，采用概算定额法。

（1）The estimate rating method is adopted when the PD has reached a certain depth and the building structure is relatively clear, and the quantities of floors, ground, walls, doors and windows, roof and other divisional works can be basically calculated according to the PD drawings.

（2）当初步设计深度不够，不能准确计算工程量，有类似概算指标可以利用时，采用每 $100m^2$ 或 $1000m^3$ 为计量单位来规定人工、材料和施工机械台班消耗量以及价值表现标准。

（2）When the depth of PD is not enough to accurately calculate the quantities, but similar budget estimate indexes can be used, the consumption and value performance criteria of labor, materials and construction machinery per $100m^2$ or $1000m^3$ shall be specified.

（3）类似工程预算法适用于拟建工程初步设计与已建工程或在建工程的设计类似，却没有可用的概算指标时。

（3）The similar project budget method is applicable when the PD of the proposed project is similar to the design of the existing project or the project under construction, but there is no budget estimate index available.

（4）设备安装工程概算编制方法有预算单价法、扩大单价法、概算指标法。

（4）Preparation methods adopted for budget estimates of device installation works include unit price budget method, extended unit price method and estimate index method.

2.4.4.3　必须在初步设计图纸中将拟建工程所有交钥匙的功能要求表达在设计文件中，务必完整全面，不漏项。业主方进行 EPC 项目招标时，必须在招标文件中明确：项目功能描述、主材品牌推荐表以及合同主要条款。

2.4.4.3　All turnkey functional requirements of the proposed project must be described in the PD drawings of the design documents, and must be complete without missing items. When bidding for EPC project, the owner must specify in the bidding document: project function description, recommended main material brand list and main terms of the contract.

1. 功能描述：业主方目的与需求应与设计任务书一致，初步设计图应全部反映业主需求。

1. Function description: The owner's purpose and requirements shall be consistent with the requirements specification, and the PD drawings shall fully reflect the owner's requirements.

2. 主材品牌推荐表：应能控制项目整体品质。

2. The recommended main material brand list shall be able to control the overall quality of the project.

3. 合同内的主要条款：应约定双方权责，包括概算表与初步设计图纸文件。

3. Main terms of the contract: Rights and responsibilities of both parties shall be agreed, including budget estimate table and PD drawing documents.

2.4.5 设计牵头 EPC 设计管理

2.4.5 EPC design management led by the design institute

2.4.5.1 发挥前期主导作用

2.4.5.1 Play the leading role in the early stage

设计是所有工程项目的起点，设计企业处于工程建设产业价值链上游，起龙头作用，工艺复杂的医疗保健设施建筑技术核心把握在设计院手中。以设计单位为主承揽 EPC 项目具备先天的技术主动权，在招投标阶段应从设计施工全过程和整体上考虑问题，发挥设计主导作用。

Design is the starting point of all projects. The design institute is in the upstream of the value chain of engineering construction industry and plays a leading role. The construction core technology of complex health care facilities is mastered by the design institute. EPC projects contracted mainly by the design institute have inherent technical initiative. In the bidding stage, the project design should be considered from the whole design and construction process and the whole project for playing a leading role of the design.

设计单位与施工单位相比，更能理解业主方的设计意图、建筑结构特点、机电系统配套关系和项目所在地自然地理条件，对展开各分项工程的时机也能把握恰当，这都利于施工方案的确定。而因施工单位自身的专业特点，决定其在咨询、设计方面有局限性，设计单位作为牵头人，可使 EPC 联合体降低决策成本和实施成本。

Compared with the construction company, the design institute can better understand the design intent, architectural structure characteristics, matching relationship of the mechanical and electrical system, the natural geographical conditions of the project site of the owner, and can also properly grasp the timing for carrying out various sub-divisional works, which are conducive to the determination of the construction scheme. However, due to the discipline characteristics of the construction unit itself, it is limited in consultation and design. As the leader, the design institute can reduce the decision-making cost and implementation cost of the EPC consortium.

2.4.5.2 健全人才培养机制

2.4.5.2 Perfect the talents cultivation mechanism

设计单位是知识密集型的机构，相比于施工单位员工而言可以更好地了解业主方的建设需求，在进行建设需求调研时更具备专业经验与科学方法，在 EPC 过程中与业主之间在协调和沟通、处理矛盾时一般采取技术性方式，有利于提高工程项目的进度和效率。但不能只有从事设计的人才，必须拥有和培养设计管理与项目管理的复合型人才。

The design institute is a knowledge-intensive organization. Compared with the employees of the construction unit, they can better understand the construction requirements of the owner, and have more professional experience and scientific methods in the investigation of construction requirements. In the EPC process, they usually adopt technical methods in the process of coordination, communication and conflict resolution with the owner, which is conducive to improving the progress and efficiency of the project. However, it is necessary to have and cultivate versatile talents who are engaged in design management and project management instead of only talents who are engaged in design.

2.4.5.3 发展技术管理优势

2.4.5.3 Develop technical management advantages

设计人员结合行业技术发展趋势和工程实际情况，更倾向采用新技术、新材料、新工艺和新设备，做到功能适用、技术先进、经济合理，取得最佳投资效果。设计单位牵头 EPC 时，为了实现联合体的整体利益和设计师的利益，应致力于通过设计技术手段进行施工方案优化，实现缩短工期、保证质量、控制投资；获得业主方对技术创新应用的支持，以有效减少实施过程中可能发生的变更和索赔情况，降低监督、制裁违约行为等的成本；通过新技术方案或路线整合联合体内部资源，获得联合体内部更强的主导权。

Designers are more inclined to adopt new technologies, new materials, new processes and new devices in combination with the development trend of industry technology and the actual situation of the project, striving to be applicable in function, advanced in technology and reasonable and effective in cost and obtain the best investment effect. When the design institute takes the lead in EPC, in order to realize the interests of the consortium and the designers, it should devote itself to optimizing the construction scheme through technical design means to shorten the construction period, ensure the quality and control the investment; obtain the support of the owner for the application of technological innovation to effectively reduce the possible changes and claims in the implementation process, and reduce the costs of supervising and sanctioning violations, etc; obtain the stronger domination within the consortium through the integration of internal resources of the consortium by using new technical proposals or routes.

2.4.5.4 建立数据支持平台

2.4.5.4 Establish data support platform

设计企业具有前期策划能力，其设计项目多、覆盖面广、周期短，因而比施工单位掌握更多、更丰富的项目估算指标、概算指标、技术经济指标比较等数据库，在 EPC 招投标阶段，设计企业在业主方建设需求相对模糊的前提下，应总结同类项目相对准确的经济技术基础数据，包括运用大数据与 AI 技术，分析指导工程建设项目总承包的竞标策划，为联合体成员筛选和投标报价阶段提供强大的数据支撑。

The design institute has the ability of preliminary planning, and has numerous design projects, wide coverage and short cycle, so, compared with the construction unit, it has more and richer databases such as project estimate indexes, budget indexes and technical and economic indexes comparison. In EPC bidding stage, the design institute should summarize the relatively accurate economic and technical basic data of similar projects if the construction requirements of the owner are relatively vague, including using big data and AI technology to analyze and guide the bidding planning of the general contract of engineering construction projects, so as to provide strong data support for the selection of consortium members and the bid quotation.

2.4.5.5 把握全程统筹优势

2.4.5.5 Grasp the advantage of overall planning in the whole process

设计牵头 EPC 联合体在前期工作中就应提早展开项目实施的统筹，在政策允许时提前进行施工组织设计或提前开工，边设计边施工或边设计边指导施工准备工作，提早进行专项工程深化施工图设计及设备产品采购，有利于交叉作业的施工组织设计，缩短施工周期。

Led by the design institute, the EPC consortium shall carry out the overall planning of project implementation in advance in the preliminary work, carry out the construction organization design or start construction in advance if the policy allows, start the construction while designing or guide the construction preparation work while designing, and deepen the design of the construction drawing for special projects and purchase devices and products in advance, which are conducive to the cross construction organization design and reduction of the construction period.

2.4.5.6 运用主观意识优势

2.4.5.6 Use the advantage of subjective consciousness

设计人员掌握现行工程设计标准、规范等，在工程质量中实行各专业注册工程师终身负责制和建筑师负责制，设计中考虑设计余量、风险防范、保持建筑永久性的主观意识强，对医疗保健设施在施工期间可能发生的功能变化、运营期的改造弹性、设备材料全生命周期使用成本把握较准确，及时改变设计的主动性、灵活性强，而施工单位在此方面会有短期行为和被动意识，质量保证局限于保修期内。

Designers master the current engineering design standards and specifications, implement the lifelong responsibility system of certified engineer of each discipline and architect responsibility system in engineering quality, consider the design allowance, risk prevention, and maintain a strong subjective consciousness of building permanence in the design; have a more accurate control of the functional changes of health care facilities during the construction, the reconstruction flexibility in the operation period, the use cost of devices and materials in the whole life cycle; and have a strong initiative and flexibility in the design change in time. While the construction company has short-term behavior and passive consciousness in these aspects, and the quality assurance is limited within the warranty period.

2.4.5.7 设计牵头 EPC 风险防范

2.4.5.7 EPC risk prevention led by the design institute

1. 规避权责不匹配风险

1. Avoid the risk of mismatch between rights and responsibilities

设计单位是轻资产企业，注册资金无法与大型项目投资相匹配，融资能力和抗风险能力也较低。一旦项目出现问题，设计单位没有能力赔付，须由联合体其他成员承担连带责任，因此在联合体合约中要健全责权对等机制，明确抗风险对策。

The design institute is an asset-light enterprise, the registered capital cannot be matched with the investment of large-scale projects and the financing capacity and risk resistance capacity are also relatively low. In case of any problems of the project, the design institute is unable to compensate and other members of the consortium need to undertake the joint liability. Therefore, it is necessary to improve the mechanism that rights match with responsibilities and to define the countermeasures against risks in the consortium contract.

2. 提高项目管理能力

2. Improve the project management capability

EPC 最基本的管理要求是：设计管理能力、采购能力、资源整合能力、计划管控能力和专业管理能力，设计单位不能只懂设计，还需引进资源整合人才、工程项目管理人才。不得已时将以设计牵头联合体的 EPC 模式分解为“设计总承包 + 全程项目管理 + 施工总承包”。

The most basic management requirements of EPC include: design management capability, procurement capability, resource integration capability, plan control capability and professional management capability. The design institute not only understands the design, but also needs to introduce resource integration talents and engineering project management talents. If necessary, the EPC mode of consortium led by the design institute will be broken down into “general design contract + overall project management + general construction contract”.

3. 转变设计单位内部运行机制

3. Change the internal operation mechanism of the design institute

设计单位内应建立 EPC 专职管理部门对承揽的 EPC 项目进行统一协调与管理；应将设计师的收入分配体系与限额设计水平、项目盈利水平、工程创优、项目全生命周期成本挂钩，使设计师不满足于只出一版图、交差了事，有主动伴随施工进程优化图纸、关注建成后运营效益的主观愿望。

The design institute shall establish a special EPC management department for unified coordination and management of the contracted EPC projects. The income distribution system of designers should be linked to the quota design level, project profitability, project excellence creation and full life cycle cost of the project, so that designers are not satisfied with only one design drawing and skimp through the work, but have the subjective consciousness to actively optimize the drawings along with the construction process and pay attention to the operating benefits after completion.

4. 重视合同制度

4. Pay attention to the contract system

合同是工程建设中一切工作的纲领性文件，合同应涉及：

The contract is a programmatic document for all work in the construction of the project. The contract shall cover:

（1）明确监理与 EPC 之间的界面，对于工程监理的工作依据之一即图纸的交接过程要做阶段性规定，例如：分成初步设计阶段和施工图阶段。

（1） Define the interface between supervisor and EPC, and make a staged plan for the handover process of drawings (which are one of the work basis of the supervisor), for example, the handover process is divided into the PD stage and construction drawing stage.

（2）明确对施工许可制度与 EPC 管理环节错时的处理办法，设计单位应建立与施工单位固定的长期的战略伙伴关系，包括内部分配机制，做到商务先行。

（2） Clarify the method to deal with the construction permit system and the EPC management link in staggered time. The design institute shall establish a fixed long-term strategic partnership with the construction company, including internal distribution mechanism, so as to achieve business first.

（3）明确五方责任制度与 EPC 之间矛盾的处理办法。建筑工程五方责任主体指：建设单位项目负责人、勘察单位项目负责人、设计单位项目负责人、施工单位项目经理、监理单位总监理工程师。设计牵头联合体 EPC 时，勘察、设计、施工合并为同一责任人，与工程另两方责任主体的矛盾解决办法要在合同中确定，包括解决政策矛盾问题。

（3） Clarify the method to deal with the contradiction between the five-party liability system and EPC. The five subjects of the construction engineering project refer to the project principal of the project owner, the project principal of the survey institute, the project principal of the design institute, the project manager of the construction company and the chief supervision engineer of the supervision company. When the design institute takes a lead in consortium EPC, the same responsible person is responsible for the survey, design and construction, and the method to handle with the contradiction with the other two responsible subjects of the project shall be determined in the contract, including the resolution of policy contradiction.

（4）事先统一结算审计口径与方式

（4） Unify settlement auditing standards and methods in advance

EPC 模式采用总价包干的合同计价方式，合同价变更仅限投资方或经投资许可的业主方提出的实施范围变化、建设标准变化两种，没有上述两种变更情况即按合同价结算，无须审计。实行这样的模式必须落实限额建设的目标，及时调整设计、控制和减除造成突破造价的诉求，使工程造价控制在合同范围内。

EPC mode adopts the lump sum contract pricing method. The contract price can only be changed in case of the change of implementation scope and construction standard proposed by the investor or the owner approved by the investor. In the absence of the above two changes, the settlement will be made according to the contract price without auditing. The implementation of such a mode must implement the goal of quota construction, and must adjust the design, control and reduce the appeal causing the breakthrough of cost in a timely manner to make the project cost within the scope of the contract.

（5）对不同阶段启动 EPC 实行限额设计

（5） Implement the quota design for starting EPC in different stages

对政府投资在初步设计与概算批复后才能介入的 EPC 项目，在已经完成的初步设计限制之下，发挥设计主导优势，对施工图设计的施展空间或优化措施严格把控，务必使施工图预算低于概算，为施工和采购留有弹性，发挥设计主导优势，严格把控施工设计变更不突破概算。对非政府投资项目 EPC 宜从方案设计阶段介入，且建筑策划先行，医疗工艺设计先行，完整获取建设需求与确定造价的可行性，控制规模、量力而为。设计主导的 EPC 应设立全程跟踪的审计服务，强化过程中动态造价控制，重视合同审查与合同交底、多环节技术交底，严格审查合同各项条款用词准确性、承包范围界线，合同双方的义务、责任划分、联合体协议条款分工与权责，必要时应聘请专业律师把关。

For EPC projects in which government investment can only be introduced after the PD and budget estimate approval, under the restrictions of the completed PD, give full play to dominant advantages of design, strictly control the construction drawing design space or optimization measures to ensure that the construction drawing budget is lower than the budget estimate, so as to reserve flexibility for construction and procurement, play the dominant advantage of design and strictly control the changes in construction design without exceeding the budget estimate. For non-government invested EPC projects, it is advisable to intervene from the stage of SD, give priority to architectural planning and design for medical process, so as to fully obtain the construction requirements and determine the cost feasibility, control the scale and do what it can. For EPC led by the design institute, set up an audit service to track the whole process, strengthen the dynamic cost control in the process, attach importance to contract review, contract disclosure and multi-link technical disclosure, strictly review the accuracy of the terms and conditions of the contract, the boundaries of the contract scope, the division of obligations and responsibilities of both parties to the contract, the division of responsibilities and rights in the terms of the consortium agreement, and employ professional attorneys to control if necessary.

第3章

医疗工艺设计与咨询

Chapter Ⅲ

Design for Medical Process and Consultation

徐小田　朱　希

Xiaotian Xu　Xi Zhu

3.1 医疗工艺设计内容

3.1 Design contents for medical process

3.1.1 术语与一般规定

3.1.1 Terms and general provisions

3.1.1.1 医疗工艺是医疗流程和医疗设备的匹配，以及其他相关资源的配置。

3.1.1.1 Medical process is the matching of medical flows and medical devices, as well as the allocation of other related resources.

3.1.1.2 医疗流程是医疗服务的程序和环节。

3.1.1.2 Medical flows are procedures and links of medical services.

3.1.1.3 医疗工艺设计应确定医疗业务结构、功能和规模，以及相关医疗流程、医疗设备、技术条件和参数。

3.1.1.3 The design for medical process shall determine the structure, function and scale of medical services, as well as relevant medical flows, medical devices, technical conditions and parameters.

3.1.1.4 医疗工艺设计应进行前期设计和条件设计；前期设计应满足编制可行性研究报告、设计任务书及建筑方案设计的需要；条件设计应与医院建筑初步设计同步完成，并应与建筑设计的深化、完善过程相配合，同时应满足医院建筑初步设计及施工图设计的需要。

3.1.1.4 The design for medical process shall carry out the PD and conditional design; the PD shall meet the needs of preparing feasibility study report, requirements specification and architectural SD; the conditional design shall be completed simultaneously with the PD of hospital buildings, and shall be matched with the deepening and perfection process of architectural design. At the same time, it shall meet the needs of the PD and construction drawing design of hospital buildings.

3.1.1.5 医疗工艺流程应分为医院内各医疗功能单元之间的流程和各医疗功能单元内部的流程。

3.1.1.5 The medical process flows shall be divided into the flow between each medical functional unit in the hospital and the flow within each medical functional unit.

3.1.1.6 医疗工艺设计应根据实际需求对医疗功能单元进行选取和细分，并对各单元的功能设置、医疗流程、用房需求和卫生学要求进行设计，确定各科门诊、住院的比例及总量。

3.1.1.6 The design for medical process shall select and subdivide medical functional units according to actual needs, and design the functional settings, medical flows, medical room demand and hygiene requirements of each unit to determine the proportion and total amount of outpatients and hospitalization in each department.

3.1.1.7 医疗工艺设计应对医院主要诊断治疗项目的任务量进行策划和计算，根据医院特点确定以下设计参数：门诊诊室间数、急救抢救床数、每个护理单元的病床数、各类手术室间数、重症监护病房（ICU）床位数、各类影像设备台数、介入治疗设备台数、放射治疗设备台数、消毒灭菌设备台数、物流站点数、信息点位数等。

3.1.1.7 The design for medical process shall plan and calculate the tasks of the main diagnosis and treatment items in the hospital, and determine the following design parameters according to the characteristics of the hospital: number of consulting rooms of outpatient department, number of beds of emergency department, number of beds in a nursing unit, number of various operating rooms, number of beds in ICU, number of various imaging devices, number of interventional therapeutic devices, number of radiotherapy devices, number of disinfection and sterilization devices, number of logistics stations, number of information points, etc.

3.1.2 医疗工艺设计深度

3.1.2 Design depth of medical process

3.1.2.1 成果文件

3.1.2.1 Deliverables

应提供医疗工艺系统说明、医疗任务量计算书、医疗工艺流程设计、医疗设备、装备、配置及说明、医疗用房配置要求、医疗相关系统配置等文件。

Provide documents such as medical process system specification, calculations of medical tasks, medical process flow design, medical devices, apparatus, configuration and specification, medical room configuration requirements, relevant medical system configuration.

3.1.2.2 医疗工艺前期（方案）设计深度

3.1.2.2 Depth of preliminary (proposal) design of medical process

1. 医院性质及医疗任务量；

1. Nature of the hospital and medical tasks;

2. 医疗结构设计；

2. Medical structure design;

3. 医疗功能单元设置与任务量设计；

3. Medical functional unit setting and tasks design;

4. 一级医疗工艺流程设计；

4. Design of primary medical process flow;

5. 医疗设备配置计划；

5. Medical devices allocation plan;

6. 医疗装备配置计划；

6. Medical equipment allocation plan;

7. 医疗用房配置要求；

7. Medical room allocation requirements;

8. 医疗工艺相关专业设计方案（水、电、医用气体、净化等）；

8. Professional design proposals related to the medical process (water, electricity, medical gas, purification, etc.);

9. 医疗物流的要求；

9. Requirements for medical logistics;

10. 信息流；

10. Information flow;

11. 初步分析、评价。

11. Preliminary analysis and evaluation.

3.1.2.3 医疗工艺条件设计深度

3.1.2.3 Depth of medical process condition design

1. 医疗任务量细化设定（门诊、住院、手术等）；

1. Detailed setting of medical tasks (outpatient, hospitalization, operation, etc.) ;

2. 医疗结构与功能设计；

2. Medical structure and function design;

3. 医疗功能单元设置与任务量优化设计；

3. Medical functional unit setting and tasks optimized design;

4. 一级医疗工艺流程优化设计；

4. Optimized design of primary medical process flow;

5. 二级医疗工艺流程设计；

5. Design of secondary medical process flow;

6. 医疗设备配置标准及设备选型、技术规格，设备所需水、电、空调等条件要求；

6. Medical devices configuration standards and devices selection, technical specifications, water, electricity, air conditioning and other conditions required by the devices;

7. 医疗装备配置标准、种类、规格等参数；

7. Medical devices configuration standards, types, specifications and other parameters;

8. 医疗用房配置要求及房间条件要求；

8. Requirements for medical room allocation and conditions;

9. 医疗工艺相关专业配置标准及技术参数；

9. Configuration standards and technical parameters of disciplines related to the medical process;

10. 综合分析、结论。

10. Comprehensive analysis and conclusion.

3.1.3 医疗工艺设计目标

3.1.3 Design objective of medical process

促进患者自身拥有的自然治愈力，并排除阻碍这种自愈的要素，实现功能、安全、效率、经营、发展。

Promote the natural healing power of patients, and remove the factors hindering the self-healing, so as to realize functions, safety, efficiency, management and development.

3.1.3.1 功能方面：医疗规划，功能设置与发展，部门相互关系，配套设施及技术参数落实。

3.1.3.1 Function: implementation of medical sources planning, function setting and development, relationships among departments, auxiliary facilities and technical parameters.

3.1.3.2 安全方面：院感安全、病人行为安全、医护人员行为安全、诊疗环境安全、废弃物管控安全。

3.1.3.2 Safety: hospital infection control safety, patient behavior safety, medical care staff behavior safety, diagnosis and treatment environment safety, waste management and control safety.

3.1.3.3 效率方面：急救效率、就诊效率、科室运营效率。

3.1.3.3 Efficiency: first aid efficiency, medical treatment efficiency and department operation efficiency.

3.1.3.4 经营方面：服务产品规划、流程改善与服务品质、规模与成本控制。

3.1.3.4 Management: service product planning, flow improvement and service quality, scale and cost control.

3.1.3.5 发展方面：医疗空间的可变性、医疗方式不断改进、未来可能发生重大改变的区域、对未来发展规模变化的应对。

3.1.3.5 Development: variability of medical space, continuous improvement of medical methods, areas that may undergo major changes in the future, and responses to changes in the scale of future development.

3.1.4 工艺设计表单

3.1.4 Process design form

3.1.4.1 面积分配统计表

3.1.4.1 Statistical table of area distribution

1. 总体规划技术指标表：总建筑面积（地上、地下）、总床位数量、绿地率、停车位（救护车、运输保障货车、自有大型车辆、职工车位、外来车位）、停机坪；

1. Table of technical indicators for overall planning: total building area (aboveground and underground), total number of beds, greening rate, parking spaces (ambulance, transportation truck, large vehicle, staff vehicles, foreign vehicles), helipad;

2. 七项基本设施，如科研、教学、规培、预防保健、康复、养老等设施面积与分配比例；

2. Area and distribution ratio of seven basic facilities and other facilities, such as scientific research, teaching, normalized training, prevention and health care, rehabilitation, endowment;

3. 公共空间面积，含门厅、公共通道、公共垂直交通与疏散设施、机电设备用房及管道间、商业服务、保洁、公共厕所及无障碍设施、消防和人防设施、停车及落客港湾空间等；

3. Area of public spaces, including entrance hall, public passageway, public vertical trans-

portation and evacuation facilities, mechanical and electrical equipment rooms and plumbing rooms, commercial services, cleaning, public toilets and barrier-free facilities, fire and civil air defense facilities, parking and drop-off bay space, etc. ;

4. 各科室总面积，含科室内部交通与疏散设施、内部厕所、保洁间等；

4. Total area of each department, including internal transportation and evacuation facilities, internal toilets, cleaning room of each department, etc. ;

5. 各科室主要房间面积与数量。

5. Area and quantity of main rooms in each department.

3.1.4.2 大中型医疗设备统计表：占用房间的设备、台数、大型设备荷载、相关产品机电参数。

3.1.4.2 Statistical table of large-size and medium-sized medical devices: devices occupying the room, number of devices, load of large devices, and mechanical and electrical parameters of related products.

3.1.4.3 医用气体设备终端点位表。

3.1.4.3 Table of terminal points of medical gas devices.

3.1.4.4 自动化物流站点位表。

3.1.4.4 Table of automatic logistics stations location.

3.1.4.5 智能化设备终端点位表。

3.1.4.5 Table of terminal points of intelligent devices.

3.1.4.6 信息化设备终端点位表。

3.1.4.6 Table of terminal points of information devices.

3.1.4.7 固定家具设置点位表：服务台/安保岗点位及数量、各科护士站点位数量、固定窗口（挂号、收费、取药、标本采集等）点位数量。

3.1.4.7 Table of fixed furniture setting point: setting points and number of help desks/security duty rooms, setting points and number of nurse stations in each department, setting points and number of fixed windows (registration, charging, medicine service, specimen collection, etc.).

3.1.5 工艺流程分级

3.1.5 Process flow classification

3.1.5.1 一级工艺流程：是总体医疗功能、总系统的流程，应涵盖医疗保健设施内全体部门的安排，为总平面设计、建筑单体选型、各部门在建筑内的选址提供依据。

3.1.5.1 Primary process flow: It is the flow of overall medical function and overall system, and shall cover the arrangement of health care facilities in all departments, providing basis for general layout design, single building type selection and location selection of various departments in the building.

一级流程图：表达总平面各类出入口，总体流线关系；各科室或部门在各栋建筑内的选址，科室之间的流线关系。

First-class flow chart: It shows all kinds of entrances and exits of the general plane and the overall flow relationship; location selection of each department in each building and flow relationship among departments.

3.1.5.2 二级工艺流程：是医疗功能部门、单元、分系统的流程，应覆盖基层医疗活动区域，确定基层医疗流程，制定部门、科室和单元内部作业流程，分配所需空间，为建筑设计提供依据，以确定部门/单元空间形态、房间相互关系。

3.1.5.2 Secondary process flow: It is the flow of medical functional departments, units and subsystems, and shall cover the basic medical activity area, determine basic medical flows, formulate the internal operation flow of departments and units, allocate the required spaces, providing basis for architectural design to determine the spatial form of departments/units and the relationship between rooms.

二级流程图：表达科室内部空间划分，各类出入口，各类人流、物流的流线。

Second-class flow chart: It shows the division of internal spaces of the department, all kinds of entrances and exits, and all kinds of people flows and logistics flows.

3.1.5.3 三级工艺流程：是房间功能和其中作业人员的操作流程，应覆盖医疗人员工作区域，确定医疗人员行为流程，罗列房间内的各类诊疗设备、设施，为室内设计提供依据。

3.1.5.3 Tertiary process flow: It is the function of rooms and the operation flow of the operators therein, shall cover the working area of the medical care staff, determine the behavior flow of the medical care staff, list all kinds of diagnosis and treatment devices and facilities in rooms, providing basis for the interior design.

三级流程图：表达一个开敞式大空间内和房间内的流程和机电终端点位要求，如ICU大开间病房、消毒供应室区域空间、大空间实验室等；提出固定设施数量和布局要求，提出特殊荷载、水、电、暖、通风、气体、蒸汽、物流、信息等使用终端数量和点位布置要求。

Three-class flow chart: It shows the flow and requirements for mechanical and electrical terminal points in an open large space and rooms, such as ICU large-bay ward, disinfection supply room area space, large space laboratory, etc.; provides the number and layout requirements of fixed facilities, and the number and location layout requirements of terminals for special load, water, electricity, heating, ventilation, gas, steam, logistics, information, etc.

3.2 医疗工艺设计方法

3.2 Method of design for medical process

3.2.1 医疗工艺前期策划

3.2.1 Preliminary planning of medical process

3.2.1.1 设施选型：是综合性医院、教学医院、某一类专科医院、疗养院，还是其他医疗保健设施？应确定医院服务范围，区域卫生规划所赋予的预防、保健及医疗任务。不同设施的建筑设计既有共性，又有个性，医疗工艺流程设计差别很大。

3.2.1.1 Type of facilities: general hospital or teaching hospital, specialized hospital, sanatorium, or other health care facilities? Determine the scope of hospital services and the prevention, health care and medical tasks assigned by the regional health planning. The architectural design of different facilities has both commonness and individuality, and the design of medical process flow is very different.

3.2.1.2 组织设计：明确产权关系、财政体制、组织结构、治理结构、人力资源配置、服务方式及后勤保障体系等，对新建设施应进行医疗服务流程设计，对改扩建项目要进行精益管理与流程再造。组织设计对医疗工艺流程设计影响较大。

3.2.1.2 Organizational design: Clarify the property relations, financial system, organizational structure, governance structure, human resource allocation, service mode and logistics support system, etc. Design the medical service process for new facilities, and keep lean management and flow reengineering for reconstruction and expansion projects. Organizational design has great influence on design of medical process flow.

3.2.1.3 医疗功能和规模：确定医院等级、学科、专业设置及业务能力，教学、科研设置，床位数、年门诊量、年急诊量、年手术量，是医疗工艺前期策划的核心部分。

3.2.1.3 Medical function and scale: Determining hospital level, subject, major setup and professional ability, teaching and scientific research setting, number of beds, annual outpatient quantity, annual emergency quantity and annual operation quantity is the core part of the preliminary planning of medical process.

3.2.1.4 设施要素：确定人员（管理者、医务人员、患者）、设备（建筑设备、医疗设备、后勤设备）、物资（药品、耗材、医疗消耗、能源）、经费（工资、资金）、信息（医疗信息、管理信息），是医疗工艺设计的重要前提条件。

3.2.1.4 Facilities elements: Determining personnel (managers, medical personnel, patients), device (building device, medical device, logistics device), materials (drugs and consumables, medical consumption, and energy), funds (wages, and funds), information (medical information, management information) is important prerequisite for design for medical process.

3.2.1.5 临床科室医疗工作流程：医疗工作流程设计是建筑中医疗功能用房设置、医患动线、流程设计的重要依据，是医疗工艺流程设计的输入条件。

3.2.1.5 Medical flows of clinical departments: Medical flow design is an important basis for setting up medical functional rooms, routines of doctors and patients, and process design in buildings, and is also an input condition for medical process flow design.

3.2.2 医疗工艺前期设计

3.2.2 PD of medical process

3.2.2.1 医疗流程、医疗设备、技术条件和参数设计

3.2.2.1 Medical flows, medical devices, technical conditions and parameter design

1. 医疗专业结构：医院功能单元（科室）设置和业务设定。

1. Medical organization: setting of hospital functional unit (department) and business setting.

2. 医疗工艺流程：一级流程和二级流程设定。

2. Medical process flow: setting of first-class flow and second-class flow.

3. 医疗业务标准：卫生学标准和各科业务标准的设定。

3. Medical service standards: setting of hygienic standards and service standards for various subjects.

4. 医疗指标体系：医院质量和效率各种指标的预设。

4. Medical indicator system: Presupposition of various indicators of hospital quality and efficiency.

3.2.2.2 自动化系统工艺设计

3.2.2.2 Process design of automatic system

1. 医疗工艺前期设计应包括药学部药品自动分配系统、消毒供应自动物流及物联网系统、影像信息自动传输系统的策划和方案；应包括后勤支持系统中的物流传输系统、标识导向系统、无障碍辅助系统、医用气体系统及医用家具系统。应从医疗工艺的角度对上述内容进行策划、设计，使这些系统的策划和设计结果符合医疗工艺设计的各个步骤和环节；这些策划和设计可以分散在各个专项工艺前期设计中，但应由医疗工艺总体设计统筹实现。

1. The PD for medical process includes the planning and proposal of automatic drug distribution system of pharmaceutical department, automatic logistics and Internet of Things system of disinfection supply, and automatic image information transmission system, it shall include the logistics transmission system, identification and guidance system, barrier-free auxiliary system, medical gas system and medical furniture system in the logistics support system. Plan and design the above contents from the perspective of medical process so that the planning and design results of these systems conform to all steps and links of design for medical process. These plans and designs can be scattered in the PD of each special process, but they should be realized by the overall design of medical process.

2. 在工艺设计前期应确定上述各类系统的自动化程度，以作为建筑方案设计的重要前提。各类保障系统自动化程度的高低，或是否为自动化系统，不仅会影响到医疗工艺设计中的功能设置、流程设计、点位设置以及人员安排，还影响建筑面积和层高的分配。

2. In the early stage of design for medical process, the degree of automation of the above-mentioned systems should be determined as an important prerequisite for architectural SD. The degree of automation of various security systems, or whether such systems are automatic, will not only affect the function setting, process design, point setting and personnel arrangement in design for medical process, but also affect the allocation of building area and floor height.

3. 应将医疗工艺设计延伸至医用家具设计，通过灵活分隔的医用家具来划分医疗和感染控制的功能分区、空间和进行流程、流线设计，以减少建筑土建墙体间隔，使室内设计与家具设计一体化，使医疗设施、系统终端、卫生器具、控制面板与家具一体化，减少建材使用和污染，利于保洁，便于灵活改造、扩建，是绿色的设计。

3. Extend the design for medical process to the medical furniture design, and divide the functional zones and spaces of medical treatment and infection control and design processes and streamline through the flexible separation of medical furniture, so as to reduce the separation of civil construction walls, integrate interior design with furniture design, integrate the medical facilities, system terminals, sanitary fixtures, control panels and furniture, reduce the use and pollution of building materials, facilitate cleaning, and facilitate flexible reconstruction and expansion as well as keep green design.

3.2.2.3 信息化系统工艺设计

3.2.2.3 Process design of information system

1. 收费点设置

1. Charging point setting

以最少的患者移动距离、减少排队等候次数或零排队、工作点位及时间为医疗工艺设计目标。根据医嘱和收费的对应关系，在医嘱系统中把收费和医嘱融合为一个系统，在医嘱区域设计收费点。

The goal of design for medical process is to minimize the patient moving distance, reduce number of times of queuing or reach zero queue, work points and time. Integrate the charge and the medical advices into one system in the medical advice system according to the relationship between the medical advices and the charge, and set the charge points in the medical advice area.

2. 分系统设置检查区

2. Examination areas set by system

在内科系统、外科系统、妇科系统等分区或模块内设置各个系统专用的检查区，通过信息交换和共享实现信息联通，患者通过自助发卡和自助挂号进入候诊程序，方便执行医嘱，实现患者一站式服务。

Set special examination areas for each system in divisions or modules of internal medical system, surgical system, gynecology system, etc. and keep information communication through information exchange and sharing. Patients access the waiting procedure through self-service card issuing and self-service registration, which facilitates the implementation of medical advices and realizes one-stop service for patients.

3. 关联设计

3. Associated design

将就诊区与药房、检验中心的信息互联，在医嘱下达的同时药房和检验科做好相关准备，在费用等相关信息确认后，通过自动物流传输系统完成药品和血样等标本的传递，患者通过自动打印机提取检验报告，医生通过信息系统直接看到患者的检验结果，这个过程中患者无须移动。通过关联设计，医院不需要设检验报告发放窗口和患者取报告的等候区，只需设置自助打印机和将信息发送到医生的电脑里。

Interconnect the information of the treatment area with the pharmacy and inspection center. When the medical advices are issued, the pharmacy and the laboratory department make relevant preparations. After the confirmation of relevant information such as charge, the transfer of samples such as drugs and blood samples is completed through the automatic logistics transmission system. The patient collects the inspection report through the automatic printer. The doctor could directly observe the patient's inspection results through the information system, and the patient does not need to move during this process. Through correlation design, the hospital does not need to set up an inspection report issuing window and a waiting area for patients to take reports, but only needs to set up a self-service printer and transfer the information to the doctor's computer.

4. 基于信息化的第三方服务

4. Information-based third-party services

包括检验、病理、影像、消毒供应、药库、洗衣等由第三方提供服务的情况。有第三方服务的医疗保健设施内不需要提供这些内容的建筑空间，不需要空间流程设计，只需要信息化联通。

Including inspection, pathology, imaging, disinfection supply, pharmacy, laundry and other services provided by third parties. Health care facilities with third-party services do not need to provide building space for these aspects, nor do they need related flow design, but only need information connectivity.

5. 信息化急救模式

5. Information first-aid mode

基于信息化的院前急救模式，是通过信息系统实现专家远程指导，提高现场救护水平，设置监护设备远程数据传递功能，医疗救治前移。应将医院或设施的 PACS 系统和设备纳入急救物联网，实现大数据共享，供医院之间会诊、共享影像数据、及时获取使用。

The information-based pre-hospital first-aid mode is to realize the remote guidance of experts through the information system, improve the on-site rescue level, set up the remote data transmission function of monitoring device, and perform medical treatment as early as possible. PACS systems and devices of hospitals or facilities should be incorporated into the first-aid Internet of Things to realize big data sharing for consultation, image data sharing and timely access between hospitals.

6. 基于物联网的社区与医院联网医疗流程

6. Internet of Things-based networking medical flows for communities and hospitals

将各个医院作为网络上的各个存储节点，把物联网的信息系统推广到各个医院和各个社区健康服务中心，将医院门诊下放到社区门诊，以缓解医院门诊的压力，减少门诊空间，提高救治效率。

Take each hospital as each storage node on the network, extend the information system of the Internet of Things to each hospital and each community health service center, and transfer the hospital outpatient service to the community outpatient service, so as to relieve the pressure of the hospital outpatient service, reduce the outpatient service space and improve the treatment efficiency.

理想的大型医院应只包括急诊、急救、医技、住院、保障系统，门诊功能只设日间治疗、日间手术和 MDT 综合会诊，把一般门诊下放到一级社区医疗点或分布匀质的分散门诊部去，从而使医院内的流程变得更加简单。

The ideal large hospitals should only include emergency treatment, first-aid, medical skills, hospitalization and guarantee system. Outpatient services only include daytime treatment, day surgery and MDT comprehensive consultation. Transfer the general outpatient service to the first-class community medical point or the uniformly distributed decentralized outpatient service, thus making the flows in the hospital simpler.

7. 信息化所需建筑空间与预留

7. Building space and reservation for informatization

在设施建筑面积分配时，应有信息化系统所需要的建筑空间，包括信息机房、布线管道井、进线室、吊顶层高、检修口。若最初没有信息化系统投资，则应考虑预留。

When allocating the construction area of the facilities, there should be the construction space required by the information system, including the information machine room, cabling pipe well, incoming line room, ceiling height and access holes. Consider reservation if there is no information system investment initially.

3.2.2.4 智能化系统工艺设计

3.2.2.4 Process design for intelligent system

1. 应编制项目智能化系统需求方案，做出系统构成规划；应明确项目中使用哪些计算机技术、通信技术、网络技术、自动化控制技术；设计目标是节省人力资源，降低能源消耗，提高设施内工作效率。

1. Prepare the requirement proposal for the intelligent system of the project and plan the system composition. It should be clear which computer technology, communication technology, network technology and automatic control technology are used in the project; the design goal is to save human resources, reduce energy consumption and improve working efficiency in the facility.

2. 医院智能化系统应满足医院内高效、规范与信息化管理的需要；应向医患提供“有效地控制医院感染，节约能源，保护环境，构建以人为本的就医环境”的技术保障。

2. The hospital intelligent system should meet the needs of high efficiency, standardization and information management in the hospital. Provide the technical guarantee to doctors and patients to “effectively control hospital infection, save energy, protect the environment and build people-oriented medical environment”.

3. 医疗工艺智能化条件设计

3. Design for intelligent conditions of medical process

应提供与建筑室内各界面末端装饰设计相关的智能化系统建设技术条件。

Provide technical conditions for the construction of intelligent systems related to the end decoration design of each interface within the buildings.

4. 综合布线系统点位配置

4. Point location of generic cabling system

应根据项目需求提供全部点位设计清单。

All points design lists shall be provided according to project requirements.

5. 确定综合布线系统主干带宽

5. Determination of backbone bandwidth of generic cabling system

主干带宽根据医院或设施的规模确定。如大型医院采用万兆主干，规模较小的医院采用千兆主干。

Determine the backbone bandwidth according to the size of the hospital or facility. For example, large hospitals adopt 10-gigabit backbone network, while smaller hospitals adopt gigabit backbone network.

6. 提出进线要求

6. Proposal of requirements for incoming line

（1）明确是每个楼层均设弱电间与进线室，还是隔层设置；

（1）Make clear whether low voltage room and incoming line room are set in each floor or alternate floor;

（2）提出楼层管理间的通风（自然或机械）、室内温度（10～30℃）、相对湿度（20%～80%）要求；

（2）Propose the requirements for ventilation (natural or mechanical ventilation), indoor temperature (10°C to 30°C), and relative humidity (20% to 80%) of floor administration room;

（3）明确是否设备网与内外网弱电间分开；

（3）Make clear that whether the device network and the internal and external low voltage rooms are separated or not;

（4）是否采用双电源供电（常规电源、UPS电源、两个不同电站供电）。

（4）Whether dual power supply is adopted (the conventional power supply, UPS, providing power supply by two different power stations).

7. 确定无线网络系统应用范围

7. Determination of application range of the wireless network system

应作出对安装工程及装饰效果有影响的应用选择，包括：

The selection of application which can influence the installation works and the decoration effect should be made, it includes:

（1）移动查房、移动护理、移动心电采集、移动输液室管理、移动点餐、各类移动工作站。

（1）Mobile ward round, mobile nursing, mobile electrocardiogram data acquisition, arrangement of mobile infusion rooms, mobile ordering, various mobile workstations.

（2）结合物联网定位技术和传感技术的母婴匹配与婴幼儿防盗，手术室、ICU、供应室设备清点与跟踪系统、移动设备定位监控、医疗废弃物管理、生命体征实时侦测等。

（2） The maternal and infant matching and infant security, operating rooms, ICU, device inventory and tracking system of supply room, position monitoring of mobile devices, medical waste management, real-time detection of vital signs are achieved by Internet of Things-based positioning and sensing technology.

（3） AP 选型和安装方式：外置吸顶天线，将 AP 隐藏到天花板后面，或直接安装在吊顶下或是走廊墙上。

（3） The selection and installation mode of AP: Put the AP of external ceiling antenna behind the ceiling in a concealed manner, or install it directly under the ceiling or on the corridor wall.

8. 提出工艺流程相关安保系统终端点位

8. Proposal of terminal points of security system related to process flow

（1） 安防监控系统视频监控点位

（1） Video monitoring points of security monitoring system

① 医院周界/围墙警戒区及医院各出入口；

① Warning areas in hospital perimeter or enclosure and each entrance and exit of the hospital;

② 医院公共区域（监视区），如停车库（场）、广场道路、楼道、电梯、挂号区域、各功能科室及诊疗科室出入口；

② Public areas of hospital (monitoring areas), such as parking garage, square roads, corridors, elevators, registration areas, each functional department and entrance and exit of diagnosis and treatment department;

③ 在医院防护区域，如收费区、住院区、办公区、贵重医疗设备应用室、妇产、传染、危重急救病房、手术室等区域或出入口、通道等处；

③ Protective areas of hospital. For example, charging area, inpatient area, office area, valuable medical device application rooms, maternity wards, infectious disease wards, critical emergency wards, operating rooms and other areas, or entrance and exit and other access;

④ 医院内不允许自由出入的禁区，如药品、器材、设备存放库，财务、档案室，安全技术防范系统中心控制室等内外部、通道、出入口；

④ Restricted zones in hospital where free access is not allowed, such as the storage warehouse which stores medicine, equipment, and device, the financial office, archives room, the central control room with security & technology protection system and other internal and external access and entrances and exits;

⑤ 各病区护士站、医患沟通室、ICU、纠纷接待室、各类会议室，其中医患沟通室、纠纷接待室增加音频同步采集，用于证据保全以及远程监控。

⑤ Nurse stations in each ward, doctor-patient communication rooms, ICU, dispute resolution rooms, and all kinds of meeting rooms, with audio devices set in doctor-patient communication rooms and dispute resolution rooms for data collection in a synchronization manner in order to preserve evidence and remotely monitor.

（2）入侵防盗报警系统监控点位

(2) Monitoring points of anti-theft intrusion alarm system

① 医院周界的围墙、栅栏、河道以及容易攀爬的屋顶；

① Enclosures, fences, river channels and the roofs which are easy to climb around the hospital;

② 财务室、现金结算处、血库、药品库房、毒麻及精神类药品存放地、实验室、化验室、存放致病微生物、放射性物品的场所、病史室、病理档案室、重要设备仪器存放处、计算机网络管理中心、中心供氧站、配电站（间）；

② Financial office, cash settlement office, blood bank, drug storehouse, narcotic drug and psychotropic drug storehouse, laboratory, testing laboratory, rooms storing pathogenic microorganism and objects with radioactivity, medical records room, pathology archives rooms, important device and instruments storehouse, computer network management center, central oxygen supply station, and power distribution station (room);

③ 门卫室、挂号处、收费处、院领导办公室、现金结算处、财务室、医患纠纷接待室、收费处、各护士站、各急诊诊室、急诊抢救室。

③ Guard room, registration office, charging office, offices of hospital leaders, cash settlement office, financial office, doctor-patient dispute resolution room, charging office, each nurse station, each emergency consulting room, and emergency room.

（3）门禁管理系统点位

(3) Points of access control system

对设施内人员的流动进行合理的监管和控制。一般在公共场所和非公共场所交界处、病区进出通道、病区治疗室、重要场所进出通道等地布点，安装门磁开关、电控锁及读卡器等门禁控制装置，进行身份识别，设置不同的权限，限制人员的进出，保障医院的人员及设备安全。

Make reasonable regulation and control for the flow of personnel within the facilities. Generally, such points are set at the junction of public places and non-public places, access to wards, therapy rooms in wards, entrances and exits of important areas and other places, with magnetic door switches, electric control locks, card readers and other access control apparatus installed for identification, and different authorities can be set to restrict access, thus ensure the safety of people and device in the hospital.

① 重要场所的出入口（挂号及收费处、现金结算处、手术室、ICU病区、血库、药品库房、毒麻及精神类药品存放地、存放致病微生物的场所、放射性物品存放处、重要设备仪器存放处、传染病院的清洁区、半污染区和污染区等）应设置出入口控制；

① Arrange access control at entrances and exits of important places (registration and charging office, cash settlement office, operating room, ICU wards, blood bank, drug storehouse, narcotic drug and psychotropic drug storehouse, rooms storing pathogenic microorganism and objects with radioactivity, important device and instruments storehouse, clean areas in infectious disease hospital, semi-contaminated area and contaminated area, etc.);

② 限制性（禁区）出入口（医院领导办公室、重要办公场所、实验室、化验室、安防控制室等）宜设置出入口控制装置，并设置可视对讲门禁装置；

② Set access control apparatus and visual intercom control apparatus at restricted entrances and exits (restricted zones) (offices of hospital leaders, important offices, laboratory, testing laboratory and security control rooms, etc.);

③ 专用电梯安装读卡器和门禁控制器，刷卡后电梯启动根据卡的权限抵达授权楼层；

③ Install card reader and access controller in specialized lifts, which can reach to the authorized floor according to the permission of cards after punching;

④ 挂号及收费处、现金结算处、药品库房、ICU 病区等区域可视情况采用双向出入口控制装置；

④ Arrange two-way access control apparatus as appropriate at registration and charging office, cash settlement office, drug storehouse, ICU wards and other areas;

⑤ 医院病区和手术室使用与视频呼叫相结合的可视对讲门禁系统；

⑤ Keep visual intercom access control system in wards and operating rooms in combination with video call;

⑥ 病区治疗室安装中距无线射频门禁系统，解决夜间护士在治疗过程中，治疗室无人而造成的药品遗失等安全隐患。中距无线射频门禁系统或手机蓝牙 app 功能可在护理人员距离门禁 30～50cm 时自动打开门，解决护理人员手上带有医疗污染或持有托盘等物品不便刷卡的问题。

⑥ Install middle-distance radio frequency access control system at the therapy room in wards, so as to prevent drug loss and other safety hazards when no one is in the treatment room in the night while nurses are busy with treatment. The middle-distance radio frequency access control system or mobile phone bluetooth app can open the door automatically for nurses in a distance of 30 to 50 centimeters from the door, thus solves the problem that it is not convenient to punch the card for nurses with medical pollution or trays and other things in their hands.

（4）电子巡查系统点位

(4) Points of electronic patrol system

应根据医院布局、特点和安全防范工作的要求，合理选用电子巡查系统，规划出保安巡查路线，依据巡查路线配置电子巡查信息装置（巡查点）点位，根据实际需要选择采用在线式或离线式系统。

Select electric patrol system rationally according to the requirements of the layout, feature and safety precautions of the hospital, plan the patrol routes for security guards, collocate electric patrol information apparatus at corresponding points (patrol points) as per patrol routes, and choose online or off-line system based on actual requirements.

（5）探视对讲系统点位

(5) Points of visitation intercom system

医院隔离病房因病情或病房管理原因，家属与亲友不能直接探视病人，一般采用探视走廊的方式，亲属隔着玻璃探视病人，包括：血液移植病房、传染病房、ICU、新生儿病房、烧伤病房、高压氧舱、手术室家属谈话间和病房非探视时间的联系对讲。

Due to the illness of patients and the ward management, family members and friends can't visit patients directly in the isolation wards, they can generally visit the patients through windows at the visiting corridor, such wards include blood transplantation ward, infectious diseases ward, ICU, neonatal ward, burn ward, hyperbaric oxygen chamber, family communication room in operating room, including the intercom of wards during non-visiting hours.

① 探视对讲系统有网络型和非网络型两种。在建立家属和患者音视频对讲的同时，还可以附加探视计时功能、监听插话功能、图像查看功能、主机呼叫分机功能、分机呼叫主机功能、多方通话等功能；

① Visitation intercom system includes network system and non-network system. In the meantime of setting voice video intercom between family members and patients, additional visiting chronograph function, monitor interposing function, image viewing function, host calling extension function, extension calling host and multi-party calls and other functions can be added;

② 通过音视频网络远距离进行探视交流，可以在全院范围内选择建设探视间，灵活性大，家属可以在医院有网络的任何地方，甚至将来在家中，也可以通过网络与病人进行视频和语音交流，实现隔离探视；

② Communication through voice video network from a great distance can be realized, and visitation rooms can be flexibly constructed in the whole hospital, family members can communicate with the patients in the manner of video and audio at anywhere in the hospital or even at home in the future through the network for isolation visitation;

③ 隔离病房的摄像机等前端设备宜考虑使用固定和移动方式；

③ Front-end device such as cameras in isolation wards may be fixed and movable type;

④ 被探视的患者有的有自主对讲能力，有的没有，在方案选择上，根据实际需要婴儿或重症病人考虑单向的探视方案，传染病或一般病人考虑双向的探视方案；

④ Some of the patients visited can speak autonomously, while others cannot speak autonomously. In terms of proposal selection, infants or severe patients are visited as required according to the one-way visitation proposal, and infectious diseases or general patients are visited according to the two-way visitation proposal;

⑤ 在隔离病房病人端设移动前端设备，家属方设固定后端设备，以较少的前端设置实现较大范围的探视对讲。

⑤ Movable front-end device is provided for the patient in the isolation wards, and fixed back-end device is provided for the family, so that a large range of visitation intercom can be realized with less front-end devices.

3.2.3 医疗设备与装备规划

3.2.3 Planning of medical devices and equipment

不同等级的医院有国家规定的各级设备配置的要求，也有医院不同个性化的需求，不同医院设备配置的档次不一，在前期方案规划中要体现出来。既要防止医院大型设备超负荷运转甚至不能满足诊疗需求、利用过度的情况出现，也要防止医院为了提高综合竞争实力，在没有进行详细调查分析的前提下，盲目购置设备的情况。

There are national requirements for the hospitals at different levels to allocate the device at all levels and personalized requirements for some hospitals. The difference among device allocation levels of different hospitals shall be reflected in the preliminary planning. It is necessary to not only prevent the overload operation of large devices in hospital, which even cannot meet the diagnosis and treatment demands, and over-utilization, but also to prevent the hospitals from blindly purchasing devices without detailed investigation and analysis in order to improve their comprehensive competitive strength.

3.2.3.1 医疗设备

3.2.3.1 Medical devices

包括医疗设备规划配置及机房条件。由国家卫生健康委员会发布的《医疗机构基本标准》规定了十一类医疗机构必须设置的医疗设备，比如口腔医院、美容医院、儿童医院、村卫生室等都有相应的标准。在每一种医疗建筑的建设标准中均有相关规定。

This part includes medical device planning and allocation and machine room conditions: The Basic Standards for Medical Institutions issued by the National Health Commission of the People's Republic of China stipulates eleven categories of medical devices that medical institutions must provide. There are corresponding standards for stomatological hospitals, beauty hospitals, children' s hospitals and village clinics. The construction standards of all healthcare buildings are stipulated.

1. 大型医用设备配置

1. Allocation of large medical devices

（1）医疗设备配置方案应根据医院总体医疗能力和各医疗专业需求而定。设备配置计划应根据各专业技术能力进行设备选型，并结合投资能力来确定。根据医院科室、负荷、发展等情况，制定明确的设备计划及预算。

(1) The medical device allocation proposal shall be determined according to the overall medical capacity of each hospital and the demands of each medical discipline. The device allocation plan shall be determined according to the technical capability of each discipline for device selection and the investment capability of each hospital. A clear device plan and budget are formulated according to the departments, load, development and others of the hospital.

（2）应根据医疗工艺的要求，设计与医疗设备使用有密切关系的医疗工艺流程。需明确大型医疗设备机房的空间、平面、通道、防护、屏蔽、承重、配套设施等方面的要求。

(2) Medical process flow closely related to the use of medical device shall be designed according to the requirements of the medical process. It is required to define the requirements for space, plane, passage, protection, shielding, bearing, auxiliary facilities and others in large medical device rooms.

（3）制定房间清单面积需求表，明确各项医疗设备使用所需水、电、气、温度、湿度等建筑要求，及其他建筑必须考虑的问题。配合进行建筑平面的调整，确定重要的建筑尺寸，完成医疗设备和装备的典型房间的平面布置，明确医疗装备计划。

（3）Make a list of room area demands, and define the water, electricity, gas, temperature, humidity and other building requirements for the use of medical device and other building issues that must be considered. Adjust the building plane properly, determine important building dimensions, complete the plane layout of typical rooms of medical device and equipment, and define medical device plan.

2. 手术室设备配置及规划

2. Allocation and planning of devices in operating room

（1）手术室区域功能划分

（1）Functional division of operating room area

① 将手术室、介入治疗室、ICU 三者临近布局，让麻醉资源得以共享。

① The OR, DSA and ICU are arranged adjacently, so that the anesthesia resources can be shared.

② 兼顾各学科不同功能的手术间，明确临床功能需求。将日间手术室、Ⅰ级手术室、铅防护骨科手术室、术中核磁、腔镜手术室、放疗手术室、感染手术室、杂交/复合手术室、心导管室的布局和相互之间的流程、安全防护、空气净化组织、污物流线做到最佳；心血管杂交、肿瘤介入、冠脉、冠脉＋电生理 DSA 手术室布局及预留机位，符合介入治疗岛临床功能需求。

② Give the operating rooms with different functions of various disciplines into consideration, and define clinical functional requirements. Optimize the layout and mutual flows, safety protection, air purification and sewage streamlines of day operating room, class Ⅰ operating room, lead-protected orthopedic operating room, intraoperative nuclear magnetism operating room, endoscopic operating room, radiotherapy operating room, infection operating room, hybrid / compound operating room and cardiac catheterization room, and make the layout and reserved device positions of cardiovascular hybridization operating room, tumor intervention operating room, coronary artery operating room, coronary artery + electrophysiology DSA operating room meet the clinical functional requirements of interventional therapy.

③ 确定复合手术室类型：

③ Determine the type of compound operating room:

一体化复合手术室：腔镜手术、外科手术、介入手术、影像、声音、摄像、设备控制等信息全面整合，外科吊塔、麻醉吊塔、手术灯。

Integrated compound operating room: It is used for endoscopic surgery, surgical operation, interventional operation and the comprehensive integration of information such as image, sound, video and device control signals. There are surgical ceiling pendants, anesthesia ceiling pendants and operating lamps.

数字化复合手术室：外科手术、介入手术，多种影像信息的整合，外科吊塔（可选）、麻醉吊塔（可选）、手术灯。

Digital compound operating room: It is used for surgical operation, interventional operation and image information integration. There are surgical ceiling pendants (optional), anesthesia ceiling pendants (optional) and operating lamps.

普通复合手术室：外科手术、介入手术，影像信息独立显示，麻醉机、外科吊塔（可选）、手术灯。

General compound operating room: It is used for surgical operation, interventional operation and independent display of image information. There are anesthesia machines, surgical ceiling pendants (optional) and operating lamps.

术中 DSA: DSA 有天吊式和落地式，两种类型要根据多种因素进行选型设计。

Intraoperative DSA: The device is vascular machine of a ceiling type or a floor type, which can be selected according to various factors.

术中 CT: 主要用于骨科脊椎类手术，可以设计成一拖二，即两间手术间可共享一台 CT 设备。

Intraoperative CT: It is used for orthopedic spinal surgery. It can be designed as one CT device for two operating rooms, that is, one CT device can be shared in two operating rooms.

术中 MRI: 主要用于脑外科手术，MRI 可在一个手术间固定使用，患者不需移动。移动式 MRI 分两类：吊轨式、地轨式，两种场地要求不一样。一拖二患者移动式术中 MRI, 可以两个或多个手术间共享扫描间，患者移至扫描间做检查。

Intraoperative MRI: It is used for brain surgery. MRI may be fixed in an OR with no need to move the patients. There are two types of movable MRI: overhead rail type and ground rail type. They have different requirements for the sites. In terms of the "One-MRI-two-ORs", one MRI can be shared by two or more ORs, and the patients are moved to the scanning room for examination.

（2）手术室单元装备配置

(2) Allocation of devices in operating room units

① 结合不同学科发展与高新设备引进科学配置，符合临床需求，包括：手术床、手术无影灯、吊塔、高频电刀（或电外科工作站）、麻醉机、监护仪、输液泵、注射泵、观片灯、吸引器。

① High-tech device is introduced and allocated scientifically according to the development of different disciplines and clinical requirements, including: operating bed, shadowless lamp, ceiling pendant, high-frequency electrotome (or electrosurgical workstation), anesthesia machine, monitor, infusion pump, injection pump, film viewer and aspirator.

② Ⅲ级（一般洁净）手术间内的医疗设备：手术床、无影灯（子母）、吊塔、麻醉机、监护仪、高频电刀、腔镜、观片灯、输液泵、注射泵、吸引器；适用手术种类：胸外科、泌尿外科、妇产科、耳鼻咽喉科、普外科。

② Medical devices in class Ⅲ (generally clean) operating room include operating bed, shadowless lamp (master-slave), ceiling pendant, anesthesia machine, monitor, high-frequency electrotome, cavity mirror, film viewer, infusion pump, infusion pump and aspirator. They are applicable to thoracic surgery, urological surgery, obstetrics and gynecology surgery, otorhinolaryngology surgery and general surgery.

③ Ⅰ级（特别洁净）手术间内的医疗设备：除万级手术间内的设备外，增设骨科手术配置C型臂、铅衣、体外循环机；适用手术种类：瓣膜置换、心脏手术器官移植、血管外科、人工关节置换、神经外科、全身烧伤、感染率大的手术。

③ Medical devices in class Ⅰ(specially clean) operating room includes C-arm used for orthopedic surgery, lead clothing and extracorporeal circulation machine in addition to those in the 10000 level operating room. They are applicable to the surgeries such as valve replacement surgery, heart transplantation surgery, vascular surgery, artificial joint replacement surgery, neurosurgical surgery, systemic burn surgery and surgeries with high infection rate.

(3) 麻醉准备及术后恢复室装备配置

(3) Allocation of devices in anesthesia preparation and recovery rooms

电动病床、麻醉机、监护仪、电动吸引器。

Electric beds, anesthesia machines, monitors and electric aspirators.

(4) 医疗设备配置的多参数变量回归计算法

(4) Regression calculation method of multi-parameter variables for medical device allocation

计算急诊病人检查所需的设备台数的变量系数（A），门诊病人检查所需的设备台数的变量系数（Y），住院病人检查所需的设备台数的变量系数（Z），并通过多参数变量回归法导出公式。将新建医院开业时的预计门急诊病人数和住院病人数代入公式计算出该医院所需要的设备台数。

Calculate variable coefficient (A) of the number of device required for emergency patient examination, variable coefficient (Y) of the number of device required for outpatient examination, and variable coefficient (Z) of the number of device required for inpatient examination, and derive the formula through the regression calculation method of multi-parameter variables. Substitute the expected number of emergency patients and inpatients when a new hospital opens into the formula to calculate the number of devices required by the hospital.

某新建医院所需的某设备台数（X）= 日急诊病人数 × 1‰ × A + 日门诊病人数 × 1‰ × Y + 日住院病人数 × 1‰ × Z

the number of devices required for a new hospital (X) = the number of emergency patients per day × 1‰ × A + the number of outpatients per day × 1‰ × Y + the number of inpatients per day × 1‰ × Z

每千住院病人中需要做某种机器检查的人次 = 机器平均每日检查的住院病人数/（平均每日住院病人数 × 1‰）

the number of inpatients requiring mechanical examination per 1000 inpatients = the average number of inpatients examined by the machine per day / (the average number of inpatients per day × 1‰)

每千门诊病人中需要做某种机器检查的人次 = 机器平均每日检查的门诊病人数/

（平均每日门诊病人数×1‰）

the number of outpatients requiring mechanical examination per 1000 outpatients = the average number of outpatients examined by the machine per day / (the average number of outpatients per day ×1‰)

每千急诊病人中需要做某种机器检查的人次 = 机器平均每日检查的急诊病人数/（平均每日急诊病人数×1‰）

the number of emergency patients requiring mechanical examination per 1000 emergency patients = the average number of emergency patients examined by the machine per day / (the average number of emergency patients per day ×1‰)

Y = 每千门诊病人中需要某种机器检查的人次 ÷ 单机饱和工作量

Y = the number of outpatients requiring machine examination per 1000 outpatients ÷ the saturated workload of a single machine

Z = 每千住院病人中需要某种机器检查的人次 ÷ 单机饱和工作量

Z = the number of inpatients requiring machine examination per 1000 inpatients ÷ the saturated workload of a single machine

A = 每千急诊病人中需要某种机器检查的人次 ÷ 单机饱和工作量

A = the number of emergency patients requiring machine examination per 1000 emergency patients ÷ the saturated workload of a single machine

单机饱和工作量 = 工作时间最多能检查的病人数量

the saturated workload of a single machine = the maximum number of patients examined during working hours

（5）医用设备配置相关规定

(5) Regulations on medical device allocation

《大型医用设备配置与使用管理办法》（卫规财〔2004〕474 号）；

Measures for Administration of Allocation and Use of Large Medical Devices (WGC [2004] No. 474) ;

《乙类大型医用设备配置规划指导意见》（卫办规财发〔2005〕64 号）；

Guidance on Allocation Planning of Class B Large Medical Devices (WBGCF [2005] No. 64) ;

《2008—2010 年全国正电子发射型断层扫描仪配置规划》（卫办规财发〔2008〕89 号）；

2008—2010 National Allocation Planning of Positron Emission Tomography Scanners (WBGCF [2008] No. 89) ;

《广东省基本实现卫生现代化建设纲要》（粤卫〔2003〕5 号）；

Outline for Basically Realizing Modernization of Health Facilities in Guangdong Province (YW [2003] No. 5) ;

《广东省乙类大型医用设备配置基本要求》（粤卫办〔2008〕70 号）。

Basic Requirements for Allocation of Class B Large Medical Devices in Guangdong Province (YWB [2008] No. 70).

3.2.3.2 护理装备：包括各种护理装备配置及房间布置要求。

3.2.3.2 Nursing devices: including the requirements for nursing device allocation and room layout.

3.2.3.3 医用家具：包括各种家具配置方案及标准。

3.2.3.3 Medical furniture: including various furniture allocation proposals and standards.

如诊室的基本装备：见表 3-1。

For example, basic devices in a consulting room: see Table 3-1.

表 3-1 诊室基本装备

Table 3-1 Basic devices in a consulting room

医疗用房 Medical room	装备名称 Device name	单间装备数量 Number of devices in a single room	说明 Description
诊室 Consulting room	诊查床 Examination bed	1	1950(L) ×700(W) ×700(H)，可安装一次性床垫卷筒纸 1950 (L) × 700 (W) × 700 (H) Disposable mattress and roll papers can be installed
	诊桌 Diagnostic table	1	1200(L) ×750(W) ×800(H) 1200 (L) ×750 (W) ×800 (H)
	患者圆凳 Round stool for patients	1	可升降，无靠背 The round stool is liftable without backrest
	屏风 Screen	1	1980(L、三折) ×1800 (H) 1980 (L, three folds) ×1800 (H)
	观片灯 Film viewer	1	单联 Single copy
	脚蹬 Foot pedal	1	200 高度 200 (H)

3.2.4 信息系统 (HIS) 策划

3.2.4 Hospital information system (HIS) planning

3.2.4.1 HIS 结构：包括 HIS、CIS[临床信息系统，如图像传输系统 (PACS)、电子病历、LIS 等]、OA(自动办公系统) 和移动医疗系统等。

3.2.4.1 HIS structure: including HIS, CIS [clinical information system, such as image transmission system (PACS), electronic medical records and LIS], OA (automatic office system) and mobile medical system and others.

3.2.4.2 HIS 作业流程：包括与医院服务流程和医疗流程相匹配的 HIS 作业流程设计等。

3.2.4.2 HIS operation process: including HIS operation process design matching with hospital service process and medical flows and others.

3.2.4.3 HIS 网络结构：包括 HIS 综合布线方案及信息点位设置等。

3.2.4.3 HIS network structure: including HIS integrated cabling proposal and information point setting and others.

3.2.4.4 HIS 管理：包括编码管理、权限及数据管理等。

3.2.4.4 HIS management: including code management, authority and data management and others.

3.2.5 专项医用设施

3.2.5 Special medical facilities

3.2.5.1 洁净室：包括洁净手术部、ICU、血液移植病房、生殖中心等配置方案。

3.2.5.1 Clean room: including the allocation plan of clean operating department, ICU, blood transplant ward, reproduction center and others.

3.2.5.2 医用气体：包括气体种类、终端设置及技术要求等方案。

3.2.5.2 Medical gas: including gas types, terminal settings and technical requirements.

3.2.5.3 辐射防护与屏蔽

3.2.5.3 Radiation protection and shielding

3.2.5.4 物流系统：包括物资存储、物流传输方式及点位设置方案。

3.2.5.4 Logistics system: including material storage, logistics transmission mode and point setting proposal.

3.3 医疗工艺设计咨询

3.3 Consultation on design for medical process

3.3.1 医疗工艺设计咨询必要性

3.3.1 Necessity of consultation on design for medical process

3.3.1.1 医疗工艺设计应由医疗保健设施的使用方自行完成，若无法完成，可委托他方进行设计咨询，需支付设计咨询费用，并应予以审定；确认后作为使用方的设计，并承担医疗工艺设计责任。医疗保健设施的使用方不应以建筑设计代替医疗工艺设计，两种设计具有本质的不同，前者是对房建工程的设计，后者是对医疗流程、医疗设备、资源配置的设计。

3.3.1.1 The design for medical process shall be completed by the user of the health care facilities. If the user cannot complete the design by himself, he can entrust a third party to conduct the design consultation and pay corresponding fees, and finally reviews the design. The design can be taken after determination and confirmation and used for design for medical process. The user of health care facilities shall not replace design for medical process with architectural design. Both types of design are essentially different. The former is the design of buildings, while the latter is the design of medical flows, medical devices and resource allocation.

3.3.1.2 医疗工艺设计咨询应就“医疗业务结构、功能和规模，以及相关医疗流程、医疗设备、技术条件和参数”全部内容形成设计咨询文件，包括文字、流程图、清单表格等；医疗工艺设计咨询应持续追求工艺的合理性和先进性。

3. 3. 1. 2 During design for medical process consultation, prepare the design consultation document as per all contents of “structure, function and scale of medical services as well as relevant medical flows, medical device, technical conditions and parameters”, including text, flow chart, list and forms. The rationality and advancement of the process shall be kept continuously through design for medical process consultation.

3. 3. 1. 3 医疗工艺前期设计咨询成果经建筑方案设计方使用时，若出现前期设计指标与规划许可建筑技术条件不匹配、超出投资计划等不可行情况，咨询方应对不适用的医疗工艺前期设计做出调整，使之可行。在建筑初步设计及施工图设计方使用中，若出现不符合建筑规范、不适用或超投资的情况，咨询方应对不适用的医疗工艺条件设计做出调整，使之符合建筑规范、可以实现。

3. 3. 1. 3 When the building SD institution uses the preliminary consultation results of design for medical process, if the PD indicators do not match with the planned and permitted technical conditions of the buildings and exceed the investment plan, or other infeasible situations occur, the consultant shall adjust the inapplicable PD for medical process to make it feasible. In case of non-compliance with building specifications, inapplicability or over-investment when the design institution uses the PD and construction drawing of buildings, the consultant shall adjust the design of inapplicable medical process conditions, so as to make it meet the building specifications and feasible.

3. 3. 1. 4 医疗工艺设计允许发生变更，咨询方应对变更做出回应，变更内容应及时通知各使用方。若变更过大，造成较大返工、突破造价投资、延误建设工期，应充分论证变更的必要性和可行性，减少各方损失。医疗工艺设计咨询成果和设计变更文件应作为工程的重要档案予以保存，并用于设计评审和审查、施工文件审核、工程竣工验收、试车运营。

3. 3. 1. 4 The design for medical process can be changed, but the consultant shall explain the changes and notify all users of the changes in time. If the change is too large, causing large rework, exceeding cost investment or delaying the construction period, the necessity and feasibility of the change shall be fully demonstrated, so as to reduce the losses of all parties. The consultation results of design for medical process and design change documents shall be kept as important files of the project and the basis of design review, construction document review, project completion acceptance and trail operation.

3. 3. 2 医疗工艺设计咨询重要性

3. 3. 2 Importance of consultation on design for medical process

3. 3. 2. 1 医疗建筑是最为复杂的民用建筑，作为医疗建筑的设计方应对医院的系统、功能、流程、运营管理以及常用的医疗专业术语有所了解，以便于日后与业主和各参建方能进行有效的交流与沟通。其特殊性和复杂性，决定了使用方对建筑设计的更高要求。

3. 3. 2. 1 The healthcare buildings are the most complicated civil buildings. The design institution of healthcare buildings shall understand the system, functions, flow, operation and management of the hospital as well as common medical terms, so as to effectively communicate with the owner and all participating parties in the future. The specificity and complexity of such buildings determine the higher requirements of the user for building design.

3.3.2.2 医疗建筑的设计内容比较多，除常规的主体建筑、结构、给排水、暖通、电气、消防设计外，还有基坑围护、人防、幕墙、钢结构、智能化、室内装饰（含医用家具设计）、室外景观绿化、室外配套、泛光照明、夜景亮化等专业工程设计，以及医疗工艺和医疗专项工程（包括但不限于医用洁净工程、医用气体、放射防护和电磁屏蔽、实验室工艺、污水处理、营养厨房、医院智能弱电、医用指示标识、医院文化标识、医用纯水、物流传输、污物被服智能收集等，其中前十项是所有医院工程必备的，后两项由业主根据医院项目的定位、整个项目的投资等情况综合考虑设置）设计。尤其医疗工艺和医疗专项工程，是完整实现医疗建筑功能的重要组成部分，也是医疗建筑设计的重点。

3.3.2.2 As for the design of healthcare buildings, in addition to the conventional main building, structure, water supply and drainage, HVAC, electrical works and fire-fighting, it also covers professional works such as foundation pit enclosure, civil air defense, curtain wall, steel structure, intelligence, interior decoration (including medical furniture), outdoor landscape greening, outdoor supporting, floodlighting, night view lighting, as well as medical process and special medical works (including but not limited to medical cleaning works, medical gas, radiology protection and electromagnetic shielding, laboratory technology, sewage treatment, nutritious kitchen, hospital intelligent low voltage, medical instruction sign, hospital cultural sign, medical pure water, logistics transmission, intelligent collection of sewage and bedding and clothing, etc.). The first ten items are necessary for all hospital projects, and the last two items shall be determined by the owner upon comprehensive consideration of the positioning of the hospital project and the investment of the whole project. In particular, medical process and special medical works are an important part to completely realize healthcare building functions, and also the design of healthcare buildings.

3.3.2.3 医疗工艺设计是对医院内部医疗系统活动过程及程序的策划，是在整合了医院管理、经济、信息、卫生、疾病预防控制、保健、医疗、护理、康复、教学、科研、生活、保障等需求后，对其业务结构、功能、规模，以及相关医疗服务流程、医疗设备技术条件、参数等进行功能性设计，为医疗建筑设计提供依据。其目的是让设计人员对医院的经营模式、服务模式、医院的空间布局、组合、单元、特色等方面有一个比较明确的认识，进而发挥设计人员的水平，尽可能地设计出具有满足医院经营要求和医疗康复功能的建筑。

3.3.2.3 The design for medical process is to plan the activity process and procedure of the internal medical system of the hospital, conduct functional design for its business structure, function, scale, relevant medical service flow, technical conditions, parameters and so on of medical devices after integrating the needs of hospital management, economy, information, health, disease prevention and control, health care, medical treatment, nursing, rehabilitation, teaching, scientific research, life, security, etc., and provide the basis for the design of healthcare buildings. It aims to give the designer a clear understanding of the hospital's operation mode, service model, spatial layout, combination, unit, characteristics and other aspects, and then give full play to the level of the designer, and try to design buildings that can meet the hospital's operation requirements and medical rehabilitation functions.

3.3.2.4 只有将所有涉及的专业及专项系统设计考虑周全，梳理好和各专业之间的界面与范围，才能尽可能地避免在施工阶段的变更与返工。

3.3.2.4 Only when all the involved disciplines and special system design are considered comprehensively, and the interface and scope with each discipline are sorted out, can the change and rework during construction be avoided as much as possible.

3.3.2.5 《设计任务书》是业主使用方明确使用功能、表达需求，为设计方提供设计依据的重要文件，是使用方和设计方之间沟通的桥梁。它是将设计意图转化为图纸的中间形式，建立在功能调研、需求分析的基础之上，是对设计进行交底的工作。在多达30项的专业设计里，没有科学、有效的《设计任务书》，设计方很难进行有序、系统的设计及设计协调工作。医疗工艺设计是很多建筑设计单位的最大短板。要想做好医疗建筑设计及设计管理，设计方除了需熟悉公共建筑的相关技术规范外，还要研究相关医疗技术和规范，同时还需要掌握系统的项目设计管理方法。使用方、设计方，双方各自的需求识别是一个逐渐深入、逐渐明了的过程。要让使用方或投资方在建筑规划设计前，完成一份涵盖所有专业、内容的《设计任务书》，给设计方及时开展工作，也是不太现实的。应借助医疗工艺设计咨询第三方分层次、分阶段、分专业编制《设计任务书》。

3.3.2.5 The Requirements Specifications are an important document for the owner to define the use function and express demand, providing the design basis for the design institution and also a bridge between the user and the design institution. It is an intermediate form to transform the design intent into drawings, based on the functional survey and demand analysis, for design disclosure. In professional design of as many as 30 items, without scientific and effective Requirements Specifications, it is difficult for the design institution to carry out orderly and systematic design and its coordination. The design for medical process is the biggest weakness of many building design institutions. To do well in the design and design management of healthcare buildings, the design institution should not only be familiar with the relevant technical specifications of public buildings, but also study the relevant medical technologies and specifications, and master the systematic project design management methods. Identification of requirements of the user and the design institution is a gradually in-depth and clear process. It is also unrealistic for the user or investor to complete Requirements Specifications covering all disciplines and contents before the architectural planning and design, so that the design institution timely carries out the work. The third party of medical process design consultation is asked to prepare the Requirements Specifications by levels, phases and specialties.

3.3.2.6 咨询是为决策服务的，咨询是决策的参谋。咨询服务有助于使用方或投资方或设计方能够从紧张的工程建设进程中脱开身、省下时间，集中精力高效地做好自身领域内力所能及的事情。

3.3.2.6 Consultation serves decision-making and acts as an adviser of decision-making. Consultation services help the user or the investor or the design institution to get away from the tense project construction process, save time, and focus on doing things within their own domain effectively, so as to create the greatest value.

3.3.3 咨询范围及服务目标

3.3.3 Consultation scope and service objectives

3.3.3.1 对医院建设前期的医院总体规划、门急诊、医技、住院、后勤保障、科研、教学等的主要功能布局、医疗工艺、流程和动线设计等进行设计咨询；对各专项工程设计和初步设计的评审等提供技术咨询服务。依据国家相关政策、相关规范及建设方的要求，协助建设方、设计方努力打造出一座功能布局科学、流程合理、运行效率高、用户体验好、保障安全节能、可持续性的医疗建筑。

3.3.3.1 Provide design consultation for the overall planning of the hospital in the early stage of the hospital construction, the main functional layout of outpatient and emergency, medical technology, inpatient, logistics support, scientific research, teaching, etc. , and the design of medical process, flow, moving line, etc. ; provide technical consultation services for the review of the special engineering design and PD. Follow relevant national policies, specifications and requirements of the project owner to assist the project owner and the design institution to strive to build a healthcare building with scientific functional layout, reasonable flow, high operation efficiency, good user experience, safety, guaranteed energy conservation and sustainability.

3.3.3.2 协助业主制定科学合理的建设目标。

3.3.3.2 To assist the owner to formulate scientific and reasonable construction objectives.

1. 设计进度目标：项目的设计进度受政策、环境、程序等多种因素影响，在制订总控进度计划时，根据项目的特点综合考虑，以总控计划作为进度控制基准，其合理性是进行进度控制的基础；

1. Design schedule objectives: The design schedule of the project is affected by various factors such as policies, environments and procedures. When the master control schedule is formulated comprehensive consideration needs to be made according to the characteristics of the project, the master control plan should be taken as the schedule control benchmark. Its rationality is the basis for schedule control;

2. 投资控制目标：投资控制的基准一般为初步设计概算，其合理性是实现控制目标的关键因素；

2. Investment control objectives: The benchmark of investment control is generally the PD budget estimate, and its rationality is the key factor to achieve control objectives;

3. 设计质量目标：在设计过程中要充分考虑新工艺、新技术、新材料的应用和绿色医院的要求，可持续性医疗建筑的设计质量好坏将对项目投资和工期产生巨大影响。

3. Design quality objectives: During design, the application of new processes, new technologies and new materials, and the requirements of green hospital should be fully considered. The

design quality of sustainable healthcare buildings will have a huge impact on project investment and construction period.

3.3.4 咨询工作程序及成果

3.3.4 Consultation procedures and results

3.3.4.1 接受使用方或建设方或设计方委托后签订设计咨询服务合同，收集咨询服务所需的资料，踏勘现场，了解项目情况，制订咨询服务实施方案。

3.3.4.1 Accept the entrustment of the user of the project owner or the design institution, and then sign the design consultation service contract, collect the information needed for consultation services, survey the site, understand the project situation, and formulate the implementation proposal of the consultation services.

3.3.4.2 根据咨询实施方案，按照医院基本建设程序、重点工作及在设计工作中需注意的管理难点和要点，协助使用方或建设方或设计方进行梳理。

3.3.4.2 Assist the user or the project owner or the design institution to sort it out according to the consultation implementation plan, the hospital's infrastructure procedures, key work and the management difficulties and key points to be focused on in the design.

3.3.4.3 提供项目规划阶段的咨询、方案设计的初步审核咨询、方案设计的详细审核咨询、初步设计及投资概算的咨询。

3.3.4.3 Provide consultation during project planning, preliminary review consultation of SD, detailed review consultation of SD, PD and investment budget estimate.

3.3.5 咨询服务理念与方法

3.3.5 Concept and method of consultation services

3.3.5.1 运用国际化先进的医院设计理念、国内现代化医院建设的经验与项目的具体情况相结合，以获得医疗建筑设计的最佳成果。

3.3.5.1 To obtain the best design results of healthcare buildings, the international advanced hospital design concept and domestic modern hospital construction experience are combined with the specific situation of the project.

3.3.5.2 组织知名的医院建设管理专家、专业从事医院设计工作的建筑师、工程师，实地踏勘、调研、了解医院现状和发展需求，了解当地医政、规划建设、市政等有关部门的政策规定。

3.3.5.2 Organize well-known hospital construction management experts, architects and engineers specializing in hospital design to conduct field survey, research, understand the current situation and development needs of the hospital, and understand the policies and regulations of local medical administration, planning and construction, municipal administration and other relevant departments.

3.3.5.3 与院方领导、各相关业务科室领导、专家以及项目设计单位、设计师进行充分沟通，交换意见、讨论设计、优化方案，提出书面咨询建议，供项目建设方和设计方参考采纳。

3.3.5.3 Fully communicate with leaders of the hospital, leaders of relevant business departments, experts, project design institution, designers, exchange opinions, discuss design and optimize proposals, and put forward written consultation suggestions for reference and adoption by the project undertaker and design institution.

3.3.6 咨询服务步骤

3.3.6 Consultation service steps

设计阶段在整个医院建设中尤为重要，影响80%的投资因素，并对项目的质量与工期有着重要的影响，此阶段的设计咨询工作尤为重要，建筑设计阶段咨询服务分为四个阶段：

The design phase is particularly important in the whole hospital construction, affecting 80% of the investment factors, and has an important impact on the quality and construction period of the project. The design consultation in this phase is particularly important. Consultation services during architectural design can be divided into four phases:

3.3.6.1 规划阶段的咨询服务

3.3.6.1 Consultation services during planning

1. 建设规模、条件的初步核定

1. Preliminary verification of construction scale and conditions

医院性质、定位、规模和特点，主要医疗科室设置、重点专科和特色科室设置、主要医疗功能单元的设置；承担的医疗、教学、科研、预防、保健、康复任务量；医院医疗模式、建筑风格建议、地方就诊习惯；床位数、日门诊量、日急诊量，各科病房床位规模的组成、各科门诊诊室数量的组成、各种手术间的数量、各种ICU床位数量、大型医疗设备的配置规划、重点医技科室的规模等。

It covers the nature, orientation, scale and characteristics of the hospital, setting of main medical departments, key specialties and special departments, and setting of main medical functional units; the tasks of medical treatment, teaching, scientific research, prevention, health care and rehabilitation undertaken by the hospital; the medical treatment model of the hospital, suggestions on architectural style, and local medical treatment habits; the number of beds, daily outpatient visits, daily emergency visits, composition of beds of ward of each department, composition of consulting room of outpatient department in each department, number of various operating rooms, number of beds in various ICUs, configuration planning of large-scale medical devices, scale of key medical technology departments, etc.

2.. 建筑面积的核定

2. Verification of total building area

七项设施（急诊、门诊、医技、住院、后勤保障、行政办公、院内生活）建筑面积、单列项目建筑面积、科研用房建筑面积、教学用房建筑面积、预防保健用房建筑面积、核增建筑面积。

It covers building area of seven facilities (emergency, outpatient, medical technology, inpatient, logistics support, administrative office and hospital life), building area of separate project, scientific research room, teaching room, and prevention and health care room, as well as increasing building area.

3. 建设条件核定

3. Verification of construction conditions

征地拆迁、市政供电、道路交通、给排水、蒸汽等建设条件落实情况或建议。

It covers implementation or suggestions of construction conditions such as land requisition and resettlement, municipal power supply, road traffic, water supply and drainage, and steam.

4. 院区总体规划设计方案布局的合法性与合理性

4. Legality and rationality of the layout of the overall planning and design proposal of the hospital

主要功能分区的合理性，院区主要出入口设置的合理性；院区道路、消防线路、人流、车流、物流动线及污物汇集处理、运出动线设置的合理性；规划设计未来扩建的功能衔接等。

It covers rationality of main functional areas; rationality of setting of main entrances and exits in the hospital ; rationality of setting of roads, fire-fighting lines, people flow, vehicle flow, logistics moving lines, sewage collection and disposal, transportation moving lines; planning and design of the functional connection of future expansion, etc.

3.3.6.2　方案设计阶段的初步审核咨询

3.3.6.2　Preliminary review consultation during SD

七项设施、健康管理、科研、教学等用房面积设计的合理性；各主要功能位置设计的合理性；医疗区主要工艺流程、动线设计的合理性；医疗区水平动线距离的合理性、竖向运输能力的核定；大型影像诊断、检验、放射治疗设备配置数量的核定等。

It covers rationality of room area design for seven facilities, health management, scientific research, teaching, etc. ; rationality of main functional location design; the rationality of main process flow and moving line design of medical area; rationality of horizontal moving line distance and vertical transportation capacity of medical area; verification of number of large-scale image diagnosis, examination and radiotherapy devices.

3.3.6.3　方案设计阶段的详细审核咨询

3.3.6.3　Detailed review consultation during SD

完善各使用功能单元内医疗工艺、流程、动线设计的合理性，确定各医疗用房构成、房间名称、定位分布、相邻关系、数量、面积、层高；确立医疗机构重点部门（手术部、重症医学科、产房、母婴室、无陪护新生儿病房、新生儿沐浴室、新生儿配奶室、导管室、血液透析室、内镜室、口腔科、消毒供应中心、感染性疾病科门诊、隔离病房、器官移植病房、骨髓移植病房、血液病房、烧伤病房、急诊科、洗衣房、生物安全实验室等）建筑布局和设施管理、环境管理、人员管理、物品清洁消毒、污物收集转运流线等，为项目初步设计打好基础。

Improve the rationality of medical process, flow and moving line design in each functional unit, determine the composition, room name, location distribution, adjacent relationship, quantity, area and floor height of each medical room, and determine building layout and facility management, environmental management, personnel management, cleaning and disinfection of articles, sewage collection and transfer flow of the key departments of medical institutions (operating department, intensive care unit, delivery room, mother-child room, unattended neonatal ward, newborn shower room, newborn milk distribution room, catheter room, hemodialysis room, endoscopy room, stomatology department, disinfection supply center, outpatient department of infectious diseases department, isolation ward, organ transplantation ward, bone marrow transplantation ward, hematology ward, burn ward, emergency department, laundry room, biosafety laboratory, etc.), so as to lay a good foundation for the PD of the project.

3.3.6.4 初步设计及投资概算的咨询

3.3.6.4 Consultation on PD and investment budget estimate

1. 全面评价初步设计的先进性、可行性、实用性和投资概算的可控性，为上级主管部门最终批准建设规模、投资总额做好准备。

1. Comprehensively evaluate the advancement, feasibility and practicability of the PD as well as the controllability of the investment budget estimate, so as to prepare for the final approval of the construction scale and total investment by the superior competent department.

2. 审核初步设计说明、审核建筑专业设计图纸，提出合理化建议和具体意见，以便在施工图设计阶段予以落实。侧重于明确、细致地指导具体的设计，包括：确定医疗单间用房内装备名称、数量、规格说明、点位；各房间对承重、热水、冷水、纯水、排水、消防、喷淋、烟感、照明、强电插座、弱电末端、监控、投影、指示标识、医用气体、紫外线消毒灯（吸顶式、壁挂式、立式）空气消毒器、空气品质、压力梯度和温湿度要求；病床的输液、床帘轨道、空调新风送风口、回风口、排风口位置；物流传输、污物被服智能收集站点等；建筑和装饰等专业之间的设计界面、大型医疗设备房间的结构降板、吊装、运输路径荷载、病区不同地面材料之间的结构层高差、钢筋含量、空调系统选型、建筑材料选择的建议等。

2. Review the PD specification, review the architectural design drawings, and put forward reasonable suggestions and specific opinions for implementation during CD. Should focus on clear, detailed and specific guidance design, such as: determination of the name, quantity, specification and point location of devices in the single medical room; requirements of each room for load bearing, hot water, cold water, pure water, drainage, fire protection, spray, smoke sensing, lighting, high-voltage socket, low-voltage terminal, monitoring, projection, indication sign, medical gas, ultraviolet disinfection lamp (ceiling, wall-mounted and vertical) air sterilizer, air quality, pressure gradient, temperature and humidity; location of infusion, bed curtain track, fresh air supply outlet of air conditioner, air return outlet and air exhaust outlet of the hospital bed; logistics transmission, intelligent collection station for sewage, bedding and clothing; design interface between architecture and decoration, structural floor dropping of large-scale medical devices, hoisting, transportation path load, height difference of structural layer among different ground materials in the ward, reinforcement content, and selection of air conditioning system and building materials.

3. 审核投资概算的编制深度，包括与立项批准投资的符合性分析、概算工程费用与建筑市场的符合性分析、其他费用与预备费用的编制标准等。重点在于深度够、不漏项、不以估算代概算。

3. Review the preparation depth of the investment budget estimate, including the analysis on compliance with the initiated and approved investment, the analysis on compliance of the estimated project cost with the construction market, and the preparation standards of other costs and preparatory costs. The focuses are enough depth, no missing items and no estimation instead of budget estimate.

3.3.6.5 作为业主方管理的延伸，咨询方可为业主提供专业化的服务，向业主提供可实施性的咨询方案与计划，供业主决策。并按业主决策的方案与计划进行实施，接受业主的监督与检查，协助业主达成既定的建设目标。医院项目的建设是一个复杂、长期的过程，使用方或投资方应做好长远规划与计划，按照项目建设的科学时序，及时采购咨询服务。

3.3.6.5 As an extension of the owner's management, the consultant provides the owner with professional services and feasible consultation proposals and plans for decision-making. In addition, the consultant puts it into practice according to the decision-making proposal and plan of the owner, accepts the owner's supervision and inspection, and assists the owner to achieve the construction objectives established. The construction of the hospital project is a complex andlong-term process. The user or the investor should make a long-term planning, and timely purchase consultant services based on the scientific sequence of the project construction.

3.3.6.6 医疗工艺优化咨询

3.3.6.6 Consultation on medical process optimization

1. 任何已经做出的医疗工艺设计都不是绝对正确的，工艺设计方案具有多样性。在同时理解医疗工艺设计、建筑设计和经营者愿望的基础上，进行工艺优化咨询，可以提高建筑空间、人力资源、医疗设备的利用率，缩短服务流线，降低成本。

1. Any design for medical process that has been made is not absolutely correct, and the process design proposal is diverse. On the basis of understanding the design for medical process, architectural design and the wishes of operators, the process optimization consultation can improve the utilization rate of building space, human resources and medical devices, shorten the service flow line and reduce the cost.

2. 对工艺流程设计从医疗服务量分析、研究，做出精细化决策。

2. Make a refined decision for the design of process flow from the analysis and research of medical service volume.

（1）一级流程：放射设备门诊病人和病房病人都要使用，确定在什么程度下实行资源共享，什么程度下实行一站式就近服务。

(1) First-class flow: Both patients from outpatient department and ward should use radiology devices. To what extent should resource sharing be implemented, and to what extent is one-stop nearest service implemented.

（2）二级流程：从科室、患者、运营成本多维度考虑，确定在多大的服务流量下，门诊设备放在门诊区，医技设备放在医技区，或分散随科室设置。

(2) Second-class flow: Upon consideration of the department, patient and operation cost, under what service flow, the devices for outpatient department are placed in the outpatient area, the medical technology devices are placed in the medical technology area, or scattered with the department.

（3）三级流程：对标准化房间和非标准房间的尺度、内部布局做出安全性策划。

(3) Three-class flow: Make safety planning for the scale and internal layout of standardized rooms and non-standardized rooms.

3. 基于理疗流程、服务流程，考虑将荷载大、体积大的大型医疗设备安装在低层，既要方便使用，又易于运输、安装，如直线加速器、钴 60 等，将设备远离办公区、病区和生活区。

3. Based on the physiotherapy flow and service flow, large-scale medical devices with large load and volume should be installed on the lower floor, which should be convenient for use, transportation and installation, such as linear accelerator and Co60, making devices be far away from the office area, ward and living area.

3.4 综合医院医疗工艺设计

3.4 Design for medical process of general hospital

3.4.1 制定医疗规模

3.4.1 Development of medical scale

3.4.1.1 医疗指标

3.4.1.1 Medical indicators

明确学科设置、门急诊规模测算（确定诊床比）、手术间数配置（医院手术间的配置根据外科系统业务量、业务水平和手术室管理水平综合确定）、大型医疗设备配置（根据门诊量、床位数、业务水平并考虑未来发展，估算大型放射影像、核医学、核磁影像、超声影像、内镜等设备配置数量）、检验及病理样本量预测、消毒器械数量预测、门诊处方量及病房配剂量预测等。

Make clear the discipline setting, the measurement of outpatient and emergency scale (determine the ratio of daily outpatient/emergency visits), the number of operating rooms (the configuration of hospital operating rooms are determined according to the business volume, business level and operating room management level of the surgical system), the configuration of large-scale medical devices (estimate the number of large-scale radiography, nuclear medicine, magnetic resonance imaging, ultrasound image, endoscope and other devices according to outpatient visits, number of beds, business level and development), the prediction of test and pathological sample size, the prediction of the number of disinfection devices, the prediction of outpatient prescription and ward dosage, etc.

3.4.1.2 功能单位指标

3.4.1.2 Functional unit indicators

明确急诊、门诊、医技、住院、行政、后勤保障、院内生活七项设施的面积指标，根据运营服务需要增加预防保健、教学、科研、健康体检、机动车和非机动车停车场（库）、直升机停机坪和其他单列项目的内容和面积、数量指标。应细分每个功能单元内的床位数量、各科医技设备配置数量、科室人员岗位设置。

Make clear the area indicators of emergency, outpatient, medical technology, inpatient, administration, logistics support and hospital life facilities, and increase the content, area and quantity indicators of prevention and health care, teaching, scientific research, health examination, parking garage for motor vehicles and non-motor vehicles, helipad and other single items according to the needs of operation services. Subdivide the number of beds in each functional unit, the number of medical technology devices in each department and the post setting of personnel in each department.

综合医院医疗功能单元划分表：见表3-2。

Division of medical functional units of general hospital: see Table 3-2.

表3-2 综合医院医疗功能单元划分表

Table 3-2 Division of medical functional units of general hospital

分类 Classification	门诊、急诊 Outpatient and emergency	预防、保健管理 Prevention and health care management	临床科室 Clinical department	医技科室 Medical technology department	医疗管理 Medical management
各功能单元 Functional units	分诊、挂号、收费、各诊室急诊、急救输液、留院观察等 Triage, registration, charging, emergency treatment in each consulting room, first-aid infusion, observation in hospital, etc.	儿童保健、妇女保健等 Children's health care, women's health care, etc.	内科、外科、眼科、耳鼻喉科、儿科、妇产科、手术部、麻醉科、重症监护科（ICU、CCU等）、介入治疗、放射治疗、理疗科等 Internal medical department, surgical department, ophthalmology department, otolaryngology department, pediatric department, obstetrics and gynecology department, operating department, anesthesiology department, intensive care unit (ICU, CCU, etc.), interventional therapy department, radiotherapy department, physiotherapy department, etc.	药剂科、检验科、医学影像科（放射科、核医学、超声科）、病理科、中心供应、输血科等 Pharmacy department, laboratory department, medical imaging department (radiology department, nuclear medicine department and ultrasonic department), pathology department, central supply department, blood transfusion department, etc.	病案、统计、住院管理、门诊管理、感染控制等 Medical records, statistics, inpatient management, outpatient management, infection control, etc.

各科门诊量占总门诊量比例应根据医院统计数据确定，当无统计数据时根据表 3-3 确定。

The proportion of outpatient visits of each department to the total outpatient visits shall be determined according to the statistical data of the hospital. If no statistical data is available, it can be determined according to Table 3-3.

表 3-3 各科门诊量占总门诊量比率

Table 3-3 The proportion of outpatient visits of each department

科别 Category	占门诊总量比率（%） Proportion of total outpatient visits (%)	科别 Category	占门诊总量比率（%） Proportion of total outpatient visits(%)
内科 Internal medicine department	28	儿科 Pediatric department	8
外科 Surgical department	25	耳鼻喉、眼科 Otolaryngology department and ophthalmology department	10
妇科 Gynecology department	15	中医 Chinese medicine department	5
产科 Obstetrics department	3	其他 Others	6

各科住院床位数占医院总床位数比率应根据医院统计数据确定，当无统计数据时根据表 3-4 确定。

The proportion of inpatient beds of each department to the total beds of the hospital shall be determined according to the statistical data of the hospital. If no statistical data is available, it can be determined according to Table 3-4.

表 3-4 各科住院床位数占医院总床位数比率

Table 3-4 The proportion of inpatient beds of each department to the total beds of the hospital

科别 Category	占医院总床位比率（%） Proportion of total hospital beds (%)	科别 Category	占医院总床位比率（%） Proportion of total hospital beds (%)
内科 Internal medicine department	30	儿科 Pediatric department	6
外科 Surgical department	25	耳鼻喉 Otolaryngology department	6
妇科 Gynecology department	8	眼科 Ophthalmology department	6
产科 Obstetrics department	6	中医 Chinese medicine department	6
—	—	其他 Others	7

3.4.2 大型医疗设备规划

3.4.2 Planning of large-scale medical devices

详见 3.2.3。

See 3.2.3 for details.

3.4.3 确定部门核算要求

3.4.3 Determination of department accounting requirements

包括科室独立计量独立核算的模式，在建筑中划分时区（7 小时、12 小时、24 小时）、环境控制区（采暖、非采暖，空调、非空调，净化、非净化）、照明分区、信息管控区、冷热水配置分区、直饮水配置分区等。

It includes the mode of independent measurement and independent accounting of departments, such as time zones (7h, 12h and 24h), environment control area (heating, non-heating, air conditioning, non-air conditioning, purification, non-purification), lighting area, information control area, hot and cold water configuration area, direct drinking water configuration area, etc. are divided in the building.

3.4.4 确定医疗工艺设计参数

3.4.4 Determination of design parameters of medical process

应根据不同医院的要求研究确定，当无相关数据时应符合下列要求：

The determination is based on research according to the requirements of different hospitals. The following requirements shall be met when there is no relevant data:

1. 门诊诊室间数可按日平均门诊诊疗人次/（50 ~ 60 人次）测算；

1. The number of consulting rooms in outpatient department can be calculated by the daily average number of outpatient visits (50 ~ 60) ;

2. 急救、抢救床数可按急救通过量测算；

2. The number of emergency or first aid beds can be calculated according to the emergency throughput;

3. 1 个护理单元宜设 40 ~ 50 张病床；

3. 40 ~ 50 beds should be set up in each nursing unit;

4. 手术室间数宜按病床总数每 50 床或外科病床数每 25 ~ 30 床设置 1 间；

4. Operating rooms should be set based on the situation that one operating room is set for every 50 beds or 25 ~ 30 surgical beds;

5. 重症监护病房（ICU）床数宜按总床位数的 2% ~ 3% 设置；

5. The number of beds in the intensive care unit (ICU) should be 2% ~ 3% of the total number of beds;

6. 心血管造影机台数可按年平均心血管造影或介入治疗数/（3 ~ 5 例 × 年工作日数）测量；

6. The number of cardiovascular imaging system can be measured according to the annual average number of cardiovascular imaging or interventional therapy/(3 ~ 5 cases × annual working days) ;

7. 日拍片人次达到 40 ~ 50 人次时，可设 X 射线拍片机 1 台；

7. One X-ray machine can be set when the daily number of patients undergoing X-raying reaches 40 ~ 50;

8. 日胃肠透视人数达到 10 ~ 15 例时，可设胃肠透视机 1 台；

8. One gastrointestinal fluoroscopy machine can be set when the number of daily patients undergoing gastrointestinal fluoroscopy examination reaches 10 ~ 15;

9. 日胸透视人数达到 50 ~ 80 人次时，可设胸部透视机 1 台；

9. One chest fluoroscopy machine can be set when the number of daily patients undergoing chest fluoroscopy examination reaches 50 ~ 80;

10. 日心电检诊人次达到 60 ~ 80 人次时，可设心电检诊间 1 间；

10. One ECG consulting room can be set when the number of daily patients undergoing ECG examination reaches 60 ~ 80;

11. 日腹部 B 超人数达到 40 ~ 60 人次时，可设腹部 B 超机 1 台；

11. One abdominal B ultrasound machine can be set when the number of daily patients undergoing abdominal B ultrasound examination reaches 40 ~ 60;

12. 日心血管彩超人数达到 15 ~ 20 人次时，可设心血管彩超机 1 台；

12. One cardiovascular color ultrasound machine can be set when the number of daily patients undergoing cardiovascular color ultrasound examination reaches 15 ~ 20;

13. 日检诊人数达到 10 ~ 15 例时，可设十二指肠纤维内窥镜 1 台。

13. One duodenal fiber endoscope can be set when the number of daily patients undergoing examination reached 10 ~ 15.

3.4.5 医疗工艺流程设计

3.4.5 Design of medical process flow

3.4.5.1 一级流程设计

3.4.5.1 First-class flow design

1. 一级流程设计对总平面规划和建筑单体选型、建筑面积分配、建筑占地面积、建筑层数影响很大，应与总体规划相结合。

1. The first-class flow design has a great influence on the master planning, selection of individual building, distribution of building area, floor area and number of floors. It should be combined with the master planning.

2. 确定医院出入口数量、医疗功能单元之间的医疗服务流程、医技平台共享关系、物流方式、智能化方案、信息化方案；

2. Determine the number of hospital entrances and exits, medical service flow between medical functional units, sharing relationship of medical technology platform, logistics mode, intelligent proposal and information-based proposal;

3. 明确医疗模式是按病种、器官、系统划分，还是按治疗手段划分；

3. Clarify whether the medical model is divided by disease, organ and system, or by means of treatment;

4. 确定护理单元数量及每个单元床位规模；确定各护理单元内特殊病房数量（如ICU、净化病房、负压病房）；

4. Determine the number of nursing units and the size of beds in each unit; determine the number of special wards in each nursing unit (such as ICU, purification ward, and negative pressure ward) ;

5. 对药房、大型医技设备、日间和门诊手术、消毒供应、餐饮、保障等部门进行规划，确定集中、分散、分层、分期建设要求；

5. Plan the pharmacy, large-scale medical technology devices, daytime and outpatient surgery, disinfection supply, catering, security and other departments, and determine the requirements for centralized, decentralized, tiered and phased construction;

6. 规划除病区外其他有人力配备的服务台、护士站、标本采集点、服务窗口；

6. Plan other service counters, nurse stations, specimen collection sites and service windows equipped with manpower in addition to the ward;

7. 确定是否有中心化、一站式诊疗单位，如心血管中心、儿科中心、创伤中心；

7. Determine whether there is a centralized and one-stop diagnosis and treatment unit, such as cardiovascular center, pediatric center and trauma center;

8. 对于急诊、门诊和住院病人都要使用的各类医技用房、垂直交通、公共空间、生活服务、营养供给等，一般宜以500床、1500人/日门急诊量为基本模数，当设施总床位规模在500床及以下时宜实行资源共享，当床位数量超过500床时宜考虑实行部分科室一站式就近服务，小型设备分散配置，大型设备共享；对于800床及以上的大型、超大型设施，宜普遍采取就近配置的一站式服务方式，减少人群聚集、防止院内感染、缩短服务距离。对于要求服务标准较高或较低的设施或机构，可以调整此模数。

8. For all kinds of medical technology rooms, vertical transportation, public space, life service, nutrition supply, etc. used by patients from emergency, outpatient and inpatient departments, 500 beds and 1500 people/daily outpatient and emergency visits should be generally taken as the basic module. When the total bed size of the facility is 500 beds or below, resources shall be shared. When the number of beds exceeds 500, one-stop nearby services for some departments should be considered. Small-scale devices should be decentralized and large-scale devices should be shared. For large and super large facilities with 800 beds or above, one-stop nearby service mode should be generally adopted, to reduce crowd aggregation, prevent hospital infection and shorten the service distance. For facilities or institutions requiring higher or lower service standards, this module can be adjusted.

3.4.5.2 二级流程设计

3.4.5.2 Second-class flow design

1. 总则：应确定各医疗功能单元内部的工艺流程，确定单元是否允许跨楼层、跨栋、跨院区（平台化、集团化）；从医疗、患者、运营成本多维度考虑，一般宜以500床、1500人/日门急诊量或接待量为基本模数，小于此规模的，医技用房、垂直交通、公共空间、生活服务、营养供给等集中设置，超过此规模的按单栋楼座、科室或按急诊、门诊、住院分散设置；对于要求服务标准较高或较低的设施或机构，可以调整此模数。

1. General rules: Determine the internal process flow of each medical functional unit, determine whether the unit allows crossing floors, buildings and hospital (platform and group); considering from the multiple dimensions such as medical treatment, patient and operation cost, 500 beds and 1500 people/daily outpatient and emergency visits or reception volume are generally taken as the basic module. In case of being less than such scale, medical technology room, vertical transportation, public space, living services, nutrition supply, etc. should be centralized. Those beyond this scale should be decentralized with individual building, department or according to emergency, outpatient and inpatient departments; for facilities or institutions requiring higher or lower service standards, such module can be adjusted.

2. 急诊部

2. Emergency department

(1) 院前：确定是否有直升机、院前急救区是否有平战/平疫结合场地要求、院前科规模、是否有急救中心指挥部要求、救护车数量、救护车是否进入室内换床；

(1) Pre-hospital: Determine whether there are helicopters, whether there are requirements for the combination of peacetime and wartime/epidemic prevention in pre-hospital first-aid area, the scale of pre-hospital department, whether there are requirements for the headquarters of emergency center, the number of ambulances, whether ambulances enter the room to change beds;

(2) 分诊：分设急救和急诊两组护士站，还是一组，急缓Ⅰ、Ⅱ、Ⅲ、Ⅳ分区要求；

(2) Triage: one or two groups of nurse stations are provided for first-aid and emergency treatment respectively according to the requirements of emergency degree Ⅰ, Ⅱ, Ⅲ and Ⅳ;

(3) 挂号收费：独立还是与门诊部共享；

(3) Registration and charging: Be independent or shared with outpatient department;

(4) 抢救室：床位数量；

(4) Emergency room: number of beds;

(5) 诊室：开放拉帘式、半隔断式、单间式、带值班卫生间的单间式；

(5) Consulting room: open curtain type, semi partition type, single medical room type, single medical room type with duty toilet;

(6) EICU: 床位数量，与中央 ICU 合用还是独立；

(6) EICU: number of beds, be shared or independent with central ICU;

(7) 医技：药房、检验、手术室、影像检查配备，是否与全院医技共享；

(7) Medical technology room: whether to equip pharmacy, laboratory, operating room, imaging examination equipment, whether to share with the whole hospital's medical technology rooms;

(8) 病房：留观床位、急诊科住院床位，是否设隔离感染护理单元；

（8） Ward: beds for observation, inpatient beds in emergency department, isolation and infection unit with a nurse station is provided or not;

（9） 治疗：儿科是否独立，输液是否集中配药，设哪些治疗项目；

（9） Therapy room: whether the pediatric department is independent, whether the transfusion is subject to centralized dispensing, and what treatment items are provided;

（10） 应设计的二级医疗流程包括：卒中中心流程、胸痛中心流程、急诊手术流程、创伤救治流程、紧急分娩流程、急诊检验流程、感染患者转运流程等。

（10） The second-class health care flows that should be designed include: stroke center process, chest pain center process, emergency operation process, trauma treatment process, emergency delivery process, emergency laboratory process, infectious patient transfer process, etc.

3. 门诊部

3. Outpatient department

（1） 挂号、分诊及分级诊疗流程设计；

（1） Process design for registration, triage and tiered diagnosis and treatment;

（2） 是否设儿科、体检科、产科、辅助生殖医学以及特需医疗的独立出入口；

（2） Whether there are independent entrances and exits for pediatric department, physical examination department, obstetrics department, assisted reproductive medicine and special medical treatment department;

（3） 就诊、叫号、收费、取药、用药咨询信息流程设计；

（3） Information process design for medical treatment, queuing, charging, medicine receiving and consultation on medicines;

（4） 分散到各门诊科室的专用医技用房/设备配置规划；

（4） Allocation planning of special medical technology rooms/devices scattered to each outpatient department;

（5） 对进行日间治疗、日间手术、门诊手术、人流室的患者和医护流程设计，对麻醉师岗位和疼痛门诊的统筹安排；

（5） Process design for patients and medical care staff of day treatment, day surgery, outpatient operation and induced abortion room as well as overall arrangement for anesthetist and pain clinic;

（6） 对内镜、DSA、口腔科等器械消毒供应的安排；

（6） Arrangement for disinfection supply of endoscopy, DSA and medical apparatus of stomatology department;

（7） 门诊中西药房集中还是分散，是否全自动发药；

（7） Whether the traditional Chinese medicine pharmacy and western medicine pharmacy are centralized or decentralized, and they are fully automatic in dispensing;

（8） 门诊检验是否与中心检验或与急诊检验合并，是否设分散采集点，儿科采样是否独立设置；

（8） Whether the outpatient laboratory is merged with the central laboratory or the emergency laboratory, whether the decentralized collection site is provided, and whether the pediatric sampling site is provided independently;

（9）感染门诊设备、留观、病房要求，是否独立成栋；

（9）Whether the requirements for equipment, observation and wards in the infection outpatient department are independent;

（10）门诊支持服务区：安防、保洁、医疗协调办公室、门诊部办公室、警务办公室、医保办等；

（10）Service areas supported by outpatient department: security and protection office, cleaning office, medical coordination office, outpatient office, police office, medical insurance office, etc;

（11）应设计的二级医疗流程包括：分诊与感染性疾病诊疗流程、检验-报告流程、日间手术流程、门诊治疗流程、配发药流程、特需服务流程、日间病房流程、体检流程、各科独立消毒供应流程等。

（11）The second-class health care flows that shall be designed include: triage and infectious disease diagnosis and treatment flow, examination-report flow, day surgery flow, outpatient treatment flow, dispensing flow, VIP flow, daytime ward flow, physical examination flow, independent disinfection supply flow of each department, etc.

4. 医技部

4. Medical technology department

（1）确定设置以下哪些医技：检验、医学影像（含普通放射、核磁共振、介入放射、核医学、内窥镜、超声、放射治疗）、物理诊断、功能检查、病理、输血、高压氧、生殖医学、生物细胞治疗、消毒供应、救护车/床消毒；

（1）Determine which of the following medical technologies should be provided: examination, medical image (including ordinary radiology, MRI, interventional radiology, nuclear medicine, endoscope, ultraphonic and radiotherapy), physical diagnosis, functional examination, pathology, blood transfusion, hyperbaric oxygen, reproductive medicine, biological cell therapy, disinfection supply, and ambulance/bed disinfection;

（2）确定分散设置的医技设备，如在心血管科及 CCU 内配置 DSA，设复合手术室；

（2）Determine medical technology devices provided separately, such as DSA in cardiovascular department and CCU, and set up compound operating room;

（3）确定大型影像设备所在楼层，如地下室、地面、楼上；超过 800 床的大型医院急诊、门诊、住院、特需、体检设备如何分设。

（3）Determine the floor where the large-scale imaging devices are located, such as basement, ground and upstairs; determine how to set up the devices respectively for emergency, outpatient, inpatient, VIP and physical examination departments in large hospitals with more than 800 beds.

应设计的二级医疗流程包括：影像及功能检查——诊断流程、介入放射诊疗感控流程、中心消毒供应流程、检验实验室和病理实验室流程等。

The second-class health care flows that should be designed include: image and functional examination-diagnosis flow, interventional radiology diagnosis treatment and infection control flow, central disinfection supply flow, examination laboratory and pathology laboratory flow, etc.

5. 药学及营养部

5. Medicine and nutrition department

（1）对门诊药房（含中、西药房）、急诊药房（含中、西药房）、住院药房（中心药房）、儿科门诊药房、感染疾病科门诊药房、静脉用药调配中心（室）、中药煎药室、仓储库房、临床药学工作室、办公生活用房等做规划和流程设计；确定是否设制剂室、临床药学检测实验室，具有临床药物试验机构资质的医院设 GCP 管理部门和配套试验用房。

（1） Make planning and flow design for outpatient pharmacy (including traditional Chinese and western medicine pharmacies), emergency pharmacy (including traditional Chinese and western medicine pharmacies), inpatient pharmacy (central pharmacy), pediatric outpatient pharmacy, outpatient pharmacy of infectious diseases department, intravenous drug dispensing center (room), traditional Chinese medicine decoction room, warehouse, clinical pharmacy room, office and living room, etc.; determine whether there are GCP management department and supporting laboratory with the qualification of clinical drug testing institution in the hospital, drug manufacturing laboratory and clinical pharmacy testing laboratory.

（2）若采用门诊药房自动化系统，应根据储存药品种类、储药量、日均处方量、发药加药速度、发药窗口数、药房面积、设备尺寸、质量等因素确定设备选型，提供设备尺寸、安装要求。

（2） If the outpatient pharmacy automation system is adopted, the type selection of devices shall be determined according to factors such as type of stored drugs, quantity of stored drugs, daily average prescription dosage, dispensing speed, number of dispensing window, pharmacy area, dimension of devices, weight, and the dimension of equipment and installation requirements shall be provided.

（3）对营养厨房及成套设备、营养配置室、营养代谢实验室、营养门诊及辅助用房等做流程设计，对营养发餐、餐食配送等确定自动物流站点。

（3） Make flow design for nutritious kitchen and packaged devices, nutrition collocation room, nutrition and metabolism laboratory, nutrition outpatient department and auxiliary room, and determine the automatic logistics stations for nutrition and food delivery.

（4）为运药、运送食品的货车规划专用货运路径和港湾式停靠车位、卸货台。

（4） Plan the dedicated freight path, bay-type stop space and unloading platform for the trucks transporting medicine and food.

6. 住院部

6. Inpatient department

（1）确定标准临床护理单元和特殊临床功能单元与数量，包括：麻醉手术部、导管介入、重症监护单元（ICU、CCU、NICU、PICU）、分娩室、血液净化、人类辅助生殖、高压氧、核素病房、烧伤单元、器官和骨髓移植等。

(1) Determine the number of standard clinical nursing units and special clinical function units, including anesthesiology department, catheter intervention, intensive care units (ICU, CCU, NICU, PICU), delivery room, blood purification, human assisted reproduction, hyperbaric oxygen chamber, radionuclide therapy ward, burn unit, organ and bone marrow transplantation, and other units.

(2) 对500床或以上规模亚专科较多的医院，安排单元之间的协作关系，形成中心化，如：心血管中心、妇儿中心、骨科中心、五官中心、肿瘤中心等；每个中心的床位总量达到200～300床，相当于一个二级专科医院住院部的规模；若达到400～500床，应为其配备专用的特殊临床功能单元。

(2) For hospitals with 500 beds or above and with more subspecialties, arrange the cooperative relationship between units, to form centralization, such as cardiovascular center, gynecology and pediatrics center, orthopedic center, five sense organs center, and oncology center. Each center has a total of 200～300 beds, equivalent to the size of the inpatient department of a second-class specialized hospital. If a center has 400～500 beds, special clinical function units shall be provided.

(3) 感染病房、核素病房、精神病房要考虑病人路径的单向流动和隔离性，尽量放在住院部底层，设独立的楼梯、电梯、出入口，避免占用上层的建筑空间。

(3) Consider the one-way flow and isolation of the patient's path for the infection wards, radionuclide therapy wards, and psychiatric wards, which shall be arranged on the ground floor of the inpatient department as far as possible, with independent stairs, elevators and entrances and avoiding occupying the building space of upper floors.

(4) 手术科室应靠近手术部，应设计外科医生和患者到达手术部的最短流线。

(4) Arrange the operating department close to the operating department, and design the shortest flow line for the surgeon and the patient to reach the operating department.

(5) 护理单元内部根据护理等级进行规划，围绕护士站、治疗区依病情的护理等级（分Ⅰ、Ⅱ、Ⅲ、Ⅳ级）安排病房，抢救及重症病房靠近护士站，病情较轻的逐渐远离护士站，往末端布置。发挥专病诊疗优势，在各病种的护理单元设该病种的重症监护室，以便随时将患者在不同护理等级的病房之间进行转移，如心内科护理单元可附设CCU、DSA介入、重病房、普通病房、临终关怀。

(5) Plan the nursing units according to the grade of nursing; arrange the wards around the nurse stations and treatment areas according to the grade of nursing (Grade Ⅰ, Ⅱ, Ⅲ and Ⅳ); arrange the emergency rooms and ICUs close to the nurse stations, and arrange the patients with mild conditions gradually far from the nurse station and toward the end. Give full play to the advantages of diagnosis and treatment of disease species, set the ICU of each disease species in the nursing unit of the disease, to facilitate transfer of patients between wards in different nursing grades at any time. For example, attach the CCU, DSA intervention, intensive ward, general ward and hospice care to the cardiology nursing unit.

（6）应对各个特殊临床功能单元做出符合感染控制要求的二级医疗流程设计。

（6）Provide the second-class medical process design complying with infection control requirements to each special clinical functional unit.

3.4.5.3 医疗工艺条件设计

3.4.5.3 Design for medical process conditions

1. 一般情况下医院的水、电、暖通、空调按照《综合医院建筑设计规范》（GB 51039—2014）中的规定及各专业本身的设计要求进行顶棚、墙面、地面上的终端布点、布线，医疗工艺只提对运营有特殊要求的条件，工艺条件涉及的内容包括：

1. For the water, electricity, heating, ventilation and air conditioning of hospitals in general, the stationing and wiring of terminals on ceiling, wall and floor are carried out in accordance with the provisions of Architectural Design Code for General Hospital and the design requirements of each discipline in general. Only the conditions with special operation requirements are presented for medical process, and the technological conditions involve the following contents:

（1）土建：大重量库存荷载，重型和特殊尺度的医疗设备自重、安装口及通道预留要求，两层或两层以上高大空间要求，楼地面降板要求，设备管道穿墙洞口预留要求。

（1）Civil engineering: requirements for heavy inventory load, self-weight of heavy and special-size medical devices, reserve of access opening and passage, large space in two or more floors, floor dropping of buildings, and reserve of wall holes for device pipeline.

（2）防水、防鼠、防尘、隔声、隔振要求。

（2）Requirements for preventing water, rats, dust and insulating sound and vibration.

（3）特殊温湿度要求。

（3）Requirements for special temperature and humidity.

（4）医用气体：氧、负压、压缩空气、CO_2、氧化亚氮（笑气）等。

（4）Medical gases: oxygen, negative pressure, compressed air, CO_2, nitrous oxide (laughing gas), and others.

（5）洁净室与特殊用房：手术部、ICU、生物实验室、生殖中心、血液干细胞移植病房等。

（5）Clean rooms and special rooms: operating department, ICU, biological laboratory, reproductive center, blood stem cell transplantation ward, and others.

（6）医院信息系统综合布线：HIS、PACS 网络等。

（6）Hospital information system integrated wiring: HIS, PACS network, and others.

（7）大型医疗设备机房工程：放射、放疗、核医学、实验室等。

（7）Large medical machine room engineering: radiation, radiotherapy, nuclear medicine, laboratory, and others.

（8）物流传送系统：气动物流、轨道物流、箱式物流、智能车传送、自动污物收集、自动传送带、非载人小型货运电梯、全自动配药发药系统等。

（8）Logistics delivery system: pneumatic logistics, rail logistics, box-type logistics, intelligent vehicle delivery, automatic waste collection, automatic conveyor belt, unmanned small freight elevator, full-automatic dispensing system, and others.

(9) 医用装备：护理站、实验台、护理用车、台、架、床等。

(9) Medical devices: nursing station, laboratory bench, nursing vehicle, bench, frame, bed, and others.

(10) 医院标识系统：VI 布点中与医疗工艺流程、医疗服务、生物安全、防护隔离、身体隐私、感染控制、婴儿防盗、病案管理等有关的点位。

(10) Hospital identification system: point location related to medical process and flows, medical services, biosafety, protective isolation, body privacy, infection control, infant security, medical record management and others in VI stationing.

(11) 医用纯水系统：如 5 兆欧和 10 兆欧纯净水制备供应。

(11) Medical pure water system: supply as preparation of 5 megohm and 10 megohm purified water.

2. 工艺条件可以用清单形式提出，也可以标注在建筑方案图中，条件一经提出，不仅要落实在土建和机电等专业的扩充、施工图纸中，还应在现场土建预埋、室内家具订货、专项设备采购、装修工程过程中结合订货、施工组织方案和验收规范加以调整，避免僵化执行。

2. Process conditions can be presented in the form of lists, and can be marked in the building proposal drawings. Once the conditions are put forward, they shall be implemented in the PD of building expansion and construction drawings of civil engineering and mechanical and electrical disciplines, and shall be adjusted during civil embedding on site, indoor furniture ordering, special device procurement and decoration project in combination with ordering, construction organization proposal and acceptance specifications, to avoid rigid implementation.

3.5 平疫结合医疗工艺设计

3.5 Medical process design in combination of normal time and epidemic time

3.5.1 院感控制工艺设计

3.5.1 Medical process design of hospital sensing control

3.5.1.1 应按疾病病种的病源传播性强度不同分区建设，以防止医院建筑成为疫情传播的高危公共场所，导致患者、医务人员在院内交叉感染。

3.5.1.1 The hospital shall be constructed with zoning according to different infectious intensities of disease species, so as to prevent the hospital building from becoming a high-risk public place for the spread of epidemic, leading to cross-infection of patients and medical care staff in the hospital.

3.5.1.2 为不同的人群分别设计医院出入口通道，把急救患者、病源传播性风险较强的感染患者、普通感染疾病患者、非感染患者和医院员工的出入口通道分开设计；应强化出入口标识，可以借助色彩如：急救患者出入口为绿色通道，发热门诊及肠道病门诊患者出入口为红色通道，普通感染疾病患者出入口为黄色通道，非感染患者出入口为蓝色通道，员工出入口为白色通道等。

3.5.1.2 Design the entrance and exit passages of the hospital for different groups of people, that is, separately design the entrance and exit passages for first-aid patients, infected patients with high risk of disease propagation, patients with common infectious diseases, non-infected patients and hospital staff; strengthen the entrance and exit signs, for example, make use of colors, design the entrance and exit for first-aid patients as green channels, design the entrance and exit for fever outpatient and intestinal disease outpatient departments as red channels, design the entrance and exit for patients with common infectious diseases as yellow channels, design the entrance and exit for non-infected patients as blue channels, and design the entrance and exit for staff as white channels.

3.5.1.3 尽可能减少患者与患者之间、患者与陪诊者、患者与医院员工等不同人之间的密切接触机会。把发热门诊等高传染的危险性诊疗区域独立分栋、分区设计，按照所在地主导风向由非感染到感染进行布局，各区域有醒目的标识系统；有独立的出入口、挂号收费处、发药窗口、检查设备及等候区等，严禁共用通道、电梯、扶梯、餐厅、厕所、污物间，把感染患者、疑似患者和非感染患者的流线和就诊空间分开。

3.5.1.3 Minimize the chances of close contact between patients, between patients and their companions, and between patients and hospital staff and others as far as possible. Design the dangerous diagnosis and treatment areas with high infection risk, for example, fever department, in independent buildings and zones, and arrange them from non-infection zone to infection zone according to the local prevailing wind direction, with eye-catching identification system in each zone; design separate entrance and exit, registration and charging office, dispensing window, examination device and waiting area; sharing passages, elevators, escalators, dining room, toilets and waste rooms shall be strictly prohibited, and the flow lines and treatment spaces for infected patients, suspected patients and non-infected patients shall be separated.

3.5.1.4 把传染性疾病诊疗区（含门诊和住院）、其他感染性疾病诊疗区和非感染性疾病诊疗区进行空间和空气的隔绝设计，除了一些大型昂贵的医疗设备共用外，其余均运用信息技术和物流技术提供分散服务。

3.5.1.4 Conduct space and air isolation design for the infectious disease diagnosis and treatment area (including outpatient and inpatient areas), other infectious and non-infectious disease diagnosis and treatment areas. Except for some large and expensive medical devices that have to be shared, the other areas are provided with decentralized services through information and logistics technologies.

3.5.1.5 传染性疾病诊疗区在传染病疫情发生后，不应影响非感染性疾病诊疗区的日常运行，不应因疫情而停止其他基础疾病的诊疗，一旦这些科室停止运行，其造成的死亡人数可能比因传染病而去世的人数更多。

3.5.1.5 After the outbreak of infectious diseases epidemic, the infectious disease diagnosis and treatment areas shall not impact the daily operation of non-infectious disease diagnosis and treatment areas, and the diagnosis and treatment of other basic diseases shall not be stopped due to the epidemic. Once these departments stop operation, the death toll may be higher than that due to infectious diseases.

3.5.2 门诊工艺流程设计

3.5.2 Design of outpatient process flow

3.5.2.1 改变医院建筑公用大厅、共享空间设计模式，消除无自然通风的公共大厅；依托分级诊疗、5G、物联网、智能物流技术，通过网络预约、无线缴费等，取消挂号、缴费窗口，减少门诊人群聚集，降低感染患者和非感染患者及其他人群密切接触的机会。

3.5.2.1 Change the design mode of public lobby and shared space in hospital buildings, eliminate the public hall without natural ventilation; cancel the registration and charging windows to reduce the gathering of outpatients and reduce the chances of close contact between infected patients, non-infected patients and other groups relying on tiered diagnosis and treatment, 5G, the Internet of things, and intelligent logistics technology, through online appointment and wireless payment.

3.5.2.2 通过网络智能预诊咨询服务平台和网络预约服务平台，结合医院智能导引系统，将病源传播性风险程度不同的就医人群引导走不同的出入口，减少感染患者和非感染患者及其他人群密切接触的机会，疏解医院出入口的高峰期交通拥堵问题。

3.5.2.2 Guide the patients with different levels of risk of disease transmission to different entrances and exits combining with the hospital intelligent guidance system through the network intelligent pre-diagnose consultation service platform and the network appointment service platform, to reduce the close contact chances between infected and non-infected patients and other groups, and ease the traffic congestion at the entrance and exit of the hospital during peak hours.

3.5.2.3 改变门诊药房窗口集中模式。

3.5.2.3 Change the centralized mode of outpatient pharmacy window.

1. 医院建筑按疾病病种的病源传播性强度不同而分区建设，在传染性疾病诊疗区、其他感染性疾病诊疗区和非感染性疾病诊疗区分设不同的药房。

1. Construct the hospital buildings in zones according to different infectious intensities of disease species, and set different pharmacies in the infectious, non-infectious diseases, and other infectious diseases diagnosis and treatment areas.

2. 将医生的处方用信息系统远程传递到门诊药库的智能发药机，将智能发药机发出的药物通过智能物流系统输送到各个专科门诊的护士站，派送给患者，使患者不需要到门诊药房窗口取药。

2. Deliver the doctor's prescription to the intelligent dispensing machine of outpatient dispensary through the information system, and deliver the outpatient drugs dispensed by the intelligent dispensing machine to the nurse stations of each specialist outpatient through the intelligent

logistics system, so that the patients do not need to go to the outpatient pharmacy window to get the drugs.

3. 通过采取急速快递等物流配送的方式将药物配送到患者家里，中药可以代煎后通过物流配送到患者家里。

3. Deliver the drugs to the patient's home through logistics modes, for example, rapid delivery, in which the traditional Chinese medicine can be boiled and delivered to the patient's home through logistics.

3.5.2.4 一站式诊疗服务流程设计

3.5.2.4 Design of one-stop diagnosis and treatment service flow

1. 通过信息化、物联网、智慧物流、无现金支付和人工智能等，采取门诊精准预约和无现金支付，不设挂号、收费窗口；使用智能化发药机、物流系统，不设门诊药房集中发药窗口；应用5G、物联网等信息传输和远程操控科技，把医技检查治疗设备分散配置到各个专科诊疗区，实现一站式诊疗服务。

1. Through information-based, Internet of Things, smart logistics, cashless payment, artificial intelligence and others, precise outpatient appointments and cashless payment are adopted, and no registered fee windows are set; intelligent dispensing machines and logistics systems are used, and no centralized dispensing windows in the outpatient pharmacy; 5G, Internet of Things and other information transmission and remote control technologies are used to decentralize and allocate the medical technology examination and treatment devices to various specialist diagnosis and treatment areas, to achieve one-stop diagnosis and treatment service.

2. 专科门诊诊疗单元基本配置包括一次候诊区、护士站（包括标本采集、物流站）、二次候诊区、诊室、专科检查设备、专科治疗室等，患者在该诊疗区完成除了大型设备检查和治疗外的所有普通专科诊疗服务。

2. The basic configuration of specialist outpatient diagnosis and treatment unit includes primary waiting area, nurse station (including specimen collection and logistics station), secondary waiting area, consulting room, specialist examination devices, specialist therapy room and others. Patients can complete all the general specialist diagnosis and treatment services in this diagnosis and treatment area, except the large-scale equipment examination and treatment.

3.5.2.5 门诊各级医疗工艺流程设计

3.5.2.5 Design of all levels of medical process flows for outpatient;

1. 按病种进行一级流程设计，让患者以最短的流程完成全部就诊活动。

1. The first-class flow design is carried out according to the disease species, so that patients can complete all outpatient activities in the shortest flow.

如心悸患者诊疗流程：

For example, diagnosis and treatment flows for patients with palpitation:

（1）分诊、导诊：通过网络预诊分诊平台指引到心内科专科门诊就诊；

(1) Triage and leading examining: Guide patients to the specialist outpatient department of cardiology through the network pre-diagnosis and triage platform;

（2）挂号、缴费、叫号：通过预约挂号平台精准预约心内科门诊号并缴费及提醒按预约时间就诊；

(2) Registration, payment and queuing: accurately make an appointment with the outpatient of cardiology department through the registration platform, pay the fee and be reminded to go to see the doctor according to the appointment time;

（3）入院、就诊：通过导引系统直接把患者导引到心内科专科诊区候诊、就诊；

(3) Admission and outpatient: directly guide patients to the department of cardiology to wait for diagnosis through the guidance system;

（4）检查：在心内科门诊护士站完成抽血检验和心电图、心脏超声等检查，标本通过智能物流系统传输到实验室；

(4) Examination: complete the blood examination, electrocardiogram, echocardiography and other examinations in the outpatient nurse station of Cardiology Department; transfer the specimens to the laboratory through the intelligent logistics system;

（5）诊断治疗：医生在电脑上查看检查结果、开处方；

(5) Diagnosis and treatment: the doctor checks the examination results on a computer and write out the prescription;

（6）取药：药物通过智能物流系统运送到心内科门诊护士站发给患者，患者在护士站通过视频系统咨询药师用药方式、离开医院。

(6) Taking medicine: deliver the medicines to the outpatient nurse station of the Cardiology Department for dispensing to patients through the intelligent logistics system. The patients consult the pharmacist on the use of drugs through the video system at the nurse station and then leave the hospital.

2. 门诊单元二级工艺流程设计

2. Design of second-class process flow of outpatient unit

建立二次候诊和医患双通道，一次候诊区和护士站在前面，就诊者通过闸门进入患者通道，患者通道内每个诊室门口设二次候诊区，医生的诊室后面有医护专用通道；一次候诊区为部分提早到的患者和陪诊者的等候区，按预约时间前 10～20 分钟到达的患者通过刷预约挂号凭证二维码或就诊卡或人脸识别通过闸门进入患者通道，在诊室前两次候诊区等候；除了儿童、老人和行动不方便或沟通不方便的患者允许一名陪诊者进入患者通道之外，一般不允许陪诊者进入患者通道，医护人员通过员工通道进入诊室。

Set the secondary waiting and the doctor-patient double channel. Set the primary waiting area and the nurse station in the front, so the patients can enter the patient channel through the gate. In the patient channel, set a secondary waiting area at the entrance of each consulting room, and a special channel for medical care staff at the back of the doctor's consulting room. The primary waiting area is the waiting area for the patients who arrive in advance and the accompanying personnel. Patients who arrive 10 ~ 20 minutes before the appointment time can enter the patient channel through the gate by scanning the QR code of the appointment registration certificate or the medical card or face recognition, and wait in the secondary waiting area in front of the consulting room. Except children, the elderly and patients with inconvenient actions or communication who are allowed to enter the patient channel with an accompanying person, generally, the accompanying persons are not allowed to enter the patient channel; medical care staff enter the consulting room through the staff channel.

3. 诊室三级工艺流程设计

3. Design of tertiary process flow of consulting room

以医护通道一侧作为相对洁净的区域，以患者通道一侧作为相对污染的区域，用诊桌将医患相对分开，拉开医患距离，把医生洗手盆布置在医护通道一侧，把医生的活动范围安排在相对洁净的区域，把体检床布置在患者通道一侧，把患者活动范围安排在相对污染的区域；因在体检床上进行体格检查时医生头部在患者上方，体检床区域应采用上进风下排风的模式，控制气流从上到下，减少医生接触患者的呼吸飞沫，降低医务人员被感染的风险。

Taking the medical channel side as the relatively clean area, and the patient channel side as the relatively contaminated area, separate the patients from the doctors with diagnostic tables to widen the distance between them; set the doctor's wash basin on the medical channel side; arrange the doctor's activity range in the relatively clean area; arrange the physical examination bed on the patient channel side, and arrange the patient's activity range in the relatively contaminated area. Because the doctor's head is above the patient during the physical examination on the physical examination bed, the medical examination bed area shall use the mode of upper-air supply and down-air exhaust to control the air flow from top to bottom, so as to reduce the doctor's contact with the patient's respiratory droplets, and reduce the risk of infection of medical care staff.

3.5.3　医技检查工艺流程

3.5.3　Process flow of medical technology examination

3.5.3.1　对于超过 1000 床的医院，按照每 500 张床为一个基本单元，分区、分散布局，将采集标本窗口分散在各个门诊科室护士站，通过物流系统将标本运送到中央实验室，减少患者在院内走动的次数和路程。

3.5.3.1　For hospitals with more than 1000 beds, carry out decentralized layout in zones, taking 500 beds as a basic unit. Decentralize the specimen collection windows in nurse stations of each outpatient department, and transfer specimens to the central laboratory through the logistics system, so as to reduce the number of times and distance of patients walking in the hospital.

3.5.3.2　在感染性疾病门诊区域配置独立的大型检查设备，如 CT、MR 等，让患者可以在一个诊疗区内完成所需要的基本医技检查服务。

3.5.3.2　In the outpatient area of infectious diseases, set independent large-scale examination devices, such as CT and MR, to enable patients to complete the basic medical technology examination services in a diagnosis and treatment area.

3.5.3.3　根据使用患者病源传播性风险程度的不同，对分区、分散布局的医疗检查设备采取不一样的消毒措施，降低院内感染的风险。

3.5.3.3 According to different degrees of disease transmission risk from patients who use the devices, take different disinfection measures for the medical examination devices that are distributed dispersedly, to reduce the risk of infection in hospital.

3.5.4 住院工艺流程设计

3.5.4 Design of process flow of inpatient department

新建、改扩建医院项目要以防疫思维和平疫结合的方式进行医院住院病区的医疗工艺流程设计。

For the project of new construction, reconstruction and expansion of hospital, the design for medical process flows of the hospital inpatient area shall be based on the epidemic prevention thinking and the combination of normal time and epidemic time.

3.5.4.1 在自然通风状态下，医护人员工作区不应位于患者区的下风向；在空调状态下，应让新风气流从清洁区向污染区流动；清洁区应布置在主导风向和气流上游方向。

3.5.4.1 Under natural ventilation condition, the working area of medical care staff shall not be located downwind to the patient area; under air conditioning condition, fresh air flow shall be arranged to flow from the clean area to contaminated area; the clean area shall be arranged in the leading wind direction and the upstream direction of the air flow.

3.5.4.2 应设医护人员及其他工作人员专用电梯和通道，医护人员工作的清洁区和病房污染区相对隔离，预留穿脱防护服的缓冲区空间，缓冲区应配备更衣淋浴室、消毒间。

3.5.4.2 Special elevators and channels for medical care staff and other staff shall be set up. The clean area for medical care staff shall be relatively isolated from the contaminated area of ward. The uncontaminated transition area for wearing and taking off protective clothing shall be reserved, and the uncontaminated transition area shall be equipped with locker room, shower room and disinfection room.

3.5.4.3 每个住院病区应规划少量规范的感染隔离病房，以便隔离疑似感染的病人；减少 3 张及以上床位病房的数量，增加单床病房数量。

3.5.4.3 A small number of standardized infection isolated wards shall be planned in each inpatient area to isolate the suspected infected patients; reduce the number of wards with 3 or more beds and increase the number of wards with single bed.

3.5.4.4 病房使用单体式空调终端，防止病毒细菌从空调系统传播；部分普通病区病房预留安装负压通风设备的接口条件，当疫情爆发时，可以快速改造为符合规范的传染病住院病区，减少医院工作人员院内感染及疫情传播。

3.5.4.4 Use single air-conditioning terminals in wards to prevent virus and bacteria from spreading through the air conditioning system; reserve the interfaces for installation of negative pressure ventilation devices in some inpatient ward areas, so that when the epidemic outbreaks, the wards can be rapidly transformed into a standard infectious disease inpatient area, to reduce the infection and epidemic transmission of hospital staff within the hospital.

第4章

医疗保健设施通用设计

Chapter Ⅳ

General Design of Health Care Facilities

朱 希　张伯林　范美态　谭晚芳

Xi Zhu　Bolin Zhang　Meitai Fan　Wanfang Tan

骆 艳　罗 红　张振华

Yan Luo　Hong Luo　Zhenhua Zhang

4.1 通用设计规定

4.1 General design regulations

4.1.1 适用性规定

4.1.1 Applicability regulations

本通用设计规定适用于无论类型、不分大小的每一座医疗保健设施，是最基本的规定。设计应满足所有现行规范、规定和标准的要求，如有背离的情况，应在设计任务书中加以说明并证明其是正确的，以获得主管部门特别批准。

The general design regulations apply to all health care facilities, regardless of type or size, and are the most basic regulations. The design shall meet the requirements of all current codes, regulations and standards. In case of any deviation, it shall be explained in the design specifications and proved to be correct, so as to obtain the special approval of the competent authority.

4.1.2 功能大纲

4.1.2 Functional outline

每一座医疗保健设施都应该有功能大纲，应表述：

Each health care facility shall have a functional outline, to state:

4.1.2.1 规模和功能规划

4.1.2.1 Scale and functional planning

设施中各部门的大小与净面积取决于任务要求与医院中各项服务的组织情况，如果功能规划不能兼顾安全标准、医疗与护理活动，那么有些功能允许组合或共享。

In the facility, the size and net area of each department depend on the task requirements and the organization of various services in the hospital. If the functional planning fails to take into account safety standards, medical and nursing activities, some functions can be combined or shared.

4.1.2.2 机动床位

4.1.2.2 Mobile beds

在护理单元里有待转运患者的机动床位设施时，需要另加走道的门，并根据使用要求设置由一个护士站转至另一个护士站的护士呼叫开关。

If there are mobile bed facilities for transfer of patients in a nursing unit, a passageway door shall be added, and the nurse call switch from one nurse station to another nurse station shall be set according to the operating requirement.

4.1.3 停车要求

4.1.3 Parking requirements

4.1.3.1 应对停车和交通进行专项研究，确保有足够的停车量与交通流量，以适应住院病人、门诊病人、员工与访客的需要。

4.1.3.1 Special research on parking and traffic shall be carried out to ensure sufficient parking volume and traffic flow, to meet the needs of inpatients, outpatients, staff and visitors.

4.1.3.2 如一座新的医院在功能方面有较多的增加或重要变更，应考虑停车位问题，以满足病人、员工及公众的需要。

4.1.3.2 If a new hospital has more additions or significant changes in function, parking spaces shall be considered to meet the needs of patients, staff and the public.

4.1.3.3 若未进行正式的停车研究，可以假定每张床一个车位再加任何一个工作班次中正常上班员工每人一个车位，在能方便到达公共交通或公共停车设施的地区，或在能减少交通量的合伙用车、网约车或其他用车安排已得到发展的地区，此车位数比例可以减少。

4.1.3.3 If no formal parking research is carried out, it can be assumed as one parking space per bed plus one parking space per regular staff in any work shift; this ratio of parking spaces can be reduced in areas where public transport and public parking facilities can be easily reached, or where car-pooling, online car-hailing or other vehicle use arrangements have been developed.

4.1.3.4 若无法分级诊疗导致门诊量特别大，应设法增加门诊病人或其他服务的车位。

4.1.3.4 If the outpatient amount is too large due to the failure of tiered diagnosis and treatment, it is necessary to increase the number of parking spaces for outpatients and other services.

4.1.3.5 应为服务性运输车和急诊病人用车提供单独另加的车位。

4.1.3.5 Additional parking spaces shall be provided for service carrier vehicles and emergency patient vehicles.

4.1.4 病房或护理区域

4.1.4 Ward or nursing area

4.1.4.1 厕所

4.1.4.1 Toilet

每位病人拥有不必经公共走道区而进入厕所的通道；一间厕所服务不超过两间病房和不多于四张病床；厕所内应有一件冲便器和一个洗手处，厕所门应向外开或双向开，在局部需要允许时，如有足够的隔声和防目视措施，则可采用折叠门。

The toilets shall have the passage that is accessible to each patient without passing through the public passage area; one toilet shall serve not less than two wards and not more than four beds; there shall be a toilet flusher and a hand washing place in the toilet, and the toilet door shall be opened outwards or in two directions. If required in part area, folding door can be used on the premise of adequate sound insulation and watch prevention measures.

4.1.4.2 病人储物处

4.1.4.2 Storage place of patient

每位病人在其病房内应有单独的衣橱、锁柜或壁橱，适用于挂长外衣与存放个人财务。

The ward shall have the separate closets, lockers or closets for each patient for hanging gowns and storing personal property.

4.1.4.3 病人和医护支持区域

4.1.4.3 Patient and medical support area

1. 管理中心或护士站：此区域应放置柜台与供储藏用的房间，应能方便地通向洗手处；此区域可以与接待、通信中心相结合或将它们包括在内。

1. Management center and nurse station: This area shall be equipped with counters and storage rooms leading to the hand washing place conveniently; this area can be combined with reception and communication centers or include them.

2. 文件制作区：图表绘制设施应有长条形工作面，足以确保员工、医生绘制图表和同时利用信息及通信系统。

2. Document making area: long and narrow working surfaces shall be equipped for Chart making facilities, to enable the staff and doctors to draw charts and utilize the information and communication system at the same time.

3. 多功能室：为员工、病人、病人家属使用的多功能室，可供召开病人研讨会、做报告、教学、进行培训和咨询活动，这些房间应能通达每一个护理单元；如这些房间能方便地供正常使用，它们也可以设在其他楼层；一间这类房间可以服务几个护理单元与部门。

3. Multi-function room: The multi-function rooms for staff, patients and patients' families can hold seminars, deliver speeches, and carry out teaching, training and consultation, and these rooms shall be accessible to each nursing unit; on the premise of convenient and normal usage, these rooms can also be arranged on other floors; one such room can serve several nursing units and departments.

4. 药品配送站：药品应由药品准备室分配，用整体式药品分送装置如全自动药品配送机，或其他可靠的物流系统来配送。

4. Drug supply station: Drugs shall be distributed by the drug preparation room, and delivered with automatic drug supply devices such as automatic drug supply machine, or other reliable logistics systems.

5. 药品准备室：此房间应处于护理员工可见控制下，房间内应有工作台、洗手处、可锁冰箱和受控上锁的药品存放柜；当药品准备室用于储存一个或多个整体式药品分送装置时，房间应设计有足够大的空间，以放置药品与整体式药品分送装置。

5. Drug preparation room: This room shall be under the visible control of nursing staff. There shall be a workbench, a hand washing place, a lockable refrigerator and a controlled and locked drug storage cabinet in the room. When the drug preparation room is used to store one or more integrated drug dispensing apparatus, the room shall be designed with sufficient space to place drugs and integrated drug dispensing apparatus.

6. 整体式药品分送装置：如整体式药品分送装置对受控药品有足够的安全性，并有足够的照明可很容易地识别药品，则它可位于护士站、清洁工作室内，应有方便的通道至洗手处，洗手盆应使用鹅颈水龙头和感应开关，以便可以冲洗到肘部。

6. Integrated drug dispensing apparatus: If the integrated drug dispensing apparatus is safe enough for controlled drugs, and has sufficient lighting to easily identify the drugs, it can be arranged in the nurse station and the cleaning room, which shall have a convenient passage to the hand-washing place. The wash basin shall be equipped with a gooseneck faucet and an inductive switch so that it can flush to the elbow.

7. 营养食品区域：配餐室或营养食品区域应配有洗涤盆、工作台、冰箱储藏柜，以及在规定用餐时间之间能提供冷热营养品的设备，该区域还应有非规定用餐服务时使用的盘、盆的放置房间。

7. Nutrition food area: The catering room or nutrition food area shall be equipped with washing basins, workbenches, refrigerator storage cabinets, and devices that can provide hot and cold nourishment during the specified meal time. The area shall also have rooms for placing the plates and pots used for non-specified dining service.

8. 清洁工作室或清洁物供应室：这类房间应单独设置，不可与污物工作室或污物直接相通。

8. Cleaning room or cleaner supply room: These rooms shall be set up separately, and shall not be directly connected to the waste work room or the wastes.

（1）清洁工作室：如房间是为病人的护理内容做准备的，则它应有工作台、洗手处以及清洁与消毒供应物的储藏设施。

（1）Cleaning room: If the rooms are used for preparing the nursing of patients, they shall have workbenches, hand washing places and storage facilities for cleaning and disinfection supplies.

（2）清洁物供应室：如房间是作为清洁与消毒用品分配系统的一部分，只做储藏与存放用，则不用工作台与洗手处。

（2）Cleaner supply room: If the rooms are used as a part of the distribution system of cleaning and disinfection supplies, they shall be used only for storage, and workbench and hand washing place are not required.

9. 污物工作室或污物存放室：这类房间应单独设置，不可与清洁工作室或清洁物供应室直接相通。

9. Waste work room or waste storage room: These rooms shall be set up separately and shall not be directly connected to the cleaning room or cleaner supply room.

（1）污物工作室：应有临床用洗涤盆与洗手处，这两项装置均应有冷热水混合龙头；也应有工作台和存放脏布件及各种废物的有盖、分装容器的放置空间。

（1）The waste work room: It shall have clinical washing basin and hand washing place, both of which shall be equipped with cold and hot water mixing faucets, workbench, and space for storage of containers with covers containing dirty cloth and various wastes.

（2）污物存放室：在只用于暂存污物的房间内，应设用于冲洗便盆的冲洗池。

（2）Waste storage room: A flushing pool for flushing the bedpan shall be set in the room only used for temporary storage of wastes.

10. 设备及供应物储藏室。

10. Storage room for devices and supplies.

（1）清洁布件储藏：应允许在每层楼面的清洁工作室内的指定地点、单独的橱柜内或经批准的配送装置内储藏；如果采用封闭的手推车装置，则应允许将清洁布件车存放在库房内；这样的小车必须储藏在正常的交通路径之外，并处于员工的控制下。

（1）Storage of cleaning cloths: storage of cleaning cloths shall be allowed in designated locations in the cleaning rooms of each floor, in separate cabinets, and in approved supply devices; if enclosed trolley apparatus is used, the trolleys shall be allowed to be stored in warehouse; such trolleys must be stored out of normal traffic path and under the control of staff.

（2）设备储藏室：应提供合适的房间或库房，存放病人、医护人员所需和功能大纲所要求的设备；病房里的每个病床单元应提供足够的储藏面积，病房内的通道宽度能让所有设备和供应物顺畅通过；每张床位应有的储藏面积不少于 $1m^2$。

（2）Storage room for devices: Appropriate rooms or warehouses shall be provided to store the devices demanded by patients and medical care staff and required by functional outline; each bed unit in the ward shall be provided with sufficient storage area, and the passage width in the ward shall allow all devices and supplies to pass smoothly; the storage area for each bed shall not be less than $1m^2$.

（3）担架与轮椅储藏室：应位于重要位置处，且不使正常交通受限。

（3）Storage room for stretchers and wheelchairs: this room shall be arranged in an important position without restricting normal traffic.

（4）抢救设备储藏：应提供处于护理人员直接控制下的抢救设备如心肺功能复苏（CPR）车的储藏用房，这类房间应位于功能大纲要求的区域内，但应在正常的交通路线之外。

（4）Storage room for emergency devices: The storage rooms for emergency devices under the direct control of nurses, for example, CPR trolley, shall be provided, which shall be located in the area specified in the functional outline, but outside the normal traffic route.

11. 整理服务用房：应能直接通达它们所服务的单元或楼层；在护理区每一楼层至少有一间整理服务用房，它应有服务用盆或楼层中使用的收受器具，并具备供应物与整理设备的储藏条件。

11. Make-up service room: the room shall lead to the units or floors they serve; there shall be at least one make-up service room on each floor of the nursing area, which shall have the service basin or receiving appliances used in the floor, and have the storage conditions of supplies and sorting devices.

4.1.5 员工支持区域

4.1.5 Staff support area

4.1.5.1 员工休息设施：休息设施的大小应按功能大纲确定，不应小于 $10m^2$，员工厕所位置应方便员工使用，可以不分男女。

4.1.5.1 Staff rest facility: The size of the rest facilities shall be determined according to the functional outline and shall not be less than 10m^2. The staff toilets shall be in the position convenient for staff, which can be unisex.

4.1.5.2 员工储藏设施：供护理员工安全存放个人用品的壁柜或储藏柜，应设在护士站内或临近护士站，这些橱柜应足够大，至少能存放小包与钱包，如设置外套存放间，则外套可存放在每一楼层的加锁壁柜内。

4.1.5.2 Staff storage facilities: Wall cabinets or storage cabinets for safe storage of personal items of nurses shall be located in or near the nurse stations. These cabinets shall be large enough to store at least small bags and wallets. If a coat storage room is set, the coat can be stored in the locked closet on each floor.

4.2 总体规划

4.2 Overall planning

4.2.1 规划设计条件

4.2.1 Planning and design conditions

4.2.1.1 对建设目的、目标、重点和战略地位的形势评估，项目发展的需要和资金来源。

4.2.1.1 Evaluate the construction purpose, objectives, key points and situation of strategic position, project development needs and the source of funds.

4.2.1.2 对临床需求的量化分析，与现有的功能设施共享的可能性和运行效率判断等可操作性评估。

4.2.1.2 Evaluate the operability of quantitative analysis of clinical needs, the feasibility of sharing existing functional facilities, operational efficiency judgment and others.

4.2.1.3 评估区位和规模、分析场所的作用条件因素。

4.2.1.3 Evaluate the location and scale of areas, and analyze the function and condition factors of the areas.

4.2.1.4 医院功能脉络的复合度分析，包括医院人流、物流分析，新设施和技术的运用计划；水平交通和垂直交通规划分析，临时设施和措施设定条件。

4.2.1.4 Analyze the multiplicity of the hospital's functional context, including the analysis of hospital's people flow and material flow, the application plan of new facilities and technologies, the planning analysis of horizontal and vertical transportation, and the setting conditions of temporary facilities and measures.

4.2.1.5 市政设施和能源供应条件。

4.2.1.5 Municipal facilities and energy supply conditions.

4. 2. 1. 6 对现有设施（包括设备和家具）清单分析，以便确定升级、改造的程度和策略。

4. 2. 1. 6 Analyze the list of existing facilities (including devices and furniture) in order to determine the upgrade and renovation degree and strategy.

4. 2. 1. 7 对未来需求作分析判断，对院区功能分区和流程的容量评估做出规划的弹性。

4. 2. 1. 7 Analyze and judge the future demand, and make the flexible planning by evaluating the capacity of functional zones and flows in the hospital.

4. 2. 1. 8 投资分析和分期实时策略，未来建设对日常运营的影响。

4. 2. 1. 8 Investment analysis and phased real-time strategy, impact of future construction on daily operations.

4. 2. 1. 9 制订一个符合医院长期发展的园区用地计划，以符合功能更新和容量需求。

4. 2. 1. 9 Develop a land use plan in line with the long-term development of the hospital, so as to meet the function upgrade and the capacity requirements.

4. 2. 1. 10 新建园区的总体规划应充分考虑医院建筑的发展和分期建设的可能性，采取适当的组合体系，有计划地留有发展余地。

4. 2. 1. 10 The overall planning of the newly-built area shall fully consider the possibility of hospital building development and phased construction, adopt an appropriate combination system, and leave room for development in a planned way.

4. 2. 2 前期策划

4. 2. 2 Preliminary planning

4. 2. 2. 1 进行 1. 1、1. 2、1. 3 章节中提到的前期咨询中所需概念性规划设计，研究可行的规划指标和规划方案。

4. 2. 2. 1 Carry out the conceptual planning and design required in the preliminary consultation of sections 1. 1, 1. 2 and 1. 3, and make a research on feasible planning indicators and proposals.

4. 2. 2. 2 以规划条件和可行性研究的建设方案为基础进行建筑策划，不仅依赖于经验和规范，更以实际调查为基础，运用计算机等近现代科技手段对研究目标进行客观的分析，最终定量地得出实现既定目标所应遵循的方法及程序；为建筑设计能够充分实现总体规划的目标，保证项目在设计完成之后具有较高的经济效益、环境效益和社会效益而提供科学的依据。包括下列成果：

4. 2. 2. 2 The architectural planning carried out based on planning conditions and construction proposals for feasibility studies depends not only on experience and specifications, but also on real survey. Carry out the objective analysis of research objectives by using modern scientific and technological means such as computers, and finally obtain methods and procedures quantitatively that shall be followed to achieve the established objectives, providing scientific basis for architectural design to fully realize the overall planning goal and ensure that the project has higher economic, environmental and social benefits after the design is completed. Including the following results:

1. 项目构成

1. Project composition

2. 城市空间关系

2. Urban spatial relationship

3. 建筑规模与空间组成

3. Building scale and space composition

4. 医疗工艺与流程验证

4. Verification of medical process and flow

5. 建筑功能与标准

5. Building functions and standards

6. 结构选型

6. Structure form selection

7. 设备系统

7. Device system

8. 环境保护

8. Environmental protection

9. 节能策略

9. Energy saving strategy

10. 建筑经济分析

10. Economic analysis of buildings

11. 工程投资

11. Project investment

12. 建设周期

12. Construction period

4.2.3 总平面规划

4.2.3 General plane planning

4.2.3.1 应满足医疗保健功能的要求，创造人性化的就医、疗愈、康养环境，并使之有利于使用者的工作和患者的健康。

4.2.3.1 It shall meet the requirements of health care functions, create a humanized environment for medical treatment, healing, rehabilitation and endowment, and make it conducive to the work of users and the health of patients.

4.2.3.2 功能分区合理，各种流线组织清晰。

4.2.3.2 Functional zone is reasonable, and various streamline organizations are clear.

4.2.3.3 洁污、医患、人车等路线清楚，避免院内交叉感染。

4.2.3.3 The route for cleaning and waste, doctors and patients, people and vehicles and other routes are clear to avoid cross infection in the hospital.

4.2.3.4 建筑布局紧凑，交通便捷，管理方便，减少能耗，最大可能保持可持续发展的空间，病房楼应获得良好朝向或景观。

4.2.3.4 The building is characterized by compact layout, convenient traffic, convenient management and energy consumption reduction, so that the space for sustainable development is maintained to the greatest extent. The ward building shall have a good orientation or landscape.

4.2.3.5 应保证住院、手术、功能检查和教学科研等用房的环境安静。

4.2.3.5 It shall ensure a quiet environment around rooms for hospitalization, operation, functional examination, teaching and scientific research.

4.2.3.6 应留有发展或改扩建余地，并作出拟发展或扩建规划。

4.2.3.6 Room for development or expansion shall be left and plans for future development or expansion shall be made.

4.2.3.7 应有完整的绿地规划。

4.2.3.7 There shall be complete green space planning.

4.2.3.8 对废弃物的处理应作出妥善安排，并应符合有关环境保护法令、法规的规定；提倡城市或地区集中处理，减少对环境的污染。

4.2.3.8 Wastes shall be properly disposed and shall conform to the provisions of relevant environmental protection laws and regulations. It is advocated to dispose wastes in a centralized way in cities or regions to reduce environmental pollution.

4.3 建筑设计

4.3 Architectural design

4.3.1 公共空间与通用空间

4.3.1 Public spaces and universal spaces

4.3.1.1 公共空间

4.3.1.1 Public spaces

1. 公共空间与功能分区合理，人和物路线清楚，布局紧凑。

1. Public spaces and functional zones are reasonable, the routes for people and materials are clear and the layout is compact.

2. 医疗医技区应在主要中心位置。其中急诊部、门诊部应面对交通干道，在大门入口处。

2. The medical and technical zones shall be arranged in the main central position. The emergency department and outpatient department shall be arranged at the entrance of the gate, facing the main trunk road.

3. 后勤供应区应在下风向，与医疗区应有一定距离或线路互不交叉干扰。

3. The logistics supply zone shall be arranged in the downwind direction and shall keep a certain distance from the medical zone or the routes do not interfere with each other.

4. 公共空间环境绿化景观优美。

4. Environmental green landscape in public space is beautiful.

4.3.1.2　通用空间

4.3.1.2　Universal space

1. 通用空间应开畅明亮，具有足够的高度。

1. The universal space shall be open and bright, and have sufficient height.

2. 空间符合功能要求，且温馨和舒畅。

2. The space meets the functional requirements and is warm and comfortable.

3. 满足最小的空间（宽度、高度、面积）要求：

3. Meet the minimum space (width, height, area) requirements:

最小空间尺度：见表 4-1。

Minimum space scale: see Table 4-1.

表 4-1　最小空间尺度

Table 4-1　Minimum space scale

使用空间 Space	宽度/m Width/m	高度/m Height/m	面积/m^2 Area/m^2
推床的通道 Passage for pushing beds	2.4m	2.3	—
双人诊查室 Double consulting room	开间不宜小于 3.0m The width should not be less than 3.0m	2.6	使用面积不应小于 12.0m^2 The effective use of space shall not be less than 12.0m^2
单人诊查室 Single consulting room	开间不宜小于 2.5m The width should not be less than 2.5m	2.6	使用面积不应小于 8.0m^2 The effective use of space shall not be less than 8.0m^2
抢救室 Emergency room	门净宽不应小于 1.4m The clear width of the door shall not be less than 1.4m	—	30.0m^2/每床 30.0 m^2/bed
每患儿候诊室 Waiting room for each child patient	—	—	不应小于 1.50m^2 The area shall not be less than 1.50m^2
口腔诊查单元 Oral examination unit	椅中距不应小于 1.80m, 椅中心距墙不应小于 1.2m The center-to-center distance between two chairs shall not be less than 1.80m and the distance from the chair center to the wall shall not be less than 1.2m	2.6	—
卫生间隔间 Toilet cubicle	不应小于 1.1m×1.4m It shall not be less than 1.1m×1.4m	—	—

续表

使用空间 Space	宽度/m Width/m	高度/m Height/m	面积/m^2 Area/m^2
门诊手术室 Outpatient operating room	平面尺寸不宜小于 3. 6m ×4. 8m The plane dimension should not be less than 3. 6m ×4. 8m	2. 7 ~3. 0 2. 7 ~3. 0	—
病房 Ward	平行的两床净距不应小于 0. 8m, 靠墙病床床沿与墙面的净距不应小于 0. 6m, 单排病床通道不小于 1. 1m, 双排病床通道不小于 1. 4m, 房门净宽不小于 1. 1m The clear distance between the two parallel beds shall not be less than 0. 8m, and the clear distance between the edge of the bed against the wall and the wall shall not be less than 0. 6m. The width of the passage with a single row of ward is not less than 1. 1m, the width of the passage with double rows of ward is not less than 1. 4m and the door clear width is not less than 1. 1m	2. 8	—
监护病房 ICU	床间净距不应小于 1. 2m The clear distance between beds shall not be less than 1. 2m	2. 8	单床间不应小于 12. 0m^2 The area of the single-bed room shall not be less than 12. 0m^2
分娩室 Delivery room	平面净尺寸宜为 4. 2m ×4. 8m The clear plane dimension should be 4. 2m ×4. 8m	—	—
剖宫产手术室 Cesarean operating room	宜为 5. 4m ×4. 8m It should be 5. 4m ×4. 8m	2. 7 ~3. 0 2. 7 ~3. 0	—
血液透析室 Hemodialysis room	治疗床（椅）之间的净距不宜小于 1. 2m, 通道净距不宜小于 1. 3m The clear distance between treatment beds (chairs) should not be less than 1. 2m, and the clear distance between passages should not be less than 1. 3m	—	—
放射照相室 Radiograph room	最小净尺寸宜为 4. 5m ×5. 4m The minimum net dimension should be 4. 5m ×5. 4m	2. 8	—
透视室 Fluoroscopic room	最小净尺寸宜为 6. 0m ×6. 0m, 机房门的净宽不应小于 1. 2m, 控制室门净宽宜为 0. 9m The minimum net dimension should be 6. 0m ×6. 0m, the clear width of the machine room door shall not be less than 1. 2m and the net width of the control room door should be 0. 9m	2. 8	—

手术室平面最小净尺寸：见表 4-2。

Minimum net plane dimension of operating Room: see Table 4-2.

表 4-2　手术室平面最小净尺寸

Table 4-2　Minimum net plane dimension of operating Room

手术室类型 Operating room type	平面净尺寸/m Net plane dimension/m	高度/m Height/m
特大型 Extra-large	7.50 × 5.70	2.7 ~ 3.0
大型 Large	5.70 × 5.40	
中型 Medium	5.40 × 4.80	
小型 Small	4.80 × 4.20	

4. 人防医疗工程规模：见表 4-3。

4. Medical project scale for civil air defense: see Table 4-3.

表 4-3　人防医疗工程规模

Table 4-3　Medical project scale for civil air defense

工程名称 Project name	掘开式工程 Excavation type project		坑、地道式工程防护区有效面积/m^2 Effective area of the defense area for the pit type and tunnel type project/m^2	人员数/人 （含伤员） Number of personnel (including the wounded)	床位数/张 Number of beds /pieces
	防护区最大建筑面积/m^2 Maximum building area of the defense area/m^2	防护区有效面积/m^2 Effective area of the defense area/m^2			
中心医院 Central hospital	4500	2500 ~ 3300	3300 ~ 4500	390 ~ 530	150 ~ 250
急救医院 Emergency hospital	3000	1700 ~ 2000	2200 ~ 2600	210 ~ 280	50 ~ 100
救护站 First-aid station	1500	900 ~ 950	1170 ~ 1250	140 ~ 150	15 ~ 25

4.3.2　急诊急救

4.3.2　Emergency treatment and first aid

4.3.2.1　急诊部：通常设在综合性医院内；专科医院设置急诊室，有条件的可设急诊科。

4.3.2.1　Emergency department: It is usually arranged in general hospitals; the specialized hospitals are provided with the emergency room, and are also provided with the emergency department if conditions permit.

1. 科室设置：设急诊科、车管科；急诊科应包含院前功能，大型医院可设独立的院前科；院前科位于 120 车辆进出车最为便利的位置，靠近急诊科。

1. Setting of departments: The emergency department and vehicle management department are set; the emergency department shall have the pre-hospital function and the large hospitals can be provided with an independent pre-hospital department; the pre-hospital department is located at the most convenient location for 120 ambulances to drive in and out of the hospital, closing to the emergency department.

2. 急救车辆：应有独立停车位，靠近急救入口；寒冷地区救护车应可以进入室内。

2. Ambulance: Ambulances shall be provided with an independent parking space near the emergency entrance. Indoor access shall be available for ambulances in cold regions.

3. 建筑面积：一般综合医院急诊科的面积占全院总建筑面积的3%。

3. Building area: The area of the emergency department in general hospitals accounts for 3% of the total building area of the whole hospital.

4. 通信：应开通急救120专线电话。

4. Communication: 120 special emergency telephone line shall be opened.

4.3.2.2 急救站：通常独立设置，也可以设在综合医院内。

4.3.2.2 Emergency station: It is usually set independently or set in general hospitals.

1. 科室设置：至少设有急救科、通信调度室、车管科。

1. Setting of departments: The emergency department, communication and dispatching room and vehicle management department are set at least.

2. 急救车辆：按每5万人口配1辆急救车，至少配备5辆急救车、5名司机、5名急救医护人员。

2. Ambulance: One ambulance is set per 50000 people, 5 ambulances, 5 drivers and 5 medical care staffs for emergency are provided at least.

3. 建筑面积：不少于$400m^2$。

3. Building area: It is not less than $400m^2$.

4. 通信：应开通急救120专线电话，与该市担任急救医疗任务医院的急诊科之间建立急救专用电话。

4. Communication: 120 special emergency telephone line shall be opened, and the dedicated emergency telephone line shall be set to contact with the emergency department of hospitals that perform emergency medical tasks in the city.

4.3.2.3 急救中心：是向100万人口以上区域提供高水平院前院内急救服务的医疗机构，并承担相应的高等医学院校教学和科研任务，是国家高层次的医疗机构；是省内或全国急救医疗、教学、科研相结合的技术中心。常独立设置，也可以设在综合医院内，或与其他机构如中心血站、医学信息中心联合建设。

4.3.2.3 Emergency center: It is a medical institution that provides regions with more than one million of people with the high-level pre-hospital and in-hospital emergency services and undertakes the teaching and scientific research tasks of corresponding medical colleges and universities. It is a national high-level medical institution, and a technical center combining with emergency and medical treatment, teaching and scientific research in the province or the whole country. It is usually set up independently, can also be set in general hospitals, or built jointly with other institutions such as central blood station, medical information center.

1. 功能设置：设急救科、通信调度室、车管科。指挥调度用房主要包括调度室、会商室、信息机房等，面积可按每辆车 10m^2 计，最低建筑面积不小于 200m^2。业务用房主要包括车库和隔离用房，其面积可按表 4-4 计算，大于 60 辆，大于部分可按每辆 30m^2 计算。行政管理用房主要包括综合办公用房和专业办公用房。后勤保障用房包括各类物资库房和其他服务用房。

1. Function setting: The emergency department, communication and dispatching room and vehicle management department are set. Commanding and dispatching rooms mainly include the communication and dispatching room, conference and consultation room, information machine room, etc. , with an area being calculated by 10m^2 per vehicle and a minimum building area of not less than 200m^2. Business rooms mainly include garage and isolation room, with an area being calculated according to Table 4-4; the additional area can be calculated by 30m^2 per vehicle in case of more than 60 vehicles. Administrative rooms mainly include comprehensive office rooms and professional office rooms. Logistics support rooms include various material warehouses and other service rooms.

2. 急救车辆：按每 5 万人口配 1 辆急救车，至少配备 20 辆急救车，21 名司机，30 名急救医护人员。

2. Ambulance: One ambulance is set per 50000 people, 20 ambulances, 21 drivers and 30 medical care staffs for emergency are provided at least.

3. 建筑面积：见表 4-4。

3. Building area: see Table 4-4.

表 4-4 建筑面积

Table 4-4 Building area

车辆数 Number of vehicles	20 辆 20 vehicles	30 辆 30 vehicles	40 辆 40 vehicles	50 辆 50 vehicles	60 辆 60 vehicles
建筑面积/m^2 Building area/m^2	950	1150	1300	1500	1800

4. 通信：应开通急救 120 专线电话，与该市担任急救医疗任务医院的急诊科之间建立急救专用电话，建立区域急诊临床信息系统。

4. Communication: 120 special emergency telephone line shall be opened, and the dedicated emergency telephone line shall be set to contact with the emergency department of hospitals that perform emergency medical tasks in the city, and the regional emergency clinical information system shall be established.

4.3.2.4 综合医院急诊部设计

4.3.2.4 Design of emergency department in general hospitals

1. 设有急诊部的医院应有标识清楚的出入口，以方便从为该院服务的市政道路进入，其他车道和人行道不应与通往急诊服务区的通道相冲突，进入急诊部的通道应位于遭受洪水或其他自然灾害最小之处，台风、飓风等气候地区的急诊通道两侧不应有容易倒伏和遮挡视线的树木。

1. Hospitals with the emergency department shall have clearly marked entrances and exits of the emergency department to facilitate access from municipal service roads serving the hospitals. Other lanes and sidewalks shall not conflict with the access to the emergency service area. The access to the emergency department shall be located at the place with minimum risk of floods or other natural disasters. There shall be no trees prone to lodging and blocking sight on both sides of the emergency access in climatic regions such as typhoons and hurricanes.

2. 急诊部应位于救护车快速抵达处，自成一区，单独设置建筑出入口，便于急救车、担架车、轮椅车的停放；急诊部与门诊部、医技部、手术部应有便捷的联系；设置直升机停机坪时，应与急诊部有快捷的通道。

2. The emergency department shall be located at the place where ambulances can arrive quickly, and shall be set in a separate area with a separate building entrance and exit to facilitate the parking of ambulances, stretchers and wheelchairs. The emergency department shall have convenient contact with the outpatient department, medical technology department and operating department. When setting up the helipad, a quick access to the emergency department shall be set.

3. 急诊部应当设医疗区和支持区；医疗区包括分诊处、抢救室、就诊室、治疗室、处置室和观察室，有条件的可设急诊手术室和急诊重症监护室；支持区包括挂号、各类辅助检查、药房、收费等部门，若与其他部门共享，应设 24 小时急诊值班；医疗区和支持区应当合理布局，缩短急诊检查和抢救距离。

3. The emergency department shall be provided with a medical area and a support area; the medical area includes triage room, emergency room, consulting room, therapy room, medical waste storage room and observation room. The emergency operating room and emergency ICU can be set up if conditions permit. The support area includes departments for registration, various auxiliary examinations, pharmacy, charging, etc. If theses departments are shared with other departments, 24-hour emergency duty shall be set up. The medical area and support area shall be rationally distributed to shorten the distance of emergency examination and rescue.

4. 急诊部应当有醒目的标识，以方便和引导患者就诊，与手术室、重症医学科等相连接的院内紧急救治绿色通道标识应当清楚明显，在医院挂号、化验、药房、收费等窗口应当有抢救患者优先的措施。

4. The emergency department shall have eye-catching signs to facilitate and guide patients for medical treatment. The signs of the emergency treatment green channel in the hospital connecting with the operating room and ICU shall be clear and obvious. Priority measures for emergency patients shall be provided at windows of registration, testing, pharmacy, charging, etc. in hospitals.

5. 急诊部应当明亮，通风良好，候诊区宽敞，就诊流程便捷通畅，建筑格局和设施应当符合医院感染管理的要求；儿科急诊应当根据儿童的特点，提供适合患儿的就诊环境。

5. The emergency department shall be bright and well ventilated and the waiting area shall be spacious; the medical treatment flow shall be convenient and smooth; the building pattern and facilities shall meet the requirements of infection management of hospitals. The pediatric emergency department shall provide an appropriate treatment environment for children patient according to their characteristics.

4.3.2.5 传染病医院急诊部设计

4.3.2.5 Design of the emergency department in hospitals for infectious diseases

1. 急诊部应明显易找、自成一区，尽可能与综合门诊部毗邻，在可能时使化验及药房布置在二者之间，分别设置服务窗口以利资源共享。

1. The emergency department shall be clearly marked and can be easily found, and shall be set in a separate area; in addition, it shall be as close as possible to the comprehensive outpatient department, and the laboratory and pharmacy can be set between these two departments when possible, and service windows can be set respectively to facilitate resource sharing.

2. 急诊部包括入口门厅、挂号、计价收费、小型药房、小化验室、男女值班室、处置室、抢救室、手术室、医生办、护士办、移动式X射线机或小型X射线机、病人卫生间、污物间及医务人员更衣间、卫生间、重症监护区及观察区。

2. The emergency department includes outpatient hall, registration, pricing and charging, small pharmacy, small laboratory, duty room for men and women, medical waste storage room, emergency room, operating room, doctors office, nurse office room, mobile X-ray machine or small X-ray machine, patient toilet, waste room and dressing room for medical care staff, toilet, ICU and observation area.

3. 特大城市中的大型传染病医院急诊部可配置大型X射线机及CT机房，通风空调装置应按正负压可转换调配考虑。

3. The emergency department of large hospitals for infectious diseases in a megacity can be equipped with a large X-ray machine and a CT machine room, and the ventilation and air conditioning equipment shall be equipped with the conversion function of positive and negative pressure.

4. 较大规模传染病医院的急诊部可按（快速抢救）绿色危急抢救区及一般急诊区划分。

4. The emergency department of large hospitals for infectious diseases can be divided into the (rapid rescue) green emergency rescue area and general emergency area.

5. 大型传染病医院的观察病区，重症监护病区中应设置负压隔离小间，每间1床，并附设缓冲过渡小间。

5. In observation wards of large hospitals for infectious diseases, the ICU shall be provided with negative pressure isolation cubicles, which are equipped with one bed and a buffer transition cubicle.

6. 大型传染病医院急诊部的X射线检查室及抢救室、手术室应考虑各有1~2间负压检查室或抢救手术室，可考虑平时采用正压，需要时改为负压。

6. The X-ray examination room, emergency room and operating room of the emergency department of large hospitals infectious disease shall consider to be equipped with 1 ~ 2 negative pressure examination rooms or emergency operating rooms respectively. These rooms can be considered to adopt the positive pressure in normal times and to adopt the negative pressure when necessary.

4.3.3 门诊

4.3.3 Outpatient service

4.3.3.1 门诊分类设计

4.3.3.1 Design of outpatient service classification

1. 门诊类型包括综合医院门诊部、专科医院门诊、社康中心、诊所、职业病防治所等，提供日间疾病诊疗、健康管理、康复、保健等服务。

1. Outpatient service types include outpatient department of general hospitals, outpatient department of specialized hospitals, community health service center, clinic, and occupational diseases prevention and treatment center, which provide services such as diseases diagnosis and treatment in the daytime, health management, rehabilitation, and health care, etc.

2. 门诊建筑形式：

2. Outpatient building type:

（1）医院内的门诊楼、门急诊医技楼、门急诊医技住院楼；

（1）Outpatient building, medical technology building for outpatient and emergency, medical technology inpatient building for outpatient and emergency in hospitals;

（2）有独立用地的门诊楼；

（2）Outpatient building with independent land;

（3）在居住社区、工业厂区、学校、部队营房等的配套用房内；

（3）Set in residential communities, industrial plants, schools, army barracks and other supporting rooms;

（4）在商业、办公等建筑综合体内的独立门店。

（4）Independent storefronts in commercial, office and other building complexes.

3. 门诊建筑应满足通用设计规定，按日门诊量和服务人群特点配置适宜的公共空间、商业或生活配套；诊疗用房设置和布局需满足门诊医疗工艺设计要求；门诊结构形式应考虑使用的灵活性和改造的可能性。

3. Outpatient buildings shall meet the general design requirements, and shall be equipped with appropriate public space, commercial or living facilities according to the daily outpatient visits and the characteristics of the patient. The setting and layout of diagnosis and treatment rooms shall meet the design requirements of medical process for outpatient. The structure of the outpatient building shall consider the flexibility of use and the possibility of reconstruction.

4.3.3.2 综合医院门诊部

4.3.3.2 Outpatient department of general hospitals

1. 门诊部占医院总建筑面积的比率为15%，日门（急）诊量与编制床位数的比值宜为 3 ∶ 1，也可按本地区相同规模医院前三年日门（急）诊量统计的平均数确定。

1. The area of the outpatient department accounts for 15% of the total building area of hospitals. The ratio of daily outpatient (emergency) visits to the number of beds should be 3 ∶ 1, and can also be determined by the average number of daily outpatient (emergency) visits in the first three years of hospitals of the same scale in the region.

2. 门诊公共部分应设置门厅、挂号、问讯、病历、预检分诊、记账、收费、药房、候诊、采血、检验、输液、注射、门诊办公、卫生间等用房和为患者服务的公共设施；各科应设置诊查室、治疗室、护士站、污洗室等；可设置换药室、处置室、清创室、X 射线检查室、功能检查室、值班更衣室、杂物储藏室、卫生间等。

2. The public space of the outpatient building shall be equipped with outpatient hall, rooms for registration, inquiry, medical records, pre-examination and triage, bookkeeping, charging, pharmacy, waiting, blood collection, testing, infusion, injection, outpatient office, toilet and public facilities serving patients. Each department shall be provided with consulting room, therapy room, nurse station, cleaning room, etc.; it can also be provided with the dressing room, medical waste storage room, debridement room, X-ray examination room, functional examination room, duty room and locker room for staffs, sundry storage room, toilet, etc.

3. 门诊宜分科候诊，门诊量小时可合科候诊；利用走道单侧候诊时，走道净宽不应小于 2.40m，两侧候诊时，走道净宽不应小于 3.0m；可采用医患通道分设、电子叫号、预约挂号、分层挂号收费等方式。

3. Outpatients waiting for diagnosis shall be classified, or may not be classified when there is a small amount of outpatients. When waiting on one side of the passageway, the clear width of the passageway shall not be less than 2.40m, while waiting on both sides, the clear width of the passageway shall not be less than 3.0m; the doctor-patient double channel, electronic queuing, appointment registration, tiered registration and charging and other types can be adopted.

4. 门诊手术可与手术部合并设置，门诊手术用房应由手术室、准备室、更衣室、术后休息室和污物室组成。

4. The outpatient operating room can be combined with the operating department, and rooms for outpatient operations shall consist of operating room, preparation room, locker room, postoperative rest room and waste room.

5. 门诊卫生间宜按日门诊量计算，男女患者比例宜为 1 ∶ 1；男厕每 100 人次设大便器不应小于 1 个、小便器不应小于 1 个，女厕每 100 人次设大便器不应小于 3 个。

5. Outpatient toilets should be set according to the daily outpatient visits, and the ratio of toilets for male to female patients should be 1 ∶ 1. The closet pans and urinals shall be no less than 1 for every 100 people in the men's toilet, the closet pans shall be no less than 3 for every 100 people in the women's toilet.

6. 各门诊专科设计

6. Design of individual outpatient departments

（1）内科：病人数量较多，为减轻垂直交通压力尽可能设于低层，采用通用诊室；

各专科中可以配套专用的小型医技用房，如功能检查室。

（1） Internal medical department: There are a large number of patients. In order to reduce the vertical transportation pressure, the department is set in the lower floors as far as possible and a general consulting room is adopted. Small special medical technology rooms, such as functional examination room, can be provided for each department.

（2） 外科：病人通常带有创伤疼痛，身体虚弱需躺卧，多人陪同，行动非常不便，应减少垂直流线行动，设计中应尽可能设于低层，采用通用诊室；换药室、处置室是外科门诊治疗用房，应划分清洁区、污染区；骨科门诊需要设石膏室。

（2） Surgical department: Patients usually suffer from trauma and pain, need to lie down due to physical weakness, and are very inconvenient to move and accompanied by many people, therefore, the vertical transportation shall be reduced. The departments shall be set in the lower floors as low as possible when designing and a general consulting room is adopted. Rooms for treatment of surgical outpatient department include the dressing room and medical waste storage room, which shall be divided into the clean area and contaminated areas. The plaster room needs to be set for the orthopedics outpatient department.

（3） 妇产科和计划生育科：妇科应自成一区，可设单独出入口，应设隔离诊室、妇科检查室及专用卫生间，宜采用不多于 2 个诊室合用 1 个妇科检查室的组合方式，检查区一般设有 B 超室、阴道镜室等，治疗区设宫颈治疗室、细胞学检查室、门诊手术室、休息室等；孕、产妇为健康人群，为避免交叉感染宜自成一区，设独立候诊区、诊室、检查室、处置室、B 超室、健康教育室、产后康复室、胎心监护室、休息室和孕妇专用卫生间；计划生育门诊及手术区设置在产科附近。

（3） Obstetrics and gynecology department and family planning: department The gynecology department shall be set in a separate area with separate entrances and exits, and shall be provided with the isolated consulting room, gynecology examination room and special toilet. It is advisable to adopt a combined use mode (that is, 1 gynecology examination room can be shared by no more than two consulting rooms) . The examination area is generally equipped with B-ultrasound room, colposcopy room, etc. The therapy area is equipped with cervical therapy room, cytological examination room, outpatient operating room, rest room, etc. Because the pregnant women of the obstetrics department are a healthy population, to avoid cross infection, it shall be set in a separate area with separate waiting area, consultation room, examination room, medical waste storage room, B-ultrasound room, health education room, postpartum recovery room, fetal heart rate monitoring room, rest room and toilet for pregnant women. The family planning department and operating area are set near the obstetrics department.

（4） 儿科：应自成一区，可设单独出入口，应增设预检、候诊、儿科专用卫生间、隔离诊查和隔离卫生间等用房，隔离区宜有单独对外出口，可单独设置挂号、药房、注射、检验和输液等用房；儿科门诊和急诊可以统筹安排，区域内应充分考虑儿童童趣化的活动空间和家属陪护空间的设置。

（4） Pediatric department: The department shall be set in a separate area with separate entrances and exits, and rooms for pre-examination, waiting, special toilet for pediatrics, isolation diagnosis and isolation toilet, etc. shall be set. The isolation area should have separate exits, and rooms for registration, pharmacy, injection, laboratory and infusion, etc. can be set separately. The pediatric outpatient department and emergency department can be arranged as a whole, and the setting of children' s fun activity space and family accompanying space shall be fully considered in the area.

（5）耳鼻喉科：应设诊室、内镜检查（包括食道镜等）、治疗室、换药室、高频测听室、听觉诱发电位检查室、耳声发射、声阻抗、助听器分析室、前庭检查室、内窥镜检查室、门诊手术室等，特殊检查房间要做隔声、屏蔽处理；可设置手术、测听、前庭功能、内镜检查（包括气管镜、食道镜等）等用房；诊区划分为候诊区、诊疗区、检查区、手术区、医护工作区、助听器验配区等。

（5） Otolaryngology department: The consulting room, endoscope examination (including esophagoscope, etc.) room, therapy room, dressing room, high frequency audiometry room, auditory evoked potential examination room, otoacoustic emission, acoustic impedance, hearing aids analysis room, vestibule examination room, endoscope examination room, outpatient operating room, etc. shall be set up. Special examination rooms shall be soundproofed and shielded. Rooms for operation, audiometry, vestibular function, endoscopy examination (including bronchoscope, esophagoscope, etc.) can also be set; the diagnosis area is divided into waiting area, diagnosis and treatment area, examination area, operation area, medical work area, hearing aid fitting area, etc.

（6）眼科：应设初检（视力、眼压、屈光）、诊查、治疗、检查、暗室、换药室、验光室、眼底照相室、AB超检查室、视野检查室、角膜内皮室、超声显微镜室、眼科OCT室、眼底电生理室、眼底血管造影室、眼激光治疗室、门诊手术室等用房，初检室和诊查室宜具备明暗转换装置，宜设专用手术室；诊区划分为：候诊区、诊疗区、检查区、准分子及飞秒激光治疗中心、手术区、医护工作区，验光区、配镜区通常设在眼科门诊同层区域。

（6） Ophthalmology department: Rooms for primary examination (vision, intraocular pressure, refraction) , diagnosis, treatment, examination, darkroom, dressing room, optometry room, fundus photography room, A-B ultrasound examination room, perimetry examination room, corneal endothelium room, ultrasonic microscope room, ophthalmic OCT room, fundus electrophysiology room, fundus angiography room, eye laser therapy room, outpatient operating room and other rooms shall be set up. The primary examination room and diagnosis room shall be equipped with light-dark conversion device, and should be equipped with the special operating room. The diagnosis area is divided into waiting area, diagnosis and treatment area, examination area, excimer and femtosecond laser therapy center, operation area, medical work area, optometry area and lens matching area, which are usually set in the same floor with the ophthalmology outpatient department.

（7）口腔科：有口腔颌面外科、口腔内科、口腔修复和口腔正畸等亚专科，包括患者候诊区、诊室、专家诊室、处置室、口腔技工室、牙片机室、数字口腔全景X光机室、口腔CT、设备机房等；诊区划分为候诊区、诊疗区、检查区、加工区和医护工作区；口腔诊室可以是单间，也可以是开放空间分出的隔断诊间，诊室楼面须降板15cm或设管道沟，用以敷设压缩空气及负压吸引气体管道和水电管道。

(7) Stomatology department: Sub-departments such as oral and maxillofacial surgery department, internal medical department for stomatology, prosthodontics and orthodontic department are set, and the patient waiting area, consulting room, expert consulting room, medical waste storage room, dental technician room, dental film room, digital oral panoramic X-ray machine room, oral CT, device room, etc. are arranged. The diagnosis area is divided into waiting area, diagnosis and treatment area, examination area, processing area and medical work area. The dental consulting room can be a separate room or a partitioned consulting room in an open space. The floor of the consulting room must be lowered by 15cm or provided with a pipe trench for laying compressed air and negative pressure suction gas pipes and water and electricity pipes.

(8) 皮肤科：皮肤病多带有传染性，宜设在门诊部独立僻静的地段；功能用房包括普通诊室、专家诊室、真菌检查室、免疫室、光动力室、冷冻室、激光室、微生物化验室等；诊区划分为候诊区、诊疗区、检查区、治疗区和医护工作区。

(8) Dermatology department: Most skin diseases are infectious, so the dermatology department should be set in an independent and secluded section of the outpatient department. The functional rooms include general consulting room, expert consulting room, fungus examination room, immunity room, photodynamic room, freezing room, laser room, microorganism laboratory, etc. The diagnosis area is divided into waiting area, diagnosis and treatment area, examination area, therapy area and medical work area.

(9) 中医科：功能用房包括诊室、专家诊室、针灸室、推拿室、牵引治疗室、离子导入室等；诊区划分为候诊区、诊疗区、治疗区、医护工作区。

(9) Traditional Chinese medicine department: Functional rooms include consulting room, expert consulting room, acupuncture and moxibustion room, massage room, traction therapy room, iontophoresis room, etc. The diagnosis area is divided into waiting area, diagnosis and treatment area, therapy area and medical work area.

(10) 感染性疾病科：需与其他门诊分开，要相对独立；呼吸道（发热）和肠道疾病科应设置各自挂号收费室、候诊区和诊室、治疗室、隔离观察室，以及检验室、放射检查室、药房、专用卫生间、处置室、抢救室等；严格区分污染区、半污染区和清洁区。

(10) Infectious diseases department: The department must be separated from other outpatient departments and relatively independent; departments of respiratory (fever) and intestinal diseases shall have their own registration and charging room, waiting area and consulting room, therapy room, isolation observation room, as well as laboratory, radioscopy examination room, pharmacy, special toilet, medical waste storage room, emergency room, etc.; in addition, the contaminated area, semi-contaminated area and clean area shall be separated strictly.

(11) 预防保健科：应设宣教、档案、儿童保健、妇女保健、免疫接种、更衣、办公等用房；可增设心理咨询用房。

(11) Prevention and health care department: Rooms for health education, archives, children's health care, women's health care, immunization, dressing, office and others shall be set; the psychological counseling rooms can be added.

4.3.3.3 传染病医院门诊部

4.3.3.3 Outpatient department in hospitals for infectious diseases

1. 传染病医院门诊量：门诊人次与住院床位比例1∶1～0.6∶1范围。

1. Outpatient visits in hospitals for infectious diseases: The ratio of outpatient visits to inpatient beds ranges from 1:1 to 0.6:1.

2. 大型传染病医院门诊应考虑不同传染病种的门诊区域，包括发热门诊、呼吸道传染病门诊、肝炎门诊、消化道传染病门诊、艾滋病门诊等。

2. The outpatient department in large hospitals for infectious diseases shall consider to set outpatient service areas of different infectious disease species, including fever outpatient area, respiratory infectious disease outpatient area, hepatitis outpatient area, digestive tract infectious disease outpatient area, AIDS outpatient area, etc.

3. 门诊部与接诊部的位置应靠近院区的主要出入口以方便病患者急找。

3. The outpatient department and the reception department shall be located close to the main entrance and exit of the hospital area to facilitate patients to find them in emergency.

4. 为保证传染病患者就医环境，减少与控制交叉感染风险，传染病医院门诊应有足够的室内候诊就医空间。

4. In order to ensure the medical environment for patients with infectious diseases and reduce and control the risk of cross infection, the outpatient department of in hospitals for infectious diseases shall have sufficient spaces for indoor waiting and medical treatment.

5. 门诊部一般包括入口厅、问讯处、挂号、划价交费、中西药房、各科诊室、医务人员值班更衣、医办、病人厕所。

5. The outpatient department generally includes the outpatient hall, inquiry desk, rooms for registration and charging, traditional Chinese medicine pharmacy, western medicine pharmacy, consulting rooms of various departments, duty room and dressing room for medical care staffs, doctors office and patient toilet.

6. 各类传染病门诊科室包括候诊、诊室等，均应分别自成一区，相对独立。

6. Outpatient departments of various infectious diseases, including waiting room and consulting room, shall be set in their respective separate area and relatively independent.

7. 传染病医院门诊平面布局可采用双通道布置方式，即病患者与医务人员分别使用不同通道；医务人员进出门诊工作区的口部应设置医务人员更衣室与卫生间，发烧门诊医务工作人员可按卫生通过室要求设置。

7. The plane layout of the outpatient department in hospitals for infectious diseases can adopt two-channel layout, i. e. patients and medical care staffs use different channels respectively. The entrance for the medical care staff getting in and out of the outpatient area shall be e-

quipped with locker room and toilet for medical care staffs, and the entrance for medical care staff of the fever outpatient department can be set according to the requirements of the uncontaminated transition room.

8. 候诊厅及候诊廊可根据传染病医院的规模以及预计接纳的日平均门诊人次设置，大中型医院采用厅式候诊，小型医院采用走道式候诊。

8. The waiting hall and waiting passageway will be set based on the scale of hospitals for infectious diseases and the predicted daily average number of outpatient visits. Large and medium-sized hospitals adopt the waiting hall and small hospitals adopt the waiting passageway.

4.3.4 医技

4.3.4 Medical technology

4.3.4.1 医学影像

4.3.4.1 Medical image

1. 一般规定

1. General provisions

承担门急诊及住院患者的影像检查及介入放射诊疗工作，大型影像中心可成为独立商业服务机构对外提供服务。中小型医院的大型影像设备用房通常集中布置，选址于急诊部、门诊部和住院部三者之间较低的楼层；大型医院影像设备的类型和数量可以分组分散布置在急诊部、门诊部、住院部的配套检查区；小型设备跟随相应专科布置，由分管科室统一排程管理；影像设备有最佳工作状态温、湿度控制要求，可以采用局部设备层，或给室外机组以足够的场地或建筑面积，并有取风、排风、排气口面积。

Undertake the imaging examination and interventional radiology diagnosis and treatment of outpatients, emergency patients and inpatients, and the large-scale imaging center can provide external services as an independent commercial service organization. Large-scale imaging device rooms in small and medium-sized hospitals are usually centrally arranged and set in the lower floors between the emergency department, outpatient department and inpatient department. The type and quantity of imaging devices in large hospitals can be distributed in groups in the supporting examination areas of emergency department, outpatient department and inpatient department in a decentralized manner. Small devices will be distributed in the corresponding departments and be managed by the department in charge in a unified schedule. The imaging device has the requirements of the optimum temperature and humidity control. Local device layer can be adopted, or the sufficient site or building area can be provided for outdoor units, including areas of air intake, air exhaust and gas exhaust.

2. 分类设计

2. Classification design

(1) 普通放射科

(1) Ordinary radiology department

是提供X射线检查的场所，其房屋和设施应符合国家环保标准、职业卫生标准、医院感染控制和放射防护要求。放射诊断功能用房包括：数字化X射线摄影系统（DR）、胃肠造影、乳腺钼靶、计算机 X 射线断层扫描（CT）等扫描室、控制室和配套的设备房、病员预约及候诊空间、更衣室、注射室、抢救室、卫生间、图像处理室、读片室、会诊室、医护人员办公值班及更衣室等辅助房间。

The department is a place to provide X-ray examination. Rooms and facilities of the department shall meet the requirements of national environmental protection standards, occupational health standards, hospital infection control and radiation protection. The radio diagnostic functional rooms include: rooms for digital X-ray tomography system (DR), gastrointestinal angiography, mammography, X-ray computed tomography (CT) and other scanning rooms, control room and supporting device rooms, patient appointment and waiting space, locker room, injection room, emergency room, toilet, image processing room, film reading room, consulting room, office, duty room, locker room and other auxiliary rooms for medical care staff.

（2）核磁共振

（2）MRI

核磁共振 MR 设备机房应尽量远离高低压配电房、电梯、汽车等有磁场干扰区域，同时考虑大型医疗设备之间的相互干扰，需满足专用医疗设备最小间距的技术条件；MR 机房尽量靠近外墙区域布置，满足设备吊装运输、空调水冷室外机、紧急排风口、失超管口布置等工艺条件；为进行 MR 增强检查，在患者检查等候区应设置专用注射室、观察室及抢救设施等。

MR device room shall keep as far away as possible from high and low voltage power distribution rooms, elevators, automobiles and other areas with magnetic field interference. Meanwhile, mutual interference between large medical devices shall be considered and the technical condition of minimum spacing between special medical devices shall be met. The MR device room shall be arranged as close to the outer wall area as possible so as to meet the process conditions such as equipment hoisting and transportation, arrangement of water-cooled outdoor unit of air conditioner, emergency exhaust outlet, quench pipe orifice, etc. For the purpose of MR enhancement examination, special injection rooms, observation rooms and rescue facilities shall be set up in the waiting area for patient examination.

（3）介入放射

（3）Interventional radiology

介入放射学是融医学影像学和临床治疗学于一体的交叉学科；数字减影血管造影系统（DSA）导管室是实施介入性放射诊疗的重要场所，是施术医师在医学影像系统监视指导下进行微创性操作的治疗检查室，类似于外科手术室，其机房的整体布局除了有适合 X 射线机工作的环境、放射防护，还要符合手术室的要求分区（限制区、半限制区、非限制区）。

Interventional radiology is an interdiscipline integrating medical imaging and clinical therapeutics; the catheter room of digital subtraction angiography (DSA) system is an important place for interventional radiology diagnosis and treatment, and a treatment and examination room for minimally invasive operation under the supervision and guidance of medical imaging system, similar to surgical operating room. The overall layout of its machine room is not only suitable for the working environment of X-ray machine and radiological protection but also meets the requirements for zoning of operating room (restricted area, semi-restricted area and non-restricted area).

介入中心功能用房主要有：机房兼手术操作室、控制室、设备间及辅助空间等；辅助空间包括：家属等候间、谈话间、病人换床间、观察室、无菌器械室、无菌敷料室、污物处置室、工作人员刷手间、值班室、办公室、卫生间、更衣室等。

The functional rooms of intervention center mainly include: machine room and operating room, control room, device room and auxiliary space; auxiliary space includes: waiting room for family members, communication room, patient bed changing room, observation room, sterile device room, sterile dressing room, waste storage room, staff scrub room, duty room, office, toilet, locker room, etc.

（4）核医学科

（4）Nuclear medicine department

是用放射性核素诊断、治疗疾病和进行医学研究的医学学科；在临床上的应用主要是脏器显像、脏器功能测定、核素治疗和体外分析；其选址应在医院内部偏僻的区域，与非放射性工作场所隔开，放射源应有单独的出入口。

It is a medical discipline concerning diagnosis with radionuclide, treatment of diseases and medical research; its clinical application mainly includes organ imaging, organ function measurement, radionuclide treatment and in vitro analysis; its location should be in the remote area inside the hospital, separated from the non-radioactivity workplace, and the radioactivity source should have a separate acess.

主要功能用房有：病员预约及候诊空间、门诊诊室、SPECT/CT 设备（在有条件情况下可配置 PET/CT、PET/MR 及提供正电子药物生产、制备与研究的回旋加速器室等配套设施的医疗设备）、配套设备房及控制室、功能测定室、敷贴治疗室、粒子植入室、体外分析室、贮源间、核废物暂存间、图像处理室、读片室、会诊室、医护人员办公值班及更衣室等辅助房间，同位素衰变池污水处理设施、特殊通排风及处理设施等。各个功能间的设计应遵循清洁区向污染区逐步过渡，控制区、监督区、非限制区分开设置，并在适当位置设立标志；医患分流，使患者和医务人员、公共区域相对分离，减少接触放射线的机会。

The main functional rooms include: patient appointment and waiting room, outpatient consulting room, SPECT/CT devices (medical equipment of auxiliary facilities such as PET/CT, PET/MR and editcyclotron room for production, preparation and research, of positron medicine are equipped if possible), supporting device room and control room, function measurement room, application therapy room, particle implantation room, in vitro analysis room, source storage room, temporary storage room of nuclear waste, image processing room, film reading room, consulting room, office duty room, locker room and other auxiliary rooms for medical care staff, sewage treatment facilities of isotope decay tank, special ventilation and treatment facilities, etc. The design of each functional room should follow the gradual transition from clean area to contaminated area. The control area, supervision area and non-restricted area should be set up separately, and signs should be set up at appropriate positions. The diversion of doctors and patients can separate patients from medical care staff and public areas, so as to reduce the chance of exposure to radioactive rays.

（5）消化内镜中心

（5）Gastrointestinal endoscopy center

包括患者的接待与术前准备区、内镜诊疗区、患者术后复苏或休息区、内镜器械洗消区、内镜器械储存区以及其他辅助区域。内镜诊疗区一般包括：胃镜室、肠镜室、ERCP 诊疗室、VIP 内镜诊疗室；一般按“病人预约 → 术前检查 → 按号进入接待 → 术前用药”的流程规划；不同部位、系统内镜的诊疗工作应当分室进行；上、下消化道内镜室应分开设置，各内镜室的使用面积应≥20m^2，应根据可开展的内镜诊疗项目，合理设置相应的诊疗室。

It includes patients' reception and preoperative preparation area, endoscopic diagnosis and treatment area, patients' postoperative recovery or rest area, endoscopic instrument washing area, endoscopic instrument storage area and other auxiliary areas. Endoscopy diagnosis and treatment area generally includes: gastroscopy room, enteroscopy room, ERCP diagnosis and treatment room, VIP endoscopy diagnosis and treatment room; it is generally planned according to the flow of “patient appointment → preoperative examination → reception by number → preoperative medication”; diagnosis and treatment of different parts and systems of endoscopy should be carried out in different rooms; upper and lower gastrointestinal endoscopy rooms should be set up separately, the use area of each endoscopy room should be≥20m^2, and the corresponding diagnosis and treatment rooms shall be set up reasonably according to the available endoscopic diagnosis and treatment items.

（6）超声医学

（6）Ultrasound medicine

包括三维超声成像技术、介入性超声成像技术、组织弹性超声成像技术等；其规模应视医院专业特点、科室分工，根据门诊、住院综合业务量、设备配置发展远景等综合因素来确定；诸多器官与疾病（如心、脑、肾、输尿管、膀胱、尿道、前列腺、肝、胆、脾、胰、子宫及其附件、甲状腺、乳腺、肿瘤疾病、外周血管疾病、围产医学检查等）检查与超声成像设备的使用关系密切，其位置的选择应考虑与上述相关联科室的门急诊部和住院部有便捷的联系。宜自成一区，或者根据专业科室的运营需求而分散设置；超声科内可设置成三个区域，即患者候诊区、检查区与工作人员办公生活区。

It includes 3D ultrasound imaging technology, interventional ultrasound imaging technology, tissue elastic ultrasound imaging technology, etc. ; its scale should be determined according to discipline characteristics of the hospital, division of departments, comprehensive business volume of outpatient and inpatient departments, development prospect of device configuration and other comprehensive factors; many organs and diseases (such as heart, brain, kidney, ureter, bladder, urethra, prostate, liver, gallbladder, spleen, pancreas, uterus and its accessories, thyroid, breast, tumor diseases, peripheral vascular diseases, and perinatal medical examination) examinations are closely related to the use of ultrasound imaging devices. The location should be selected in consideration of the convenient connection with the emergency department and inpatient department of the above-mentioned related departments. It should be set in a separated or decentralized according to operation needs of professional departments; the ultrasonic department can be set up in three areas, namely the waiting area for patients, the examination area and the office and living area for the staff.

4.3.4.2 检验科

4.3.4.2 Laboratory department

1. 检验科用房位置及平面布置应符合下列要求：

1. The site selection and plane layout of the laboratory room shall meet the following requirements:

（1）对于规模较大的医院，急诊检验区域面积不应小于 100m^2，门诊检验区域面积不应小于 300m^2，中心检验区域面积应为 1500～2000m^2。

(1) For large-scale hospitals, the area of laboratory in the emergency department, outpatient department shall not be less than 100m^2, 300m^2 and the center department should be 1500～2000m^2 respectively;

（2）应自成一区，微生物学检验应与其他检验分区布置；

(2) It should be set in a separate area, and the microbiological laboratory should be arranged with other laboratory zones;

（3）微生物学检验室应设于检验科的末端。

(3) Microbiological laboratory should be located at the end of laboratory department.

2. 检验科用房设置应符合下列要求：

2. Laboratory room shall be set in line with the following requirements:

（1）应设临床检验、生化检验、微生物检验、血液实验、细胞检查、血清免疫、洗涤、试剂和材料库等用房；

(1) There should be rooms for clinical laboratory, biochemical laboratory, microbiological laboratory, blood laboratory, cell examination, serum immunity, washing, reagent and material bank, etc.;

（2）可设更衣、值班和办公等用房。

(2) Rooms for dressing, duty, office, etc. can be set up.

3. 检验室应设通风柜、仪器室（柜）、试剂室（柜）、防振天平台，并应有储藏贵重药物和剧毒药品的设施。

3. The laboratory shall be equipped with ventilation cabinet, instrument room (cabinet), reagent room (cabinet), anti-vibration table, and facilities for storing valuable medicine and highly toxic medicine.

4. 细菌检验的接种室与培养室之间应设传递窗。

4. A transfer window shall be set between the vaccination room and the culture room for bacterial laboratory.

5. 检验科应设洗涤设施，细菌检验应设专用洗涤、消毒设施，每个检验室应装有非手动开关的洗涤池。检验标本应设废弃消毒处理设施。

5. The laboratory department shall be equipped with washing facilities, bacteria laboratory shall be equipped with special washing and disinfection facilities, and each laboratory shall be equipped with a washing pool with a non-manual switch. The laboratory specimens shall be equipped with waste disinfection facilities.

6. 危险化学试剂附近应设有紧急洗眼处和淋浴处。

6. Emergency eye wash device and shower shall be provided near hazardous chemical reagents.

7. 实验室工作台间通道宽度不应小于1.20m。

7. The width of the aisle between the laboratory workbenches shall not be less than 1.20m.

4.3.4.3 输血科

4.3.4.3 Blood transfusion department

1. 输血科（血库）是进行采血工作和储备血浆、血液以供手术部、急诊部及病房等输血之用。在150床以上的医院因采血工作较多，输血科（血库）可以独立设置。

1. Blood transfusion department (blood bank) is to collect blood and reserve plasma and blood for transfusion in operating department, emergency department and ward. In hospitals with more than 150 beds, the blood transfusion department (blood bank) can be set up independently for heavy blood collection workload.

2. 输血科（血库）的位置

2. Location of blood transfusion department (blood bank)

（1）规模较小的医院在手术部内设“储血冰箱”，以供应急诊手术之用，供应方便。

(1) The smaller hospitals set up “blood refrigerators” for emergency operation and convenient supply in the operating department.

（2）将输血科（血库）设置在检验科旁边，由检验科管理，因配血前血液检验、血型交配等工作均有化验人员担任，这样工作较方便。但输血科（血库）与检验科应有适当的隔离，以保证血库不受感染。同时应适当考虑与手术室的联系，亦可在手术室与输血科（血库）之间设置物流传送血液和血浆。

(2) The blood transfusion department (blood bank) is set next to the laboratory department, which is managed by the laboratory department. Before blood matching, blood laboratory, blood type mating and other work are performed by laboratory personnel, so the work is more convenient. However, the blood transfusion department (blood bank) shall be properly isolated from laboratory department to ensure that the blood bank is not infected. At the same time, the relationship with the operating room shall be properly considered. Logistics may also be set between the operating room and the blood transfusion department (blood bank) to transfer blood

and plasma.

（3）在城市中心设置中心输血科（血库），由卫生机构主办，或几个医院合办。其优点是血液可以统筹分配，避免有的缺少，有的积压的现象，并可节约人力物力。但在运输时要妥善解决，才能使血液供应安全及时。

(3) A central blood transfusion department (blood bank) shall be set up in the city center, which shall be hosted by the health institution or jointly run by several hospitals. Its advantage is that the blood can be distributed overall, to avoid shortage and overstocking, and save manpower and material resources. But it should be solved properly during transportation so that the blood supply is safe and timely.

3. 输血科（血库）的房间组成：由储血、配血、清血、消毒等房间组成。有采血要求的应增设配液，采血、献血人员等候和休息的房间。

3. The room of blood transfusion department (blood bank) consists of rooms for blood storage, blood matching, blood clearing, disinfection and others. If blood collection is required, the dosing should be added. A room for waiting and rest of blood collectors and blood donors shall be provided.

4. 输血科（血库）各功能房间之间的关系见图 4-1。

4. The relationship between the functional rooms of blood transfusion department (blood bank) is shown in Figure 4-1.

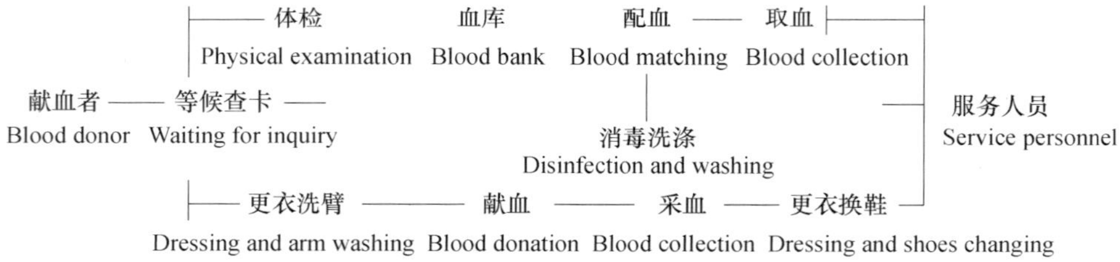

图 4-1 输血科各功能房间之间的关系

Figure 4-1 The relationship between the functional rooms of blood transfusion department

4.3.4.4 病理科

4.3.4.4 Pathology department

1. 病理科选址与平面布置

1. Site selection and plane layout of pathology department

（1）小型医院病理一般与检验合并，大型医院宜单独设置，用房总面积一般为 600 ~ 1000m²，二级医院一般不应小于 300m²。

(1) The pathology department is generally combined with laboratory in small-scale hospitals, and should be set up separately in large-scale hospitals. The total area of rooms should not be 600 ~ 1000m², and that of second-class hospitals should not be less than 300m².

（2）可成为独立商业机构对外提供服务。

(2) It can become an independent business organization to provide external services.

（3）病理科用房由切片（取材）、制片、镜检、标本整理和陈列、洗涤、消毒以及病理解剖等组成。

(3) The room for pathology department is composed of slice cutting (sampling), slice making, microscopy, specimen sorting and display, washing, disinfection, pathological anatomy, etc.

（4）病理解剖宜和太平间合建，与停尸室宜有内门相通，并应设置医务人员的更衣室和淋浴室。

(4) Pathological anatomy room should be built together with mortuary, and connected with morgue by inner door, and equipped with room for dressing and shower of medical care staff.

（5）教学医院解剖室应有观察设施，取材台与解剖台的一端应安装水池，另一端应有冲洗装置，解剖台应在距水池 0.70 米处设泄水口，且两侧均可操作；室内地面、操作台台面以及洗涤池等均应采用易于清洗、耐腐蚀的材料。

(5) The anatomy room of the teaching hospital shall be equipped with observation facilities. The sampling workbench and anatomy workbench shall be equipped with a water pool at one end and washing device at the other end. The anatomy workbench shall be equipped with a water outlet that is 0.70m away from the water pool, and can be operated at both sides; the indoor ground, console, washing pool, etc. shall be made from materials that are easy to clean and resistant to corrosion.

2. 病理科与其他科室的关联

2. Association between pathology department and other departments

（1）应与中心手术室、门诊手术室、产房、解剖室、病案室之间有水平或垂直通道，以传送病理标本，也可以采取自动物流系统传输。

(1) There should be horizontal or vertical channels between the center operating room, outpatient operating room, delivery room, anatomy room and medical record room to transmit pathological specimens, and automatic logistics system can also be adopted for transmission.

（2）应满足门诊、住院各相关临床科室病人标本送检、打印、咨询、会诊、借阅等功能，通道便捷。

(2) It should meet the requirements of sending, printing, consulting, joint consulting, borrowing and other functions of patients' specimens from outpatient, inpatient and relevant clinical departments, with convenient channels.

4.3.4.5 电生理科

4.3.4.5 Electrophysiology department

1. 根据医疗工艺设计，有的医院独立成一个医技科室，有的医院分散在相应的门诊专科内。

1. According to the design for medical process, some hospitals are independent into a medical technology department, and other hospitals are scattered in the corresponding outpatient specialty.

2. 用房包括：脑电图、肌电图、脑血流图、心电图室。

2. The room for EEG, EKG, REG and ECG is provided.

3. 各检查室因检查设备仪器较敏感，选址时应注意避免外来强电磁波的干扰，远离大型放射影像、放射治疗、磁共振等设备，且与使用关系较为密切的相关联科室的门急诊部、住院部有便捷的联系。

3. Due to the sensitivity of the examination devices and instruments, the examination room should be away from the interference of external strong electromagnetic waves, large-scale radiography imaging, radiotherapy, MRI and other devices, and have a convenient connection with the outpatient, emergency and inpatient departments of relevant departments that are closely related to the use.

4.3.4.6 麻醉手术部

4.3.4.6 Anesthesiology department

1. 手术部环境应符合现行国家标准《医院消毒卫生标准》(GB 15982—2010)的有关规定，手术部应分为一般手术部和洁净手术部，洁净手术部应按现行国家标准《医院洁净手术部建筑技术规范》(GB 50333—2013)的有关规定设计。

1. The environment of the operating department shall comply with the relevant provisions of the current national standard Hygienic Standard for Disinfection in Hospitals (GB 15982—2010). The operating department shall be divided into general operating department and clean operating department. The clean operating department shall be designed in accordance with the relevant provisions of the current national standard Architectural Technical Code for Hospital Clean Operating Department (GB 50333—2013).

2. 手术部选址和平面布置要求：

2. Requirements for site selection and plane layout of operating department:

(1) 手术部应自成一区，宜与外科护理单元邻近，并宜与相关的急诊、介入治疗科、重症监护科(ICU)、病理科、中心(消毒)供应室、血库等路径便捷；

(1) The operating department should be separated, close to the surgical nursing unit, and convenient to the relevant emergency department, interventional therapy department, intensive care unit (ICU), pathology department, central (disinfection) supply room, blood bank, etc.;

(2) 手术部不宜设在首层；

(2) The operating department should not be located on the first floor;

(3) 平面布置应符合功能流程和洁污分区要求；

(3) The plane layout shall meet the requirements of functional flow and division of clean and contaminated areas;

(4) 入口处应设医护人员卫生通道，且换鞋处应采取防止洁污交叉的措施；

(4) The entrance should be equipped with facilities for uncontaminated transition for the medical care staff, and measures should be taken to prevent the cross between cleaning and contamination at the shoes changing place;

(5) 通往外部的门应采用弹簧门或自动启闭门。

(5) The door to the outside shall be spring door or automatic opening and closing door.

3. 手术部用房设置：

3. Setting of room for operating department:

（1）应设手术室、刷手、术后苏醒、换床、护士室、麻醉师办公室、换鞋、男女更衣、男女浴室和卫生间、无菌物品存放、清洗、消毒、污物和库房等用房；

（1）Operating room, room for hand scrub, room for wake-up after operation, room for bed changing, nurse room, anesthesiologist office, room for shoes changing, locker room for men and women, bathroom and toilet for men and women, room for storage of aseptic supplies, cleaning room, disinfection room, dirt and warehouse room shall be set;

（2）可设洁净手术室、手术准备室、石膏室、冰冻切片、敷料制作、麻醉器械储藏、教学、医护休息、男女值班和家属等候等用房。

（2）Clean operating room, operation preparation room, plaster room, frozen slicing room, dressing production room, room for storage of anesthesia devices, teaching room, rest room for medical care staff, duty room for men and women and family waiting room can be set.

4. 手术部平面设计要求：

4. Graphic design requirements of operating department:

（1）应根据需要选用手术室平面尺寸。

（1）The plane size of the operating room shall be selected as required.

（2）每 2 ~4 间手术室宜单独设立 1 间刷手间，可设于清洁区走廊内。刷手间不应设门。洁净手术室的刷手间不得和普通手术室共用。每间手术室不得少于 2 个洗手水龙头，并应采用非手动开关。

（2）Every 2 ~ 4 operating rooms should be equipped with a separate scrub room, which can be located in the corridor of the clean area. No door shall be set in the scrub room. The scrub room of the clean operating room shall not be shared with the common operating room. Each operating room shall be equipped with no less than 2 hand washing faucets, and non-manual switch shall be used.

（3）推床通过的手术室门，净宽不宜小于 1.40m, 且宜设置自动启闭装置。手术室可采用天然光源或人工照明，当采用天然光源时，窗洞口面积与地板面积之比不得大于 1∶7，并应采取遮阳措施。

（3）The clear width of operating room door through which the bed is pushed should not be less than 1.40m, and should be equipped with automatic opening and closing devices. Natural light source or artificial lighting can be used in operating room. When natural light source is used, the ratio of window opening area to floor area shall not be greater than 1∶7, and sun-shading measures shall be taken.

5. 手术间室内设计要求：

5. Indoor design requirements of operating room:

（1）手术台长向宜沿手术室长轴布置，台面中心点宜与手术室地面中心点相对应，患者头部不宜置于手术室门一侧；

（1）The length direction of the operating table should be arranged along the long axis of the operating room, the center of the table should be corresponding to the ground center point of the operating room, and the head of the patient should not be placed on the side of the door of the operating room;

（2）观片灯应设置在手术医生对面墙上；

（2）The film viewer should be set on the opposite wall of the surgeon;

（3）设置医用气体终端装置；

（3）Set up medical gas terminal devices;

（4）采取防静电措施；

（4）Take anti-static measures;

（5）不应有明露管线；

（5）There shall be no exposed pipelines;

（6）吊顶及吊挂件应采取固定措施，吊顶上不应开设人孔；

（6）Fixed measures shall be taken for ceiling and hangers, and manholes shall not be opened on the ceiling;

（7）手术室内不应设地漏。

（7）Floor drains shall not be set in operating room.

4.3.4.7 放射治疗科

4.3.4.7 Radiotherapy department

1. 是治疗恶性肿瘤的临床科室，治疗设备有质子、直线加速器、伽马刀、后装机、放疗定位仪、重粒子加速器等，辅助设备有模拟定位机、治疗计划系统、挡铅制作系统、患者定位体架等。

1. It is a clinical department for the treatment of malignant tumors. The treatment devices include proton, linear accelerator, gamma knife, afterloader, radiotherapy localizer, heavy particle accelerator, etc. The auxiliary devices include analog localizer, treatment planning system, lead blocking production system, patient positioning frame, etc.

2. 功能用房包括：放疗门诊、病患预约及治疗前等候区、各类放射治疗机房、控制室、放疗模拟定位室、放疗计划室及讨论室、技师办公室、医师办公室、信息机房、医护区（更衣室、休息室）、公共卫生间、设备层等。

2. Functional rooms include: radiotherapy consulting room, patient appointment and pre-treatment waiting area, various radiotherapy rooms, control room, radiotherapy simulation positioning room, radiotherapy planning room and discussion room, technician office, doctor office, information room, medical care area (locker room and rest room), public toilet, device floor, etc.

3. 用房设计要求：

3. Room design requirements:

（1）科室位置要方便门诊和住院放疗患者与家属到达，其候诊厅、门诊区、办公区应考虑自然采光，可设会诊和值班用房。

（1）The location of the department should be convenient for patients of outpatient department, inpatient radiotherapy patients and their families. The waiting hall, outpatient area and office area should consider the natural lighting, and can be equipped with consultation and duty rooms.

（2）放射治疗机房应自成一区，应进行放射防护，多台放射治疗机房宜并排布置，以节约防护墙体；治疗室的出入口应设迷路，且有用线束照射方向应尽可能避免照射在迷路墙上，防护门和迷路的净宽均应满足设备要求。

（2）The radiotherapy room should be separated for radiology protection. Multiple radiotherapy rooms should be arranged side by side to save the protective wall. The entrance and exit of the therapy room should be set with labyrinth, and the irradiation direction of the useful wire harness should be avoided on the labyrinth wall as far as possible, and the clear width of the protective door and labyrinth should meet the device requirements.

（3）机房四周、地面及顶棚防护墙体很厚（如直线加速器主防护带厚度为 2. 6 ~ 3. 0m），结构荷载很大，应位于地下室底板上，宜布置在地下室尽端贴挡土墙的部位；应有放疗设备运输通道和设备吊装口。

（3）The protective walls around the machine room, on the ground and on the ceiling are very thick (for example, the thickness of the main protective belt of linear accelerator is 2. 6 ~ 3. 0 m), and the structural load is very large. It should be located on the basement floor, and should be arranged at the place where the end of the basement is close to the retaining wall; there should be a transport channel for radiotherapy devices, and a hoisting opening for devices.

（4）放射防护应按国家现行有关后装 γ 源近距离卫生防护标准、γ 远距治疗室设计防护要求、医用电子加速器卫生防护标准、医用 X 射线治疗卫生防护标准等的规定设计。

（4）Radiology protection shall be designed in accordance with the current national standards for the short-distance health protection of after-loaded γ source, the requirements for the design of γ long-distance therapy rooms, the standards for the health protection of medical electron accelerators, and the standards for the health protection of medical X-ray therapy.

（5）治疗室内噪声不应超过 50dB（A）。

（5）The noise in the therapy room shall not exceed 50dB (A).

4. 3. 4. 8　生殖医学中心

4. 3. 4. 8　Reproductive medicine center

1. 是妇产科的一个分支，也可于妇产科之外成为一个独立的科室，治疗不孕症；是人类辅助生殖技术包括人工授精和体外受精、胚胎移植及其衍生技术工作的场所，宜自成一区，位于医院内相对最干净的部位。

1. It is a branch of obstetrics and gynecology department, and can also become an independent department outside obstetrics and gynecology department to treat infertility; it is a work place for human assisted reproduction technologies including artificial insemination and in vitro fertilization, embryo transplantation and its derivative technologies, which should be separated and located in the relatively cleanest part of the hospital.

2. 基本分区：诊察区、手术试验区、手术留观区、医护工作区；根据人员和物品不同区分医生流线、男女患者流线、洁净物品流线和污物流线。

2. Basic division: examination area, operation trial area, operation observation area and medical work area; according to different personnel and objects, streamlines of doctors, men and women patients, clean objects and dirt can be distinguished.

用房组成：见表4-5。

Room composition: see Table 4-5.

表4-5 用房组成

Table 4-5 Room composition

功能分区 Functional area	房间名称 Room name
门诊诊断用房 Outpatient diagnosis room	候诊区、护士站、咨询室、检查室、B超室、诊断室、注射、抽血、样本采集、取精室、精液处理室、清洗室 Waiting area, nurse station, consulting room, examination room, B-ultrasound room, diagnosis room, injection, blood collection, sampling, sperm collection room, semen processing room, and cleaning room
手术实验室用房 Operation laboratory room	缓冲区（包括更衣室）、取卵手术室、胚胎移植室、术后休息室、胚胎培养室、胚胎学实验室、精子库、精液化验室、体外授精实验室、细胞培养、冷冻室、妇科内分泌测定室、试剂库、耗材库、荧光显微镜、更衣室、药品库、辅料间、污物间等 Buffer zone (including locker room), operating room for ovum pickup, embryo transplantation room, postoperative rest room, embryo culture room, embryology laboratory, sperm bank, semen laboratory, in vitro fertilization laboratory, cell culture, freezing room, gynecological endocrine test room, reagent bank, consumables bank, fluorescence microscope, locker room, drug bank, auxiliary material room, waste room, etc.
留观用房 Observation room	护理单元 Nursing unit
管理用房 Management room	办公室、资料档案室及其他辅助用房 Office, data archive room and other auxiliary rooms

3. 建筑空间宜明快舒展，空间适度放松、色彩柔和，使患者感到宽慰和温暖，环境安静，配置家具应避免锐角、眩光和冷硬，候诊厅室内设计和布置不可太过严肃整齐，以免引起情绪紧张，室内装修避免出现父母、儿童、肥胖等题材，以免患者触景生情、引起伤感。

3. The building space should be bright and spread, the space should be moderately relaxed and the color should be soft, so that the patients feel relieved and warm, the environment should be quiet, the furniture should be equipped to avoid sharp corners, glare and cold and hard states, the interior design and layout of the waiting hall should not be too serious and neat, so as to avoid emotional tension, the interior decoration should avoid the occurrence of parents, children, obesity and other themes, avoiding the scene brings back memories to the patients, and they feel sad.

4. 受精卵在强光照射下易被杀死，胚胎实验室的墙面宜设亮度可调式灯具，其安装最低点为 1. 80m 以上，采用嵌入式不受限制；胚胎实验室靠外窗设置时要采取遮阳设施；取卵室、移植室、胚胎实验室等应满足生物净化要求。

4. The fertilized ovum is easily killed under strong light. The wall of embryo laboratory should be equipped with lamps with adjustable brightness. The lowest point of installation is more than 1. 80m, and the use of "embedded" lamps is not limited. When the embryo laboratory is set near the external window, sunshade facilities should be equipped. The ovum pick-up room, transplantation room, embryo laboratory, etc. should meet the requirements of biological purification.

5. 室内装修材料应无毒、无味、不产生有害物质，避免受精卵的变异或死亡。

5. Interior decoration materials shall be nontoxic, tasteless and free from generation of harmful substances, so as to avoid variation or death of fertilized ovum.

4. 3. 4. 9　高压氧科

4. 3. 4. 9　Department of hyperbaric oxygen

1. 高压氧舱类型：见表 4-6。

1. Type of hyperbaric oxygen chamber: see Table 4-6.

表 4-6　高压氧舱类型

Table 4-6　Department of hyperbaric oxygen

类型 Type	适用人数 Applicable number of people	科室建筑面积/m^2 Department construction area /m^2
小型 Small	1 ~ 2 人	170
中型 Medium	8 ~ 12 人	400
大型 Large	12 ~ 20 人	600

高压氧舱分空气加压舱和氧气加压舱两种；有单人舱、双人舱、多人舱及婴儿氧舱；舱体有过渡舱、治疗舱、手术舱、急救运输舱、动物实验舱及潜水减压病的特殊治疗舱；小型舱可以不设过渡舱，治疗舱分为立体式和卧式；大中型舱根据舱室、舱门的数量可分为三舱七门、两舱两室四门、一舱两室四门、一舱两室二门等五种类型。

The hyperbaric oxygen chamber is divided into two types: air-pressurized chamber and oxygen-pressurized chamber; there are single-person chamber, double-person chamber, multi-person chamber and infant oxygen chamber; the chamber is divided into transition chamber, treatment chamber, operation chamber, first-aid transportation chamber, animal experiment chamber and special treatment chamber for diving decompression sickness; the small chamber can be without transition chamber, and the treatment chamber can be divided into three-dimensional and horizontal chambers; the large and medium chamber can be divided into five types according to the number of chambers and doors: three chambers and seven doors, two chambers, two rooms and four doors, one chamber, two rooms and four doors, and one chamber, two rooms and two doors.

2. 高压氧科设计要点

2. Key points of design of hyperbaric oxygen department

(1) 科室选址宜靠近急诊科、感染科，并有推床通道，方便门诊和住院患者抵达。

(1) The location of the department should be close to the emergency department and the infection department. There should be a channel for bed pushing to facilitate the arrival of patient of outpatient department and inpatient department.

(2) 功能房间包括等候厅、诊室、抢救室、高压氧舱、控制台、男女更衣室、医生办公室、会议室、休息室、库房、卫生间等。

(2) The functional rooms include waiting hall, consulting room, emergency room, hyperbaric oxygen chamber, console, locker room for men and women, doctor office, meeting room, rest room, warehouse, toilet, etc.

(3) 氧舱宜设置于独立建筑中或位于建筑底层，沿建筑外墙放置；因舱体较重、体积较大，为便于运输安装，墙体需预留安装洞口。

(3) Oxygen chambers should be set in independent buildings or at the bottom of buildings and placed along the outer wall of buildings; because of heaviness and large volume of the chambers, in order to facilitate the transportation and installation, the wall should be reserved with installation holes.

(4) 自带制氧机和设备的氧舱，其机房若设置于氧舱地下，通过楼梯到达，在地面上应有开口通风泄瀑。

(4) For the oxygen chamber with oxygen generator and devices, if the machine room is set underground and accessible by stairs, there should be an opening for ventilation and leakage and explosion on the ground.

3. 舱体房间设计要求

3. Design requirements of chamber room

(1) 高压氧舱间设计

(1) Design of hyperbaric oxygen chamber

高压氧舱间内安装高压氧舱舱体前需要在地面预留一定的安装深度，舱体需设置结构基础承重；控制台下需预留楼板洞，以便于与舱体各类管线相联系；高压氧舱间与其他房间之间的分隔墙体、楼板及门窗需满足防火设计要求。

Before the hyperbaric oxygen chamber is set in the room for hyperbaric oxygen chamber, a certain installation depth shall be reserved on the ground, and the chamber shall be set with structural foundation for load bearing; a floor hole shall be reserved under the console to facilitate the connection with various pipelines of the chamber; the partition walls, floors and doors and windows between the hyperbaric oxygen chamber and other rooms shall meet design requirements of fire protection.

（2）辅助间设计

（2）Design of auxiliary rooms

候诊面积可按每人 $2m^2$ 设置，人数按照舱体核定人数的两倍考虑；诊室按照单人诊室设计，一般 $8 \sim 12m^2$，小型高压氧舱可不考虑诊室的设置。

The waiting area can be set as $2m^2$ per person, and the number of people is considered as twice the number of people approved for the chamber; the consulting room is designed as a consulting room for single person, generally $8 \sim 12m^2$, and the consulting room is not considered for small hyperbaric oxygen chamber.

（3）机房设计

（3）Design of machine room

机房一般包括空压站、储氧罐间、氧气间等；如使用医院主供氧系统，可不设制氧间；空压站内装修考虑吸声、隔声处理，设备基础考虑减震构造；机房与其他房间之间的分隔墙体、楼板及门窗需满足防火规范设计要求。

The machine room generally includes the air compressor station, oxygen storage tank room, oxygen room, etc. ; if the main oxygen supply system of the hospital is used, the oxygen generation room may not be set; the sound absorption and sound insulation shall be considered for the decoration of the air compressor station, and the shock absorption structure shall be considered for the device foundation; the separation walls, floors and doors and windows between the machine room and other rooms shall meet the design requirements of the fire protection code.

4.3.5 药学部

4.3.5 Pharmacy department

4.3.5.1 药房

4.3.5.1 Pharmacy

1. 有门诊药房（含中、西药房）、急诊药房（含中、西药房）、住院药房、儿科门诊药房、感染疾病科门诊药房、静脉用药调配中心（室）、中药煎药室、制剂室、仓储库房、临床药学工作室、办公生活用房等。

1. There are outpatient pharmacy (including traditional Chinese and western medicine pharmacies), emergency pharmacy (including traditional Chinese and western medicine pharmacies), inpatient pharmacy, pediatric outpatient pharmacy, outpatient pharmacy of infectious diseases department, intravenous drug dispensing center (room), traditional Chinese medicine decoction room, drug manufacturing room, warehouse, clinical pharmacy room, office and living room, etc.

2. 中、西药房规模、配置、分布和流程应遵循医疗工艺设计要求；两者可分可合，当中药和西药设置在一起时，两药房在空间区划上应加以适当的分隔；大型医院常设门诊药房自动化系统，药房有集中也有分散在各楼层、各诊疗单元甚至各科室的情况，应根据设备条件为药房留有足够的空间尺度。

2. The scale, configuration, distribution and flow of Chinese and western medicine pharmacies should meet the requirements of design for medical process; the two pharmacies can be divided and combined. When Chinese medicine and western medicine are set together, the two

pharmacies should be properly separated in the space division; large-scale hospitals always set automation systems of outpatient pharmacy, and the pharmacies are centralized or scattered in each floor, each diagnosis and treatment unit and even each department. According to equipment conditions, enough space is reserved for the pharmacy.

3. 西药房：根据药品的性质、特点分别设置冷藏库、阴凉库、常温库；化学药品、生物制品、中成药、中药饮片分别储存，分类定位存放；麻醉药品、精神药品、医疗用毒性药品、放射性药品、易燃易爆、强腐蚀性高危等药品，应按有关规定分别设库，单独储存，存放区域应远离污染源（区）；药品库应设置验收、退药、发药等功能区域；应避免阳光直射，保持室内干燥和良好通风条件。

3. Western medicine pharmacy: According to the nature and characteristics of medicine, refrigerated warehouse, cool warehouse and normal temperature warehouse shall be set up respectively; chemical medicines, biological products, Chinese patent medicines and traditional Chinese medicine decoction pieces shall be stored separately, classified and positioned; narcotic medicines, psychoactive medicines, toxic medicines for medical treatment, radioactive drugs, inflammable, explosive, highly corrosive and high-risk medicines shall be stored separately according to relevant regulations. The storage area shall be far away from the contamination source (area) ; the medicine warehouse shall be set with functional areas for acceptance, withdrawal and distribution; direct sunlight shall be avoided, and the indoor dry and properly ventilated conditions shall be maintained.

4. 中药房：由调剂室、煎药室、切片、磨粉、制丸、炒药等室、中药库、生药库、成药库、原货仓库及晒场等组成。

4. Traditional Chinese medicine room: It is composed of dispensing room, decoction room, slicing room, grinding room, pill making room, medicine roasting room, traditional Chinese medicine warehouse, crude medicine warehouse, finished medicine warehouse, original goods warehouse, drying yard, etc.

（1）调剂室：可设在西药调剂室的旁边，共用候药厅，收方、发药柜台应在门诊大厅内；配方分生药和成药配方，生药配方较复杂，设有斗橱的调剂台，调配台对面设药橱，使工作人员坐在旋转凳上能取药调配，成药配方如设置转盘的配方台较方便，药架可低些，以利于通风采光。

（1）Dispensing room: It can be located next to the dispensing room of western medicine, sharing the waiting room, and the formula receiving and medicine distribution counters should be set in the outpatient hall; the formula is divided into formulas for crude medicines and compounded medicines, and the formula of crude medicines is more complex. There is a dispensing table with a cupboard, and a medicine cabinet is set on the opposite side of the dispensing table, so that the staff can take and dispense medicines on the rotating stool. The formula table with a turntable is more convenient for formulas of patent medicine, and the medicine rack can be lower, so as to be conducive to ventilation and lighting.

（2）煎药室：一般负责病房患者煎药，有的也给门诊患者代煎药。由于一般煎药室

煎制中药时，最好选择柴火、炭火，或者煤气灶点燃的明火，因此煎药室有明火和灰尘，要注意防火以及对其他科室的影响，要独立设置。

(2) Decoction room: It is generally for decoction for ward patients, and some for outpatients. As it is recommended to use open fire ignited by wood, coal and gas stove for decoction of traditional Chinese medicine for general decoction rooms, so attention shall be paid to fire prevention due to open fire and dust, and the impact on other departments, and decoction room should be set independently.

(3) 中药炒锅制剂：用超导电热锅将原材料经过加工炒制制成丸、散、膏、片、药片、针剂、冲剂等成品药。

(3) Traditional Chinese medicine wok preparation: The raw materials are processed and infused into pills, powder, paste, tablets, injections, granules and other finished medicines by using superconducting electric cookers.

(4) 中药整理加工室：由原药库、整理加工和室外晾晒场组成。

(4) Finishing and processing room of traditional Chinese medicine: It is composed of crude medicine warehouse, finishing and processing room, and outdoor drying site.

5. 药房调剂、制剂均要求朝北向，尤其在南方尽量避开东西向，避免阳光直射，以保持光线均匀，保存怕光的药品；药房调剂室内地面、墙面，宜采用易于洗刷、不易油垢和耐酸的材料，一般选择耐酸碱的油漆和瓷砖等材料。

5. Room for dispensing and preparation in pharmacy is required to face north, especially in the south, try to avoid east-west direction and direct sunlight, so as to keep the light even and protect the medicine sunlight; for dispensing room in pharmacy, indoor floor and wall shall be made from materials that are easy to wash, uneasy to generate oil scale and resistant to acid, generally paint, ceramic tile and other materials resistant to acid and alkali.

4.3.5.2 药学实验室

4.3.5.2 Pharmaceutical laboratory

1. 用房位置与平面布置应符合下列要求：

1. The site selection and plane layout of the room shall meet the following requirements:

(1) 中药实验室和西药实验室应分别设置，规模较小的可以合并在一处。

(1) The traditional Chinese medicine laboratory and the western medicine laboratory should be set up separately, and the smaller one can be combined.

(2) 应接近医院中心实验室。

(2) They shall be close to the hospital's central laboratory.

(3) 应符合实验室设计规范和实验室工艺设计要求。

(3) They shall meet laboratory requirements specifications and requirements of laboratory process design.

2. 实验室用房：

2. Laboratory room:

主要有生物/植物检验室、冰箱与烘箱室、天平室、暗室、显微镜室、库房、清洗室、污物处理室、检材处理室、更衣室、卫生间，应设空调、通风、工作台排风、通风柜。

They mainly include biological/plant laboratory, refrigerator and oven room, balance room, darkroom, microscope room, warehouse, cleaning room, waste therapy room, test material therapy room, locker room and toilet, which shall be equipped with air conditioner, ventilator, workbench ventilator and ventilation cabinet.

4.3.5.3 制剂室

4.3.5.3 Drug manufacturing room

1. 包括配制、检验两部分，是医疗单位配制制剂的地方，其周围环境必须保证制剂质量的要求，距制剂室 30m 以内不得有公厕、锅炉房、太平间、传染病房、动物房以及其他污染源，10m 以内不得有露土地面，外部环境要保持清洁。

1. It covers two parts: preparation and examination. It is the place where medical units prepare preparations. The surrounding environment must meet the requirements of preparation quality. No public toilets, boiler rooms, mortuaries, infectious wards, animal rooms and other contamination sources shall be set within 30m from the preparation room. No ground shall be exposed within 10m. The external environment shall be kept clean.

2. 药检室使用面积与设施应与其所开展的检验任务相适应，药检室应按制剂规模、品种设立化学、生物、仪器等检测室和留样观察室等。

2. The use area and facilities of the drug control room should be suitable for the examination tasks it carries out. The drug control room should set up chemical, biological, instrument and other testing rooms and sample observation rooms according to the scale and variety of the preparations.

3. 制剂室负责配制大量的协定处方剂、各种储备剂以及科研需要的或市面上供应不足的药品；制剂室的位置可接近调剂室、灭菌制剂室和洗涤室；如门诊验房和中心药房分开，制剂室可设在中心药房；制剂室面积较大时，可隔成小间，以便药水、药粉、药膏等分别操作，避免粉尘飘扬，影响制剂纯度。室内地坪、墙面可用易于清洗的材料，色彩宜用浅色，使光线明亮。主要家具有操作台、油膏台、药架、制剂台等，还有烘箱、冰箱、蒸馏水过滤器、搅拌箱、水池等设备。

3. The preparation room is responsible for preparing a large number of prescription agents agreed, various reserve agents and drugs needed for scientific research or in short supply in the market; the location of the preparation room can be close to the dispensing room, the sterilization preparation room and the washing room; for example, the outpatient laboratory and the central pharmacy are separated, and the preparation room can be located in the central pharmacy; when the preparation room is large, it can be divided into small rooms, for separate operation of the liquid medicine, powder medicine, unguent, etc. , thus avoiding dust flying and affecting the purity of the preparation. The indoor floor and wall can be made from materials that are easy to clean. Light color should be used so that the light is bright. The main furniture includes console, ointment table, medicine rack, preparation table, etc. , as well as oven, refrigerator, distilled wa-

ter filter, mixing tank, water tank and other devices.

（1）普通制剂室：配制常用的内服、外用药水、药粉、药膏等，可分别成小间，将药水、药粉分开制作，以避免互相影响。

（1）General preparation room: It is equipped with commonly used internal and external liquid medicine, powder medicine, unguent, etc. , which can be divided into small rooms to separate the liquid medicine and powder, thus avoiding mutual influence.

（2）灭菌制作间：配制糖、盐水、蒸馏水等大量应用注射液，以及临床科研所需的药剂；300床以上的医院均需设置灭菌制剂室，灭菌制剂室严格要求无菌，其部位宜隐蔽，避免与交通走廊直接联系，也应与门诊部等人流较多的地方有所隔离，应设有前室，供工作人员更衣消毒；本室要求双层窗、两道门，并要求有空调，墙面、地面、吊顶、水池、制剂台等分别用瓷砖、油漆、玻璃、搪瓷等易于清洗消毒的建筑材料，室内用紫外线照射等消毒；为便于消毒灭菌，该室的面积不宜过大，但也不宜小于6m^2；室内装置上下水管，水盆、电插头、管道均应暗设；要求光线明亮，多设窗户，墙面、家具等均宜采用浅色。

（2）Sterilization workshop: Prepare a large number of injections such as sugar, brine and distilled water, as well as the medicament for clinical research; hospitals with more than 300 beds need to be equipped with sterilization preparation rooms, which are strictly sterile, and their location should be concealed to avoid direct contact with the traffic corridor, and should also be isolated from the outpatient department and other places with large flow of people. The front room should be set up for the staff for dressing and disinfection; double windows and two doors are required in this room, and air conditioner is required. Walls, floors, ceilings, pools, preparation tables and others are made from building materials that are easy to clean and disinfect, such as ceramic tiles, paint, glass and enamel. Indoor disinfection is conducted by ultraviolet radiation; to facilitate disinfection and sterilization, the area of this room should not be too large, but should not be less than 6m^2; indoor devices such as water pipes, basins, plugs and pipes shall be concealed; the light shall be bright, with more windows, and walls and furniture shall be in light color.

（3）分析室：分析鉴定药房自制药品的性能、质量。

（3）Analysis room: Analyze and identify the performance and quality of self-made drugs in pharmacy.

4.3.5.4 静脉用药调配室（中心）

4.3.5.4 Intravenous drug dispensing room (center)

1. 是将分散在各病区治疗室开放环境下进行调配的静脉用药，由药学专业技术人员按照无菌操作要求在洁净环境下集中进行加药混合调配。

1. It is a kind of intravenous drug that is distributed in the open environment of therapy room in each ward, and the pharmaceutical professionals and technicians will centrally mix the drugs in the clean environment according to the sterile operation requirements.

2. 应当设置于人员流动少的安静区域，且便于与医护人员沟通和成品的运送；设置地点应远离各种污染源，禁止设置于地下室或半地下室；周围30m内环境应清洁、无污染、周边路面、植被等不会对调配过程造成污染。

2. It shall be set in a quiet area with less flow of people, and convenient for communication with medical care staff and transportation of finished products; the location shall be far away from all kinds of contamination sources, and shall not be set in the basement or semi-basement; the surrounding environment within 30m shall be clean and contamination-free, and the surrounding roads, vegetation, etc. shall not cause contamination to the deployment process.

3. 设计应符合医疗工艺流程设计要求，划分为洁净区、辅助工作区和生活区，不同区域之间人流、物流、出入走向合理，不同洁净级别区域间应有防治交叉污染的相应设施。

3. The design shall meet the design requirements of medical process flow, and be divided into clean area, auxiliary work area and living area. The flow of people, logistics and access trend between different areas shall be reasonable. The corresponding facilities for prevention and control of cross contamination shall be provided between areas of different clean levels.

4.3.5.5 营养科

4.3.5.5 Nutritional department

1. 用房包括营养咨询门诊、营养办公、肠内营养配制室、肠外营养制剂库房、营养实验室、营养厨房、营养餐厅、回民餐厅等。

1. The rooms include room for nutrition counseling outpatient department, nutrition office, enteral nutrition preparation room, parenteral nutrition preparation warehouse, nutritious laboratory, nutritious kitchen, nutrition dining room, Muslim restaurant, etc.

2. 营养厨房位置与平面布置应符合下列要求：

2. The site selection and plane layout of the nutritious kitchen shall meet the following requirements:

(1) 应自成一区，宜邻近病房，并与之有便捷联系通道；

(1) It should be separated and close to the ward, with convenient access to it;

(2) 配餐室和餐车停放室（处）应有冲洗和餐车消毒的设施；

(2) The catering room and dining car parking room (area) shall be equipped with facilities for washing and disinfection of the dining car;

(3) 应避免营养厨房的蒸汽、噪声和气味对病区的窜扰；

(3) The steam, noise and smell of nutritious kitchen should be avoided to disturb the ward;

(4) 平面布置应遵守食品加工流程；

(4) The plane layout should follow the food processing flow;

(5) 营养厨房应设置主食制作、副食制作、主食蒸煮、副食洗切、冷荤熟食、回民灶、库房、配餐、餐车存放、办公和更衣等用房。

(5) The nutritious kitchen shall be equipped with rooms for staple food making, non-staple food making, staple food stewing, non-staple food washing and cutting, cooked cold meat,

Muslim kitchen, warehouse, catering, dining car storage, office and dressing.

3. 营养门诊可设在门诊部标准门诊单元里，通常只有1～2间诊室，与其他科室共用会诊室、宣教室。

3. The nutrition outpatient department can be set up in the standard outpatient unit of the outpatient department. Usually, only 1～2 consulting rooms are available, and consulting rooms and health education rooms are shared with other departments.

4. 营养实验室可与其他各科实验室一并设在中心实验室附近，共用实验室辅助用房、通风系统等。

4. The nutrition laboratory can be located near the central laboratory together with other laboratories, sharing auxiliary rooms of laboratory, ventilation system, etc.

4.3.6　住院

4.3.6　Inpatient department

4.3.6.1　通用护理单元

4.3.6.1　General nursing unit

1. 护理单元的规模

1. Scale of nursing unit

(1) 一个护理单元宜设40～50张病床。专科病房或因教学科研需要者可少于30床。

(1) 40～50 ward beds should be equipped in each nursing unit. Specialized wards may be equipped with less than 30 beds, which may be required by teaching scientific research.

(2) 一个护理单元宜为同一病科；性质相近、病床数较少的可合并为一个护理单元。

(2) One nursing unit should be of the same pathology. The pathologies with similar nature and fewer beds can be combined into one nursing unit.

(3) 传染病科应单独设置护理单元，并应自成一区。

(3) The infectious diseases department shall be designed with a separate nursing unit, forming a separate zone.

2. 护理单元用房设置应符合下列要求：

2. The design of rooms for nursing unit shall meet the following requirements:

(1) 病房、重病房；病人卫生间、盥洗室、浴室；配餐室、库房、污洗室；护士室、医生办公室、治疗室、男女更衣值班室、医护人员卫生间。

(1) Wards and intensive disease wards; patient toilet, wash room and bathroom; catering room, warehouse, cleaning room; nurse room, doctor office, therapy room, locker room for men and women, duty room, medical care staff toilet.

(2) 根据需要配备的：重点护理病房、病人餐室兼活动室；主任医生办公室、换药室、处置室；勤杂人员更衣休息室；教学医院的示范教室、小化验室。

(2) Rooms to be provided as needed: ICUs, patient dining rooms doubled as activity rooms; chief physician office rooms, dressing rooms, medical waste storage rooms; dressing and

rest rooms for office assistants; demonstration classrooms and small laboratories in teaching hospital.

3. 病房

3. Ward

（1）病床的排列应平行于采光窗墙面。单排一般不超过 3 床，特殊情况不得超过 4 床；双排一般不超过 6 床，特殊情况不得超过 8 床。

（1）The arrangement of ward beds shall be parallel to the wall space of lighting window, generally, set no more than 3 beds for single row wards, and no more than 4 beds in special circumstances; set no more than 6 beds for double row ward, and no more than 8 beds in special circumstances.

（2）平行两床的净距不应小于 0. 80m，靠墙病床床沿同墙面的净距不应小于 0. 60m。

（2）The clear distance between the two parallel beds shall not be less than 0. 80m, and the clear distance between the edge of the bed against the wall and the wall shall not be less than 0. 60m.

（3）单排病床通道净宽不应小于 1. 10m，双排病床（床端）通道净宽不应小于 1. 40m。

（3）The clear width of single row bed passage shall not be less than 1. 10m, and the clear width of double row bed (bed end) passage shall not be less than 1. 40m.

（4）病房门应直接开向走道，不应通过其他用房进入病房。

（4）The ward door shall open directly to the corridor, and the ward shall not be accessed through other rooms.

（5）重点护理病房宜靠近护士室，不宜超过 4 床；重病房宜靠近护士室，不得超过 2 床。

（5）The ICUs should be close to the nurse room, with no more than 4 beds; the intensive disease ward should be close to the nurse room, with no more than 2 beds.

（6）病房门净宽不得小于 1. 10m，门扇应设观察窗。

（6）The clear width of the ward door shall not be less than 1. 10m, and a sight glass shall be set on the door leaf.

4. 护士室宜以开敞空间与护理单元走道连通，到最远病房门口不应超过 30m，并宜与治疗室以门相连。

4. The nurse room should be connected with the nursing unit corridor by an open space, and should not exceed 30m from the door of the furthest ward, and should be connected with the therapy room by a door.

5. 配餐室应近餐车入口处，并宜有烧开水和热饭菜的设施。

5. The catering room shall be near the entrance of the dining car, and should have the facilities for boiling water and heating meals.

6. 护理单元的盥洗室和浴厕。

6. The wash room, bathroom and toilet in the nursing unit.

（1）设置集中使用卫生间的护理单元，男女病人比例一般为 6：4，男卫生间每 16

床设 1 个大便器和 1 个小便器；女厕每 12 床设 1 个大便器。

(1) When setting the nursing unit with centralized use of toilet, since the ratio of male to female patients is generally 6 : 4, for the male toilet, set 1 closet pan and 1 urinal every 16 beds, and for the female toilet, set 1 closet pan every 12 beds.

(2) 医护人员卫生间应单独设置。

(2) The toilets for medical care staff shall be set separately.

(3) 当设置集中使用盥洗室和浴室的护理单元，每 16 床各设 1 个大便器和 1 个小便器。女卫生间每 16 床应设置 3 个大便器。

(3) When setting the nursing unit with intensively used wash rooms and bathrooms, set 1 closet pan and 1 urinal every 16 beds. The number of closet pans for female toilets shall be determined according to 3 closet pans every 16 beds.

(4) 附设于病房中的浴厕面积和卫生洁具的数量，根据使用要求确定，并宜有紧急呼叫设施和输液吊钩。

(4) The area of the bathroom and toilet and the quantity of sanitary ware attached to the ward shall be determined according to the demand of use. The wards should be equipped with emergency call facility and transfusion hook.

7. 污洗室应近污物出口处，并应有倒便设施和便盆、痰杯的洗涤消毒设施。

7. The cleaning room shall be close to the sewage outlet, and there should be defecation disposal facilities and bed pan and sputum cup washing and disinfection facilities.

8. 病房楼不宜设置垃圾管道；护理单元内不得设置垃圾管道。

8. Ward buildings should not be equipped with waste pipeline; nursing units shall not be equipped with waste pipeline.

9. 监护病房。

9. ICU

(1) 重症监护病房 (ICU) 宜与手术部、急诊部临近，并应有快捷联系；

(1) Intensive care unit (ICU) should be close to the operating department and emergency department and shall have quick contact;

(2) 心血管病房 (CCU) 宜与急诊部、介入治疗科室邻近，并应有快捷联系；

(2) The cardiovascular care unit (CCU) should be adjacent to the emergency department and interventional treatment department, and shall have quick contact;

(3) 应设监护病房、治疗、处置、仪器、护士站、污洗等用房；

(3) The ICU, therapy room, medical waste storage room, instruments room, nurse stations, cleaning room and other ICU accessories shall be set up;

(4) 护士站的位置宜便于直接观察患者；

(4) The position of the nurse station should be convenient for direct observation of patients.

(5) 监护病床的床间净距不应小于 1.2m；

(5) The clear distance between the ICU beds shall not be less than 1.2m;

(6) 单床间不宜小于 12.0m^2。

(6) The area of single bed room should not be less than 12.0m^2.

4.3.6.2 儿科单元

4.3.6.2 Pediatric unit

1. 儿科室设置应符合下列要求：

1. Pediatric department shall be arranged to meet the following requirements:

(1) 应自成一区，宜设在首层出入方便之处，并应设单独出入口。

(1) Pediatric department shall be a separate zone, and should be located in the first floor where is convenient to come in and go out, with separate entrance and exit.

(2) 入口应设预检处，并宜设挂号处和配药处。

(2) The entrance shall be set with a pre-examination office, and should be set with a registration office and a dispensing office.

(3) 候诊处每病儿面积不宜小于1.50m^2。

(3) The waiting area should not be less than 1.50m^2 per sick child.

(4) 设置仅供一病儿使用的隔离诊查室，并宜有单独对外出口。

(4) Set an isolated consulting room for the use of a sick child, and set a separate exit.

(5) 应分设一般厕所和隔离厕所。

(5) General toilets and isolated toilets shall be set up separately.

(6) 儿科门诊程序。

(6) Pediatric outpatient procedures.

2. 儿科门诊设置

2. Pediatric outpatient setting

挂号处、预诊室、病历室、药房、候诊室、诊室、治疗室、隔离诊室以及隔离病儿专用卫生间。

Registration office, pre-examination room, medical record room, pharmacy, waiting area, consulting room, therapy room, isolated consulting room and isolated special toilet for sick children.

4.3.6.3 产科单元

4.3.6.3 Obstetrics unit

1. 妇、产科病房

1. Gynecology and obstetrics wards

(1) 妇、产二科合为一个单元时，妇科的病房、治疗室、浴厕应与产科的产休室、产前检查室、浴厕分别设置。

(1) When the gynecology and obstetrics departments are combined into one unit, the wards, therapy room and bathroom and toilet of gynecology department shall be set separately with the parturition rest room, prenatal examination room and bathroom and toilet.

(2) 产房应自成一区，入口处应设卫生通过室和浴厕。

(2) The delivery room shall be a separate zone, and the entrance shall have an uncontaminated transition room and a bathroom and toilet.

(3) 待产室应邻近产房，宜设专用厕所。

（3）The predelivery room shall be close to the delivery room and should have a special toilet.

（4）应设隔离待产室和隔离产房，如条件限制，两者可兼用。应设产期监护室。

（4）Isolated predelivery rooms and isolated delivery rooms shall be set. In the case of limited conditions, both can be doubled. A perinatal period ICU shall be set up.

（5）一般产房平面净尺寸宜为 4.20m×5.10m，剖宫产产房宜为 5.40m×5.10m。两者的室内装修和设施均应与“无菌手术室”相同，但无观片灯装置。

（5）The general delivery room and the cesarean delivery room should have the clear plan view size of 4.20m×5.10m and 5.40m×5.10m respectively. The interior decoration and facilities of the two should be the same as that of “aseptic operating room”, except film viewer apparatus.

（6）洗手池的位置必须使医护人员在洗手时能观察临产产妇的动态。

（6）The wash basin must be positioned so that medical care staff can observe the status of parturient women while washing their hands.

2. 产科的婴儿室

2. Infant room of the obstetrics department

（1）应贴近产房区和产休室。

（1）The infant room shall be close to the delivery room area and parturition rest room.

（2）婴儿室宜朝南，应设观察窗，并应有防鼠、防蚊蝇等措施。

（2）The infant room should face south, and shall have sight glass, and rat proof, flies proof and other measures.

（3）洗婴池应贴邻婴儿室，水嘴离地面高度为 1.20m 并应有防止蒸汽窜入婴儿室的措施。

（3）The infant bath basin should be next to the infant room, with the water nozzle 1.20m above the floor, and the measures preventing steam from entering the infant room.

（4）宜设隔离婴儿室和隔离洗婴设施。

（4）Isolated infant room and infant bath facilities should be provided.

（5）配乳室与奶具消毒室不得同护士室合用。

（5）The milk kitchen and disinfection room shall not be doubled with the nurse room.

4.3.6.4 洁净病房

4.3.6.4 Clean ward

是指无菌病房、洁净护理单元、移植仓，用于收治白血病移植患者、癌症晚期化疗、重度烧伤、严重呼吸器官疾病和器官移植等患者。其特点是患者身体抵抗力差、住院周期长，特别是白血病患者治疗周期长达 1～2 个月，此期间不能离开病房，活动受限。

It refers to the aseptic wards, clean nursing units and transplantation chambers, which are used to treat leukemia patients in transplantation, terminal cancer patients in chemotherapy, severe burn, severe respiratory organ disease and organ transplantation patients. Its characteristic is that the patients have poor body resistance and long-time hospitalization cycle, especially, the leukemia patients have the treatment cycle as long as 1 ~ 2 months, and cannot leave the ward during this period, and are restricted in activity.

洁净病区设在各病种对应的护理单元内，通常位于末端，自成一区，区域内划分清洁区和洁净区，医护人员在清洁区工作，设男女患者恢复室兼准备室、药浴室、治疗室、无菌间、处置室等。净化病房位于洁净区，病房为单人间，血液病房符合Ⅲ级净化等级，骨髓移植病房符合Ⅰ、Ⅱ级净化等级，带负压卫生间，有窗户可以观看室外，以不超过4间病房为一组运营较为高效。

The clean ward area is arranged in the nursing units corresponding to various disease species, usually at the end of a nursing unit, forming a separate zone. The zone is divided into a cleaning area and a clean area. The medical care staff works in the cleaning area. The zone is arranged with recovery room doubled as preparation room for male and female patients, medicated bathroom, therapy room, aseptic room and medical waste storage room. Purification wards are located in clean areas, the wards are single rooms, the hematopathy wards meet class Ⅲ purification, the bone marrow transplantation wards meet class Ⅰ and Ⅱ purification, with negative pressure toilet, and with windows to view outside. For more efficiency, the wards should be operated with no more than 4 wards as a group.

患者、医护及服务工作者进出该区需经卫生通过室，设更衣间和缓冲区，药食物品传递采用传递窗。区域内宜采用通透性的隔墙，使患者与护士站之间有视线相通，增加室内空间的视觉明亮性和通透性，让患者和护士与外界有更多的沟通，改善心情、增进疗效。

Patients, medical care staff and service workers who enter this area shall pass through the sanitary uncontaminated transition room. The area has locker rooms and buffer zone, and has transfer windows to deliver food and medicine. Transparent partition walls should be used in the area, to enable sight contact between the patients and nurse station, increase the visual brightness and transparency of the indoor space, allow the patients and nurses have to communicate with the outside world, and improve their mood and the curative effect.

根据平面条件可设置家属探视廊，并设对讲电话，但不应阻挡病房望向室外的视线，有保证房间隐私的设施，也可采用视频探视。应控制洁净区内的面积和层高，以降低造价和运营费用。

Family visitation corridor can be set up according to the layout condition, with intercom telephone equipped, which shall not block the sight line from wards to the outside. The facilities guaranteeing the privacy of the room shall be provided, and video visitation can also be used. The area and story height of the clean area shall be controlled to reduce the cost and operating expenses.

4.3.6.5 重症单元 ICU

4.3.6.5 Intensive care unit (ICU)

1. 重症单元 ICU 组成：以护士站为中心，另设隔离室、电子仪器室、化验室、处置室、器材室、休息室、卫生间等主要护理单元，重点监护和抢救重症疾病，以提高抢救成功率。

1. Composition of ICU: taking the nurse station as the center, set the isolated room. electronic instrument room, laboratory, medical waste storage room, apparatus room, rest room, toilet, and other ICU accessories, to care and rescue intensive disease patients emphatically, so as to improve the rescue success rate.

2. 手术区的组成与要求和一般手术室相同，但要有至少两间与手术部合用。

2. The composition and requirements of ICU operating zones are the same as that of the general operating room, but at least two rooms shall be doubled with the operating department.

3. 恢复治疗区，是专门供患者在恢复与观察、治疗期间使用，待病情稳定后，再送到有关普通病房继续治疗。

3. The recovery treatment area is specially used for patients in recovery and observation period, and the patients can be sent to the general wards for further treatment after their condition is stabilized.

4. 重症单元 ICU 设计要求

4. Design requirements for ICU

（1）ICU 设计应设置在医院适中之处，以利于病人和物资的运行，便于医护人员快速集中。其内部布置应有最短的抢救治疗距离，并有清洁区和半清洁区之分。

（1） The ICU shall be set in a centrical place in the hospital, so as to facilitate the operation of patients and supplies, and facilitate the rapid concentration of medical care staff. The shortest rescue and treatment distance shall be arranged in ICU, with the clean area and semi-clean area divided.

（2）ICU 的床位应根据工作的重点来确定，占医院总床位的 1% ~2% ，最多可达 5% 左右。一般每张床位占使用面积 12 ~ 20m^2, 以 14 ~ 18m^2 为宜。辅助用房和床位相等，也可超出 50% 。

（2） The number of ICU beds shall be determined according to the focus of work, accounting for 1% ~2% of the total number of hospital beds, up to about 5%. Each bed covers an usable area of 12 ~ 20m^2 in general, and 14 ~ 18m^2 is advisable. Auxiliary rooms have the same area as the beds, and may exceed 50%.

（3）ICU 可根据需要设置单间或大病房。大病房宜设置 4 ~ 12 床，病床之间可以用玻璃隔断成半封闭式单间，病人之间的视线应予阻挡，仅使医护人员能观察到每一个病人，因此大多数 ICU 的病床以监护站为中心呈辐射状排列。

（3） Single room or large ward can be set for ICU as needed. It is advisable to set 4 ~ 12 beds in large wards, and set glass partition between the beds to form semi-enclosed single rooms. The sight line between patients shall be blocked, so that only medical care staff can observe each patient. Therefore, most ICU beds are arranged radially with the ICU station as the center.

（4）设有医用氧气、吸引、压缩空气等设施的机房，必须有急用电源，空调系统宜单独分区设置，送风口除有净化和灭菌装置。室内要求温度 25 ± 5℃，湿度 60%

±20%。

(4) Set the machine room for medical oxygen, suction, compressed air and other facilities, with emergency power supply necessarily; the air conditioning system should be set up in separate zones, with the purification and sterilization devices at air supply outlet. The required indoor temperature and humidity are 25±5℃ and 60%±20% respectively.

(5) ICU 入口处要设工作人员更衣室、盥洗室。在工作区内需有足够数量的洗手池。在2~4张床位之间应有一个备有浸手消毒液的洗手池。

(5) Staff locker room and wash room shall be set at the entrance of ICU. Sufficient amount of wash basins shall be set in working area. A wash basin with hands soaking disinfectant shall be provided among 2~4 beds.

4.3.6.6 烧伤单元

4.3.6.6 Burn unit

1. 用房组成

1. Room composition

由单人隔离病房、重点护理病房及康复病房（均不得超过3床）以及处理室、治疗室、护士室、洗涤消毒室、消毒品储藏室（柜）、器械室（柜）、化验室、扩创室、手术室等组成。

Burn unit consists of single isolated wards, ICUs and rehabilitation wards (each type of wards shall not exceed 3 beds) and medical waste storage rooms, therapy rooms, nurse rooms, washing and disinfection rooms, disinfectant item storage rooms (cabinets), apparatus rooms (cabinets), laboratories, debridement room, operating room, and others.

2. 抢救室、扩创室、手术室的医护人员和患者出入口分开、流线分开。

2. The entrance/exit and flow for medical staff and patients in emergency rooms, debridement rooms and operating rooms must be differentiated.

3. 设计要求

3. Design requirement

(1) 灼伤患者易感染，病床少者可设于外科护理单元的尽端，自成一区。病床多者可建立灼伤病护理单元（符合25床位的标准）。每一护理单元都应独立成区，避免穿行。

(1) Burn patients are susceptible to infection, in the case of insufficient ward beds, the wards can be located at the end of nursing unit of the surgery department, forming a separate zone. In the case of many ward beds, the burn nursing unit can be set (25 beds as the standard). Each nursing unit shall be a separated zone, to avoid walking through.

(2) 应设置在条件良好、空气清洁之处。

(2) It shall also be arranged in a place with good conditions and clean air.

(3) 出入口应设医护人员卫生通过室，应有换鞋、更衣、卫生间、淋浴设施，宜设风淋。病房中应带有卫生间。

(3) The entrance and exit shall be equipped with uncontaminated transition rooms and the

shoes changing, changing, toilet, and shower facilities. Air shower facilities are advisable. The wards shall have a toilet.

（4）探视者不得进入病区，宜在病房外侧设置探视走道。

（4）Visitors are not allowed to enter the ward area, and visitation corridors should be provided outside the ward.

（5）宜有空调设施，并应采用直流式空调系统，排风须经初效过滤处理后排放。

（5）Air conditioning facilities should be provided, and direct flow air conditioning system shall be used. The exhaust air must be discharged after primary-efficiency filtration treatment.

4.3.7　后勤保障

4.3.7　Logistics support

包括后勤、行政办公、职工食堂、洗衣房、锅炉房、机电设备、总务仓库、垃圾、尸体等。

It includes logistics room, administrative office room, staff dining room, laundry, boiler room, mechanical and electrical devices, general affairs warehouse, garbage room, corpse room and others.

4.3.7.1　太平间设计

4.3.7.1　Mortuary design

1. 应建于隐蔽之处，还应方便尸体搬运。尸体出口周围应与其他医疗用房绿篱隔离。

1. It shall be built in a concealed place, and are convenience for corpse handling. The corpse exit periphery shall be isolated from other medical rooms by green fences.

2. 应有冷藏尸体设施，停放尸体数量一般按医院总床位数量的 2% ~3% 计，应有尸体推车的回旋面积。

2. It shall have facilities for cold storage of corpses and a turning area for corpse trolley. Generally, the number of corpses stored shall be calculated as 2% ~3% of the total number of beds in the hospital.

3. 应有防蚊、蝇、鼠、雀等侵扰的措施。室内应有水源，以便冲洗。

3. Measures shall be taken to prevent the invasion from mosquitoes, flies, rats, sparrows and others. There shall be a water source in the mortuary, to facilitate washing.

4. 应有良好的通风，地面及墙面采用耐冲洗的材料。

4. Mortuaries shall have good ventilation, and the floor and walls with washable materials.

5. 应设置亲属入口，大型太平间还可设遗体瞻仰室，附近宜设休息室。

5. The entrance for family members shall be set up in the mortuary, body viewing rooms can also be set up for large mortuaries, with rest rooms set nearby.

4.3.7.2　尸体解剖室设计

4.3.7.2　Design of autopsy room

1. 应有通向太平间的门，以便于尸体从太平间进入。

1. A door leading to the mortuary shall be set to facilitate corpses entering from the mortuary.

2. 室内要求清洁，应设置医生淋浴更衣室、卫生间、器械室、洗涤消毒室、库房等。还应附有标本陈列室（柜），以及脏器固定池。

2. The autopsy room is required to be clean, and shall have doctor shower and locker room, toilet, apparatus room, washing and disinfection room, warehouse, and others. And the specimen exhibition room (cabinet) and organ fixing pool shall be arranged.

4.3.8 行政办公

4.3.8 Administration building

4.3.8.1 综合医院行政建筑面积占建筑总面积的4%。

4.3.8.1 The area of administration buildings in general hospital accounts for 4% of the total building area.

4.3.8.2 行政办公分为单独建筑或与住院楼、门诊楼合为一栋建筑等形式。可以把综合医院的急诊部、门诊部、住院部、医技科室、保障系统、行政管理和院内生活用房等设置在一栋建筑内，或部分功能放在一栋建筑内，形成高层建筑综合体。

4.3.8.2 Administration buildings are divided into detached buildings, and comprehensive buildings integrated with inpatient buildings and outpatient buildings. For general hospitals, the emergency department, outpatient department, inpatient department, medical technology department, support system, administrative management and living rooms in hospital or part of the function rooms can be integrated in a building, forming a high-rise building complex.

4.3.8.3 行政办公一般为独栋建筑，放置在次要入口处或靠近主体建筑，也可以与员工生活合建成综合楼。

4.3.8.3 In general, administration buildings are detached buildings, which are located at secondary entrances or close to main buildings, and can be integrated with staff living buildings, forming comprehensive buildings.

4.3.9 员工生活

4.3.9 Staff living building

4.3.9.1 大型综合医院应设置员工生活建筑，主要是宿舍与食堂等。

4.3.9.1 Large general hospitals shall have staff living buildings, mainly dormitories and dining rooms.

4.3.9.2 综合医院员工生活建筑面积占建筑总面积的4%。

4.3.9.2 The area of staff living buildings in general hospital accounts for 4% of the total building area.

4.3.9.3 员工生活一般为独栋建筑，放置在次要入口处或靠近主体建筑，也可以与行政办公合建成综合楼。

4.3.9.3 In general, staff living buildings are detached buildings, which are located at secondary entrances or close to main buildings, and can be integrated with administration buildings, forming comprehensive buildings.

4.4　建筑安全与风险控制

4.4　Building safety and risk control

4.4.1　建筑防火

4.4.1　Building fire protection

4.4.1.1　医院建筑耐火等级一般不应低于二级，当为三级时，不应超过三层。

4.4.1.1　The fire resistance rating of hospital buildings shall not be lower than Level Ⅱ in general. In the case of Level Ⅲ, the building shall not be more than three floors.

4.4.1.2　医院建筑的防火分区应结合建筑布局和功能分区划分。

4.4.1.2　The fire compartment of hospital buildings shall be divided in combination with the building layout and function division.

4.4.1.3　防火分区的面积除按建筑耐火等级和建筑物高度确定外；病房部分每层防火分区内，尚应根据面积大小和疏散路线进行防火再分隔；同层有两个及两个以上护理单元时，通向公共走道的单元入口处，应设乙级防火门。

4.4.1.3　In addition to determining the area of the fire compartment according to the fire resistance rating and height of the building, within the fire compartment on each floor of the ward area, fire compartment shall be carried out once again according to the area size and evacuation route; when there are two or more nursing units on the same floor, B-grade fireproof door shall be set at the unit entrance leading to the public passageway.

4.4.1.4　防火分区内的病房、产房、手术部、精密贵重医疗装备用房等，均应采用耐火极限不低于 1h 的非燃烧体与其他部分隔开。

4.4.1.4　The wards, delivery rooms, operating department, and precise and valuable medical device rooms in the fire compartment shall be separated from other parts by noncombustible materials with a fire resistance limit of no less than 1 hour.

4.4.1.5　楼梯、电梯防火

4.4.1.5　Fire prevention for stairs and elevators

1. 病人使用的疏散楼梯至少应有一座为天然采光和自然通风的楼梯。

1. At least one of the emergency staircases for patients shall have natural lighting and ventilation.

2. 病房楼的疏散楼梯间，不论层数多少，均应为封闭式楼梯间；高层病房楼应为防烟楼梯间。

2. Regardless of the number of floors, the emergency staircases of ward buildings shall be enclosed staircases; the high-rise ward buildings shall have smoke-proof staircases.

3. 每层电梯间应设前室，由走道通向前室的门，应为向疏散方向开启的乙级防火门。

3. The elevator room on each floor shall be equipped with a front chamber. The door leading to the front chamber door from the passageway shall be B-grade fireproof door opened in the direction of evacuation.

4.4.1.6 安全出口

4.4.1.6 Emergency exit

1. 在一般情况下，每个护理单元应有两个不同方向的安全出口。

1. In general, each nursing unit shall have two emergency exits in different directions.

2. 尽端式护理单元或自成一区的治疗用房，其最远一个房间门至外部安全出口的距离和房间内最远一点到房门的距离，如均未超过建筑设计防火规范规定时，可设一个安全出口。

2. For the dead-end type nursing unit or therapy rooms forming a separate zone, if the distance between the furthest room door and the external emergency exit and the distance between furthest point in the room and the door do not exceed the fire protection code for building design, an emergency exit can be set.

4.4.2 建筑防灾

4.4.2 Building disaster prevention

4.4.2.1 灾难包括：火灾、暴雨、洪水、地震、台风、战争、瘟疫、矿坑坍塌、群体暴力事件、重大事故等自然灾害和人为造成的危害。

4.4.2.1 The disasters include: fire, rainstorm, flood, earthquake, typhoon, war, plague, mine collapse, mass violence, major accidents and other natural and man-made hazards.

4.4.2.2 医院应对突发的上述灾难采取积极有效的措施，包括建筑防灾、灾难人员救治、院内人员保护与疏散、应急物资储备等。

4.4.2.2 The hospitals shall take active and effective measures in response to the above sudden disasters, including building disaster prevention, disaster personnel treatment, hospital personnel protection and evacuation, emergency supplies storage, and others.

4.4.2.3 医院建筑设计应满足各类有关防灾、减灾的设计规范。

4.4.2.3 The architectural design for hospitals shall meet all the design specifications for disaster prevention and mitigation.

4.4.3 建筑局部改造与临时建筑

4.4.3 Partial reconstruction of building and temporary building

4.4.3.1 医院是永续经营的建筑类型，任何医院在不断的发展中常常需要改建、扩建，这是医院发展建设必不可少的方式。

4.4.3.1 Hospitals are sustainable-operation buildings. Hospitals often need reconstruction and expansion in the continuous development, which is an essential way of hospital development and construction.

4.4.3.2 医院可以利用旧房屋、将过去旧医院扩建、改建，也可以利用原有的民房、学校改建成门诊、住院部甚至整座医院。

4.4.3.2 The old houses can be used to expand and reconstruct the old hospitals, and the existing houses and schools can be reconstructed into outpatient and inpatient departments and even a whole hospital.

4.4.3.3 进行医疗保健设施改造应做到以下工作：

4.4.3.3 The following shall be done in the reconstruction of health care facilities:

1. 对原有建筑进行原始资料收集。

1. Collect the original data of the original building.

2. 应对改造范围、扩建用地、选址、规模、投资等进行前期策划或可行性论证。

2. Carry out preliminary planning or feasibility demonstration for the scope of reconstruction, land for expansion, site selection, scale, investment, and others.

3. 加高和改造的建筑设计应符合现行建筑、结构、机电、消防等规范。

3. The architectural design of expansion and reconstruction shall conform to the current building, structure, mechanical and electrical, fire protection and other specifications.

4. 改造建筑外观应考虑室内外交界面处理、协调新旧建筑关系、与其他建筑的通道或连接、变形缝构造等。

4. The indoor and outdoor interfacial surface treatment, the coordination of relationship between the new and old buildings, the passage or connection with other buildings, and the deformation joint structure shall be considered for the appearance of reconstruction buildings.

5. 造价应考虑拆除、改建、新建、腾挪设备及家具、设计、咨询、管理周期等各方面的费用。

5. The costs for dismantling, reconstruction, new construction, moving devices and furniture, design, consultation, management cycle and other aspects shall be considered.

4.4.4 感染控制与风险评价

4.4.4 Infection control and risk assessment

本条款适用于所有医疗保健设施工程。此处所指感染，是指进入医疗保健设施内的人员，包括病人、医务人员、探视或陪护人员，在设施内停留期间获得的一切感染性疾病，也称医源性感染，在医院里发生的感染称为院内感染。

This clause applies to all projects of health care facilities. The infection here refers to all infectious diseases acquired by the personnel who enter the health care facility, including patients, medical care staff, visitors or accompanying personnel, during the stay in the health care facility, which is also called iatrogenic infection. The infection occurred in hospitals is called hospital infection.

4.4.4.1 在新建和改扩建工程规划设计阶段，业主应提供感染控制风险性评价标准，并将其纳入设计任务书，明确通过空气、水传播各种微生物污染潜在的风险以及相关措施。

4.4.4.1 In the planning and design stage of new construction, reconstruction and expansion project, the employer shall provide the risk assessment standard of infection control, and write into the design specifications, to clarify the potential risks of various microbial contamination through air and water and relevant measures.

4.4.4.2 在新建和改扩建施工期间，设施内部护理环境受到影响，会对患者及员工产生有害的环境影响，在施工过程中要采取感染控制、保障生命安全和保护室内人员的措施。

4.4.4.2 During construction, reconstruction and expansion, the internal nursing environment of the facilities is affected, which will have harmful environmental impact on patients and staff. During the construction, measures shall be taken to control infection, ensure life safety and protect indoor personnel.

4.4.4.3 感染控制设计与实施

4.4.4.3 Infection control design and implementation

该项工作有赖于各专业设计团队协作完成，其中包括但不局限于：医院管理者、临床医生、感染控制专家、医疗工艺咨询师、政府卫生主管部门和施工技术人员。

This work depends on the cooperation of design teams of each discipline, including but not limited to: hospital managers, clinicians, infection control experts, medical process consultants, government health authorities and construction technicians.

4.4.4.4 感染控制风险评价组

4.4.4.4 Infection control risk assessment group

工作人员应有感染控制管理者，直接参加病人护理的护理师，风险管理、工程设计、总包及分包施工、监理、流行病学等专家，组成的小组在各个对应阶段做出评价。评价阶段包括建设策划、医疗工艺、规划方案、施工图、施工组织验收。

The risk assessment personnel include infection control managers, nurses directly participate in patient care, experts of risk management, engineering design, general contracting and subcontracting of construction, supervisors, epidemiology and others, which forms a group to make assessment at each corresponding stage. The assessment stage includes construction planning, medical process, planning proposal, construction drawing and construction organization acceptance.

4.4.4.5 感染控制规范与监督

4.4.4.5 Infection control standard and supervision

我国《医院消毒卫生标准》(GB 15982—2012)将医院内的环境分为四类：

The environment in hospitals is divided into four categories according to China's Hygienic Standard for Disinfection in Hospitals:

1. 一类环境：采用空气洁净技术的诊疗场所，分洁净手术室和其他洁净场所。

1. Class Ⅰ environment: diagnosis and treatment places adopting air cleaning technology, including clean operating room and other clean places.

2. 二类环境：非洁净手术室、产房、导管室、血液病病区、烧伤病区等保护型隔离病区、重症监护病区、新生儿室。

2. Class Ⅱ environment: protective isolation wards (such as non-clean operating room, de-

livery room, catheter room, blood disease ward, burn ward), ICU and newborn room.

3. 三类环境：母婴同室、消毒供应中心的检查、包装灭菌区和无菌物品存放区血液透析室、其他普通住院病区；

3. Class Ⅲ environment: mother-child room, examination and packaging sterilization area and aseptic supplies storage area in disinfection supply center, hemodialysis room, and other general inpatient ward;

4. 四类环境：为普通门急诊及其检查诊疗室，感染性疾病科门诊和病区。

4. Class Ⅳ environment: general outpatient department and emergency department, examination room and diagnosis and treatment room, outpatient department and inpatient ward of infectious diseases department.

在工程施工过程中，业主应对应用感染控制风险性推荐措施的有效性进行监督。

In the process of project construction, the owner shall supervise the effectiveness of the application of recommended measures for the infection control risk.

在改建或者新建工程中，应保持使用区原有的空气品质要求与其他公用设施要求。

In the reconstruction project or new project, the original air quality requirements and other public facilities requirements of the use area shall be maintained.

4.4.4.6 感染控制措施

4.4.4.6 Infection control measures

《医院隔离技术规范》（WS/T 311—2009) 根据患者获得感染危险的程度，将场所分为四个区域。

Places are divided into four areas based on the degree of infection risk for patients according to the Technical Specification for Hospital Isolation.

1. 低危险区域：包括行政管理、教学、图书馆、生活服务区。

1. Low risk areas: include administration, teaching, library and living service areas.

2. 中等危险区域：包括普通门诊、普通病房。

2. Moderate risk areas: include general outpatient departments and general wards.

3. 高危险区域：包括感染疾病科的门诊和病房。

3. High risk areas: include outpatient departments and wards of infectious diseases department.

4. 极高危险区域：包括手术室、重症监护病房、器官移植病房等。

4. Extremely high risk areas: include operating room, ICU, organ transplantation ward, etc.

4.4.4.7 感控设计

4.4.4.7 Infection control design

空气感染隔离病房和护理防护病房的数量、位置及形式：在急诊部的等候区与进入区，通风和过滤的位置；在外科服务用房，如手术室、消毒供应室、感染性疾病科及其空气感染隔离病房、外科重症监护病房、环境防护病房、各类有创检查室、有害媒介物局部排风系统以及其他特殊区域中，需要进行空气处理与通风。

Number, location and form of air-borne infection isolation wards and nursing protection wards: Air treatment and ventilation are required in the waiting area and entry area of emergency department, ventilation and filtration positions, surgical service rooms such as operating rooms, disinfection supply rooms, infectious diseases departments and their air-borne infection isolation wards, surgical ICU, environmental protection wards, various invasive examination rooms, local exhaust systems for harmful media and other special areas.

4.4.4.8 装饰面与表面

4.4.4.8 Decorative surface and surface

首选表面特征包含以下性能的面材：

Surface materials with the following features and properties are preferred:

1. 维护/维修和清洗简单

1. Maintenance/repair and cleaning are simple

2. 不支持微生物生长

2. Microbial growth is not supported

3. 无孔-平滑

3. Non-porous and smooth

4. 在适用处，吸声/隔声功能

4. Sound absorption/insulation function, where applicable

5. 易燃性—耐燃等级为一级或更高

5. Flammability—fire resistance of grade I or higher

6. 耐用

6. Durability

7. 可持续性

7. Sustainability

8. 挥发性有机化合物含量低（非放散形）

8. Low content of volatile organic compounds (non-diffuse)

9. 烟气毒性低

9. Low smoke toxicity

10. 初投资与寿命周期—性价比

10. Initial investment and life cycle—cost performance ratio

11. 防滑—适当的摩擦系数

11. Skid resistance-appropriate friction coefficient

12. 安装、拆除和替换简单

12. Installation, removal and replacement are simple

13. 无未知的基质/或合成体

13. There is no unknown matrix/synthetic

14. 无缝

14. Seamless

15. 抗回弹、抗撞击

15. Resilience and impact resistance

16. 光反射/眩光控制

16. Light reflection/glare control

17. 色彩、样式和织物的选择

17. Selection of color, style and fabric

18. 无毒/非过敏性

18. Non-toxic/non-allergic

4.5　绿色建筑

4.5　Green building

4.5.1　院址选择

4.5.1　Site selection

4.5.1.1　医院场地选址无洪灾、泥石流及含氡土壤的威胁。

4.5.1.1　The hospital site is free from risks of flood, debris flow and radon-bearing soil.

4.5.1.2　医院场地无危险品的威胁，无电磁辐射、含氡土壤等危害。

4.5.1.2　The hospital site is free from risks of dangerous articles, electromagnetic radiation and radon-bearing soil.

4.5.1.3　建筑场地安全范围内无电磁辐射危害和火、爆、有毒物质等危险源，场地建设满足不破坏当地文物、自然水系、湿地、基本农田、森林和其他保护区。

4.5.1.3　There are no electromagnetic radiation risk, fire, explosion, toxic substances and other dangerous sources within the safety scope of the construction site, and the site construction meets the requirements of not damaging local cultural relics, natural water systems, wetlands, basic farmland, forests and other protected areas.

4.5.1.4　将确保无直射光射入空中，不对行人和周边居住区域产生光污染。

4.5.1.4　No direct light will be injected into the air and no light pollution will occur to pedestrians and surrounding residential areas.

4.5.2　环境绿色

4.5.2　Green environment

4.5.2.1　满足日照间距和卫生间距，强化小区内的通风换气。

4.5.2.1　Meet the sunshine spacing and sanitary spacing and strengthen the ventilation in the community.

4.5.2.2 户外活动场地有乔木、构筑物遮荫措施的面积达到要求。

4.5.2.2 Meet the area requirement of outdoor activity site being covered by trees, structures and other shade measures.

4.5.2.3 总平面布局留有足够的自然风入口和出口，建筑群室外空间气流顺畅。

4.5.2.3 There are enough natural air inlets and outlets in the general layout, and the air flow in the outdoor space of the building complex is smooth.

4.5.2.4 病房前后尽量拉大建筑间距且应满足日照要求，且不宜小于12m，强化小区内的通风换气。

4.5.2.4 The building interval between wards shall be widen as far as possible and shall meet the sunshine requirements, but should not be less than 12m; the ventilation in the community shall be strengthened.

4.5.2.5 可再生能源的应用，如水能、风能、地热能以及地源、水源热泵技术等节能措施。

4.5.2.5 Application of renewable energy sources (including water energy, wind energy, geothermal energy), and ground source, water source heating bump technology and other energy saving measures.

4.5.3 绿色建筑

4.5.3 Green building

4.5.3.1 医院建筑的装配化。

4.5.3.1 Assembly of hospital buildings.

4.5.3.2 节能建筑，建筑屋面采用40mm厚的泡沫玻璃保温板进行保温隔热，外围护结构为全幕墙结构，幕墙玻璃采用Low-E中空玻璃。

4.5.3.2 Energy-saving buildings: 40mm thick foam glass insulation board is adopted on the building roof for heat insulation, the peripheral protection structure is a full curtain wall structure, and the curtain wall glass is Low-E insulated glass.

4.5.3.3 避免“热岛效应”，通过阳台绿化、空中花园、屋顶花园等立体绿化提高绿化率，设置遮阳利于节能。

4.5.3.3 Avoid the “heat island effect” and improve the greening rate through balcony greening, hanging garden, roof garden and other three-dimensional greening, adopt the sun shading measures to be conducive to energy conservation.

4.5.3.4 道路、地面停车场和其他硬质铺地采用透水铺装措施。

4.5.3.4 Permeable paving measures is adopted for roads, ground parking lots and other hard pavements.

4.5.3.5 强调自然通风。

4.5.3.5 Emphasize the natural ventilation.

4.5.3.6 节能建筑和节能门窗。

4.5.3.6 Energy-saving buildings and energy-saving doors and windows.

4.6　结构设计

4.6　Structure design

4.6.1　抗震设防类别及抗震设防标准

4.6.1　Seismic fortification categories and standards

4.6.1.1　抗震设防类别

4.6.1.1　Seismic fortification categories

医疗建筑是救死扶伤的生命线工程，属于防灾救灾建筑，《建筑工程抗震设防分类标准》(GB 50223—2008)规定：

Healthcare buildings are lifeline projects to heal the wounded and rescue the dying, and belong to disaster prevention and relief buildings. (GB 50223—2008) Standard for Classification of Seismic Protection of Building Construction stipulates:

1. 三级医院中承担特别重要医疗任务的门诊、医技、住院用房，抗震设防类别应划分为特殊设防类。

1. For rooms for the outpatient department, medical technology and inpatient in third-class hospitals that undertake particularly important medical tasks, the category of the seismic fortification category shall be classified as the special fortification category.

2. 三级医院的门诊、医技、住院用房，具有外科手术室或急诊科的乡镇卫生院的医疗用房，县级及以上独立采供血机构的建筑，抗震设防类别应划分为重点设防类。

2. For rooms for the outpatient department, medical technology and inpatient in third-class hospitals, medical rooms in township hospitals with surgical operating rooms or emergency departments, and buildings of independent blood transfusion services at county level or above, the category of the seismic fortification category shall be classified as the key fortification category.

3. 工矿企业的医疗建筑，可比照城市的医疗建筑示例确定其抗震设防类别。

3. For healthcare buildings in industrial and mining enterprises, the category of the seismic fortification category shall be determined according to the examples of healthcare buildings in cities.

4.6.1.2　抗震设防标准

4.6.1.2　Seismic fortification standards

医疗建筑的抗震设防标准，依据《建筑工程抗震设防分类标准》(GB 50223—2008)规定：

For seismic fortification standards of healthcare buildings, (GB 50223—2008) Standard for Classification of Seismic Protection of Building Construction stipulates:

1. 重点设防类建筑，应按高于本地区抗震设防烈度一度的要求加强其抗震措施；但抗震设防烈度为9度时应按比9度更高的要求采取抗震措施；地基基础的抗震措施，应符合有关规定。同时，应按本地区抗震设防烈度确定其地震作用。

1. Buildings classified as the key fortification category shall be designed with seismic fortification measures one degree higher than the seismic fortification intensity of the area; However, when the seismic fortification intensity is Ⅸ, seismic fortification measures that are stricter than intensity of Ⅸ shall be taken; the seismic fortification measures of foundation shall comply with relevant regulations. Meanwhile, the seismic action of buildings shall be determined according to the seismic fortification intensity of the area.

2. 特殊设防类建筑，应按高于本地区抗震设防烈度提高一度的要求加强其抗震措施；但抗震设防烈度为9度时应按比9度更高的要求采取抗震措施。同时，应按批准的地震安全性评价结果且高于本地区抗震设防烈度的要求确定其地震作用。

2. Buildings classified as the special fortification category shall be designed with seismic fortification measures one degree higher than the seismic fortification intensity of the area; However, when the seismic fortification intensity is Ⅸ, seismic fortification measures that are stricter than intensity of Ⅸ shall be taken. Meanwhile, the seismic action of buildings shall be determined according to approved seismic safety evaluation result and higher than the seismic fortification intensity of the area.

4.6.2 活荷载取值

4.6.2 Live load value

医疗建筑使用功能复杂、科室众多，各科室也有不同的医疗设备，所以，医疗建筑用房的活荷载取值也不同于普通的民用建筑，国家规范没有明确的规定，参考《全国民用建筑工程设计技术措施》中关于楼面荷载的补充规定及广东省标准《建筑结构荷载规范》(DBJ 15—101—2014)，一般医技及其他专用房间的活荷载取值见下表：

The healthcare buildings have many departments and complicated use functions, and each department also has different medical device. Therefore, the value of live load for healthcare buildings is different from that for ordinary civil buildings. But there are no clear provisions in the national codes. Based on the supplementary provisions on floor load in the National Technical Measures for Design of Civil Construction and the Guangdong provincial standard DBJ15—101—2014 Load Code for the Design of Building Structures, see the following table for values of live load for general medical rooms and other special rooms:

医技房间的活荷载取值：见表4-7。

Values of live load in technical room: see Table 4-7.

表 4-7 医技房间的活荷载取值

Table 4-7 Values of live load in technical room

项次 Item No.	类别 Category	标准值/(kN/m²) Standard value/(kN/m²)	组合值系数 ψ_c Combination value ψ_c	频遇值系数 ψ_f Frequent value ψ_f	准永久值系数 ψ_q Quasi-permanent value ψ_q
1	手术室（设 3000 型、3008 型万能手术床及 3001 型骨科手术台） Operating room (model 3000, model 3008 universal operating bed and model 3001 orthopedic operating table)	3.0	0.7	0.6	0.5
2	CT 室 CT room	8.0	0.8	0.8	0.8
3	核磁共振室 MRI room	8.0	0.8	0.8	0.8
4	消毒室 Sterilization room 1602 型消毒柜 Model 1602 sterilizer 2616 型治疗台及 3704 型椅 Model 2616 treatment table and model 3704 chair	6.0 5.0	0.8 0.8	0.8 0.8	0.8 0.8
5	血库（设 D-101 型冰箱） Blood bank (equipped with model D-101 refrigerator)	5.0	0.8	0.8	0.8
6	产房（设 3009 型产床） Delivery room (equipped with model 3009 obstetric table)	2.5	0.7	0.6	0.5
7	X 光室 X-ray room 30MA 移动式 X 光机 30MA mobile X-ray machine 200MA 诊断 X 光机 200MA X-ray machine 200kV 治疗机 200kV therapy machine X 光存片室 X-ray film storage room	2.5 4.0 3.0 5.0	0.7 0.7 0.7 0.9	0.6 0.6 0.6 0.9	0.5 0.5 0.5 0.8
8	口腔科 Stomatology department 201 型治疗台及电动脚踏升降椅 Model 201 treatment table and electrically-driven lifting chair with pedals 205 型、206 型治疗台及 3704 型椅 Model 205, 206 treatment table and model 3704 chair	3.0 4.0	0.7 0.7	0.6 0.6	0.5 0.5
9	药房、药库 Pharmacy, medicine storehouse	5.0	0.9	0.9	0.8

续表

项次 Item No.	类别 Category	标准值/kN/m² Standard value/kN/m²	组合值系数 ψ_c Combination value ψ_c	频遇值系数 ψ_f Frequent value ψ_f	准永久值系数 ψ_q Quasi-permanent value ψ_q
10	透视、MRI、DR、DSA、普通模拟机房、后装机房 Fluoroscopy, MRI, DR, DSA, general simulation room, afterloading room	6.5~7.5 6.5~7.5	0.8	0.8	0.8
11	药浴 Medicated bath	2.5	0.7	0.6	0.5
12	ICU	4.0	0.7	0.6	0.5
13	测听室 Audiometry room	4.0	0.7	0.6	0.5

注：1. 当医疗设备型号与表中不符时，应按实际情况采用。

Note: 1. When the model of medical devices is inconsistent with the table, the model shall be selected according to the actual situation.

2. 第1、12项顶板宜额外考虑2.0kN/m² 的活荷载；第2、10项顶板宜额外考虑4.0kN/m² 的活荷载。

2. For roofs of items 1 and 12, the additional 2.0kN/m² live load should be considered. For roofs of items 2 and 10, the additional 4.0kN/m² live load should be considered.

4.6.3 钢筋混凝土结构

4.6.3 Reinforced concrete structure

4.6.3.1 结构体系及结构类型

4.6.3.1 Structure system and structure type

现代医疗建筑多采用钢筋混凝土结构体系，常用的结构类型有框架结构、框剪结构和框架-核心筒结构，如多层门诊楼多采用框架结构；高层的医技楼、住院楼等多采用框剪结构和框架-核心筒结构。

The reinforced concrete structure system is commonly used in modern healthcare buildings. The commonly used structure types include frame structure, frame-shear structure and frame-core tube structure. For example, the frame structure is commonly used for multi-storey outpatient buildings. The frame-shear structure and frame-core tube structure are commonly used for high-rise medical technology buildings and inpatient buildings.

4.6.3.2 楼盖结构

4.6.3.2 Floor structure

医疗建筑的柱网比较规则，柱网尺寸以7.5~8.5的居多，两个方向基本差不多。考虑到墙体分隔的灵活，楼盖结构以井字梁居多，楼板厚度一般取100~120mm。由于医疗建筑的设备管线较多，会占用一定的净空高度，所以结构梁高会受到限制，可采用宽扁梁并应注意控制好裂缝及挠度。

The column grid of healthcare buildings is relatively regular, the column grid size mainly ranges from 7.5 to 8.5 and the two directions are basically the same, in addition, considering the wall separation is relatively flexible and the floor structure of healthcare buildings is mostly cross beam, the floor thickness is generally between 100 ~ 120mm. Due to the large number of device pipelines in healthcare buildings, it will occupy a certain clearance height and the height of structural beams will be limited, so wide flat beams can be adopted and cracks and deflections shall be properly controlled.

4.6.3.3 结构缝

4.6.3.3 Structural joints

医疗建筑平面往往比较细长，有时长宽比达到6以上，但因医疗工艺要求结构不能设缝，在设计时应充分考虑温度应力的影响，采用有限元分析计算温差应力作用效应，同时采取以下构造措施：

The plane of healthcare buildings is often relatively slender, sometimes the length-to-width ratio reaches more than 6. However, joints are not allowed in the structure according to the medical process requirements. Therefore, the influence of temperature stress shall be fully considered in the design, the effect of temperature difference stress shall be calculated by using the finite element analysis. At the same time, the following structural measures shall be taken:

1. 每隔30 ~ 40m设置一条后浇带或膨胀加强带；

1. A post-cast strip or an expansion reinforcing band is arranged every 30 ~ 40m;

2. 加强屋面保温措施；楼板、屋面板中配温度应力钢筋；

2. Strengthen roof insulation measures; add temperature stress steel reinforcements in the floor slab and roof slab.

3. 采用低水化热水泥，在混凝土中增加抗裂纤维。

3. Use the low hydration heat cement and add the anti-crack fiber in the concrete.

4.6.3.4 基础沉降控制

4.6.3.4 Foundation settlement control

综合医院的门诊楼一般为多层，住院楼多为高层，且建筑功能上多要求住院楼和门诊楼相连，为避免过大的差异沉降最好设置沉降缝；如果门诊楼和住院楼之间不能设缝，高层和多层之间的沉降协调至关重要；如果有地下室，可以加强基础底板及地下室结构的刚度来协调，同时可以采用先施工主楼，再施工裙楼、设置后浇带等措施来减小差异沉降。

The outpatient buildings in general hospitals are mostly multi-storey building while the inpatient buildings are mostly high-rise buildings. If the inpatient buildings shall be connected with the outpatient buildings and it is best to set settlement joints to avoid excessive differential settlement. However, according to the requirements of medical process and building function, no joints can be set between outpatient buildings and inpatient buildings. If no joints are set, the settlement coordination between high-rise buildings and multi-storey buildings is crucial. If there is a basement, the stiffness of foundation slab and basement structure can be strengthened to coordinate. At the same time, measures such as constructing the main building first, and then constructing the skirt building and setting up the post-cast strip can be adopted to reduce the differential settlement.

4.6.4 钢结构

4.6.4 Steel structure

钢结构具有自重轻、强度高、可实现大跨度、构件可工厂化施工、现场安装等特点，在医疗建筑中也常有采用，常用的结构类型有框架结构、框架—中心支撑结构等，同时一些大跨度的屋盖也采用桁架结构、空间网架结构等。为避免钢结构防腐、防火等后期维护，同时增强结构刚度，也可采用钢与混凝土组合结构的型钢混凝土梁柱。

Steel structure has the characteristics of light weight, high strength, large span, factory construction of components, on-site installation, etc. It is also commonly used in healthcare buildings. Common structure types include frame structure, frame-center supporting structure, etc. At the same time, the truss structure and space grid structure are also used in some roofs with large span. In order to avoid the later maintenance of steel structure such as corrosion prevention and fire prevention and enhance the structural stiffness, section steel-reinforced concrete beams and columns with the steel and concrete composite structure can also be used.

4.6.5 装配式结构

4.6.5 Prefabricated structure

装配式结构具有标准化设计、工厂制作、现场安装、施工噪声小、对环境污染小、可实现绿色施工等特点，是未来建筑的发展趋势，在医疗建筑中常有采用；对一些抢险应急的医疗建筑可实现快速施工，快速投入使用，如为抗击新型冠状病毒建成的雷神山医院、火神山医院。

The fabricated structure has the characteristics of standardized design, factory production, on-site installation, low construction noise and little environmental pollution, and can realize green construction. This structure is the future development trend of building structure. It is often used in healthcare buildings, and can ensure that some healthcare buildings for emergency can be constructed and put into use quickly. For example, the Leishenshan and Huoshenshan hospitals built to fight against COVID-19.

4.6.6 人防结构

4.6.6 Civil air defense structure

现代综合医院一般都设有 2 ~3 层地下室，人防地下室一般设在底部的 1 ~2 层，根据当地民防主管部门关于人防工程建设的规定、措施及建筑规模，可能设置多个防护单元，涉及二等人员掩蔽所、防空专业队员掩蔽部、人防医疗救护工程、电站等。人防等级通常为甲类核五级、核六级。以平战结合为原则，平时用作地下车库。人防医院平时一般都是围合成封闭空间不得使用。

Modern general hospitals are generally equipped with 2 ~ 3 storeys of basements, while civil air defense basements are generally set in the underground 1 ~ 2 storeys. According to the regulations, measures and construction scale of the local department in charge of civil air defense on the construction of civil air defense projects, multiple protection units may be set up, including second-class refuge room, refuge department for professional air defense team members, medical aid works for civil air defense, power stations, etc. Civil air defense grade is usually Class-A nuclear grade Ⅴ and nuclear grade Ⅵ. Based on the principle of combining peacetime with wartime, it is usually used as an underground garage. Civil air defense hospitals are usually enclosed as closed spaces and cannot be used.

4.6.7　民用建筑改造为医院建筑

4.6.7　Civil buildings are reconstructed into hospital buildings

一般民用建筑抗震设防类别为丙类，医院建筑抗震设防类别多为乙类，所以原结构的抗震措施不能满足医院建筑的要求，同时伴随着使用功能的改变、荷载的增加，原结构一般不能满足规范要求。改造前应注意收集原结构的设计资料、竣工图、地勘报告、检测报告等原始资料作为改造设计的依据，对原结构复核计算，并根据计算结果与原结构的差距采取相应的加固措施。

Generally, the seismic fortification category of general civil buildings is Class C, and the seismic fortification category of hospital buildings is mostly Class B. Therefore, the seismic measures of the original structure cannot meet the requirements of hospital buildings. At the same time, with the change of use function and the increase of live load, the original structure cannot generally meet the specification requirements. Before reconstruction, the design data, as-built drawings, geological exploration reports, testing reports and other original data of the original structure shall be collected as the basis for reconstruction design; after rechecking and calculating the original structure, corresponding reinforcement measures shall be adopted based on the difference between the calculation results and the original structure.

4.7　机电设计

4.7　Mechanical and electrical design

4.7.1　给水排水设计

4.7.1　Water supply and drainage design

4.7.1.1　给水排水

4.7.1.1　Water supply and drainage

1. 一般规定

1. General provisions

（1）给水、排水管道不应从洁净室、强电和弱电机房，以及重要医疗设备用房的室内架空通过，必须通过时应采取防漏措施。

(1) Water supply and drainage pipelines shall not pass overhead through the clean rooms, high-voltage and low-voltage machine rooms and rooms for important medical devices. Leak-proof measures shall be taken if unavoidable.

（2）传染病医院室内外给水、热水的配水主管、支管应设置检修阀门，阀门宜设在工作人员的清洁区内。

（2） The main and branch pipes for the indoor and outdoor water supply and hot water in hospitals for infectious diseases shall be equipped with service valves, which should be set in the clean area of the staff.

（3） 在有磁屏蔽的场所应采用紫铜管、塑料管等非磁性管材，排放含有放射性污水的管道应采用机制含铅的铸铁管道，水平横管应敷设在垫层内或专用防辐射吊顶内，立管应安装在壁厚不小于 150mm 的混凝土管道井内。

（3） Non-magnetic pipes such as copper pipes and plastic pipes shall be used for places with magnetic shielding. Pipes discharging radioactive sewage shall be cast iron pipes containing lead, and the horizontal pipes shall be laid in cushion layer or special radiation-proof ceiling and vertical pipes shall be installed in concrete pipe shafts with the wall thickness not less than 150mm.

2. 给水

2. Water supply

（1） 供给洁净手术部用水应有两路进口，由处于连续正压状态下的管道系统供给。

（1） The water supplied to the clean operating department shall be supplied by the pipeline system under continuous positive pressure through two inlets.

（2） 医院生活用水量定额宜符合下表的规定。

（2） The domestic water consumption quota in hospitals should comply with the provisions of the table below.

医院生活用水量定额：见表 4-8。

Domestic water consumption quota in hospitals: see Table 4-8.

表 4-8 医院生活用水量定额

Table 4-8 Domestic water consumption quota in hospitals

建筑物名称 Name of building		单位 Unit	生活用水定额/L Domestic water quota /L		使用时数/h Hours of use/h	最高日小时变化系数/kh Hourly variation coefficient of maximum water consumption day/kh
			最高日 Maximum water consumption day	平均日 Average water consumption day		
医院住院部 Hospital inpatient department	设公用卫生间、盥洗室 Set up public toilets and wash rooms	每床位每日 Daily/bed	100 ~ 200	90 ~ 160	24	2. 5 ~ 2. 0
	设公用卫生间、盥洗室、淋浴室 Set up public toilets, wash rooms and shower room		150 ~ 250	130 ~ 200		
	设单独卫生间 Set up the separate toilet		250 ~ 400	220 ~ 320		
	医务人员 Medical care staff	每人每班 Shift/person	150 ~ 250	130 ~ 200	8	2. 0 ~ 1. 5

续表

建筑物名称 Name of building		单位 Unit	生活用水定额/L Domestic water quota /L		使用时数/h Hours of use/h	最高日小时变化系数/kh Hourly variation coefficient of maximum water consumption day/kh
			最高日 Maximum water consumption day	平均日 Average water consumption day		
门诊部、诊疗所 Outpatient department, diagnosis and treatment department	病人 Patients	每病人每次 Patient/time	10 ~ 15	6 ~ 12	8 ~ 12	1.5 ~ 1.2
	医务人员 Medical care staff	每人每班 Shift/person	80 ~ 100	60 ~ 80	8	2.5 ~ 2.0
疗养院、休养所住房部 Inpatient department of sanatorium, convalescent hospital		每床位每日 Daily/bed	200 ~ 300	180 ~ 240	24	2.0 ~ 1.5

注：医疗建筑用水中已含医疗用水。

Note: The medical water is included in water for healthcare buildings.

（3）下列场所的用水点应采用非手动开关，并应采取防止污水外溅的措施：

(3) Non-manual switches shall be adopted at water consumption points in the following places, and measures shall be taken to prevent sewage from splashing out:

① 公共卫生间的洗手盆、小便斗、大便器；

① Wash basin, urinal and closet pan in public toilet;

② 护士站、治疗室、中心（消毒）供应室、监护病房等房间的洗手盆；

② Wash basins in nurse stations, therapy rooms, central (disinfection) supply rooms, ICU, etc.;

③ 产房、手术刷手池、无菌室、血液病房和烧伤病房等房间的洗手盆；

③ Wash basins in rooms such as delivery room, scrub basin for operation, sterilizing room, hematology ward and burn ward;

④ 诊室、检验科等房间的洗手盆；

④ Wash basins in consulting rooms, laboratory departments and other rooms;

⑤ 有无菌要求或防止院内感染场所的卫生器具。

⑤ Sanitary fixtures that have sterile operation requirements or prevent infection in hospitals.

（4）传染病医院生活给水系统宜采用断流水箱供水方式供给，且供水系统宜采用断流水箱加水泵的给水系统；当改造项目采用断流水箱供水确有困难时，应依据现行国家标准《建筑给水排水设计标准》（GB 50015—2019）的有关规定，分析供水系统产生回流污染的危险等级，并应符合下列规定：

(4) The domestic water supply system of hospitals for infectious diseases should adopt the water supply mode of break tank, and the water supply system should be water supply system of break tank and water pump. When it is really difficult to supply water by using the break tank in the reconstruction project, the risk level of backflow pollution in the water supply system shall be

analyzed according to the relevant provisions of the current national standard GB 50015—2019 Code for Design of Building Water Supply and Drainage, and shall meet the following requirements:

① 当产生回流污染的风险较低，且供水压力满足要求时，供水系统应设置减压型倒流防止器；

① When the risk of backflow pollution is low and the water supply pressure meets the requirements, the water supply system shall be provided with a reduced-pressure type backflow preventer;

② 当风险较高时，仍应采用断流水箱供水方式。

② When the risk is relatively high, the water supply mode of break tank shall still be adopted.

3. 排水

3. Drainage

（1）下列场所应采用独立的排水系统或间接排放，并应符合下列要求：

（1）The following places shall adopt independent drainage system or indirect drainage, and shall meet the following requirements:

① 传染病门急诊和病房的污水应单独收集处理；

① Sewage from the outpatient and emergency departments and wards for infectious diseases shall be collected and treated separately;

② 呼吸道发热门（急）诊内应设独立卫生间，排水管及通气管不宜与其他区域的管道连接，排水管应单独排出；

② The independent toilet shall be set up for respiratory (fever) outpatient and emergency departments. Drainage pipes and ventilation pipe should not be connected with pipes in other areas. Drainage pipes shall be discharged separately;

③ 放射性废水应单独收集处理；

③ Radioactive waste water shall be collected and treated separately;

④ 牙科废水宜单独收集处理；

④ Wastewater from the dental department should be collected and treated separately;

⑤ 锅炉排污水、中心（消毒）供应室的消毒凝结水等，应单独收集并设置降温池或降温井；

⑤ Boiler sewage and sterilized condensate from central (disinfection) supply room etc., shall be collected separately and cooling tank or cooling well shall be provided;

⑥ 分析化验采用的有腐蚀性的化学试剂宜单独收集，并应综合处理后再排入院区污水管道或回收利用；

⑥ Corrosive chemical reagents used for analysis and testing should be collected separately and shall be comprehensively treated before being discharged to the sewage pipeline in hospitals or recycled;

⑦ 其他医疗设备或设施的排水管道应采用间接排水；

⑦ The drainage pipes of other medical devices or facilities shall adopt indirect drainage;

⑧ 太平间和解剖室应在室内采用独立的排水系统，且主通气管应伸到屋顶无人处。

⑧ The morgue and autopsy room shall be equipped with independent drainage system, and the main ventilation pipe shall be extended to the roof where no one is around.

（2）中心（消毒）供应室、中药加工室、口腔科等场所的排水管道的管径，应大于计算管径 1 ~ 2 级，且不得小于 100mm，支管管径不得小于 75mm。

（2）The diameter of drainage pipes of Central (disinfection) supply room, traditional Chinese medicine processing room, stomatology department and other places shall be 1 ~ 2 level greater than the calculated pipe diameter, and shall not be less than 100mm; the diameter of branch pipes shall not be less than 75 mm.

（3）存水弯的水封高度不得小于 50mm，且不得大于 100mm。

（3）The height of the water seal of the trap shall not be less than 50mm and shall not be greater than 100mm.

（4）传染病医院排水应符合下列要求：

（4）The drainage of hospitals for infectious diseases shall meet the following requirements:

① 排水立管的最大设计排水能力取值不应大于现行国家标准《建筑给水排水设计标准》(GB 50015—2019) 规定值的 0.7 倍；

① The maximum design drainage capacity of the drainage riser shall not be 0.7 times greater than the value specified in the current national standard GB 50015—2019 Code for Design of Building Water Supply and Drainage;

② 地漏应采用水封补水措施，并宜采用洗手盆排水给地漏水封补水的措施；

② The water-seal water supplement measure shall be adopted for the floor drain, and the wash drain should be adopted to supplement water to the floor drain water seal;

③ 应急医疗设施室外污水排水系统应采用无检查井的管道进行连接，通气管的间距不应大于 50m，清扫口的间距应符合国家现行标准《室外排水设计规范》（GB 50014—2006) 的有关规定；

③ The outdoor sewage drainage system of emergency medical facilities shall be connected by pipes without inspection wells. The spacing of ventilation pipes shall not be more than 50m and the spacing of cleanout shall conform to the relevant provisions of the current national standard Code for Design of Outdoor Wastewater Engineering(GB 50014—2006);

④ 隔离区排水系统的通气管出口应设置高效过滤器过滤或采取消毒处理；

④ The outlet of ventilation pipes of the drainage system in the isolation area shall be provided with a HEPA filter for filtration or provided with disinfection treatment measures;

⑤ 排水管道应进行闭水试验，且应采取防止排水管道内的污水外渗和泄漏的措施；

⑤ The closed water test shall be carried out for drainage pipes, and measures shall be taken to prevent sewage leakage in drainage pipes;

⑥ 当改造项目污水处理无法满足现行国家标准《传染病医院建筑设计规范》（GB 50849—2014) 二级生化处理的有关规定时，污水处理应采用强化消毒处理工艺。

⑥ When the sewage treatment of the reconstruction project cannot meet the relevant provisions of secondary biochemical treatment in the current national standard GB 50849—2014 Architectural Design Code for Infectious Diseases Hospital, the enhanced disinfection treatment process shall be adopted for sewage treatment.

4. 热水

4. Hot water

(1) 医院生活热水用水量定额宜符合表4-9的规定，水温按60℃设计。

(1) The domestic hot water consumption quota in hospitals should comply with the provisions of the table 4-9 and the water temperature shall be designed at 60℃.

医院生活热水（60℃）用水量定额表：见表4-9。

Domestic hot water consumption quota in hospitals (60℃): see Table 4-9.

表4-9 医院生活热水（60℃）用水量定额表

Table 4-9 Domestic hot water consumption quota in hospitals (60℃)

建筑物名称 Name of building		单位 Unit	生活用水定额/L Domestic water quota/L		使用时数/h Hours of use/h
			最高日 Maximum water consumption day	平均日 Average water consumption day	
医院住院部 Hospital inpatient department	设公用盥洗室 Set up public wash rooms	每床位每日 Daily/bed	60～100	40～17	24
	设公用盥洗室、淋浴室 Set up public wash rooms and shower room		70～130	65～90	
	设单独卫生间 Set up the separate toilet		110～200	110～140	
	医务人员 Medical care staff	每人每班 Shift/person	70～130	65～90	8
门诊部、诊疗所 Outpatient department, diagnosis and treatment department	病人 Patients	每病人每次 Patient/time	7～13	3～5	8～12
	医务人员 Medical care staff	每人每班 Shift/person	40～60	30～50	8
疗养院、休养所住房部 Inpatient department of sanatorium, convalescent hospital		每床位每日 Daily/bed	100～160	90～110	24

注：医疗用水定额及热水温度应根据工艺确定。

Note: Medical water quota and hot water temperature shall be determined according to the process.

(2) 医院建筑应采用无冷温水滞水区的水加热设备。

(2) The water heating equipment without cold and warm water stagnant zone shall be adopted for hospital buildings.

(3) 医院热水系统的热水制备设备不应少于 2 台。当一台检修时，其余设备应能供应 60% 以上的设计用水量。

(3) The water heating equipment of the hot water system of hospitals shall not be less than 2. When one is maintenance status, another equipment shall be able to supply more than 60% of the designed water consumption.

(4) 生活热水系统的水加热器出水温度不应低于 60℃，系统回水温度不应低于 50℃。

(4) The outlet water temperature of water heater of the domestic hot water system shall not be lower than 60℃, and the return water temperature of the system shall not be lower than 50℃.

(5) 当淋浴或浴缸用水点采用冷、热混合水温控装置时，用水点出水温度在任何时间均不应大于 49℃。

(5) When the temperature control device for water mixed with cold and hot water are used for water consumption points of shower or bathtub, the outlet water temperature of water consumption points shall not be greater than 49℃ at any time.

(6) 热水系统任何用水点在打开用水开关后宜在 5～10s 内出热水。

(6) Any water consumption points in the hot water system should discharge hot water within 5s to 10s after turning on the switch.

(7) 手术部集中刷手池的水龙头应采用恒温供水，且末端温度可调节，供水温度宜为 30～35℃。

(7) The faucet of the centralized scrub basin in the operating department shall supply water at constant temperature, and the terminal temperature is adjustable and the water supply temperature should be between 30℃ and 35℃.

(8) 洗婴池的供水应防止烫伤或冻伤且为恒温，末端温度可调节，供水温度宜为 35～40℃。

(8) The water supplied to the infant bath basin shall be kept at a constant temperature to prevent scalding or frostbite. The terminal temperature can be adjusted and the water supply temperature should be between 35℃ and 40℃.

(9) 传染病医院热水进行再循环时，对于在严重传染区下游的不带水阀门的结构，在使循环水回到蓄水箱后，应在箱内于 80℃加热 10min 以上进行杀菌，然后以供给时所需的温度进行循环。

(9) When the hot water in hospitals for infectious diseases is circulated, water flowing through the structure without water valves at the downstream of severe infection area shall be heated at 80℃ for more than 10min to sterilize after returning the circulating water to the water storage tank, and then shall be circulated at the temperature required for supply.

5. 饮用水

5. Drinking water

(1) 饮用水可采用下列方式供应：

(1) Drinking water can be supplied in the following ways:

① 当采用管道直饮水系统时，供水点宜根据需要分散设置，传染病医院每个护理单元应单独设置饮用水供水点；

① When the direct drinking water pipe system is adopted, the water supply points should be distributed in a decentralized manner as required, and each nursing unit of hospitals for infectious diseases shall be provided with a separate drinking water supply point;

② 当采用蒸汽间接加热时，蒸汽开水炉宜集中设置，饮用水供应至护理单元和科室；

② When steam is used for indirect heating, steam boilers should be centrally set. Drinking water is supplied to nursing units and departments;

③ 当采用电开水器时，可在楼层或护理单元、科室设置电开水器，传染病医院每护理单元应单独设置电开水器；

③ Electric boilers can be set on floors or nursing units and departments. Each nursing unit in hospitals for infectious diseases shall be separately equipped with electric boilers;

④ 当采用桶装水饮水机时，供水点宜根据需要分散设置。

④ Barreled water dispensers should be set in a decentralized manner as required.

（2）当采用蒸汽开水炉和电开水器时，自来水进开水器前应设置过滤器和止回阀。

（2）When steam boilers and electric boilers are used, the filter and check valve shall be set at the upstream of the point where the tap water flows into the boiler.

（3）管道直饮水应设水质分析室，直饮水水质分析每班不应少于 2 次。

（3）The water quality analysis room shall be set up for direct drinking water in pipelines, and water quality analysis for direct drinking water shall not be less than 2 times per shift.

（4）传染病医院生活饮用水管道宜避开污染区，当条件限制不能避开时，应采取防护措施。

（4）Domestic drinking water pipelines in hospitals for infectious diseases should avoid contaminated areas. If unavoidable, protective measures shall be taken.

4.7.1.2 节能与绿色建筑

4.7.1.2 Energy saving and green buildings

1. 一般规定

1. General provisions

（1）在方案设计阶段应制定水资源规划方案，统筹、综合利用各种水资源；水资源规划方案应包括市政再生水、雨水等非传统水源综合利用的内容。

（1）At the stage of SD, water resources planning scheme shall be formulated to make overall and comprehensive use of various water resources. The water resource planning scheme shall include the comprehensive utilization of unconventional water sources such as municipal reclaimed water and rainwater.

（2）生活热水系统的能源宜采用废热和太阳能等；当采用太阳能或热泵时，宜采用可自动控制的其他辅助能源。

（2）The energy of domestic hot water system should be waste heat, solar energy and others. When solar energy or heat pump is used, other auxiliary energy that can be automatically controlled should be used.

2. 给水与热水

2. Water supply and hot water

(1) 给水系统应充分利用市政供水压力；高层建筑生活给水系统合理分区。

(1) The water supply system shall make full use of municipal water supply pressure, and the domestic water supply system of high-rise buildings should be distributed by reasonable zone.

(2) 给水系统采取减压限流的节水措施，建筑用水点处供水压力不大于 0. 2MPa。

(2) The water supply system adopts the water-saving measures of reducing pressure and limiting flow, and the water supply pressure at the water consumption points of buildings is not more than 0. 2MPa.

(3) 集中热水供应系统应设置完善的热水循环系统，且用水点出水温度达到 45℃的放水时间不应大于 10s。

(3) The centralized hot water supply system shall be equipped with a complete hot water circulation system, and the drainage time shall not be more than 10s when the outlet water temperature at the water consumption points reaches 45℃.

3. 非传统水源利用

3. Utilization of unconventional water resources

(1) 医疗污水、放射性废水、生物污染废水、重金属及其他有毒有害物质超标的排水严禁作为中水原水。

(1) Medical sewage, radioactive waste water, bio-contamination waste water, heavy metals and water with other toxic and harmful substances that exceed the standard shall not be used as raw water for reclaimed water.

(2) 传染病医院的雨水收集不得采用雨水收集回用系统。

(2) Rainwater collection and reuse system shall not be used for rainwater in the hospital for infectious diseases.

4. 节水措施

4. Water-saving measures

(1) 给水系统中使用的管材、管件，必须符合现行国家标准的要求；需选用高性能的阀门。

(1) The pipes and pipe fittings that are used in the water supply system must meet the requirements of the current national standards; the valves shall be of high performance.

(2) 卫生器具、水嘴、淋浴器等应符合现行行业标准《节水型生活用水器具》(CJ/T 164—2014) 的要求。

(2) Sanitary appliances, faucets, showers, and others shall meet the requirements of the current industry standard, Domestic Water Saving Devices (CJ/T 164—2014).

4. 7. 1. 3　消防、环保、卫生、医疗救护工程（人防）

4. 7. 1. 3　Fire control, environmental protection, hygiene and medical aid works (civil air defense)

1. 消防

1. Fire control

（1）手术部的消火栓宜设置在清洁区域的楼梯口附近或走廊，必须设置在洁净区域时，应满足洁净区域的卫生要求。

（1）The fire hydrant of the operating department should be set near the stairway or corridor in the clean area, when it must be set in a clean area, it shall meet the hygiene requirements of the clean area.

（2）护士站宜设置消防软管卷盘。

（2）The nurse station should be equipped with fire hose reel.

（3）病房应采用快速反应喷头。

（3）The ward shall be equipped with quick-response sprinklers.

（4）手术部洁净和清洁走廊宜采用隐蔽型喷头，洁净手术室内不宜布置洒水喷头。

（4）Concealed sprinklers should be used for clean corridors in operating department, sprinklers should not be arranged in clean operating room.

（5）医院的贵重设备用房、病案室和信息中心（网络）机房，应设置气体灭火装置。

（5）The valuable equipment room, medical record room and information center (network) room of the hospital shall be equipped with gas fire extinguishers.

（6）血液病房、手术室和有创检查的设备机房，不应设置自动灭火系统。

（6）Automatic fire extinguishing system shall not be set in hematology ward, operating room and equipment room for invasive examination.

2. 环保

2. Environmental protection

（1）医院污水必须进行消毒处理。

（1）Hospital sewage must be disinfected.

（2）传染病医院内含有病原体的固体废弃物应进行焚烧处理，手术中产生的医疗污物应就地或集中消毒处理。

（2）Solid waste containing pathogens in the hospitals for infectious diseases shall be incinerated, the medical waste produced in the operation shall be disinfected on the spot or intensively.

（3）医院污水处理站排臭系统宜进行除臭、除味处理，处理后应达到现行国家标准《医疗机构水污染物排放标准》（GB 18466—2016）中规定的处理站周边大气污染物最高允许浓度。

（3）Deodorization and odor removal should be carried out for the odor discharge system of the hospital sewage treatment station. After treatment, the maximum allowable concentration of air pollutants around the treatment station which is specified in the current national standard, Discharge Standard of Water Pollutants for Medical Institutions (GB 18466—2016), shall be reached.

3. 卫生

3. Hygiene

军团菌繁殖和生长的适宜温度是 20 ~ 50℃，故水加热设备要求采用无冷温水滞水区的水加热设备；生活热水系统要求水加热器出水温度不应低于 60℃，这样基本可防治军团菌的滋生。

The suitable temperature for citronella's propagation and growth is 20 ~ 50℃, so the water heating equipment is required to be those without cold and warm water stagnant zone. The domestic hot water system requires that the outlet water temperature of water heater shall not be lower than 60℃, which can basically prevent legionella growth.

4. 医疗救护工程（人防）

4. Medical aid works (civil air defense)

（1）战时人员用水量标准应按下表采用。

（1）The standard of water consumption for the personnel in wartime shall be adopted according to the following table.

战时人员用水量标准表：见表 4-10。

Standard of water consumption for the personnel in wartime: see Table 4-10.

表 4-10　战时人员用水量标准表

Table 4-10　Standard of water consumption for the personnel in wartime

<table>
<tr><th colspan="3" rowspan="3">工程类别
Category of works</th><th colspan="2" rowspan="2">用水量/L/(人·d)
Water consumption /L/(person· d)</th><th colspan="4">贮水时间
Water storage time</th></tr>
<tr><th colspan="2">有可靠内水源
Reliable internal water sources</th><th colspan="2">无可靠内水源
No reliable internal water sources</th></tr>
<tr><th>饮用水
Drinking water</th><th>生活用水
Domestic water</th><th>饮用水/d
Drinking water/d</th><th>生活用水/h
Domestic water/h</th><th>饮用水/d
Drinking water/d</th><th>生活用水/d
Domestic water/d</th></tr>
<tr><td rowspan="4">医疗救护工程
Medical aid works</td><td rowspan="2">中心医院
Central hospital
急救医院
Emergency hospital</td><td>伤病员
The sick and wounded</td><td>4 ~ 5</td><td>60 ~ 80</td><td rowspan="4">2 ~ 3</td><td rowspan="4">10 ~ 12</td><td rowspan="4">15</td><td rowspan="4">有防护外水源：3 ~ 7
Protected external water sources: 3 ~ 7
无防护外水源：7 ~ 14
No protected external water sources: 7 ~ 14</td></tr>
<tr><td>工作人员
Working personnel</td><td>3 ~ 6</td><td>30 ~ 40</td></tr>
<tr><td rowspan="2">救护站
First-aid station</td><td>伤病员
The sick and wounded</td><td>4 ~ 5</td><td>30 ~ 50</td></tr>
<tr><td>工作人员
Working personnel</td><td>3 ~ 6</td><td>25 ~ 35</td></tr>
</table>

注：设置水冲厕所的医疗救护工程，水冲厕所的用水量已计入生活用水量中。

Note: For the medical aid works with water flushing toilet, the water consumption of water flushing toilet has been included in the domestic water consumption.

（2）污水处理设施宜设在防护区外。

（2）Sewage treatment facilities should be constructed outside the protection area.

（3）人员洗消方式应为淋浴洗消，洗消人员百分数应为5% ~10%。

（3）The personnel shall take a shower for disinfect, and the percentage of personnel who take a washer for disinfection shall be 5% ~10%.

4.7.1.4 污水处理

4.7.1.4 Sewage disposal

1. 一般规定

1. General provisions

（1）医疗污水排放应符合现行国家标准《医疗机构水污染物排放标准》（GB 18466—2016)的有关规定。

（1）The discharge of medical sewage shall comply with the relevant provisions of the current national standard, Discharge Standard of Water Pollutants for Medical Institutions (GB 18466—2016).

（2）放射性污水的排放，应符合现行国家标准《电离辐射防护与辐射源安全基本标准》(GB 18871—2002)的有关规定。

（2）The discharge of radioactive sewage shall comply with the relevant provisions of the current national standard, Basic Standard for Protection Against Ionizing Radiation and for the Safety of Radiation Sources (GB 18871—2002).

（3）医院污水处理设施中应设置事故处理设备。

（3）The hospital sewage treatment facilities shall be equipped with accident treatment equipment.

（4）污水在化粪池中的停留时间不宜少于36h，污泥清掏周期为1年。

（4）The retention time of sewage in septic tank should not be less than 36h, sludge clearing period is one year.

（5）提升式医院污水处理设施应设调节池，其有效容积宜为5 ~6h污水平均小时流量。

（5）The lifting type hospital sewage treatment facility shall be equipped with a regulating tank with an effective volume of 5 ~6h average hourly sewage flow.

（6）医疗污水不得作为中水水源。

（6）Medical sewage shall not be used as the source of reclaimed water.

2. 消毒处理

2. Disinfection treatment

（1）宜采用液氯、商品次氯酸钠、现场制备次氯酸钠、二氧化氯、三氯异氰尿酸、漂粉精粉、漂粉精片作为消毒剂。

（1）Liquid chlorine, commercial sodium hypochlorite, sodium hypochlorite prepared in situ, chlorine dioxide, trichloroisocyanuric acid, bleaching powder concentrate, calcium hypochlorite tablets should be used as disinfectant.

（2）当用液氯消毒时，必须采用真空加氯机并设置必要的安全装置；加氯机宜设置两套，其中一套备用。

（2） When liquid chlorine is used for disinfection, a vacuum chlorinator must be used with necessary safety devices provided. Two sets of chlorinators should be provided, one of which is for standby.

（3） 严禁将加氯设备设置在各类建筑物的地下室。

（3） It is strictly prohibited to install the chlorinators in the basement of various buildings.

（4） 液氯容器宜采用容积为 40L 的氯瓶，氯瓶一次使用周期不得大于 3 个月。

（4） The liquid chlorine container should be a chlorine bottle with a volume of 40L, and the one-time service cycle of the chlorine bottle shall not be more than 3 months.

（5） 加氯系统的管道材料应按下列规定选择：

（5） The pipe materials of the chlorination system shall be selected according to the following provisions:

① 输送氯气的管道应使用紫铜管、无缝钢管，严禁使用聚氯乙烯管；

① Copper pipes and seamless steel pipes shall be used for the chlorine pipeline, but PVC pipes are strictly prohibited;

② 输送氯溶液的管道宜采用硬质聚氯乙烯管、工程塑料管、聚四氟乙烯管，严禁使用铜、铁等不耐氯溶液腐蚀的金属管。

② Rigid PVC pipes, engineering plastic pipes and polytetrafluoroethylene pipes shall be used as the liquefied chlorine pipeline, however, copper pipes, iron pipes and other metal pipes that are not resistant to chloride solution corrosion are strictly prohibited.

（6） 加氯系统的管道宜明装，埋地管道应设在管沟内，管道应有良好的支撑和足够的坡度。

（6） The pipeline of the chlorination system should be surface-mounted, and the buried pipeline shall be set in the trench, with good support and enough slope.

3. 污泥处理

3. Sludge treatment

（1） 医院化粪池和处理构筑物内的污泥应由具有相应资质的单位或部门定期掏取；所有污泥必须经过有效的消毒处理，在符合有关标准的规定后，方可消纳。

（1） The sludge in the septic tanks and structures of hospitals shall be cleared and removed regularly by the unit or department with corresponding qualification. All sludge must be effectively disinfected and can be consumed only after it meets the requirements of relevant standards.

（2） 污泥的处理和处置方法，应根据场地条件、投资与运行费用、操作管理和综合利用的可能性等因素综合考虑。

（2） The treatment and disposal methods of sludge shall be comprehensively considered according to the site conditions, investment and operation costs, operation management and the possibility of comprehensive utilization.

（3） 当污泥采用氯化法消毒时，加氯量应通过试验确定；当无相关资料时，可按单位体积污泥中有效氯投加量为 2.5g/L 设计；消毒时应充分搅拌混合均匀，并保证有不少于 2h 的接触时间。

（3） When the sludge is disinfected by chlorination pulping process, the amount of chlorination shall be determined by testing. When there is no relevant data, the effective chlorine dosage per unit volume of sludge can be designed as 2. 5g/L. During disinfection, the mixture shall be fully stirred and evenly mixed, and the contact time of sludge and chlorine shall be no less than 2h.

（4） 当采用高温堆肥法处理污泥时，应符合下列要求：

（4） When high-temperature composting is used to treat sludge, the following requirements shall be met:

① 合理配料，就地取材；

① The ingredients shall be reasonable and raw materials be obtained locally;

② 堆温保持在60℃以上且不应少于1d;

② The composting temperature shall be kept above 60℃ and not be less than 1 day;

③ 保证堆肥的各部分都能达到有效消毒；

③ Ensure that all parts of compost can be effectively disinfected;

④ 采取防止污染人群的措施。

④ Take measures to prevent pollution of people.

（5） 当采用石灰消毒污泥时，污泥的 pH 值不得小于 12，并应存放 7d 以上。石灰的设计投加量可采用 15g/L[以 $Ca(OH)_2$ 计]。

（5） When lime is used to disinfect sludge, the pH value of sludge shall not be less than 12 and be stored for more than 7 days. The design dosage of lime can be 15g/L [calculated by $Ca(OH)_2$].

（6） 经消毒处理后的污泥不得随意弃置，也不得用作根块作物的肥料。

（6） The sludge after disinfection shall not be discarded at will or used as fertilizer for root crops.

4. 放射性污水处理

4. Treatment of radioactive sewage

（1） 医院总排出口污水中的放射性物质含量高于现行国家标准《电离辐射防护与辐射源安全基本标准》(GB 18871—2002)规定的浓度限值时，应进行处理。

（1） When the content of radioactive substances in the sewage discharged from hospitals is higher than the concentration limit specified in the current national standard, Basic Standard for Protection Against Ionizing Radiation and for the Safety of Radiation Sources (GB 18871—2002), the sewage shall be treated.

（2） 当医院的放射性污水排入江河时，应符合下列要求：

（2） When the radioactive sewage from hospitals is discharged into rivers, the following requirements shall be met:

① 经处理后的污水不得排入生活饮用水集中取水点上游 1000m 和下游 100m 范围的水体内，且取水区的放射性物质含量必须低于露天水源中的浓度限值；

① The treated sewage shall not be discharged into the water body within 1000 meters in the upstream and 100 meters in the downstream of the centralized drinking water intake points, and the content of radioactive substance in the water intake area must be lower than the concentration limit in the open water source;

② 排放口应避开经济鱼类产卵区和水生生物养殖场；

② The discharge outlet shall be away from the spawning area of economic fish and the breeding farm of aquatic biology;

③ 在设计和控制排放量时，应取 10 倍的安全系数。

③ In the design and control of discharge, the safety factor of 10 times shall be taken.

（3）当污水中含有几种不同的放射性物质时，污水在衰变池中的停留时间应取其中最大值。

（3）When there are several different radioactive substances in the sewage, the retention time of sewage in the decay tank shall be the maximum.

（4）对同时具有病原体和放射性核素的病人，其排泄物应单独收集，经杀菌消毒再经衰变后排放。

（4）For patients with pathogens and radionuclides at the same time, their excreta shall be collected separately, and then discharged after sterilization, disinfection and decay.

4.7.2 强电弱电设计

4.7.2 Design of strong and weak current

4.7.2.1 强电供配电

4.7.2.1 Supply and distribution of strong current

1. 供配电系统

1. Power supply and distribution system

（1）医疗建筑的供配电系统应根据医疗场所的特点对供电连续性和安全性的要求以及用电容量、当地的供电条件和发展规划等进行设计，并应安全可靠。

（1）The power supply and distribution system of the healthcare buildings shall not only be designed according to the characteristics of the medical locations, the requirements for continuity and safety of power supply as well as the capacity of power consumption, local power supply conditions and development planning but also be safe and reliable.

（2）医疗场所供配电系统应具有对特别重要负荷从正常工作主电源自动转换到应急电源的功能。

（2）The power supply and distribution system in the medical locations shall have the function of automatically converting the particularly important load from the main power supply in normal operation to the electric source for safety services.

2. 医疗建筑用电负荷应根据负荷供电可靠性要求及中断供电对生命安全、人身安全、经济损失等所造成的影响程度进行分级，并应符合下表的规定。

2. The electrical load of healthcare buildings shall not only be graded according to the reliability requirements of load power supply and the impact of the suspension of power supply on life safety, personal safety, economic loss, etc. , but also conform to the regulations in the table below.

医疗建筑用电负荷分级：见表4-11。

Electric load classification of healthcare buildings: see Table 4-11.

表4-11 医疗建筑用电负荷分级

Table 4-11 Electric load classification of healthcare buildings

医疗建筑名称 Name of healthcare buildings	用电负荷名称 Name of electric load	负荷等级 Load grade
三级、二级医院 Third-class, second-class hospitals	急诊抢救室、血液病房的净化室、产房、烧伤病房、重症监护室、早产儿室、血液透析室、手术室、术前准备室、术后复苏室、麻醉室、心血管造影检查室等场所中涉及患者生命安全的设备及其照明用电；大型生化仪器、重症呼吸道感染区的通风系统 The electricity for the equipment, which is related to patients' life safety in emergency room, blood purification room in hematology ward, delivery room, burn ward, intensive care unit (ICU), premature babies, hemodialysis room, operating room, preoperative preparation room, postoperative recovery room, anesthesia room, cardiovascular contrast examination room and others, and its lighting, large biochemical instruments, and the ventilation system in severe respiratory infection area	一级负荷中特别重要的负荷 Especially important load in the first-class load
	急诊抢救室、血液病房的净化室、产房、烧伤病房、重症监护室、早产儿室、血液透析室、手术室、术前准备室、术后复苏室、麻醉室、心血管造影检查室等场所中的除一级负荷中特别重要负荷的其他用电设备； Other electrical equipment in emergency room, blood purification room in hematology ward, delivery room, burn ward, intensive care unit (ICU), premature babies, hemodialysis room, operating room, preoperative preparation room, postoperative recovery room, anesthesia room, cardiovascular contrast examination room and others except for the especially important load in the first-class load; 下列场所的诊疗设备及照明用电：急诊诊室、急诊观察室及处置室、婴儿室、内镜检查室、影像科、放射治疗室、核医学室等； The electricity for diagnosis and treatment equipment as well as their lighting in the following locations: emergency consulting room, EM observation room and debridement room, infant room, endoscopy room, imaging department, radiotherapy room, nuclear medicine clinic, etc.; 高压氧舱、血库、培养箱、恒温箱； Hyperbaric oxygen chamber, blood bank, incubator and thermostat; 病理科的取材室、制片室、镜检室的用电设备； The electrical equipment in the material room, film room and endoscopy room of the pathology department; 计算机网络系统用电； The electricity for computer network system; 门诊部、医技部及住院部30%的走道照明； 30% corridor lighting in outpatient department, medical technology department and inpatient department; 配电室照明用电 The electricity for lighting power of power distribution room	一级 First-class

续表

医疗建筑名称 Name of healthcare buildings	用电负荷名称 Name of electric load	负荷等级 Load grade
三级、二级医院 Third-class, second-class hospitals	电子显微镜、影像科诊断用电设备； Electric equipment for electron microscope and the diagnosis of imaging department; 肢体伤残康复病房照明用电； Electricity for the lighting in the rehabilitation ward for physical disabilities; 中心（消毒）供应室、空气净化机组； Central (disinfection) supply room, air purification unit; 贵重药品冷库、太平柜； The refrigerators for precious drugs and morgue refrigeration cabinet; 客梯、生活水泵、采暖锅炉及换热站等用电负荷 Electric load of passenger elevator, domestic water pump, heating boiler and heat exchange station	二级 Second and
一级医院 First-class hospitals	急诊室 Emergency room	
三级、二级、一级医院 First-class, second-class and third-class hospitals	一、二级负荷以外的其他负荷 Load other than first-class and second-class load	三级 Third-class

注：①其他医疗机构用电负荷可按本表进行分级；

Note: ①Electric load of other medical institutions can be classified according to this table;

② 本表未包含的消防负荷分级按国家现行有关标准执行。

② The fire load, which is not included in this table, shall be classified in accordance with the current relevant national standards.

3. 医用气体供应系统中的真空泵、压缩机、制氧机等设备负荷等级及其控制与报警系统负荷等级应为一级。

3. The load grade of vacuum pump, compressor, oxygen generator and other equipment in medical gas supply system and its control and alarm system shall be Class Ⅰ.

4. 医学实验用动物屏蔽环境的照明及其净化空调系统负荷等级不应低于二级。

4. The load grade of lighting and clean air conditioning system for the shielding environment of medical laboratory animals shall not be lower than Class Ⅱ.

5. 配电系统设计应符合下列规定：

5. The design of power distribution system shall meet the following requirements:

（1）当由双重电源供电，其中一路电源或变压器中断供电时，另一路电源或变压器应能承担全部一级负荷中的特别重要的负荷、一级负荷和二级负荷；

(1) When the power is supplied by dual power supply, when one power supply or transformer is suspended, the other power supply or transformer shall be able to bear all the particularly important loads, primary load and secondary load in the primary;

（2）配变电所选址应深入或接近负荷中心，并不应与诊疗设备用房、电子信息系统机房、病房等相贴邻；

（2） The location of power distribution station and powder substation shall be in or close to the load center, but not adjacent to the room for diagnosis and treatment equipment, electronic information system room, wards, etc. ;

（3） 除消防负荷外的一级负荷的两路电源宜在末端配电箱处或用电设备处自动切换；

（3） The two-way power supply of the first-class load other than the fire load shall be switched automatically at the end distribution box or electrical equipment;

（4） 除消防负荷外的二级负荷的两路电源可在配变电所或总配电箱处切换。

（4） The two-way power supply of the second-class load other than the fire load can be switched at the power distribution station and substation or the main distribution box.

6. 正常运行情况下，用电设备受电端的电压允许偏差值应符合下列规定：

6. Under normal operation, the allowable deviation value of voltage at the receiving end of electric equipment shall meet the following requirements:

（1） 室内一般照明的电压允许偏差值应为 ±5%，视觉要求较高的场所的电压允许偏差值应为 -2.5% ~ +5%；

（1） The allowable voltage deviation of general indoor lighting shall be ± 5%, and that of places with higher visual requirements shall be -2.5% ~ +5%;

（2） 应满足诊疗设备对电压允许偏差值的要求。

（2） The requirements of diagnosis and treatment equipment for the allowable deviation value of voltage shall be met.

7. 当配电变压器不能达到设备对供电电压质量要求时，可配置有载调压变压器。

7. When the distribution transformer fails to meet the quality requirements of the supply voltage for equipment, the on-load-tap-changing transformer can be configured.

8. 地震烈度为 6 度及以上地区电气设备应采取抗震措施。

8. Aseismic measures shall be taken for electrical equipment in areas with seismic intensity of 6 or above.

9. 三级医院应设置应急柴油发电机组。二级医院宜设置应急柴油发电机组。

9. The third-class hospitals ought to be equipped with emergency diesel generator sets. The second-class hospitals is advisalde to be equipped with emergency diesel generator sets.

10. 正常市电供电电源停电或故障时，应急电源的供电容量应保证一级负荷中特别重要的负荷用电，并宜保证一级负荷的用电。

10. In case of power failure or failure of normal mains power source, the capacity of emergency power supply shall ensure the power consumption of the particularly important load in the first-class load and that of the first-class load.

11. 应急电源应根据其容量、供电时间等因素选择相应的形式，并可采用独立于正常电源的柴油发电机组、蓄电池、干电池等。

11. The emergency power supply shall be selected according to its capacity, power supply

time and other factors, and the diesel generator set, battery, dry battery and others independent of the normal power supply can be used.

12. 柴油发电机房不宜与诊疗设备用房、住院部、电子信息系统机房等贴邻。当受条件限制而贴邻时，应采取机组消声及机房隔声等综合治理措施，治理后的环境噪声不应超过城市区域环境噪声Ⅰ类标准的规定，且机组的排烟不应对诊疗构成影响。

12. The diesel generator room shall not be adjacent to the diagnosis and treatment equipment room, inpatient department, electronic information system room, etc. When diesel generator room is close to the above-mentioned locations due to the restrictions of conditions, comprehensive treatment measures such as the noise control of diesel generator set and the sound insulation of computer room shall be taken. The ambient noise after treatment shall not exceed the requirements of class Ⅰ standard of environmental noise in urban areas, and the smoke exhaust of the diesel generator sets shall not affect the diagnosis and treatment.

13. 柴油发电机组的选择应符合下列规定：

13. The selection of diesel generator set shall meet the following requirements:

（1）对于柴油发电机组的供油时间，三级医院应大于 24h，二级医院宜大于 12h，二级以下医院宜大于 3h；

（1）In terms of the diesel supply time of diesel generator set, the third-class hospitals, the second-class hospitals and the below first-class hospitals shall be more than 24 hours, 12 hours and 3 hours, respectively;

（2）柴油发电机组应配有电压自动调整装置、快速自启动装置及电源自动切换装置，当正常供电电源中断供电时，应能自动启动，并应在 15s 内向规定的用电负载供电；当正常供电电源恢复供电后，应延时切换并停机。

（2）The diesel generator set shall be equipped with automatic voltage regulating device, fast self-starting device and automatic switching device for power supply. When the normal power supply is suspended, the diesel generator shall be able to start automatically and supply power to the specified electrical load within 15s; when the normal power supply is restored, it shall be switched over to the mains power and shut down after a delay.

14. 要求中断供电时间小于或等于 0.5s 的一级负荷中特别重要的负荷，应设不间断电源装置（UPS），且宜为在线式。TN-S 系统中的不间断电源装置（UPS）输出端为三相时，应加装三相隔离变压器并做重复接地。

14. For the particularly important load in the first-class load whose suspension time is less than or equal to 0.5s, on-line type UPS shall be provided. When the output terminal of the UPS in TN-S system is three-phase, a three-phase isolation transformer shall be installed and grounded repeatedly.

15. 应急电源为柴油发电机组时，不间断电源装置（UPS）应急供电时间不应小于 15min。

15. When the emergency power supply is provided by diesel generator sets, the emergency power supply time of UPS shall not be less than 15min.

4.7.2.2 弱电、智能化、信息化

4.7.2.2 Weak current, intelligence and informatization

1. 医疗建筑的智能化集成系统应按医疗建筑等级、管理水平、发展规划等进行设计，并应符合下列规定：

1. The intelligent integration system of healthcare buildings shall be designed according to the level of, management level and development planning of healthcare buildings, and meet the following requirements:

（1）三级医院宜设置智能化集成系统；

（1）The third-class hospitals shall be equipped with intelligent integration system;

（2）二级及以上医院应预留智能化系统集成接口；

（2）Intelligent system integration interface shall be reserved for second-class or above hospitals;

（3）应预留与火灾自动报警系统的接口。

（3）The interface with automatic fire alarm system shall be reserved.

2. 医疗建筑的智能化集成系统应包含信息设施系统、建筑设备及诊疗设备监控系统、公共安全系统及呼叫信号系统等。

2. The intelligent integration system of healthcare buildings shall include information facility system, building equipment and the monitoring system of diagnosis and treatment equipment, public security system and call signal system, etc.

3. 医疗建筑的智能化集成系统应实现对智能化各系统的统一监视及管理。

3. The intelligent integration system of healthcare buildings shall be able to uniformly monitor and manage intelligent systems.

4. 医疗建筑的智能化集成系统的设计，应遵循集中管理、分散控制的原则，并应满足各系统的联动要求。

4. The design of intelligent integration system of healthcare buildings shall not only follow the principle of centralized management and decentralized control but also meet the linkage requirements of each system.

5. 医疗建筑的集成系统应基于开放结构，且符合标准的通信协议和接口要求，并应为未来发展留有余地。

5. The integration system of healthcare buildings shall be based on open structure and meet the requirements of standard communication protocols and interfaces, in addition to leaving the room for future development.

6. 医疗建筑的智能化集成系统应符合下列规定：

6. The intelligent integration system of healthcare buildings shall meet the following requirements:

（1）应具有数据记录、存储和分析处理能力；

（1）Have the ability of data recording, storage, analysis and processing;

（2）应提供中文操作及图形化界面；

(2) Provide Chinese operation and graphical interface;

(3) 软件宜能由用户根据实际需要进行必要的扩充。

(3) Its software shall be able to be expanded by the user according to the actual needs.

7. 医疗建筑的信息设施系统应包括电话交换系统、计算机网络系统、综合布线系统、室内移动通信覆盖系统、医疗专用信息系统 (HIS)、卫星通信系统、有线电视及卫星电视接收系统、公共广播系统、会议系统、信息发布系统、子母钟系统等。

7. The information facility system of healthcare buildings shall include telephone switching system, computer network system, generic cabling system, indoor mobile communication coverage system, hospital information system (HIS), satellite communication system, cable and satellite television receiving system, public address system, conference system, information release system, master-slave clock system, etc.

8. 当医疗建筑设置电话交换系统时，电话交换总机应具有呼叫保留、呼叫转移、热线电话及无线通信接口等专用功能，并应具有模拟中继、数字中继接口。

8. When the healthcare building is equipped with a telephone switching system, the telephone switchboard shall have not only special functions such as call cold service, call forwarding service, telephone hotline service and wireless communication interfaces, but also interfaces of analog and digital relay.

9. 一级及以上医院应设置计算机网络系统，并应符合下列规定：

9. The computer network system shall be provided for the first-class and above hospitals, and shall meet the following requirements:

(1) 计算机网络设备应设置在专用的设备间内，并应满足设备工作环境要求；

(1) The computer network equipment shall be installed in the special equipment room and meet the requirements of equipment working environment;

(2) 医疗建筑的计算机网络系统宜设置内网和外网，并宜分别设置交换机和服务器；

(2) The computer network system of healthcare buildings shall be equipped with internal and external networks, which are furnished with switches and servers respectively;

(3) 三级医院核心交换机应采用1+1冗余设置，二级及以下医院核心交换机宜采用1+1冗余设置。

(3) 1+1 redundancy setting Ought to be adopted for the core switches of the third-class hospitals, and 1+1 redundancy setting is advisable to be adopted for the core switch of the second-class and below hospitals.

10. 一级及以上医院应设置综合布线系统，并应符合下列规定：

10. The generic cabling system shall be provided for the first-class and above hospitals, and shall meet the following requirements:

(1) 对于二级及以上医院的综合布线系统，数据传输主干应采用光纤，水平线宜采用六类及以上的4对对绞电缆；手术室、影像科室、示教室等传输信息量较大的场所，宜采用光纤到桌面的布线形式；

(1) For the generic cabling system of second-class and above hospitals, optical fiber shall be used as the main data transmission trunk, and four pairs of balanced cables of category 6 and above should be used as the horizontal line; as to the places with large amount of information transmission, such as operating rooms, imaging departments (rooms), demonstration classrooms, the cabling mode of optical fiber to desktop should be adopted;

(2) 对于一级医院的综合布线系统，数据传输主干宜采用光纤，水平线可采用六类及以上的 4 对对绞电缆；

(2) For the generic cabling system of the first-class hospitals, optical fiber should be used as the data transmission trunk, and four pairs of balanced cables of category 6 and above may be used as the horizontal line;

(3) 病房内的信息点宜按床位设置在多功能医用线槽上；

(3) The information points in wards should be set on the multi-functional medical trunking according to the beds;

(4) 信息点的设置应满足使用要求；

(4) The setting of information points shall meet the use requirements;

(5) 电信间设置应满足现行国家标准《综合布线系统工程设计规范》(GB 50311—2016) 的规定。

(5) The setting of telecommunication room shall meet the requirements of current national standard Code for Engineering Design of Generic Cabling System (GB 50311—2016).

11. 二级及以上医院应设置医疗专用信息系统，一级及以下医院宜设置医疗专用信息系统，并应符合下列规定：

11. The second-class and above hospitals ought to be equipped with special medical information system, and the first-class and below hospitals is advisable to be equipped with special medical information system, which shall meet the following requirements:

(1) 医疗专用信息系统宜由医疗建筑管理系统 (HMS) 、临床信息系统 (CIS) 等组成；

(1) Special medical information system should be composed of hospital management system (HMS) and clinical information system (CIS) ;

(2) 医疗专用信息系统的建设应包括系统的运行保障和信息安全的建设；

(2) The construction of medical special information system shall include the guarantee of system operation and the construction of information security;

(3) 医疗专用信息系统应设置操作权限并分类管理医疗建筑的管理信息和临床医疗信息，并应分级管理各科室的临床信息；

(3) The special medical information system shall not only have the operation authority but also manage the hospital management information and clinical medical information together with the clinical information of each department in different levels;

(4) 医疗专用信息系统应支持模块化方式；

(4) The medical information system shall support modularization;

（5）医疗专用信息系统前端输入识别方式，宜采用一卡通；

(5) The front-end input and identification mode of special medical information system should adopt all-purpose card;

（6）宜建立医疗建筑信息管理系统（HIMS）。

(6) A hospital information management system (HIMS) should be established.

12. 二级及以上医院可根据远程医疗的需要设置卫星通信等系统，并宜预留卫星通信系统天线基础及相关条件。

12. The second-class and above hospitals may construct satellite communication system and others according to the needs of telemedicine, and should reserve the antenna foundation and relevant conditions of satellite communication system.

13. 一级及以上医院应设置有线电视系统，涉外病房、三级医院宜设置卫星电视接收系统。设置有线电视系统的，宜预留自办节目的接口，并应符合下列规定：

13. CATV system shall be constructed in the first-class and above hospitals, and satellite TV receiving system should be constructed in foreign wards and the third-class hospitals. If a CATV system is constructed, the interface for self-run programs should be reserved, and the following provisions shall be met:

（1）会议室、示教室、医疗康复中心等教学场所宜设置有线电视插座；

(1) CATV socket should be set up in conference room, demonstration room, medical rehabilitation center and other teaching places;

（2）候诊室、输液室、休息室、活动室等公共场所宜设置有线电视插座；

(2) The waiting area, infusion room, rest room, activity room and other public places should be equipped with CATV sockets;

（3）除精神病房外，每间病房应设置一个或以上有线电视插座，多人病房宜设有电视伴音输出装置；

(3) In addition to the psychiatric ward, each ward shall be equipped with one or more CATV sockets, and the ward with several people should be furnished with TV sound output device;

（4）带套间的单人病房的有线电视系统，宜具有视频点播功能。

(4) The CATV system of single ward with en-suite should have video on demand (VOD) function.

14. 一级及以上医院应设置公共广播系统，并应符合下列规定：

14. The public address system shall be provided for the first-class and above hospitals, and shall meet the following requirements:

（1）公共广播宜与应急广播合用扬声器，平时用于业务及背景音乐广播的应具备强制切换到应急广播的功能；

(1) The public address system should share the loudspeaker with the emergency broadcast, and the system used for business and background music broadcast at ordinary times shall have the function of forcibly switching to the emergency broadcast;

（2）公共广播宜在有人值班的场所设置音量调节装置；

（2）The public address system should be equipped with a volume adjusting device in the place where someone is on duty;

（3）有独立音源及扬声器的场所宜留有应急广播接口，并应具备强制切换功能；也可专设应急广播扬声器，且火灾时应切除独立音源广播；

（3）Places with independent sound sources and loudspeakers should be equipped with emergency broadcasting interface and have forced switching function; emergency broadcasting loudspeakers may also be specially mounted, but independent sound source broadcasting should be cut off in case of fire;

（4）公共广播宜按防火分区并结合医疗功能分区，设置广播播出内容。

（4）The programs of public address system should be determined according to fire prevention zone and medical function zone.

15. 二级及以上医院宜设置会议系统、远程医疗系统，有教学要求的医疗建筑应设置手术室视频示教系统，并应符合下列规定：

15. The second-class and above hospitals should be equipped with conference system and telemedicine system. The operating room in the healthcare buildings with teaching requirements shall be furnished with video surgery demonstration system, which shall meet the following requirements:

（1）会议系统可根据需要配置多媒体投影显示、数字会议、同声传译、会议扩声、视频会议、中央控制、场景控制、数码会议桌牌、电子会标、音像资料存档查询等；

（1）The conference system can be configured with multimedia projector, digital conference, simultaneous interpretation, conference sound reinforcement, video conference, central control, scene control, table cards for digital conference, electronic conference logo, audio and video data archiving and query and others as required;

（2）远程医疗系统应具备远程诊断、专家会诊、信息服务、在线检查和远程交互功能；

（2）Telemedicine system shall have the functions of remote diagnosis, expert consultation, information service, online examination and remote interaction;

（3）手术室视频示教系统应能实现手术过程的实况观摩，并应具有双向音/视频传输、场景切换、影像录制、后期处理等功能；

（3）The video demonstration system in the operating room shall be able to realize the live observation of the operation process, and shall have the functions like bidirectional audio/video transmission, scene switching, image recording, post-processing, etc;

（4）手术室无影灯应内置摄像机或设置全景摄像机，也可配置全方位高清摄像机，并应预留 1 ~ 2 个视频输入接口；

（4）The camera or panoramic camera shall be built in the shadowless lamp in operating rooms, or the omnidirectional HD camera can be configured in operating rooms, and 1 ~ 2 video input interfaces shall be reserved;

(5) 示教室内应设置大屏幕显示、音/视频切换控制、编辑存储等装置，并宜设置专用控制设备间。

(5) The demonstration room shall be equipped with large screen display, audio/video switching control, editing storage and other devices, as well as a special control equipment room.

16. 二级及以上医院的门诊部、住院部的入口处，宜设置信息发布系统，并应符合下列规定：

16. Information release system should be set at the entrance of outpatient department and inpatient department of second-class and above hospitals, and shall meet the following regulations:

(1) 系统可采用屏幕显示装置或触摸查询装置；

(1) The system may use screen display or touch query devices;

(2) 系统应支持主流媒体格式，实现对终端的独立管理、同步播放。

(2) The system shall support mainstream media format to realize synchronous play and independent management of terminals.

17. 三级医院宜设置子母钟系统，并应符合下列规定：

17. The third-class hospitals should be equipped with a master-slave clock system, which shall meet the following requirements:

(1) 子母钟系统源信号宜采用全球定位系统（GPS）接收装置；

(1) The global positioning system (GPS) receiving device should be adopted for the source signal of the master-slave clock system;

(2) 子母钟系统母钟宜设置在一层值班室，门诊部、急诊部、医技部、手术部、住院部等处宜设置子钟。

(2) The master clock of master-slave clock system should be set in the duty room on the first floor, and the slave clock should be set in the outpatient department, emergency department, medical technology department, operating department, inpatient department, etc.

18. 在移动信号较弱的场所，应设置室内移动通信覆盖系统，并应符合下列规定：

18. The indoor mobile communication coverage system shall be set in the places with weak mobile signal and meet the following provisions:

(1) 应覆盖手机信号全频段；

(1) Cover all frequency bands of mobile phone signal;

(2) 应设置在不干扰诊疗设备正常工作的公共场所。

(2) Be set in a public place that does not interfere with the normal work of the diagnosis and treatment equipment.

4.7.2.3 绿色建筑

4.7.2.3 Green building

1. 以人为本是照明的目的，照明节能应该是在满足规定的照度和照明质量要求的前提下进行考核。

1. People foremost is the purpose of lighting. Energy saving of lighting shall be assessed on the premise of meeting the specified requirements of illumination and lighting quality.

2. 民用建筑应根据规范和绿色建筑专项规划确定的星级绿色建筑等级要求进行设计。设计自评结果至少应满足现行国家和地方绿色建筑评价标准的一星级绿色建筑要求，其中国家机关办公建筑和政府投资或者以政府投资为主的其他公共建筑，至少应满足现行国家和地方绿色建筑评价标准的二星级绿色建筑要求。

2. Civil buildings shall be designed in accordance with the requirements of codes and the star grade of green buildings which is determined by the special planning of green buildings. The self-assessment results of the building design shall at least meet the requirements of the one-star green buildings in current national and local assessment standards for green buildings. The office buildings for state organs and other public buildings invested by governments or mainly invested by governments shall at least meet the requirements of the two-star green buildings in current national and local assessment standards for green buildings.

3. 高压静电空气净化装置应设置与风机有效联动的措施，给病人一个安全、有效、无毒、无害、无臭的净化环境，且有效阻止新冠肺炎病毒的传播。

3. The high-voltage electrostatic air purifier shall have an efficient linkage measure to provide patients with a safe, effective, nontoxic, harmless and odorless purified environment and effectively prevent the spread of COVID-19 virus.

4. 相关标准规定了两种照明功率密度值，即现行值和目标值。现行值是根据对国内各类建筑的照明能耗现状调研结果、我国建筑照明设计标准以及光源、灯具等照明产品的现有水平并参考国内外有关照明节能标准，经综合分析研究后制定的，其在相关标准实施时执行。而目标值则是预测到几年后随着照明科学技术的进步、光源灯具等照明产品能效水平的提高，从而照明能耗会有一定程度的下降而制定的。目标值比现行值降低 10% ~20% 。目标值执行日期由标准主管部门决定。目标值的实施，可以由相关标准（如节能建筑、绿色建筑评价标准）规定，也可由全国或行业，或地方主管部门作出相关规定。

4. Relevant standard specifies two kinds of lighting power density (LPD) values, i. e. current value and target value. The current value is formulated after comprehensive analysis and research of the investigation results of the current situation of lighting energy consumption of various buildings in China and the current level of lighting products such as light sources and lamps, in accordance with lighting design standards of buildings in China, and by reference to the relevant lighting energy-saving standards at home and abroad. It is implemented in relevant standard. And the target value is formulated by predicting that the lighting energy consumption will decrease to a certain extent with the progress of lighting science and technology, the improvement of energy efficiency of lighting products such as light source as well as lamps and lanterns. The target value is about 10% ~20% lower than the current value. The implementation date of the target value shall be determined by the competent department of this standard. The implementation of the target value may be specified by relevant standards (such as energy-saving building and assessment standards for green buildings), or by the national, industrial or local competent departments.

4.7.2.4 消防、节能、环保、人防

4.7.2.4 Fire fighting, energy saving, environmental protection and civil air defense

1. 建筑内消防应急照明和灯光疏散指示标志的备用电源的连续供电时间应符合下列规定：医疗建筑、总建筑面积大于 100000m^2 的公共建筑和总建筑面积大于 20000m^2 的地下、半地下建筑，不应少于 1.0h。

1. The continuous power supply time of standby power supply for fire emergency lighting and light evacuation indicator signs in buildings shall meet the following requirements: health-care buildings, public buildings with a total building area of more than 100000m^2 and the underground and semi-underground buildings with a total building area of more than 20000m^2 shall not be less than 1.0h.

2. 消防用电设备应采用专用的供电回路，当建筑内的生产、生活用电被切断时，应仍能保证消防用电。

2. The special power supply circuit shall be adopted for the electrical fire-fighting equipment. When the electricity for production and living in buildings is cut off, the fire-fighting electricity shall still be used.

3. 消防控制室、消防水泵房、防烟和排烟风机房的消防用电设备及消防电梯等的供电，应在其配电线路的最末一级配电箱处设置自动切换装置。

3. The power supply of fire elevator and the electrical fire-fighting equipment in the fire control room, fire pump room, smoke control and exhaust fan room shall be equipped with automatic switching devices at the last distribution box of the distribution line.

4. 消防用电按一、二级负荷供电的建筑，当采用自备发电设备作备用电源时，自备发电设备应设置自动和手动启动装置 。当采用自动启动方式时，应能保证在 30s 内供电。

4. For the buildings whose fire control power supply is distributed according to the first-and second-class load, when the self-provided power generation equipment is used as the standby power supply, the self-provided power generation equipment shall be equipped with automatic and manual start devices. When the automatic start mode is adopted, the power shall be supplied within 30s.

5. 在医院门诊楼、病房楼、手术部、办公室等场所设置火灾自动报警系统。

5. Automatic fire alarm system is set in outpatient building, ward building, operating department, offices and other places in hospitals.

6. 电气系统的设计应经济合理、高效节能。

6. The design of electrical system shall be economical, reasonable, efficient and energy-saving.

7. 电气系统宜选用技术先进、成熟、可靠，损耗低、谐波发射量少、能效高、经济合理的节能产品。

7. The electrical system should select energy-saving products with advanced, mature and reliable technology, low loss, less harmonic emission, high energy efficiency and reasonable economy.

8. 建筑设备监控系统的设置应符合现行国家标准《智能建筑设计标准》（GB 50314—2015）的有关规定。

8. The setting of building monitoring system shall comply with the relevant provisions of the current national standard, Standard for Design of Intelligent Building (GB 50314—2015).

9. 配变电所应靠近负荷中心、大功率用电设备。

9. The distribution station and substation shall be close to the load center and high-power electrical equipment.

10. 配电系统三相负荷的不平衡度不宜大于15%。单相负荷较多的供电系统，宜采用部分分相无功自动补偿装置。

10. The unbalance of three-phase load in distribution system should not be more than 15%. The power supply system with more single-phase loads shall be equipped with automatic reactive power compensation device for partial split phases.

11. 大型用电设备、大型可控硅调光设备、电动机变频调速控制装置等谐波源较大设备，宜就地设置谐波抑制装置。当建筑中非线性用电设备较多时，宜预留滤波装置的安装空间。

11. Large-scale electrical equipment, large-scale silicon controlled rectifier dimmer, the frequency control device of motor and other equipment with large harmonic source should be equipped with local harmonic suppression device. When there are many non-linear electrical equipment in the building, the installation space should be reserved for filtering device.

12. 室内照明功率密度（LPD）值应符合现行国家标准《建筑照明设计标准》（GB 50034—2013）的有关规定。

12. The indoor lighting power density (LPD) value shall comply with the relevant provisions of the current national standard, Standard for Lighting Design of Buildings (GB 50034—2013).

13. 一般照明在满足照度均匀度条件下，宜选择单灯功率较大、光效较高的光源，不宜选用荧光高压汞灯，不应选用自镇流荧光高压汞灯。

13. Under the condition of uniform illumination, the light source with high power and light efficiency should be selected for general lighting, but the high-pressure mercury fluorescent lamp and self-ballasted high-pressure mercury fluorescent lamp should not be selected.

14. 走道、楼梯间、卫生间、车库等无人长期逗留的场所，宜选用发光二极管（LED）灯；疏散指示灯、出口标志灯、室内指向性装饰照明等宜选用发光二极管（LED）灯。

14. LED lights should be used not only in passageways, staircases, toilets, garages and other places where no one can stay for a long time but also for evacuation indicator lights, exit sign lights and indoor directional decorative lighting.

15. 照明控制应结合建筑使用情况及天然采光状况，进行分区、分组控制。

15. Lighting control shall be controlled by zone and group in combination with the use of the building and the natural lighting conditions.

16. 主要次级用能单位用电量大于等于 10kW 或单台用电设备大于等于 100kW 时，应设置电能计量装置。公共建筑宜设置用电能耗监测与计量系统，并进行能效分析和管理；公共建筑应按照明插座、空调、电力、特殊用电分项进行电能监测与计量。办公建筑宜将照明和插座分项进行电能监测与计量。

16. When the electricity consumption of primary sub-organization of energy using is more than or equal to 10kW or the electricity consumption of single electrical equipment is more than or equal to 100kW, the electric energy metering device shall be equipped. Public buildings should be equipped with power consumption monitoring and measurement system, and energy efficiency shall be analyzed and managed. The electricity consumption of public buildings shall be monitored and measured according to items of lighting sockets, air conditioning, electric power and special use of electricity. Lighting and sockets should be monitored and measured separately in office buildings.

17. 电气设计除应满足战时用电的需要外，还应满足平时用电的需要；电气设备应选用防潮性能好的定型产品。

17. The electrical design shall meet the needs of electricity consumption not only in wartime but also in peacetime. The electrical equipment shall be the approved products with good moisture-proof performance.

18. 电力负荷应分别按平时和战时用电负荷的重要性、供电连续性及中断供电后可能造成的损失或影响程度分为一级负荷、二级负荷和三级负荷。

18. The electric load shall be divided into the first-class load, the second-class load and the third-class load according to the importance of the electric load in peacetime and wartime, the continuity of power supply and the possible loss or influence that is caused by the suspension of power supply.

19. 战时电力负荷分级，应符合下列规定：

19. The classification of power load in wartime shall meet the following requirements:

（1）一级负荷：

（1）First-class load:

① 中断供电将危及人员生命安全；

① Suspension of power supply will endanger people's life;

② 中断供电将严重影响通信、警报的正常工作；

② Suspension of power supply will seriously affect the normal operation of communication and alarm;

③ 不允许中断供电的重要机械、设备；

③ Important machinery and equipment that are not allowed to suspend the power supply;

④ 中断供电将造成人员秩序严重混乱或恐慌。

④ Suspension of power supply will cause serious disorder or panic of personnel.

（2）二级负荷：

（2）Second-class load:

① 中断供电将严重影响医疗救护工程、防空专业队工程、人员掩蔽工程和配套工程的正常工作；

① Suspension of power supply will seriously affect the normal operation of medical aid works, works of service team for civil air defense, personnel shelter engineering and supporting engineering;

② 中断供电将影响生存环境。

② Suspension of power supply will affect the living environment.

（3）三级负荷：除上述两款规定外的其他电力负荷。

（3）Third-class load: other electric loads except those specified in the above two paragraphs.

20. 防空地下室的柴油电站选址应符合下列要求：

20. The site selection of diesel-driven power station in air defense basement shall meet the following requirements:

（1）靠近负荷中心；

（1）Be close to the load center;

（2）交通运输、输油、取水比较方便；

（2）Be convenient for communications and transportation, oil transportation and water intake;

（3）管线进、出比较方便。

（3）Be convenient for the entrance and exit of piping.

21. 医疗救护工程和防空专业队工程应设置与所在地人防指挥机关相互联络的直线或专线电话，并应设置应急通信设备。通信设备、电话可设置在值班室、防化通信值班室内。人员掩蔽工程应设置电话分机和音响警报接收设备，并应设置应急通信设备。

21. The medical aid works and works of service team for civil air defense shall be provided with not only a straight or special telephone line to contact with the local civil air defense commanding organ but also emergency communication equipment. Communication equipment and telephone can be set in duty room and CBR protection and communication duty room. The personnel shelter shall be provided with telephone extension, sound alarm receiving equipment, and emergency communication equipment.

22. 中心医院、急救医院内应设置电话总机，并在办公、医疗、病房、值班室、防化通信值班室、配电间、电站、通风机室等各房间内设有电话分机。

22. Central Hospital and emergency hospital shall be equipped with telephone switchboard, and telephone extensions shall be set in office, medical treatment, wards, duty room, CBR protection and communication duty room, power distribution room, power station, fan room and other rooms.

23. 救护站、防空专业队工程、人员掩蔽工程、配套工程中的值班室、防化通信值班室、通风机室、发电机房、电站控制室等房间应设置电话分机。

23. The duty room in first-aid station, works of service team for civil air defense, personnel

shelter and supporting works, CBR protection and communication duty room, fan room, generator room, the control room in power station and other rooms shall be equipped with telephone extension.

24. 战时通信设备线路的引入，应在各人员出入口预留防护密闭穿墙管，穿墙管可利用预埋备用管。当需要设置通信防爆波电缆井时，除留有设计需要的穿墙管外，还应预埋备用管。

24. For the introduction of the lines for wartime communication equipment, protective airtight through-wall pipes shall be reserved at the entrances and exits of all personnel, the through-wall pipes can use embedded spare pipes. When it is necessary to set blastproof cable well for communication, in addition to the through-wall pipes required by the design, the spare pipes shall also be embedded.

4.7.2.5　夜景亮化

4.7.2.5　Night view lighting

1. 建筑夜景照明的照明功率密度（LPD）限值应符合现行行业标准《城市夜景照明设计规范》（JGJ/T 163—2008）的有关规定。

1. The limit of lighting power density (LPD) for building night view lighting shall conform to the relevant provisions of the current industry standard Code for Lighting Design of Urban Nightscape (JGJ/T 163—2008).

2. 室外景观、道路照明应选择安全、高效、寿命长、稳定的光源，避免光污染。

2. In terms of the lighting for outdoor landscape and roads, safe, efficient and stable light sources with long service life shall be selected to avoid light pollution.

3. 建筑景观照明应设置平时、一般节日、重大节日等多种模式自动控制装置。

3. The lighting for building landscape shall be designed into an automatic control device which has various modes such as ordinary times, general festivals and major festivals.

4.7.2.6　智能化与信息化规模标准

4.7.2.6　Standards for intellectualization and informatization

1. 医疗建筑智能化系统工程应符合下列规定：

1. The intelligent system engineering of healthcare building shall conform to the following requirements:

（1）应适应医疗业务的信息化需求；

（1）Meeting the informationization requirements of medical business;

（2）应向医患者提供就医环境的技术保障；

（2）Providing the technical guarantee in medical environment for the medical patients;

（3）应满足医疗建筑物业规范化运营管理的需求。

（3）Meeting the requirements of standardized operation management of healthcare building properties.

2. 综合医院智能化系统应按下表的规定配置，并应符合现行行业标准《医疗建筑电气设计规范》（JGJ 312—2013）的有关规定。

2. The general hospital intelligent system shall be configured according to the provisions listed in the table below, and shall conform to the relevant provisions of the current industry standard Code for Electrical Design of Healthcare Buildings (JGJ 312—2013).

综合医院智能化系统配置表：见表 4-12。

Configuration of general hospital intelligent system: see Table 4-12.

表 4-12 综合医院智能化系统配置表

Table 4-12 Configuration of general hospital intelligent system

智能化系统 Intelligent system			一级医院 First-class hospital	二级医院 Second-class hospital	三级医院 Third-class hospital
信息化应用系统 Informationalized application system	公共服务系统 Public service system		⊙	●	●
	智能卡应用系统 Smart card application system		⊙	●	●
	物业管理系统 Property management system		⊙	●	●
	信息设施运行管理系统 Information facility operation management system		○	●	●
	信息安全管理系统 Information security management system		⊙	●	●
	通用业务系统 Universal business system	基本业务办公系统 Basic business office system	按国家现行有关标准进行配置 To be configured according to the current relevant national standards		
	专业业务系统 Specialized business system	医疗业务系统信息化系统 Informationalized medical business system			
		病房探视系统 Ward visitation system			
		视频示教系统 Video demonstration system			
		候诊呼叫信号系统 Waiting call signal system			
		护理呼应信号系统 Nursing response signal system			
智能化集成系统 Intelligent integrated system	智能化信息集成（平台）系统 Intelligent information integration (platform) system		○	⊙	●
	集成信息应用系统 Integrated information application system		○	⊙	●

续表

智能化系统 Intelligent system			一级医院 First-class hospital	二级医院 Second-class hospital	三级医院 Third-class hospital
信息设施系统 Information facility system		信息接入系统 Information access system	●	●	●
		布线系统 Wiring system	●	●	●
		移动通信室内信号覆盖系统 Mobile communication indoor signal coverage system	●	●	●
		用户电话交换系统 Private telephone switch system	⊙	●	●
		无线对讲系统 Wireless intercom system	●	●	●
		信息网络系统 Information network system	●	●	●
		有线电视系统 CATV system	●	●	●
		公共广播系统 Public address system	●	●	●
		会议系统 Conference system	⊙	●	●
		信息导引及发布系统 Information guidance and release system	●	●	●
建筑设备管理系统 Building management system		建筑设备监控系统 Building automation system	⊙	●	●
		建筑能效监管系统 Building energy efficiency supervision system	○	⊙	●
公共安全系统 Public security system		火灾自动报警系统 Automatic fire alarm system	按国家现行有关标准进行配置 To be configured according to the current relevant national standards		
	安全技术防范系统 Security & technology protection system	入侵报警系统 Intrusion alarm system			
		视频安防监控系统 Video surveillance & control system			
		出入口控制系统 Access control system			
		电子巡查系统 Electronic patrol system			
		停车库（场）管理系统 Parking garage (lot) management system	○	⊙	●
	安全防范综合管理（平台）系统 Integrated security management (platform) system		○	⊙	●
	应急响应系统 Emergency response system		○	⊙	●

续表

智能化系统 Intelligent system		一级医院 First-class hospital	二级医院 Second-class hospital	三级医院 Third-class hospital
机房工程 Room engineering	信息接入机房 Information access room	●	●	●
	有线电视前端机房 CATV headend room	●	●	●
	信息设施系统总配线机房 General wiring room of information facility system	●	●	●
	智能化总控室 Intelligent master control room	●	●	●
	信息网络机房 Information network room	⊙	●	●
	用户电话交换机房 Private telephone switch room	⊙	●	●
	消防控制室 Fire control room	●	●	●
	安防监控中心 Security and fire control center	●	●	●
	智能化设备间（弱电间） Intelligent device room (weak current room)	●	●	●
	应急响应中心 Emergency response center	○	⊙	●
	机房安全系统 Room security system	按国家现行有关标准进行配置 To be configured according to the current relevant national standards		
	机房综合管理系统 Integrated room management system	⊙	●	●

注：●—应配置；⊙—宜配置；○—可配置

Note: ●—Required; ⊙—Appropriate; ○—Optional

3. 信息化应用系统的配置应满足综合医院业务运行和物业管理的信息化应用需求。

3. The informationalized application system shall be configured to meet the informationalized application requirements of general hospital business operation and property management.

4. 信息接入系统应满足医疗业务信息应用的需求。

4. The information access system shall be configured to meet the requirements of medical business information application.

5. 移动通信室内信号覆盖系统的覆盖范围和信号功率应保证医疗设备的正常使用和患者的人身安全。

5. The mobile communication indoor signal coverage system shall be designed with coverage and signal power such that it could guarantee the normal service of medical device and the personal safety of patients.

6. 用户电话交换系统宜根据医院的业务需求，配置相应的无线寻呼系统或其他组群式的寻呼系统。

6. The private telephone switch system shall be configured with corresponding radio paging system or other group paging systems according to the hospital business requirements.

7. 信息网络系统应为医疗业务信息化应用系统提供稳定、实用和安全的支撑条件，并应具备高宽带、大容量和高速率，宜具备系统升级的条件。

7. The information network system shall provide stable, practical and safe supporting conditions for the informationalized medical business application system, and shall be designed with high broadband, large capacity and high speed rate, as well as the conditions for upgrading.

8. 有线电视系统应提供本地有线电视节目或卫星电视及自制电视节目。

8. The CATV system shall cover local CATV programs or satellite and home-made TV programs.

9. 信息导引及发布系统应在医院大厅、挂号及药物收费处、门急诊候诊厅等公共场所配置发布各类医疗服务信息的显示屏和供患者查询的多媒体信息查询端机，并应与医院信息管理系统互联。

9. At public places such as hospital hall, registration and charging offices, outpatient and emergency waiting areas, the information guidance and release system shall be equipped with the LCD screens for releasing various medical service information and the multimedia information inquiry terminals for inquiring by patients, and shall be interconnected with the hospital information management system.

10. 建筑设备管理系统应满足医院建筑的运行管理需求，并应根据医疗工艺要求，提供对医疗业务环境设施的管理功能。

10. The building management system shall conform to the operation management requirements of hospital buildings, and shall be responsible for managing the medical business environment facilities according to the requirements of medical process.

11. 安全技术防范系统应满足医院安全防范管理的要求。

11. The security & technology protection system shall conform to the requirements of hospital security protection management.

4.7.3 暖通空调设计

4.7.3 HVAC design

4.7.3.1 医院通风设计

4.7.3.1 Hospital ventilation design

1. 医院通风是暖通空调系统是否优良的重要评价因素，设计应注意通风与控制并重。

1. The hospital ventilation shall be deemed as an important factor for assessing the performance of HVAC system. The design shall focus on both ventilation and temperature & humidity control.

2. 医院通风设计的重点包括：送入卫生所要求的新风量，需要更加关注通风量的计算，确定合理的室内换气次数，设置局部的独立排风系统排除室内污染后的空气，以及有害物，排风必须经过处理之后才能排出，应做好气流组织设计，控制室内空气流向，控制相邻房间空气流动等。

2. The key points of hospital ventilation design are described as follows: the supply of fresh air rate required for the health, more attention required for the calculation of ventilation rate, the determination of reasonable indoor air exchange rate, and the setup of a local independent exhaust system to emit the indoor contaminated air and harmful substances. The exhaust air must be subject to treatment before emission. The design of air distribution shall be well done in order to control the indoor air flow direction, and the air flow in adjacent rooms.

3. 医疗设备机房主要分布在检验科、放射科、核医学科、放疗科等，需要更多地了解各类医疗设备的技术条件，做好医疗设备的安全通风，主要包括：生物安全通风，比如PCR生物实验通风，检验科中的生物检验通风；涉及有害有毒气溶胶的排放、处理，比如病理科通风、低温消毒通风、放疗通风等；核医学科辐射性通风；MR通风等。

3. The medical device rooms are mainly distributed in clinical laboratory, X-radiology department, nuclear medicine department, radiotherapy department, etc. It is necessary to know more about the technical conditions of various medical device and do a good job in the safe ventilation of medical device, mainly including: biological safety ventilation, i. e. PCR biological experiment ventilation, and biological inspection ventilation of clinical laboratory; emission and treatment of harmful and toxic aerosols, i. e. ventilation of pathology department, low temperature disinfection ventilation, radiotherapy ventilation, etc. ; radiation ventilation in nuclear medicine, MR ventilation, etc.

4. 医技科室实验操作中广泛使用通风柜、生物安全柜、洁净工作台，这三种设备之间最大的区别在于提供不同对象的保护，暖通设计中需注意送风口应根据通风柜、生物安全柜实际安装位置而调整，避免送风气流影响正常使用。

4. Ventilation cabinets, biosafety cabinets and clean benches are widely used for the experimental operation of medical technology departments. The biggest difference among those three devices is that they provide protection for different objects. In HVAC design, attention should be paid to the adjustment of air supply outlets according to the actual installation positions of ventilation cabinets and biosafety cabinets so as to avoid the influence of air supply flow on normal service.

4.7.3.2 综合医院暖通设计

4.7.3.2 Heating and ventilation design of general hospital

1. 采用散热器采暖时，应以热水为介质，不应采用蒸汽，供水温度不应大于85℃，散热器应便于清洗消毒。

1. In case of radiator, hot water, instead of steam, shall be used as the medium. The water supply temperature shall not be greater than 85℃. The radiator shall be easy to clean and disinfect.

2. Ⅲ级、Ⅳ级洁净用房，应采用板式或光管式散热器采暖，且应采取防护、防尘措施。

2. In terms of Grade Ⅲ and Ⅳ clean rooms, plate radiator or light tubular radiator shall be selected. Moreover, protective and dustproof measures shall be taken.

3. 凡产生气味、水蒸气以及潮湿作业的用房，应设机械排风。

3. Mechanical exhaust system shall be provided for rooms that produce odor and steam and receive operations under damp environment.

4. 空调系统应符合下列要求：

4. The air conditioning system shall meet the following requirements:

（1）应根据室内空调设计参数、医疗设备、卫生要求、使用时间、空调负荷等要求合理分区；

（1）It shall be zoned reasonably according to the indoor air conditioning design parameters, medical devices, sanitary requirements, service duration, air conditioning load and other requirements;

（2）各功能区域宜独立，宜单独成系统；

（2）All functional areas shall be independent and designed into a separate system;

（3）各空调分区应能互相封闭，并应避免通过空气传播导致院内感染；

（3）All air conditioning areas shall be enclosed to each other and shall avoid hospital-acquired infection resulting from airborne transmission;

（4）有洁净度要求的房间和严重污染的房间，应单独成一个系统。

（4）Rooms with cleanliness requirements and rooms with severe pollution shall be designed into one system.

5. 无特殊要求时不应在空调机组内安装臭氧等消毒装置，不得使用淋水式空气处理装置。

5. Ozone devices and other disinfection devices shall not be installed inside the air conditioning unit without special requirements. Sprayed air handling unit shall not be used.

6. 采用集中空调系统的医疗用房的送风量不宜低于6次/h。

6. For medical rooms equipped with centralized air conditioning system, the air supply rate shall not be less than 6 cycles/hour.

7. 集中空调系统和风机盘管机组的回风口必须设置初阻力小于50Pa、微生物一次通过率不大于10%和颗粒物一次计重通过率不大于5%的过滤设备。

7. The return air inlets of the centralized air conditioning system and fan coil unit must be equipped with a filter plant with an initial resistance less than 50Pa, primary microbial pass rate no more than 10% and primary particulate weight pass rate no more than 5%.

8. 当室外可吸入颗粒物PM_{10}的年均值未超过现行国家标准《环境空气质量标准》(GB 3095—2012)中二类区适用的二级浓度限值时，新风采集口应至少设置粗效和中效两级过滤器，当室外PM_{10}超过年平均二级浓度限值时，应再增加一道高中效过滤器。

8. If the annual mean value of outdoor inhalable particles PM_{10} does not exceed the secondary concentration limit applicable to Class Ⅱ zone stated in the current national standard Ambient Air Quality Standard GB 3095—2012, the fresh air collection port shall be provided with at

least primary and medium-effect filters. If the annual mean value of outdoor PM_{10} exceeds the annual mean secondary concentration limit, an additional HEPA and medium-HEPA filter shall be provided in addition to the above filters.

9. 医疗用房的集中空调系统的新风量每人不应低于 $40m^3/h$，或新风量不应小于 2 次/h；对人员多的场所，经过经济和技术比较，宜改为新风量运行。

9. The fresh air rate of the centralized air conditioning system in medical room shall not be less than $40m^3/h$ per person, or the fresh air rate shall not be less than 2 cycles/hour. For places with a large flow of people, according to economic and technical comparison, it is appropriate to select the fresh air rate for operation.

10. 核医学检查室、放射治疗室、病理取材室、检验科、传染病病房等含有有害微生物、有害气溶胶等污染物质场所的排风，应处理达标后才能排放。

10. The exhaust air from nuclear medicine examination room, radiotherapy room, pathology sampling room, clinical laboratory, infectious disease ward and other places containing harmful microorganisms, harmful aerosols and other pollutants shall be emitted only after reaching the standard.

11. 没有特殊要求的排风机应设在排风管路末端，使整个管路为负压。

11. Exhaust fans without special requirements shall be arranged at the terminal of exhaust duct to make the whole duct is under negative pressure.

12. 门诊部应采用自然通风；当采用采暖系统时，候诊区、办公室等的冬季采暖设计温度不应低于 18℃；当采用空调系统时，夏季空调设计温度不宜高于 26℃。

12. Natural ventilation shall be arranged for the outpatient department. If heating system is used, the design temperature for heating in winter of waiting areas, offices, etc. shall not be lower than 18℃. If air conditioning system is used, the design temperature of air conditioning in summer shall not be higher than 26℃.

13. 当医院的门厅采用空调时应减少室外空气流入，并应维持室内定向的空气流动和热环境；中庭式的门厅，宜采用分层空调，冬季可设置其他补充供暖装置。

13. If air conditioning system is used in the outpatient hall of the hospital, the inflow of outdoor air shall be reduced, and the indoor directional air flow and thermal environment shall be maintained. For atrium-type outpatient hall, the stratified air conditioning system shall be arranged, and in winter, other supplementary heating devices can be installed.

14. 候诊区的空调系统，应结合平面布局使空气从清洁区流向非清洁区。其中，小儿科候诊室和诊室对其他区域应为正压。隔离诊室及其候诊前室应采用单独的空调系统，其回风应有中效（含）以上的过滤器。当与其他诊室为同一空调系统时，应单独设回（排）风，并应维持室内负压。

14. The air conditioning system in the waiting areas shall, based on the plane layout, make the air flow from the clean area to the non-clean area. Wherein, positive pressure shall be maintained in the pediatrics waiting area and consulting room relative to other areas. For the isolated consultation room and its pre-waiting room, separate air conditioning system shall be arranged. In

terms of the return air, medium-HEPA (including) and above-stage filters shall be provided. If the isolated consultation room and its pre-waiting room share the same air conditioning system with other consulting rooms, the return air (exhaust) system shall be separately arranged and the indoor negative pressure shall be maintained.

15. 化验室、处置室、换药室等污染较严重的场所，应设局部排风。

15. Local exhaust system shall be provided for laboratories, disposal rooms, dressing rooms and other places with severe pollution.

16. 诊室的空调温度宜高于候诊区 1～2℃。

16. The air conditioning temperature in the consulting room shall be 1～2℃ higher than that in the waiting area.

17. 急诊部当采用空调系统时，应采用独立系统，可 24h 连续运行，冬季采暖设计温度不应低于 18℃，夏季空调温度不宜高于 26℃。

17. For emergency department, an independent air conditioning system shall be arranged, which shall be operated continuously for 24 hours. The design temperature for heating in winter shall not be lower than 18℃, and the design temperature for air conditioning system in summer shall not be higher than 26℃.

18. 急诊隔离区的空调系统宜独立设置，其回风应有中效（含）以上的过滤器，并应有排风系统。当与其他诊室为同一空调系统时，应单独排风，不应回风，与相邻并相通的区域应保持不小于 5Pa 的负压。

18. For the emergency isolation area, the air conditioning system shall be arranged independently. The return air system shall be equipped with medium-HEPA (including) and above-stage filters. The exhaust system shall also be provided. If the emergency isolation area shares the same air conditioning system with other consulting rooms, the exhaust system shall be separately arranged instead of return air system. A negative pressure of no less than 5Pa shall be maintained within adjacent and interconnected areas.

19. 普通病区应符合下列要求：

19. The inpatient ward shall meet the following requirements:

(1) 普通病区的病房应能开窗（有纱窗）通风；

(1) For inpatient ward, the windows (with screens) of sickrooms should be opened for ventilation;

(2) 设置普通空调时，冬季温度宜在 20℃以上，夏季温度不宜高于 27℃；应有新风和排风，系统规模不宜过大；

(2) If general air conditioning system is used, the temperature in winter shall be higher than 20℃ and the temperature in summer shall not be higher than 27℃. The fresh air system and exhaust system shall be provided, of which small size is recommended;

(3) 病区的换药室、处置室、配餐室、污物室、污洗室、公用卫生间等，应设排风，排风口的布置不应使局部空气滞留；排风换气次数宜为 10～15 次/h。

(3) The dressing room, disposal room, catering room, sewage room, cleaning room, public

toilet, etc. in the ward shall be provided with exhaust system. The exhaust outlets shall be arranged such that no local air trapping will occur; the air exchange rate of exhaust system shall be 10 ~ 15 cycles/hour.

20. 产科分娩室及准备室、淋浴室、恢复室等相关房间设置空调系统时，应能24h连续运行；分娩室宜采用新风空调系统；新生儿室室内温度全年宜保持22 ~26℃。

20. If air conditioning system is arranged for maternity delivery rooms, preparation rooms, shower rooms, recovery rooms and other related rooms, the system shall be operated continuously for 24 hours. Fresh air system shall be arranged for delivery rooms. The indoor temperature of newborn room shall be kept at 22 ~26℃ throughout the year.

21. 监护病房温度在冬季不宜低于24℃，夏季不宜高于27℃；采用普通空调系统时，宜采用连续运行，相对湿度宜为40% ~65%；噪声不应大于45dB(A)，宜采用上送下回的气流组织，送风气流不宜直接吹向头部；每张病床均不应处于其他病床的下风侧；排风（或回风）口应设在床头附近。

21. The temperature of ICU shall not be lower than 24℃ in winter and higher than 27℃ in summer; if general air conditioning system is used, the system shall be operated continuously and the relative humidity shall be 40% ~65%; the noise shall not be greater than 45dB(A). The up-supply and down-return air distribution shall be applied. The supply air flow shall not be directly blown to the head; each ward bed shall not be arranged at downward air side of other ward beds; the exhaust (or return air) outlet shall be located nearby the bed head.

22. 检验科、病理科、实验室应符合下列要求：应有单独排风系统；采用普通空调时，室内温度冬季不宜低于22℃，夏季不宜高于26℃；室内相对湿度冬季不宜低于30%，夏季不宜高于65%。

22. The clinical laboratory, pathology department and laboratory shall meet the following requirements: A separate exhaust system shall be arranged; if general air conditioning system is used, the indoor temperature shall not be lower than 22℃ in winter and higher than 26℃ in summer; the indoor relative humidity shall not be lower than 30% in winter and higher than 65% in summer.

23. 电生理、超声、纤维内窥镜等科室，宜设置独立的普通空调系统。

23. An independent general air conditioning system shall be arranged for such departments as electrophysiology, ultrasound, fibrescope, etc.

24. 听力检查室宜设置集中式空调系统，应采取消声减振措施，且噪声不应大于30dB（A）。对于有无声要求的检测，可采取暂时停止空调、隔断气流等措施。

24. For audiometry room, a centralized air conditioning system shall be arranged, and noise elimination and vibration reduction measures shall be taken. The noise shall not be greater than 30dB(A). For testing to be conducted under noiseless condition, measures such as temporarily air conditioning shutdown and air flow blocking may be taken.

25. 心血管造影室的操作区宜为Ⅲ级，洁净走廊应低于操作区一级，与相邻并相通房间应保持5Pa的正压。辅助用房应采用普通空调。

25. The operation area of the angiocardiography room shall be designed as Grade Ⅲ area, and the clean corridor shall be designed as an area lower than Grade I operation area. A positive pressure of 5Pa shall be maintained in adjacent and interconnected rooms. The general air conditioning system shall be arranged for the auxiliary rooms.

26. 放射科的检查室、控制室和机械间的空调系统和排风系统应符合下列要求：

26. The air conditioning system and exhaust system of examination room, control room and machinery room in radiology department shall meet the following requirements:

（1）应根据设备需要选择空调系统；

（1）The air conditioning system shall be selected depending on device requirements;

（2）采用半集中式空调系统时，不应在机器上方设置任何风机盘管机组等末端装置及其凝水管；

（2）If a semi-centralized air conditioning system is used, no fan coil unit and other air terminal devices and condensate pipes shall be installed above the unit;

（3）放射科的检查室、控制室和暗室应设排风系统，自动洗片机排风应采用防腐蚀的风管，排风管上应设止回阀；

（3）An exhaust system shall be arranged for the examination room, control room and darkroom in radiology department. The exhaust system of automatic film processor shall be equipped with an anti-corrosion air duct. A check valve shall be arranged on the exhaust duct;

（4）在有射线屏蔽的房间，对于穿墙后的风管和配管，应采取不小于墙壁铅当量的屏蔽措施。

（4）In rooms with X-ray shielding, shielding measures no less than the lead equivalent of the wall shall be taken for the air duct and piping passing through the wall.

27. 磁共振室宜采用独立的恒温恒湿空调系统，室内温度应为22℃ ±2℃，相对湿度应为60% ±10%。扫描间内应采用非磁性、屏蔽电磁波的风口，任何磁性管线不应穿越。磁共振机的液氦冷却系统应设置单独的排气系统，并应直接连接到磁共振机的室外排风管。管道应采用非磁性材料，管径不应小于250mm。

27. An independent constant temperature and humidity air conditioning system shall be arranged for the MRI room. The indoor temperature shall be 22℃ ±2℃ and the relative humidity shall be 60% ±10%. Non-magnetic air inlet with electromagnetic wave shielding shall be arranged within the scanning room. No magnetic pipeline shall pass through such inlet. A separate exhaust system shall be arranged for the liquid helium cooling system of the MRI machine and shall be directly connected to the outdoor exhaust duct of the MRI machine. The duct shall be made of non-magnetic materials, with a diameter no less than 250mm.

28. 核医学科所有核辐射风险的用房宜采用独立的恒温恒湿空调系统。扫描间温度应为22℃ ±2℃，且1h内的温度变化不宜大于3℃。扫描间相对湿度应为60% ±10%。其他房间可采用普通空调，但排风应按国家现行标准《临床核医学卫生防护标准》(GBZ 120—2006)和《医用放射性废物的卫生防护管理》(GBZ 133—2009)的规定处理。

28. An independent constant temperature and humidity air conditioning system shall be ar-

ranged for each room of nuclear medicine department with nuclear radiation risks. The temperature of scanning room shall be 22℃ ±2℃, of which the change within 1 hour shall not be greater than 3℃. The relative humidity of scanning room shall be 60% ±10%. For other rooms, general air conditioning system may be arranged. However, the exhaust air system shall be considered according to the provisions of the current national standards Radiological Protection Standard for Clinical Nuclear Medicine GBZ 120—2006 and Radiological Protection Standard for Management of Medical Radioactive Waste GBZ 133—2009.

29. 放射性同位素治疗用房的空调系统，应根据放射性同位素种类与使用条件确定，宜采用全新风空调方式。放射性同位素管理区域内，相对于管理区域外应保持负压，排气风管宜采用氯乙烯衬里风管，并应在排风系统中设置气密性阀门；应在净化处理装置的排气侧设置风机，并应保持排风管内负压，排风机应滞后于空调系统关闭。当储藏室、废物保管室储藏放射性同位素时，应 24h 排换气。

29. The air conditioning system of radioisotope therapy room shall be determined depending on the type and service conditions of radioisotope. A fully fresh air conditioning method shall be applied. Negative pressure shall be maintained in the radioisotope management area relative to the outside of the management area. An vinyl chloride lined air duct shall be used as the exhaust duct, and an air-tight valve shall be set in the exhaust system. An ventilator shall be installed on the exhaust side of purification apparatus. A negative pressure shall be maintained in the exhaust duct. The exhaust fan shall be shut down later than the air conditioning system. 24h air exhaust and exchange shall be maintained if radioactive isotopes are stored in storage room and waste storage room.

30. 中心（消毒）供应室应保持有序压差梯度和定向气流。定向气流应经灭菌区流向去污区。无菌存放区对相邻并相通房间不应低于 5Pa 的正压，去污区对相邻并相通房间和室外均应维持不低于 5Pa 的负压。

30. Ordered differential pressure gradient and directional air flow shall be maintained at the central (disinfection) supply room. Directional airflow shall flow to decontamination area through sterilization area. A positive pressure no less than 5Pa shall be maintained at aseptic storage area relative to the adjacent and interconnected rooms. A negative pressure no less than 5Pa shall be maintained at decontamination area relative to the adjacent and interconnected rooms.

31. 无菌存放区宜按不低于Ⅳ级洁净用房设计，并应采用独立的净化空调系统。高压灭菌器应设置局部通风，低温灭菌室应有独立排风系统，温度冬季不宜低于 18℃，夏季不宜高于 24℃；室内相对湿度冬季不宜低于 30%，夏季不宜高于 60%。

31. The aseptic storage areas shall be designed as a clean room not lower than Grade Ⅳ, and shall be equipped with an independent purifying air conditioning system. The autoclave shall be provided with local ventilation, and the low-temperature sterilization room shall be provided with an independent exhaust system. The temperature shall not be lower than 18℃ in winter and higher than 24℃ in summer. The indoor relative humidity shall not be lower than 30% in winter and higher than 60% in summer.

32. 去污区应设置独立局部排风，总排风量不应低于负压所要求的压差值风量。去污区内的回风口应设置不低于中效的空气过滤器。

32. The decontamination area shall be provided with an independent local exhaust system. The total exhaust air rate shall not be lower than the air rate under differential pressure required by negative pressure. An air filter with at least medium-HEPA shall be arranged at the return air inlet of the decontamination area.

33. 采用普通空调的区域冬季温度不宜低于 18℃，夏季温度不宜高于 26℃。

33. At the areas where the general air conditioning system is used, the temperature shall not be lower than 18℃ in winter and higher than 26℃ in summer.

34. 应根据需要和可能选用洁净用房，产科、监护用房、血液病房、烧伤病房、过敏性哮喘病室、生殖学中心、无菌存放区等对洁净用房的等级要求详见《综合医院建筑设计规范》（GB 51039—2014）的有关规定，洁净空调系统的设计应满足以下条件：

34. Clean rooms shall be selected depending on demands and possibilities. Please refer to the relevant provisions of Architectural Design Code for General Hospital (GB 51039—2014) for the grade requirements of maternity department, ICU accessories, hematology ward, burn ward, allergic asthma ward, reproductive center and sterile storage area. The clean air conditioning system shall be designed in such a way that the following conditions are satisfied:

（1）Ⅰ级洁净用房的送风末端应设高效过滤器，Ⅱ级洁净用房送风末端可设高效或亚高效过滤器，Ⅲ级洁净用房的送风末端可设亚高效过滤器，Ⅳ级洁净用房的送风末端可设高中效过滤器。

（1）A HEPA filter shall be arranged at the air supply terminal of Grade Ⅰ clean room, a HEPA or sub-HEPA filter may be arranged at the air supply terminal of Grade Ⅱ clean room, a sub-HEPA filter may be arranged at the air supply terminal of Grade Ⅲ clean room, and a HEPA and medium-HEPA filter may be arranged at the air supply terminal of Grade Ⅳ clean room.

（2）洁净用房应采用阻隔式空气净化装置作为房间的送风末端。

（2）A blocking air cleaner shall be used as the air supply terminal of the clean room.

（3）洁净用房内不应采用普通的风机盘管机组或空调器。Ⅲ级、Ⅳ级洁净用房内采用带亚高效或高中效过滤器的净化风机盘管机组或立柜式净化空调器时，新风可集中供给或设立独立的新风机组。

（3）General fan coil unit or air conditioner shall not be used in clean room. If a purifying fan coil unit with sub-HEPA or HEPA filter and medium-HEPA filter or an vertical cabinet purifying air conditioner is used in Grade Ⅲ and Ⅳ clean rooms, fresh air may be supplied centrally or an independent fresh air handling unit may be installed.

（4）洁净用房室内（不含走廊）不宜采用上送上回气流组织。洁净用房的患者通道上不应设置空气吹淋室；

（4）For clean rooms (excluding corridors), it is not appropriate to select the up-supply up-return air distribution. The patient channel of clean room shall not be provided with an air shower;

(5) 净化空调系统应在新风口、回风口和空调机组正压出风面、送风口3处设置空气过滤器。

(5) For the purifying air conditioning system, air filters shall be installed at the fresh air inlet, return air inlet, as well as positive-pressure air outlet and air supply outlet of the air conditioning unit.

35. 洁净手术部设计，应符合现行国家标准《医院洁净手术部建筑技术规范》(GB 50333—2013)的有关规定。

35. The clean operating department shall be designed in accordance with the relevant provisions of the current national standard Architectural Technical Code for Hospital Clean Operating Department GB 50333—2013.

4.7.3.3 传染病医院、应急医疗设施、急救中心等暖通设计

4.7.3.3 HVAC design for infectious disease hospitals, emergency medical facilities, first-aid centers, etc.

1. 应设置机械通风系统。机械送风、排风系统应按半清洁区、半污染区、污染区分区设置独立系统。半污染区、污染区的排风机应设置在建筑室外。机械送、排风系统应使医院内空气压力从清洁区至半污染区至污染区依次降低，清洁区应为正压区，污染区应为负压区。清洁区送风量应大于排风量，污染区排风量应大于送风量。建筑物气流组织应形成从清洁区至半污染区至污染区有序的压力梯度。房间气流组织应防止送、排风短路，送风口位置应使清洁空气首先流过房间中医务人员可能的工作区域，然后流过传染源进入排风口。送风口应设置在房间上部。病房、诊室等污染区的排风口应设置在房间下部，房间排风口底部距地面不应小于100mm。清洁区每个房间送风量应大于排风量150m^3/h。污染区每个房间排风量应大于送风量150m^3/h。同一个通风系统，房间到总送、排风系统主干管之间的支风道上应设置电动密闭阀，并可单独关断，进行房间消毒。

1. Mechanical ventilation system shall be provided. Independent mechanical air supply and exhaust systems shall be arranged in semi-clean area, semi-contaminated area and contaminated area respectively. For semi-contaminated area and contaminated area, the exhaust fans shall be installed outside the buildings. The mechanical air supply and exhaust system shall be designed to guarantee that the air pressure is sequentially reduced from clean area to semi-contaminated area to contaminated area. The clean area shall be designed into a positive pressure area and the contaminated area shall be designed into a negative pressure area. The air supply rate in the clean area shall be greater than the exhaust air rate, and the exhaust air rate in the contaminated area shall be greater than the air supply rate. The air distribution of the building shall form an ordered pressure gradient from the clean area to the semi-contaminated area to the contaminated area. The air distribution in the room shall be designed to prevent short circuit of air supply and exhaust system. The air supply outlet shall be located such that clean air flows through the possible working area of medical care staff in the room first, and then flows to the exhaust outlet through the infection source. The air supply outlet shall be arranged at the upper part of the room. The exhaust outlets of contaminated areas such as wards and consulting rooms shall be ar-

ranged at the lower part of the room, and shall be no less than 100mm from the ground. The air supply rate of each room in the clean area shall be greater than the exhaust air rate of 150 m^3/h. The exhaust air rate of each room in the clean area shall be greater than the air supply rate of $150m^3/h$. In terms of the same ventilation system, an electric airtight valve shall be provided on the branch duct from the room to the main duct of the main air supply and exhaust system, which can be separately shut off for purpose of disinfection of the room.

2. 半清洁区送风系统最低应经粗效、中效两级过滤；半污染区、污染区送风系统最低应经粗效、中效、亚高效三级过滤，排风系统应经高效过滤。送、排风系统的各级空气过滤器应设压差检测、报警装置。

2. The air supply system in semi-clean area shall be subject to at least coarse and medium-HEPA filtration. The air supply system in semi-contaminated area and contaminated area shall be subject to at least coarse, medium and sub-HEPA filtration. The exhaust system shall be subject to HEPA filtration. All stages for air filters of the air supply and exhaust system shall be equipped with differential pressure detection and alarm devices.

3. 半污染区、污染区的排风机应设在排风管路末端，排风系统的排出口不应临近人员活动区，排气宜高空排放，排风系统的排出口、污水通气管与空气压缩机房的进风口、送风系统取风口不宜设置在建筑同一侧，并应保持安全距离。

3. Exhaust fans in semi-contaminated area and contaminated area shall be located at the terminal of exhaust duct. The exhaust outlet of exhaust system shall not be close to the personnel movement area. The exhaust air shall be emitted at high altitude. The exhaust outlet of exhaust system, air inlet between sewage vent pipe and air compressor room, and air intake of air supply system shall not be located at the same side of the building and a safe distance shall be kept.

4. 医院门诊、急诊部入口处的筛查，其通风系统应独立设置。

4. The ventilation system for screening at the entrance of outpatients and emergency departments of the hospital shall be arranged independently.

5. 病房卫生间排风不宜通过共用竖井排风，应结合病房排风统一设计。

5. In ward toilet, common shaft is not appropriate for exhaust system. Instead, the exhaust system shall be designed in consideration of the exhaust system in the ward.

6. 医院各处门口不宜设置空气幕。

6. It is not recommended to install air curtains at each entrance of the hospital.

7. 有条件设置集中空调的地方，诊室、病房、医护办公等小空间可结合机械送排风系统设置风机盘管系统，机械送风系统应设计为空调新风系统。中庭、门诊大厅等大空间可设计全新风直流式空调系统。全新风直流式空调系统应采取在非呼吸道传染病流行时期回风的措施。

7. Where centralized air conditioning is feasible, small spaces such as clinics, wards and offices for medical care staff can be equipped with fan coil systems in combination with mechanical air supply and exhaust systems. The mechanical air supply system shall be designed with a fresh air system. Direct flow fresh air conditioning system can be designed for large spaces such

as atrium and outpatient hall. The direct flow fresh air conditioning system shall ensure air return during the epidemic period of non-respiratory infectious diseases.

8. 手术室、重症监护室（ICU）、负压隔离病房以及高精度医疗设备用房等，宜采用空气调节。医疗设备用房应在根据房间压差要求设置送排风系统的基础上，根据设备的温度、湿度要求设置独立的空调机组或恒温恒湿机组。

8. Air conditioning shall be used in operating rooms, intensive care units (ICU), negative pressure isolation wards and high-precision medical equipment rooms. Medical equipment rooms shall be equipped with independent air conditioning units or constant temperature and humidity units according to the temperature and humidity requirements of the equipment on the basis of air supply and exhaust systems deployed to meet pressure difference requirements of these rooms.

9. 空调的冷凝水应分区集中收集，并应随各区污水、废水排放集中处理。

9. All condensates from air conditioning shall be collected in different areas and then treated in a centralized way along with sewage and wastewater discharged from different areas.

10. 非呼吸道传染病的门诊、医技用房及病房最小换气次数（新风量），应为3次/h。污染区房间应保持负压，每间房间的排风量应大于送风量150m^3/h。

10. The minimum ventilation rate (fresh air rate) for outpatient, medical technology rooms and wards of non-respiratory infectious diseases shall be 3 AC/h. Negative pressure shall be maintained in the rooms of contaminated areas, and the exhaust rate of each room shall be greater than 150 m^3/h.

11. 呼吸道传染病的门诊、医技用房及病房、发热门诊最小换气次数（新风量），应为6次/h。

11. The minimum ventilation rate (fresh air rate) for outpatient, medical technology rooms, wards and fever clinics of respiratory infectious diseases shall be 6 AC/h.

12. 负压隔离病房（包括ICU）设计应符合下列规定：

12. The design of negative pressure isolation ward (including ICU) shall comply with the following regulations:

（1）负压隔离病房宜采用全新风直流式空调系统。最小换气次数应为12次/h。负压隔离病房的送风应经过粗效、中效、亚高效过滤器三级处理。排风应经过高效过滤器过滤处理后再排放。负压隔离病房排风的高效空气过滤器应安装在房间排风口处。

(1) Direct flow fresh air conditioning system shall be arranged for all negative pressure isolation wards. The minimum ventilation rate shall be 12 AC/h. Air supply for all negative pressure isolation wards shall be treated by coarse, medium and sub-HEPA filters for three-level filtration. Exhaust air shall be filtered by HEPA filters before being discharged. HEPA filters for the exhaust of negative pressure isolation wards shall be installed at the air outlet of each room.

（2）每间负压隔离病房的送、排风管上应设置密闭阀。负压隔离病房送风口应设在医护人员常规站位的顶棚处，排风口应设在与送风口相对的床头下侧。

(2) Sealing valves shall be set on the air supply and exhaust pipes of each negative pressure isolation ward. The air inlet of the negative pressure isolation ward shall be located at the

ceiling below which the medical care staff frequently stands, whereas the air outlet shall be set at the lower side of the bed head opposite to the air inlet.

（3）负压隔离病房与其相邻、相通的缓冲间、走廊压差应保持不小于 5Pa 的负压差。负压隔离病房的通风系统在过滤器终阻力时的送排风量，应能保证各区压力梯度要求。有条件时，可在送、排风系统上设置定风量装置。

（3）The pressure difference between a negative pressure isolation ward and its adjacent and interconnected uncontaminated transition rooms and corridors shall be kept at a negative pressure difference of not less than 5Pa. The ventilation system of all negative pressure isolation wards shall ensure the pressure gradient of each area when the final resistance of the filter is met. When it is feasible, set a constant air flow regulator on the air supply and exhaust system.

（4）负压隔离病房应设置压差传感器，门口宜安装压差显示装置。负压隔离病房送排风系统的过滤器宜设压差检测、报警装置。

（4）Install differential pressure sensors for each negative pressure isolation ward and set differential pressure display device at the entrance of each ward. The filter of air supply and exhaust system for all negative pressure isolation wards shall be equipped with differential pressure detection and alarm devices.

13. 疫情期间，为了防止集中空调系统导致的交互感染而关闭空调回风系统的被动局面，可在空调机组、风道内或送风口处安装具有杀灭病毒和细菌的空气过滤器。

13. In order to prevent the passive situation of closing the air-conditioning return air system during the epidemic period of infectious diseases and avoid cross infection caused by centralized air-conditioning system, air filters that are functionally capable of killing virus and bacteria can be installed in the air-conditioning unit, air duct or at the air inlet.

14. 为减少医护人员工作量，在急诊和门诊可以安装测温机器人，为防止医护人员感染，在半污染区和污染区安装移动式可杀灭病毒和细菌的净化器或机器人。

14. In order to reduce the workload of medical care staff, body temperature measuring robots can be installed in emergency and outpatient departments. Moreover, install mobile purifiers or robots capable of killing viruses and bacteria in semi-contaminated areas and contaminated areas so as to protect medical care staff from infections.

15. 应急医疗设施冬季和夏季送风（新风）的加热或冷却宜采用独立直膨式风冷热泵机组并根据室温控制调节送风温度，严寒地区可设辅助电加热装置。

15. The heating or cooling of air supply (fresh air) for emergency medical facilities in winter and summer shall be realized by independent direct expansion air-cooled heat pump units. It needs to control and adjust the air supply temperature according to room temperature. Auxiliary electric heating devices can be set in severe cold areas.

16. 根据当地气候条件及围护结构情况，应急医疗设施污染区可安装分体冷暖空调机，分体空调机的送风应尽量减小对室内气流方向的影响，分体空调机电源应集中管理，应根据室温需求统一管控；清洁区、半清洁区、半污染区宜设置分体冷暖空调机，CT 等散热量较大的医技机房应设置空调。严寒、寒冷地区冬季可设置电暖器供暖。

16. According to the local climate conditions and enclosure structures, split air conditioners can be installed in the contaminated areas of emergency medical facilities. As for the air supply of split air conditioners, minimize the impact on the direction of indoor air flow. The power supply of split air conditioners shall be under centralized management and control for specific room temperatures. Separate cooling and heating air conditioners shall be installed in clean areas, semi-clean areas and semi-contaminated areas. It also needs to arrange air conditioners in medical technology rooms with massive heat dissipation such as CT room. Electric heaters can be installed for heating in severe cold areas and cold areas in winter.

17. 应急医疗设施的手术室应按直流负压手术室设计，并应符合国家现行标准《医院洁净手术部建筑技术规范》(GB 50333—2013)的有关规定。

17. Operating rooms for emergency medical facilities shall be designed into negative pressure operating rooms in accordance with relevant regulations of the current national standard GB50333—2013 Architectural technical code for hospital clean operating department.

18. 急救中心隔离用房应采用独立的空调系统，送风量不宜小于每小时 10 次换气次数，新风量不宜小于每小时 3 次换气次数，并应能 24h 连续运行。隔离用房还应有独立排风系统，并应保持与其相邻、相通房间低 10Pa 的负压。急救中心也应按功能要求设置清洁区、半污染区及污染区。隔离用房的送排风系统应按区域设置，机械送排风系统应使压力从清洁区、半污染区、污染区依次降低。清洁区应为正压或 0，半污染区、污染区应为负压。设计还应符合下列规定：

18. Independent air conditioning system shall be deployed for isolation rooms of the emergency center. The air supply rate shall not be less than 10 AC/h, and the fresh air rate shall not be less than 3 AC/h. In addition, the system shall be capable of continuous operation for 24 hours. All isolation rooms shall be arranged with independent exhaust systems to maintain a negative pressure of 10Pa in the adjacent and interconnected rooms. The emergency center shall be provided with clean area, semi-contaminated area and contaminated area according to functional requirements. The air supply and exhaust system for all isolation rooms shall be set for specific areas to reduce the pressure from clean area, semi-contaminated area and contaminated area in turn. Clean area shall be under positive pressure or 0, semi-contaminated area and contaminated area shall be under negative pressure. The design shall also meet the following requirements:

(1) 隔离用房的送风口应设置于房间的上部，采用顶送或侧上送，回（排）风口应设于房间一侧下部，回（排）风口下沿距地面不应小于 100mm，回（排）风口宜设置低阻中效过滤器。

(1) The air outlet of all isolation rooms shall be set at the upper part of each room, and the top or side air supply shall be chosen. The air return (exhaust) outlet shall be set at the lower part on one side of the room, and the lower edge of the air return (exhaust) outlet shall not be less than 100mm above the floor. All air return (exhaust) outlets shall be provided with a low-resistance medium-HEPA filter.

(2) 隔离用房等排风系统应设置高效过滤器及防倒流装置，其他用房排风系统应设置防倒流装置，并均应高空稀释排放。

（2） Install HEPA filters and backflow prevention devices for the exhaust systems of isolation rooms and other rooms. Install backflow prevention devices for the exhaust system of rooms for other purposes. All exhausts shall be diluted and discharged at high altitude.

（3） 负压房间应设置压差传感器。过滤器的两端应设置压差检测报警装置。

（3） Differential pressure sensors shall be set in negative pressure rooms. Arrange differential pressure detection and alarm devices at two ends of each filter.

（4） 新风采集口距所有排风口距离不应小于10m，且应在排风口上风侧无污染干扰源的清洁区域；新风采集口下端距地面的距离不应小于3m；当新风采集口设在屋顶时，其距屋顶面的距离不应小于1m。新风系统宜设置初、中效两级过滤器，隔离用房宜再设置亚高效过滤器。

（4） The distance between fresh air collection port and all air outlets shall not be less than 10m. In addition, the fresh air collection port must be located in a clean area without contamination and interference source on the windward side of the air outlet. The distance between the lower end of fresh air collection port and the ground shall not be less than 3m. When the fresh air collection port is located on the roof, the distance to the roof surface shall not be less than 1m. The fresh air system shall be equipped with primary and medium filters, and the isolation room shall be provided with sub-HEPA filters.

19. 消毒区应设置独立送排风系统，并应保持负压，其各功能房间压力与通风应满足下表要求：

19. Independent air supply and exhaust system shall be set in the disinfection area, and negative pressure shall be maintained. The pressure and ventilation of each functional room shall meet the following requirements:

消毒区各功能房间压力和通风要求：见表4-13。

Requirements for the pressure and ventilation of each functional room in the disinfection area: see Table 4-13.

表4-13 消毒区各功能房间压力和通风要求

Table 4-13 Requirements for the pressure and ventilation of each functional room in the disinfection area

功能房间 Functional room	相对于相邻区域压力关系 Relation of pressure with adjacent areas	每小时最小换气次数（次/h） Minimum air exchange rate per hour (AC/h)	每小时新风最小换气次数（次/h） Minimum fresh air exchange rate per hour (AC/h)
消毒间 Disinfection room	负压 Negative pressure	10	—
污物间或去污间 Waste room or decontamination room	负压 Negative pressure	6	2
消毒后存放间 Post-disinfection storage room	正压（或0） Positive pressure (or 0)	4	2

20. 严寒地区和寒冷地区的精神专科医院应设置集中采暖系统，夏热冬冷地区有条件时宜设置集中采暖系统。采暖热媒应采用热水，采暖散热器宜暗装。最热月平均室外气温大于或等于25℃地区，宜设置空调降温设备。

20. Centralized heating systems shall be deployed in psychiatric hospitals in severe cold areas and cold areas. Moreover, it is recommended to set a centralized heating system when actual conditions permit in the areas with hot summer and cold winter. Hot water is proposed as the heating medium. It is recommended to conceal heat radiators. In the areas where the average outdoor temperature in the hottest month is higher than or equal to 25℃, air conditioning and cooling equipment shall be installed.

21. 精神专科医院建筑宜采用自然通风，并宜设置机械通风系统。采用空调系统时，应有新风供应和排风。

21. Natural ventilation shall be utilized for psychiatric hospital buildings and mechanical ventilation system shall be set at the same time. When air conditioning is used, fresh air supply and exhaust shall be provided.

22. 运营维护：各区域排风机与送风机应连锁，为保证各区域空气压力梯度，清洁区应先启动送风机，再启动排风机；污染区应先启动排风机，再启动送风机。各区之间风机启动顺序：清洁区、半污染区、污染区；随时关注风机故障报警信号，保证风机正常运行；随时关注送、排风系统的各级空气过滤器的压差报警，及时更换堵塞的空气过滤器，保证各区域风量；排风高效空气过滤器更换操作人员须做好自我防护，拆除的排风高效过滤器应当由专业人员进行原位消毒后，装入安全容器内进行消毒灭菌，随医疗废弃物一起处理。

22. Operation and maintenance: All exhaust fans in each area shall be interconnected with the blowers. In order to ensure the air pressure gradient in each area, it is required to start the blower at first and the exhaust fan then in the clean area. However, the exhaust fan shall be started before the blower is started in the contaminated area. Sequence of starting fans in different areas: Clean area, semi-contaminated area and contaminated area. Observe the fan fault alarm signal all the time to ensure the normal fan operation. Observe differential pressure alarms from all levels of air filters in the air supply and exhaust system all the time, and timely replace the blocked air filters to ensure the fresh air rate in each area. Take effective protective measures for the operator of replacing exhaust HEPA filters. The removed exhaust HEPA filters shall be disinfected in situ by professionals, then put into a safe container for disinfection and sterilization, and finally treated together with medical wastes.

4.7.3.4 空调负荷及冷热源规划

4.7.3.4 Air conditioning load and cooling and heating source planning

1. 医院主要医疗功能包括门诊急诊、医技、病房三大块，这三大块的需求确定了医院的负荷特点。急诊及病房要求全年24h提供通风、空调或供暖，医疗设备如MR也要求全年24h空调；多数医疗综合楼体量较大，内区房间较多，过渡季也需要供冷；洁净病房，可能处于建筑物的内区，自身存在一定的热负荷，因此在过渡季节和冬季也需要供冷。因此医院空调冷负荷具有明显的时间段和季节特点，这些特点很大程度上决定了医院冷源的配置。对于设置在建筑物内区的功能房间，当过渡季节和冬季存在热负荷时，应首先考虑采用经过过滤处理的室外新风，以便减少医院的总能耗。

1. Main medical functions of the hospital include outpatient and emergency department, technical department and ward, which determine the load characteristics of the hospital. Emergency department and wards require 24h ventilation, air conditioning or heating every year. Medical equipment such as MR also requires 24h air conditioning throughout the year. Most medical complex buildings are large in size and have many rooms inside, which need cool air during the transition season. The clean ward may be located inside the building and requires a specific heating load. In this view, cool air is also required in the transitional season and winter. Therefore, the cooling load of hospital air conditioning system is notably temporal and seasonal, which largely determine the configuration of cooling sources for the hospital. When there is heating load in the transitional season and winter for the function rooms inside the building, the filtered outdoor fresh air shall be preferred in order to reduce the total energy consumption of the hospital.

2. 冷源应优先考虑采用可靠性最高的能源，目前多数城市可靠性最高的能源仍然是电力，一般宜选用水冷冷水机组作为医院的主冷源。多联机空调系统和一体化高效冷水机组均可作为集中式冷水空调系统的补充，用于放射科、放疗科、核医学科、病案库、消防控制室、信息中心等需要24h供冷的区域。一体化高效冷水机组是国内最近几年根据市场需求新推出的一种将冷水机组、冷冻水泵、冷却水泵、冷却塔、电气控制系统等高度集成化的整体式高效制冷机房产品，完全替代了传统水冷中央空调制冷机房，具有配置灵活、高效节能、运行稳定、不占用土建机房的显著特点，特别适合用于医疗、数据中心等场合作为不间断冷源。

2. With respect to the cooling source, give priority to the most reliable energy source. At present, the most reliable energy source in most cities is still electricity. In most cases, water chillers are selected as the main cooling source for hospitals. The multi-split air conditioning system can be used as a supplement to the centralized chilled water air conditioning system, and can be used in the areas for 24h cooling such as radiology department, radiotherapy department, nuclear medicine department, medical record storeroom, fire control room and information center. Integrated high-efficiency water chiller has been developed in response to domestic market trends in recent years; it integrates the water-cooling unit, chilled water pump, cooling water pump, cooling tower, electric control system, etc. into a single highly compact package, which is an ideal substitute for the traditional water-cooled central AC mechanical room; the new product is flexible in configuration, energy-efficient, stable in operation, and does not require a mechanical room, which is particularly suitable for providing uninterruptible cooling source to medical facilities and data centers, etc.

3. 为使冷水机组高效运行，冷水机组可以采用大小搭配的组合方式。一般情况下，夏热冬暖地区的夜间负荷可以在不同时段分别由大冷水机组（17：00～22：00）和小冷水机组（22：00～7：00）来承担，夏热冬冷地区及寒冷地区则由小冷水机组承担夜间负荷、过渡季内区负荷及净化空调负荷。在北方或夏热冬冷地区的一些医院，当过渡季负荷较小，不适宜再由冷水机房供冷时，可设置风冷冷水机组作为冷源。

3. In order to ensure high-efficiency operation, the combination of high-capacity and low-capacity chillers can be selected. Under normal circumstances, the night load in the areas with hot summer and warm winter can be guaranteed by high-capacity water chillers (17:00 ~ 22:00) and low-capacity water chillers (22:00 ~ 7:00) respectively in different periods, while the low-capacity water chillers in the areas with hot summer and cold winter and in the cold areas can bear the night load, the load in transition seasons and the load of air conditioning for purification. In some hospitals in the Northern Region or in the areas with hot summer and cold winter, air-cooled chillers can be set as cooling source when the load in the transition season is too small to be guaranteed by the water chiller room.

4. 北方地区医院的热源多以市政供热和自建锅炉房为主，为了增加供热的安全，即使在市政供热保障性较高的城市，医院也会增加锅炉房作为市政供热的补充。南方地区情况比较多样化，有的自建锅炉房，有的采用热泵机组作为热源。一些城市采用风冷热泵热水机组单独供热（不作为冷源），需要注意的是，在长江中下游地区的城市，冬季湿度较大，导致风冷热泵除霜频繁且持续时间长，实际能耗未必比锅炉供热要低，而且连续供热可靠性也较差，因此这些地区采用风冷热泵供热时要慎重。

4. Most heating sources for hospitals in the Northern Region are municipal heating networks and self-built boiler rooms. In order to enhance the heating reliability, the hospitals will build boiler rooms as a supplement to the municipal heating network even in the cities with highly reliable municipal heating network. Heating sources are diversified in the Southern Region. There are self-built boiler rooms and heat pump units as heating sources. In some cities, air-cooled heat pump hot water units are used to supply heat separately (not as cooling source). However, it shall be noted that in the cities at the middle and lower reaches of Yangtze River, higher humidity in winter leads to frequent and long-time defrosting of air-cooled heat pumps. The actual energy consumption is not necessarily lower than that of boiler heating, and the reliability of continuous heating is also poor. Therefore, select air-cooled heat pumps for heating in these areas with caution.

5. 坚持适度集中、灵活分散的原则。大型医疗设备所在的诊疗间、信息中心等，为避免系统漏水风险，不宜在室内设置集中空调末端装置，建议采用多联式分体空调、独立的恒温恒湿机组来满足这些空间的温湿度要求。

5. Therefore, persevere in the principle of moderate concentration but flexible distribution. For the clinic, treatment room and information center where large medical equipment is located, it is not recommended to set centralized air conditioning terminal devices indoors to avoid the risk of system water leakage. For this reason, multi-split air conditioning and independent constant temperature and humidity units shall be set to meet the temperature and humidity requirements of these spaces.

6. 洁净手术部净化空调系统可采用独立冷热源或从医院集中冷热源能源站接入。除应满足夏、冬设计工况冷热负荷使用要求外，还应满足非设计负荷使用要求。冷热源设备不宜少于 2 台。

6. The air conditioning system for the purification of clean operating department can be pro-

vided with independent heating and cooling sources or connected to the centralized cooling and heating source energy station of the hospital. In addition to the design cooling and heating loads in summer and winter, the cooling and heating sources shall guarantee non-design loads. At least two sets of cooling and heating source equipment shall be provided.

7. 全年中需要供冷供暖运行时间较少的洁净手术部宜采用分散式冷热源。

7. Decentralized cooling and heating sources shall be used in the clean operating department that only needs a short period of cooling and heating throughout the year.

8. 洁净手术部宜采用自带冷热源的集中式净化空调系统。

8. The clean operating department shall be configured with a centralized purification and air conditioning system that has independent cooling and heating sources.

（1）对于规模较大、手术室数量较多的洁净区域可采用多个空调净化系统，以保证使用上的灵活性。

（1） For clean areas with a large size and a large number of operating rooms, multiple air conditioning purification systems can be set to ensure flexible operations.

（2）为防止交叉感染，隔离手术室应采用独立的净化空调系统，使用医用净化空调机组，可供 10 万级、1 万级的净化手术室使用；室温可保持在 23 ~ 25℃，使用面积可达 40 ~ 70m²；该机组可独立配置冷热泵，不依赖集中空调系统，也可选用外接集中空调系统的冷热源，使用灵活方便；这种方式适用、经济、节能，特别适合于区县的二、三级医院使用，也可在要求设置独立系统的急诊、感染手术室中使用。

（2） In order to prevent cross infection, all isolated operating rooms shall be provided with independent purification air conditioning system and use medical purification air conditioning units, which can be used for 100000 and 10000-purity operating rooms. The room temperature can be kept at 23℃ to 25℃, and the usable area can reach 40 – 70m². In addition, the unit can be independently equipped with a cooling and heating pump so as not to rely on centralized air conditioning system. A cooling and heating source externally connected with the centralized air conditioning system can be selected, thus ensuring flexible and convenient operations. This solution is applicable, economical and energy-saving, especially suitable for the second-class and third-class hospitals in districts and counties. It can also be used in emergency departments and infection operating rooms for an independent system.

4.8　废弃物处理

4.8　Waste disposal

4.8.1　废弃物减量分类隔离

4.8.1　Waste reduction, classification and isolation

4.8.1.1 设施应寻求各废弃物的所有成分，并进行分离，至少应考虑到普通垃圾、医疗/传染性废弃物、危害性废弃物和低放射性废弃物。

4.8.1.1 For different types of facility, it needs to analyze all components of each type of waste and isolate these wastes. At least, these wastes shall be classified into general garbage, medical/infectious waste, hazardous waste and low-level radioactive waste.

4.8.1.2 应有专门的存放和流动空间与清洁/消毒设施，以促进医疗产物、食品服务中的再利用，尽量避免清除，减少废弃物量。

4.8.1.2 For the special storage and transfer space and cleaning/disinfection facilities, it needs to promote the reuse of medical products and food services to minimize waste disposal and reduce the amount of wastes.

4.8.1.3 应在封闭、安全的区域内分段存放黄色包装袋中的废弃物。

4.8.1.3 All wastes in yellow packages shall be stored in sections in a closed and safe area.

4.8.1.4 有生物危害和环境危害的材料，包括水银、核反应废弃物或其他规定类型的废弃物，应被隔离并安全运输。

4.8.1.4 Materials with biological hazards and environmental hazards, including mercury, nuclear reaction wastes or other specific types of wastes, shall be isolated and transported safely.

4.8.1.5 应制订废弃物管理和处理意外事故计划，工人的健康与安全防护，以及所有废弃物现场处理的场所。

4.8.1.5 A waste management and emergency handling plan, the measures for the health and safety protection of workers, and the protective measures for all waste disposal sites shall be formulated.

4.8.2 规划设计

4.8.2 Planning and design

4.8.2.1 废弃物的收集与存放地点是设计任务书的重要部分，在卸货平台或其他卸货区，任务书应确定废弃物捣碎机、打包机、尖锐物与分段运输周转容器的放置地点，废弃物房间大小应按计划的容量、预期存放时间来确定。

4.8.2.1 The location of waste collection and storage is an important part of the requirements specification. In the unloading platform or other unloading areas, the requirements specification shall elaborate on the location of waste mashers, packagers, sharp objects and segmented transportation circulation containers, and the size of the waste storage room shall be determined according to the planned capacity and expected storage time.

4.8.2.2 建筑物应有足够的空间，以适应箱子/手推车的适当分类，例如：可循环的废弃物、传染性废弃物、尖锐物等。

4.8.2.2 The building shall have a sufficient space to accommodate the proper classification of boxes/trolleys, for example recyclable waste, infectious waste and sharp objects.

4.8.2.3 走道和物料处理系统、垃圾自动收集系统，应实现废弃物从产生到存放或处理的有效流动，尽量减少对人员的危险性。

4.8.2.3 Walkways, material handling systems and automatic garbage collection systems shall realize the effective flow of wastes from generation to storage or disposal to minimize the hazard brought to personnel.

4.8.2.4 应处理好高效收集、运输、虫害控制和存放，设施内产生的废弃物应有收集与存放空间，清除的废弃物应有接纳处。

4.8.2.4 Effective collection, transportation, pest control and storage shall be effectively implemented. Waste generated in the facility shall be provided with a collection and storage space, and a space shall be guaranteed to receive wastes.

4.8.2.5 废弃物在何处产生、收集与分级处理，它的优化管理对设施的计划与空间影响很大，对于需就地进行再循环处理的设施与地区，应设计分捡与分级的场地。

4.8.2.5 The optimal management on the location of waste generation, collection and classification produces a great impact on the planning and space of facilities. For the facilities and areas that need in-situ recycling, the location of waste sorting and classification shall be designed.

4.8.2.6 封闭的废弃物存放空间应远离食品准备区、个人卫生区和其他清洁功能区域。

4.8.2.6 Closed waste storage space shall be far away from food preparation area, personal hygiene area and other cleaning function areas.

4.8.2.7 高压灭菌器、垃圾捣碎机和医疗垃圾掩埋前的其他技术处理用房、分段运输周转用房，应安装紫外线消毒灯。

4.8.2.7 Autoclaves, garbage mashers and other technical treatment rooms before the burial of medical wastes, and sectional transportation circulation rooms, shall be provided with ultraviolet disinfection lamps.

4.8.2.8 禁止在设施内自行焚烧废弃物，应统一定时收集，运往城市指定的集中焚烧地点。

4.8.2.8 It is prohibited to burn wastes inside the facility. Instead, all wastes shall be collected on a regular basis and then transported to the centralized combustion site designated by the city.

4.9 室内设计与标识系统

4.9 Interior design and identification system

4.9.1 室内设计

4.9.1 Interior design

4.9.1.1 设计原则

4.9.1.1 Design principle

1. 新建或改扩建的医疗保健设施装饰装修设计关系到医疗建筑的使用、人文环境塑造，它不但要满足医疗建筑特殊的实体功能要求，同时也要满足医患人员对室内环境人性化的精神需求。

1. The decoration design of newly built or expanded medical and health facilities is related to the operation of medical buildings and the creation of humanistic environment. It needs to meet not only the special actual function requirements of medical buildings, but also the spiritual needs of doctors and patients for the humanization of interior environments.

2. 医院建筑不应给患者带来伤害，其室内设计既要满足医院内部医疗技术的合理管理、适合操作体系，又要为医护人员、病患及家属从心理学角度提供一个舒适、温馨、优美的就医、诊疗环境，同时还要表达出医院本身的人文内涵。

2. Any hospital building shall not do harm to its patients. Its interior design shall not only ensure the system of reasonable management and proper operation for various parts of the hospital's internal medical process, but also provides a comfortable, warm and beautiful environment for medical care staff, patients and their families seeking for medical treatment and diagnosis from a psychological view. At the same time, the interior design also expresses the humanistic connotation of hospitals.

3. 医院建筑空间物理环境、心理环境、人文环境对患者有一定治疗作用，可以促进患者自身拥有的自愈力，应运用环境心理学、人体工程学、信息工程学及人性化的色彩设计，创造人性化的室内外疗愈环境和信息数字化的诊疗场景。

3. The physical environment, psychological environment and humanistic environment of hospital building space bring certain therapeutic effects to patients and can promote the self-healing power of patients. It also represents the current trend and preference of hospital interior designs using environmental psychology, ergonomics, information engineering and humanized color design, and create a people-oriented interior and exterior healing environment and digitized medical scene.

4.9.1.2 医疗空间人性化设计

4.9.1.2 People-oriented design of medical space

1. 室内医疗空间设计应对的是在其空间内部工作或某一时段处于弱势地位的人群，设计应兼顾并合理地满足使用者在物质和精神方面的需求。应秉承“以人为本”的设计原则，设计出能够给使用者带来舒适、便利体验的空间环境，以体现出对患者及其家属的人文关怀。

1. The design of interior medical space aims at the people who work inside the space or are in a weak position at a certain period of time. The material and spiritual needs of users shall be considered and reasonably realized in the design. Adhere to being people-oriented and design a space environment that can bring comfortable and convenient experience to users to reflect the humanistic care for patients and their families.

2. 人性化的室内医疗空间应是具有便利性、舒适性、安全性的空间，既能方便病人就诊、舒缓患者精神压力，又能提高医务人员的工作效率、缓解工作压力。

2. People-oriented interior medical space shall be a space with convenience, comfort and safety, which can not only facilitate patients to see a doctor and relieve their mental pressure, but also improve the working efficiency of medical care staff and relieve the working pressure.

3. 通过设计的形式要素（如造型、色彩、装饰、材料等）的变化，引发人积极的情感体验和心理感受。

3. By changing the design elements (such as shape, color, decoration and material), the positive emotional experience and psychological feelings of people are triggered accordingly.

4. 设计中的造型要素是人们对设计关注点中最重要的一方面，设计的本质和特性必须通过一定的造型而得以明确化、具体化、实体化。

4. Shape elements in design play the most important role in the people's attention to design. The essence and characteristics of design must be defined, specified and materialized through certain shapes.

5. 室内造型在设计中具有重要性和引人注目之作用，应通过对建筑功能的深入探究，在日臻完善的功能中渗透人类伦理道德的优秀思想，如平等、关怀、友爱等细节设计，使人感到亲切温馨，让人感受到空间之美。

5. Interior shaping plays an important and attractive role in design. Through in-depth exploration of architectural functions, excellent ideas of human ethics and morality, such as equality, care, fraternity and other detailed designs, are integrated into the increasingly perfect functions, making people feel warmth and cordiality as well as the beauty of space.

6. 室内设计不仅要有一切为病人的服务宗旨，亦要有服务医护人员的理念，病人在身体上求助医疗的同时，又有心灵上被尊重、被理解的渴望，他们有享受就医环境的权利，设计中应体现这种指导思想；医护人员在服务病患的工作中，有压力和情绪需要缓解，因此所处的环境也应有体现；办公空间在满足功能的前提下，应创造出充满文化内涵、科室特色的环境。

6. Interior design shall not only achieve all the service objectives for patients, but also the concept of serving medical care staff. Patients have the desire to be respected and emotionally understood while seeking for medical help. They have the right to enjoy the medical environment. Therefore, the design shall embody this guiding ideology. While serving patients, the medical care staff has pressure and emotions that need to be relieved, so their environment shall also be reflected. On the premise that specific functions are realized for the office space, it needs to create an environment full of cultural connotation and department characteristics.

4.9.1.3 医疗空间色彩设计

4.9.1.3 Color design of medical space

1. 视觉信息大部分来源于色彩，应有目的、有意识地利用色彩的这一功能，让色彩成为快速传播医院形象的一种符号。

1. Most visual information comes from color, which shall be used intentionally and consciously to design colors into a symbol for the fast spreading of hospital image.

2. 应避免空间色彩单调，避免凭大致的感觉和喜好使用颜色，导致医院环境色彩整体特色不突出；应充分考虑色彩的治疗、美化和激励等使用功能，合理搭配色彩，使空间组织流线清晰明确，同时在潜意识中能平稳患者的焦虑紧张情绪。

2. The monotony of space color shall be avoided, and the use of color according to general

feeling and preference shall also be avoided since this may make the overall characteristics of hospital environment color not prominent. Full consideration shall be given to the operation functions of color such as treatment, beautification and stimulation, and the color shall be reasonably matched to make the spatial organization streamline clear and definite. At the same time, the patients can be stabilized subconsciously from anxiety and tension.

3. 在进行医疗空间色彩设计的时候要注意色调的协调统一，针对不同类别的医院、不同类型的病患群体运用不同的色调。

3. When designing the color of medical space, pay attention to the coordination and unification of colors. Different colors shall be applied to different types of hospitals and different types of patient groups.

（1）儿童医院或儿科诊室应搭配多元化色彩来渲染较为欢快的环境，可搭配相应的卡通动画来分散儿童就医的注意力，以达到减轻儿童就医的紧张、恐惧心理。

（1）Children's hospitals or pediatric clinics shall match diversified colors to render a more cheerful environment. Meanwhile, corresponding cartoons can be utilized to distract children while they are seeking for medical treatment, thus reducing their nervousness and fear.

（2）妇科医院或妇科诊所宜以清新暖色调为主，打造素雅温馨的环境。

（2）Gynecologic hospitals or clinics shall be characterized by fresh and warm colors to create an elegant and warm environment.

（3）中医院可采用展示中医文化的大自然原生态材料、中式风格、中医药元素和色系。

（3）Traditional Chinese medicine hospitals can adopt natural original ecological materials, Chinese styles and TCM elements and colors to display TCM cultures.

4. 科学合理地利用色彩能够产生对人的刺激性，以起到辅助治疗的效果。

4. The scientific and reasonable use of colors can stimulate people to achieve the effect of adjuvant therapy.

（1）室内设计应运用多种色彩来辅助治疗疾病，将色彩作为缓解肉体和精神痛苦的良方妙药。

（1）During interior design, use a variety of colors to assist in the treatment of diseases, and use colors as a panacea to relieve different physical and mental pains.

（2）科研人员应不断加大对色彩治疗的研究力度。

（2）Related researchers shall strengthen their efforts on color therapy.

（3）设计行业应不断探索将科学合理地利用色彩对人的刺激特性融于装饰设计的方法，研究出多样化和合理化的空间色彩设计手段。

（3）The design industry shall continue to explore the ways to integrate scientific and reasonable use of color in stimulation into decoration design, and to study diversified and reasonable space color design approaches.

4.9.1.4 重点空间设计

4.9.1.4 Key space designs

1. 在医疗保健设施建筑空间中，大厅、公共通道、电梯厅、接待、办公、商业服务等部分可划分为统一类型，此类空间设计与其他公共建筑相似，专业的医疗设备、机电桥架较少，因此设计受限小，是感性表达的重要发挥场地，同时也是形象展示的地方，通常会作为重点部位设计。

1. In the building space of medical and health facilities, the atrium, public passage, elevator hall, reception space, office, commercial service center and other parts can be divided into unified types. Such space design is similar to other public buildings, with fewer professional medical equipment and mechanical and electrical bridges. Therefore, the design is limited and provides an important place for emotional expression. It also provides a place for image display. These parts are usually designed as a key part.

2. 应遵循功能与美学并重的设计理念，在设计中将医院的文化空间、造型主题的设计、色彩、照明、空调节能的规划、装饰材料抗菌耐久性、标识系统与软装陈设的考虑及室内、外环境的风格整合等做整体考量，最终目标是将医院的功能诉求与人性关怀进行完美整合，空间设计与医疗工艺设计相结合，设计出来的空间能让所有的病患及家属感觉亲切舒适，也为医护行政管理人员提供一个良好的医疗办公环境，创造医疗空间的最高价值。

2. In this sense, it needs to follow the design concept of paying equal attention to function and aesthetics. In the design, the cultural space of hospital, the design of modeling theme, the color selection, the energy conservation planning of lighting and air conditioning, the anti-bacterial durability of decorative materials, the identification system and soft furnishing, and the style integration of interior and exterior environment shall be considered as a whole. The ultimate aim is to perfectly integrate the functional needs of the hospital with humanistic care. The space design shall be combined with the medical process design. The designed space shall make all patients and their family members feel cordiality and comfort. It shall also provide a good medical office environment for medical administrators and create the highest value of medical space.

3. 要保障设施满足所有人群使用的安全和方便，包括各类病患及家属、医职员工和研究者。

3. It is necessary to ensure that the facilities are safe and convenient for all people, including all kinds of patients and their family members, medical care staff and researchers.

4. 门诊及出入院大厅是各科患者、陪护人员均需要到达的空间，是建筑的交通枢纽，大厅内服务功能较多、人员密集，设计应满足净化空间、改善、合理布局、防止交叉感染，理顺患者、陪护、探视及医务工作人员的动线等要求；要处理好建筑内部、房间内部的各种动线。

4. Design of the outpatient and admission/discharge hall: The hall is the space that patients and accompanying personnel of different departments need to reach, and is also the transportation hub of the building. It has many service functions and holds a large number of people. The design shall achieve the objectives of purifying the space, improving and rationally arranging the space, preventing cross infection and straightening the flows of patients, accompanying personnel, visiting personnel and medical care staff. Furthermore, various streamlines inside the building and different rooms shall be properly handled.

5. 不仅对建筑物内部界面进行装饰，还应对医疗工艺流程和动线组织进行深入细致的规划，使内部空间顺畅、路线舒适、操作合理。

5. Not only the interior building interface shall be decorated, but also the medical process and streamline organization shall be carefully planned to make the interior space smooth, and the route comfortable and the operation reasonable.

6. 不仅要注重硬件装修，还应对软件环境加以塑造。

6. Furthermore, it needs to pay attention to not only hardware decoration, but also the shape of software environment.

7. 标识系统应与内装相结合，对家具、软装饰、无障碍设施、声光控制、色彩计划以及植物等统一规划配置，以达到预期效果。

7. The identification system shall be combined with the interior decoration. It needs to uniformly plan and configure the furniture, soft decoration, barrier-free facilities, sound and light control, colors and plants to achieve the desired effect.

4.9.1.5 材料选择

4.9.1.5 Material selection

1. 医疗空间是治疗和护理病患的空间，为了能够从物理和心理层面向病患传递出健康的信息，同时又能反映出医院国际化的水平，体现医院具有的高科技感和权威性，材料的选择上要考虑医院的特殊性，选择安全环保、耐污、易清洁、耐腐蚀、装饰性强的材料。

1. Medical space is a space for treating and nursing patients. In order to transmit health information to patients from physical and psychological aspects and reflect the internationalization, high technology and authority of the hospital, it needs to consider the particularity of hospital during material selection. Meanwhile, select the secure, environment-friendly, pollutant-resistant, easy-to-clean, corrosion-resistant and highly decorative materials.

2. 抗菌板类材料具有强度高，稳定耐久，耐水耐湿、防火、耐磨、抗菌、易清洗维护、花色多样等特性，可以用在病区的护士站、走廊、休闲活动区等公共空间的墙面。

2. Bacteria-resistant board materials have the characteristics of high strength, stability, durability, water resistance, moisture resistance, fire prevention, abrasion resistance, bacteria resistance, easy cleaning and maintenance and diverse colors. It can be used for the walls of public spaces, such as nurse stations, corridors and leisure areas of wards.

3. PVC、橡胶类同质透底卷材地板具有耐磨、抗菌、无缝易清洁、吸声、减噪、抗静电、脚感舒适、颜色多样等特性，适合用在人流量少的住院病区、医技用房及洁净室地面。

3. PVC and rubber homogeneous transparent coiled material floor has the characteristics of abrasion resistance, bacteria resistance, seamless connection and easy cleaning, sound absorption, noise reduction and electrostatic prevention, comfortable foot feeling, and various colors. It can be used as the floor of ward areas, technical rooms and clean rooms with a lower pedestrian flow.

4. 铝板、复合铝板、彩钢板具有平整度高、强度高、耐久、耐污、科技感强且可以仿木等特性，可用于大厅、走廊、洁净室、卫生间、雨篷底等空间的吊顶或墙面。

4. Aluminum plate, composite aluminum plate and color steel plate have the characteristics of high flatness, strength, durability, stain resistance, strong sense of science and technology and can be similarly used as wooden material. It can be used for suspended ceilings or wall surfaces in public space, such as hall, corridor, clean room, washroom, canopy and so on.

5. 普通水磨石、彩色水磨石具有传统工艺美感，花色多样、配色灵活、小面积无缝、大面积少缝，具有耐久、抗霉菌、可水洗、耐磨损、耐烟头烧、抗划痕、可塑造曲面、造价低等特点，适用于各类基层医疗保健设施；基于玉石级别和配比，也可成为局部空间应用的高档装饰材料。

5. Ordinary terrazzo and colored terrazzo have the aesthetic feeling of traditional process, with various colors, flexible color match, seamless connection in small area, less seam in large area and are durable, fungi resistant, washable, abrasion resistant, cigarette burn resistant, scratch resistant, curved surface moldable and low cost. Therefore, they are suitable for various primary medical and health care facilities. Based on the grade and proportion of jade, it can also become a high-grade decorative material for local space.

6. 因医院使用抗生素，不应使用易发霉、滋生细菌的材料，如石膏板、实木、普通墙纸。

6. Due to the use of antibiotics in hospitals, materials that are prone to mildew and bacteria, such as gypsum board, solid wood and common wallpaper, shall be avoided.

4.9.2　标识系统设计

4.9.2　Identification system design

4.9.2.1　医院环境导向系统是指能传递医疗功能、环境等信息以及医院服务理念的标识系统，它依据医院环境的结构特征及诊疗科室、职能科室的服务功能建立，以便能在最短时间内把患者、访客引导到目的地。

4.9.2.1　Hospital environment guidance system refers to the identification system that can transmit medical function, environment and other information as well as hospital service concepts. It is established according to the structural characteristics of the hospital environment and the service functions of diagnosis and treatment departments and functional departments, so as to guide patients to their destinations within the shortest time.

4.9.2.2　环境与道路标识的视觉传达

4.9.2.2　Visual communication of environment and road markings

“环境道路标识导向系统”构成是以提高医疗机能与效率作为最重要的追求目标。视觉传达上以文字与符号、线性方位指示、导向图标、图示等组成的标识系统。该系统设计应标准化，符合大多数设施系列性、功能性与管理要求，标准化的图示、文字表达与国际接轨，能够更好、更深刻地体现人性化，利用设计语言满足公众的需求，其中渗透对使用者的关怀。

The environment and road marking and guidance system is established to improve medical function and efficiency as its most important aim. In visual communication, the identification system is composed of characters and symbols, linear orientation indication, guide icon, graphic representation and so on. This system shall be standardized and can meet the series, function and management requirements of most facilities. Standard graphic and text expressions are in line with international standards, and can embody humanity more effectively and profoundly. Design languages are used to meet the public needs full of care for the users.

4.9.2.3 环境道路导向系统空间构成

4.9.2.3 Spatial structure of the environment and road guidance system

1. 对于空间复杂的医疗建筑建立能够易于把握的系统方法，是其空间环境构成的重要指标。良好的标识导向系统源于科学的导向分类和与之配合的设计要素。科学的导向分类是依据就诊人员的到达及其阅读习惯，由外至内、由大至小、由先至后的顺序设计。它综合了建筑环境、标识信息、专用符号、专用色彩和专用字体等应用条件，是医疗服务设施中视觉传播应用要素的重点。从分级导向以社会道路进入院区范围的引导为开始，对于不同区域、场所、目的地进行分流引导。主要形式一般从容易明确地表现标识性的地形图为主导，明确地区分各区域的标识特征，同时配合容易形成辅助记忆关系的简明图像设计。

1. It is an important indicator of the composition of the space environment for health care buildings with a complex space to establish a systematic method that can be easily grasped. An identification and guidance system originates from scientific guidance classification and corresponding design elements. The scientific guidance classification is designed according to the arrival and reading habits of medical care staff from outside to inside, from big to small, and from first to last. It integrates the application conditions such as architectural environment, identification information, special symbols, special colors and special fonts, and is the focus of elements for visual communication in medical service facilities. The system design starts with the guidance of social roads entering the hospital area, and conducts diversion for different areas, places and destinations. In general, the main form is dominated by topographic maps that easily and clearly show markings, distinguish the identification features of each area, and combine with the concise image design that easily forms assisted memories.

2. 对于要到达的不同目的地、场所，如缴费、候诊、检查、治疗、病房等地点的标识，有必要直接使用文字，明确易懂，但要注意与所具有同样性质的其他场所区别，比如另外科室的相同诊疗室等，则要考虑通过其他方法，如辅助的色彩标记或以带有同具体部位相吻合的标识性图形来区分。建筑环境应发挥出其作为背景的功能，利用空间规模、形态、地面、墙面的材料、色彩、照明要素来强调空间和场所的个性化并达到使道路容易识别理解。

2. For the markings of different destinations and places to be reached, such as payment, waiting for diagnosis, examination, treatment, ward and other places, it is necessary to directly use words and make them clear and easy to understand. However, it is necessary to observe the difference from other places with the same nature, such as the same diagnosis room in another department. Under such circumstance, other methods, such as auxiliary color markings or indica-

tive graphics matching with specific parts, shall be taken into account for discriminations. The architectural environment shall play its role as the background, and use the space scale, shape, floor, wall materials, colors, lighting elements to emphasize the individuality of space and place and to make the road easily identified and understood.

3. 在初期制订方案时必须考虑空间等级与层次，对大量患者的活动空间及与之相连的主要空间区分为主要的、辅助的不同等级。把握整体构成、总的动线、辅助动线等具体图像的区分、设定，形成重要的空间诱导等级分块地图。中央接待大厅、外来等候大厅、休息室、能看到内外景色的特定场所与总动线的关系，应在地图上明确反映。为对人们形成自然导向，首先是能看到目的地，场所的设计要求能从远处进行识别。例如：交费处位于壁面拐角处，但如果其墙壁具有独立的形式、色彩，照明也比一般空间突出，则可以形成从周围空间中“浮现”出来的效果。

3. Space level and hierarchy divide the activity space for a large number of patients and the main space connected with it into different primary and auxiliary levels, which must be taken into account during initial planning. This aims to grasp the clear distinction and setting of specific images such as overall streamline and auxiliary streamline, and form block maps of important spatial induction levels for main spaces, such as central reception hall, external waiting hall and lounge. The relation between specific places with interior and exterior sceneries and the overall streamline shall be clearly reflected on the map. In order to develop a natural guidance, people shall be firstly enabled to see the destination and the design of some places can be identified from a distance. For example: The payment office is located at the corner of the wall. If the wall has independent style and color and the lighting is more prominent than other spaces, the effect of “emerging” from the surrounding space can be created accordingly.

4.10 医疗保健设施固定家具

4.10 Fixed furniture for health care facilities

4.10.1 范围

4.10.1 Scope

是指须与土建装修及安装工程同步完成交钥匙的固定家具，不包括活动家具，是医疗工艺条件设计必须提出的定制型末端设施；应与建筑隔断、室内硬装饰、配电末端、给排水末端、通风排风口、信息点、物流站点、垃圾自动收集点等同步进行定位、尺寸、造型、选材、防火和隔声设计；应与建筑工程同步组织施工、定制、采购或现场加工。

It refers to the fixed furniture whose turnkey operation must be completed with that of the civil decoration and installation works simultaneously, excluding the movable furniture. It is customized terminal facility that must be proposed in the design of medical process conditions. It shall be positioned, sized, shaped, selected, fireproof and soundproof in synchronization with building partition, interior hard decoration, power distribution terminal, water supply and drainage terminal, ventilation and exhaust outlet, information point, logistics station and automatic garbage collection station, etc. It shall be constructed, customized, purchased or processed on site simultaneously with the construction works.

4.10.2 分类

4.10.2 Classification

4.10.2.1 综合服务接待台、护士站、开放式管理中心。

4.10.2.1 Comprehensive service reception desk, nurse station and open management center.

4.10.2.2 医疗及辅助区域所用的与照明、排风口、上下水对应的治疗台、配液台、处置台、污物回收台、清洗槽台等固定家具。

4.10.2.2 Fixed furniture for medical treatment and auxiliary areas, such as treatment tables, liquid distribution tables, disposal tables, dirt recovery tables, cleaning tank tables, etc. corresponding to lighting, exhaust outlets, and water supply and drainage.

4.10.2.3 手术部、洁净室中各类嵌入墙体内的器械柜、药品柜、麻醉柜、器械台及各类无菌物品存放、手术器械存放、麻醉物品存放的固定家具。

4.10.2.3 Various instrument cabinets, medicine cabinets, anesthetist cabinets, instrument tables embedded in the walls, as well as fixed furniture for the storage of various sterile articles, surgical instruments and anesthesia articles in the operating department and clean room.

4.10.2.4 消毒供应室内与供配电、蒸汽供配、给排水管道、通风排气管道敷设相关联的各类消毒灭菌机柜、手工清洗槽、传送带等固定家具或封口饰面组合柜。

4.10.2.4 Various disinfection and sterilization cabinets, manual cleaning tanks, conveyor belts and other fixed furniture or sealing veneer combined cabinets associated with the laying of power supply and distribution, steam supply and distribution pipes, water supply and drainage pipes, and ventilation and exhaust pipes in the disinfection supply room.

4.10.2.5 药房、静脉配制、肠外营养配置、厨房中与全自动配药机、药品发放台、派餐台、餐具自动回收设备等相关联的固定家具。

4.10.2.5 Fixed furniture associated with full-automatic dispensing machines, drug distribution tables, food delivery tables, automatic tableware recovery equipment, etc. in the pharmacy, intravenous preparation room, parenteral nutrition preparation room and kitchen.

4.10.2.6 检验科、病理科、实验室等室内与通风、排风、供电照明、给排水管道相关联的样本采集台、实验工作台、生物安全柜、洗涤台等固定家具。

4.10.2.6 Fixed furniture, such as the sample collection tables, experimental workbench, biosafety cabinets and washing tables associated with ventilation, exhaust, power supply and lighting, and water supply and drainage pipes in the laboratory department, pathology department, laboratory, etc.

4. 10. 2. 7 影像设备机房、控制室与设备管线关联的工作台。

4. 10. 2. 7 Worktables associated with device pipelines in the imaging device room and control room.

4. 10. 2. 8 病房、治疗室等房间内的固定壁柜。

4. 10. 2. 8 Fixed wall cabinets in the ward, therapy room and other rooms.

4. 10. 3 功能要求

4. 10. 3 Functional requirements

4. 10. 3. 1 减少可引起医疗感染的表面污染。

4. 10. 3. 1 Reduce surface pollution that may cause medical infection.

1. 表面易于清洁，无表面接缝或裂缝；

1. The surface is easy to clean, without surface seams or cracks;

2. 装饰材料不渗透；

2. The decorative materials are impervious;

3. 表面无孔、光滑、不反光。

3. The surface is non-porous, smooth and non-reflective.

4. 10. 3. 2 减少患者跌倒和相关伤害。

4. 10. 3. 2 Reduce falls and related injuries of patients.

所有家具不能有坚硬的、尖锐的边角，以免对跌倒的病人造成伤害。

All furniture must not have hard, sharp corners to avoid injury to the fallen patient.

4. 10. 3. 3 减少用药差错。

4. 10. 3. 3 Reduce medication errors.

1. 用药安全区的固定家具应具有内置式照明，在执行关键任务的区域使用可调节的高亮度工作灯，灯具应不发热、节能。

1. The fixed furniture in the medication safety area shall be equipped with built-in lighting. Adjustable high-brightness working lamps shall be used in the areas where key tasks are performed. The lamps shall be non-heating and energy-saving.

2. 减少药物准备、分配和管理期间，因家具而引起的视觉干扰和噪声干扰。

2. Reduce visual and noise interference caused by furniture during drug preparation, distribution and management.

4. 10. 3. 4 应满足正常人及残障人士的使用尺度，并有助于医护作业，利于医护与患者和家属之间的沟通。

4. 10. 3. 4 Meet the use standard of normal people and the disabled, and be conducive to medical work and to the communication between medical care staff and patients and their families.

4. 10. 3. 5 应符合人体工程学及舒适性要求，满足操作台上摆放设备、用具、重度肥胖病人施压等的承载力。

4. 10. 3. 5 Meet the requirements of ergonomics and comfort, and meet the bearing capacity of devices, appliances and patients with severe obesity on the operating platform.

4.10.3.6 应有助于节约室内空间面积，提高员工工作效率和沟通效率，满足嵌入式电脑、视频观察设备、电气控制设备、消火栓、标识牌、伸缩式倒便器、安装衣服物品挂钩、家具门扇开启的空间尺寸要求。

4.10.3.6 Help to save indoor space area, improve staff working efficiency and communication efficiency, and meet the space size requirements for embedded computers, video observation devices, electrical control devices, fire hydrants, signboards, retractable toilet bowls, installation of clothes hooks and opening of furniture door leaves.

4.10.3.7 确保环境安全，材料不含挥发性有机化合物（Vocs）如甲醛和苯，符合防火规范的耐燃、不燃要求。

4.10.3.7 Ensure environmental safety, and materials are free of volatile organic compounds, such as formaldehyde and benzene, and meet the requirements of fire resistance and non-flammability in the fire prevention code.

4.11 景观设计与亮化

4.11 Landscape design and lighting

4.11.1 院区环境设计

4.11.1 Environment design of hospital area

4.11.1.1 应充分利用地形、防护间距和其他空地布置绿化，并应有供病人康复活动的专用绿地。

4.11.1.1 The terrain, protective space and other open spaces shall be made full use of for greening, and there shall be special green space for patients' rehabilitation activities.

4.11.1.2 应对绿化、装饰、建筑内外空间和色彩等做综合性处理。

4.11.1.2 Comprehensive treatment shall be made for greening, decoration, internal and external space and color of the building.

4.11.1.3 在儿科用房及其入口附近，宜采取符合儿童生理和心理特点的环境设计。

4.11.1.3 In pediatric rooms and the vicinity of their entrances, it is advisable to adopt an environmental design that conforms to the physiological and psychological characteristics of children.

4.11.1.4 基地的环境设计应综合布置绿化、花坛、喷泉、坐凳、雕塑和小品建筑等各种景观内容。

4.11.1.4 The base shall be comprehensively equipped with greening, flower beds, fountains, benches, sculptures, pavilions and other landscape contents in terms of environment design.

4.11.1.5 场地建设对原有生态环境景观的有价值的树木、水塘、水系的保留，不但具有较高的生态价值，而且是传承场地所在区域历史文脉的重要载体，也是该区域重要的景观标志。

4.11.1.5 As for the construction of the site, the valuable trees, ponds and water systems in the original ecological landscape retained by the site not only have high ecological value, but also are important carriers to inherit the historical context of the site area, as well as important landscape signs of the area.

4.11.1.6 室外环境设计宜根据用地的自然条件，结合各功能区的特点，对景观、植物配置及山石水面等做出综合设计。

4.11.1.6 In terms of outdoor environment design, it is advisable to make a comprehensive design of landscape, plant configuration as well as mountain, stone and water surface according to the natural conditions of the land and the characteristics of each functional area.

4.11.2 院区环境和建筑亮化

4.11.2 Environment and building lighting of hospital area

院区入口、建筑主体和急诊急救楼部位，应写明醒目的名称，并配合鲜明的灯光照明设计，达到一目了然的效果。

The entrance of the hospital area, the main building and the emergency and first-aid building shall be marked with the striking enlarged name and be matched with bright lighting design to achieve the purpose of being clear at a glance.

第5章

医疗保健设施分类设计

Chapter V

Classified Design of Health Care Facilities

朱 希 张柏林

Xi Zhu Bolin Zhang

5.1 分级诊疗设施分类

5.1 Classification of tiered diagnosis and treatment facilities

5.1.1 分级诊疗

5.1.1 Tiered diagnosis and treatment

5.1.1.1 分级诊疗指按照疾病的轻重缓急及治疗的难易程度进行分级，不同级别的医疗机构承担不同疾病的治疗，逐步实现从全科到专业化的医疗过程。

5.1.1.1 Tiered diagnosis and treatment refers to a kind of medical process that a tiered medical service system is formed according to the priority of the disease and the difficulty of treatment, and different levels of medical institutions undertake the treatment of different diseases, gradually realizing from general practice to specialization.

5.1.1.2 应大力推行分级诊疗，把对各类医疗保健人群的服务尽量分散到基层医疗保健设施中，不断优化城市规划中医疗卫生系统的专项规划，供应充足的、布点均匀的医疗卫生用地，城乡规划部门应与卫生健康部门联合制定合理的社区医疗卫生配套建筑面积比例，控制三级医院不断扩大规模、高密度化的趋势。

5.1.1.2 It is necessary to vigorously promote tiered diagnosis and treatment, distribute the services to various medical and health care groups to the primary health care facilities as far as possible, constantly optimize the special planning of the medical and health system of urban planning, and provide sufficient and evenly distributed medical and health land. Urban and rural planning departments and health and sanitation departments shall work out a reasonable proportion of medical and health supporting building area in the community, and control the trend of expanding scale and high density of third-class hospitals.

5.1.1.3 基层医疗卫生机构门诊：家庭医生服务，老年人、孕产妇、0~6 岁儿童、慢性疾病（高血压、糖尿病、结核病等）患者和严重精神障碍患者等重点人群的签约服务、健康管理服务，贫困人口、残疾人和计划生育特殊家庭成员的签约服务工作。

5.1.1.3 Outpatient department of primary medical and health institutions: family doctor service; signing service and health management service for key groups such as the elderly, pregnant and lying-in women, children aged 0 ~ 6, patients with chronic diseases (hypertension, diabetes, tuberculosis, etc.) and patients with severe mental disorders; and signing service for the poor, the disabled and members of family planning special families.

5.1.1.4 三级医院门诊：要开展日间手术，扩大日间手术病种范围，提高日间手术占择期手术的比率，缩短患者等待住院和等候手术时间。设置日间病房、日间治疗中心等，为患者提供适宜的日间诊疗服务，提高床单元使用效率。要调整门诊病种结构，减少常见病、多发病、慢性病患者占比，增加手术、急危重症的诊疗量占比。

5.1.1.4 Outpatient department of third-class hospital: It is necessary to carry out day surgery, expand the scope of disease species of day surgery, increase the proportion of day surgery in elective surgery, and shorten the waiting time for hospitalization and surgery. Daytime wards, daytime treatment centers, etc. shall be set up to provide suitable daytime diagnosis and treatment services for patients and improve the use efficiency of sickbed units. It is necessary to adjust the structure of disease species in the outpatient department, reduce the proportion of patients with common diseases, frequently-occurring diseases and chronic diseases, and increase the proportion of diagnosis and treatment volume for surgery and acute disease or critical disease.

5.1.1.5 对分级诊疗的各类设施均应进行分类前期研究、策划和规划设计，以创造良好的各层级医疗环境，提高和改善人民就医体验。

5.1.1.5 Various facilities for tiered diagnosis and treatment shall be studied, planned and designed by classification in advance in order to create a good medical environment at all levels and improve people's medical experience.

5.1.2　医联体

5.1.2　Medical treatment alliance

5.1.2.1 以设区的地市和县域为单位，将服务区域按照医疗资源分布情况划分为若干个网格，每个网格由一个医疗集团或者医共体负责。三级医院和康复、护理等慢性病医疗机构可以跨网格服务，检查检验、消毒供应、特殊临床专科技术等可以区域内资源共享。

5.1.2.1 Taking prefecture-level cities and counties divided into districts as units, the service area is divided into several grids according to the distribution of medical resources, and each grid is the responsibility of a medical group or medical community. Third-class hospitals and medical institutions for chronic diseases involving rehabilitation and nursing can provide cross-grid services, and inspection, disinfection and supply, special clinical expertise and other resources can be shared within the region.

5.1.2.2 以医疗集团和医共体为网格内居民提供疾病预防、诊断、治疗、康复、护理等一体化、连续性医疗服务。鼓励中医医院牵头建设医联体，发挥中医药在治未病、疾病治疗和康复中的重要作用。发挥妇幼保健机构预防、保健和诊疗作用，为网格居民提供妇幼保健服务，建立防、治、康协同机制，形成以健康为中心的服务模式。

5.1.2.2 The integrated and continuous medical services such as disease prevention, diagnosis, treatment, rehabilitation and nursing are provided for the residents in the grids through medical group and medical community. Traditional Chinese medicine hospitals are encouraged to take the lead in building medical treatment alliance, so as to play an important role of traditional Chinese medicine in the preventive treatment of disease, disease treatment and rehabilitation. The prevention, health care and diagnosis and treatment functions of maternal and child health care institutions have been brought into play, providing maternal and child health care services, establishing a coordinated mechanism of prevention, treatment and health, and forming a health-centered service model for residents in the grids.

5.1.3 区域医疗中心

5.1.3 Regional medical center

符合条件的国家级、省级医院开展国家医学中心、国家区域医疗中心建设工作。

Eligible national and provincial hospitals will carry out the construction of national medical centers and national regional medical centers.

5.1.4 县医院

5.1.4 County hospital

以县医院能力建设为重点，推进分级诊疗城乡分开。

Focusing on the capacity-building of county hospitals, separately promote tiered diagnosis and treatment in urban and rural areas.

5.2 社区医疗与诊所

5.2 Community medicine and clinics

5.2.1 社康中心

5.2.1 CHSC

5.2.1.1 全称是社区健康服务中心，它是专有名称，使用统一的专用标识、机构标牌及内部环境，跟私人诊所有明显区别。

5.2.1.1 The full name is the community health service center, which is a proper name. It uses a unified special logo, institutional sign and internal environment, which is obviously different from the private clinic.

5.2.1.2 社康中心两个职能

5.2.1.2 Two functions of community health service center

1. 提供九项国家基本公共卫生服务项目：建立居民健康档案、健康教育、预防接种、传染病防治、儿童保健、孕产妇保健、老年人保健、慢性病管理、重性精神疾病管理等。

1. Nine national basic public health services are provided: establishment of resident health records, health education, preventive vaccination, prevention and treatment of infectious diseases, child health care, pregnant and lying-in woman health care, elderly health care, chronic disease management, severe mental disease management, etc.

2. 提供的基本医疗服务：包括一般常见病、多发病诊疗、护理，慢性病治疗，社区现场应急救护，家庭医疗服务，康复医疗服务等。

2. Basic medical services provided: including diagnosis and treatment and nursing of common and frequently-occurring diseases, treatment of chronic diseases, emergency rescue at community sites, family medical services, rehabilitation medical services, etc.

5.2.1.3 按规模大小一般分为三类

5.2.1.3 It is generally divided into three classes according to scale

1. 一类社康中心规模较大，有全科诊室、中医诊室、康复治疗室、抢救室、预防保健科室、检验室、B超室、心电图室、药房等。

1. Class Ⅰ community health service center is of a large scale, including general practice consulting room, traditional Chinese medicine consulting room, rehabilitation therapy room, emergency room, prevention and health care department, testing room, ultrasound room, ECG room, pharmacy, etc.

2. 二类社康中心中等规模：功能科室和一类大致相同，一般没有B超检查。

2. Class Ⅱ community health service center is of a medium scale: The functional departments are basically the same as Class Ⅰ, and there is generally no ultrasound examination.

3. 三类社康中心规模最小：一般没有中医诊室和B超检查，其他和二类大致相同。

3. Class Ⅲ community health service center is of the smallest scale: Generally, there is no traditional Chinese medicine consulting room and ultrasound examination, and the others are roughly the same as Class Ⅱ.

5.2.1.4 社康中心诊治范围：一般常见病、多发病（如感冒、发烧、腹泻等）、诊断明确的慢性病（如高血压、糖尿病）和康复期患者。

5.2.1.4 Diagnosis and treatment scope of community health service center: common diseases, frequently-occurring diseases (such as cold, fever, diarrhea, etc.), chronic diseases (such as hypertension, diabetes) with definite diagnosis and patients in rehabilitation period.

5.2.2 诊所

5.2.2 Clinic

5.2.2.1 包括：中医诊所、中西医结合诊所、口腔诊所、美容整形外科诊所、医疗美容诊所、精神卫生诊所、企业学校卫生保健所、卫生站、卫生所、医务室。

5.2.2.1 Including: Traditional Chinese medicine clinic, integrated traditional Chinese and western medicine clinic, stomatology clinic, cosmetic plastic surgery clinic, medical beauty clinic, mental health clinic, enterprise school health care center, health station, health center, medical clinic.

5.2.2.2 全科+健康管理型基于互联网医疗平台的全科诊所，主要设置内科、外科、妇科、儿科等全科科室，提供备孕计划、抗癌计划以及健康管理服务。

5.2.2.2 General practice + health management general practice clinic based on Internet medical platform, mainly provided with internal medical department, surgical department, gynecology department, pediatric department and other general departments, providing pregnancy preparation plan, anti-cancer plan and health management services.

5.2.3 村卫生室

5.2.3 Village clinic

基本标准：至少设有诊室、治疗室、药房，每室必须独立，建筑面积不少于$40m^2$。

Basic standard: at least consulting room, therapy room and pharmacy shall be set up, each room must be independent and the construction area shall not be less than $40m^2$.

5.3 综合医院

5.3 General hospital

5.3.1 综合医院定义

5.3.1 Definition of general hospital

医院是以诊治疾病、照护并认为主要目的的医疗机构，综合医院是各类型医院的主体。凡城镇以上医院，同时具备以下条件：

The hospital is a medical institution whose main purpose is to diagnose and treat diseases and care for patients. The general hospital is the main body of all kinds of hospitals. Hospitals above the town level combine the following conditions:

5.3.1.1 应设置包括大内科、大外科、妇产科、儿科、五官科等三科以上疾病科。

5.3.1.1 Hospitals with more than three disease departments, including comprehensive internal medical department, comprehensive surgical department, obstetrics and gynecology department, pediatric department, and ophthalmology and otolaryngology department.

5.3.1.2 应设置门诊部及24小时服务的急诊部和住院部。

5.3.1.2 Hospitals with outpatient department, 24-hour emergency department and inpatient department.

5.3.1.3 设有药剂、检验、放射等医技部门，拥有相应人员、设备的医院。

5.3.1.3 Hospitals with pharmacy department, laboratory department, radiology department and other medical technology departments and corresponding personnel and devices.

5.3.2 综合医院分类

5.3.2 Classification of general hospital

采用“三级医疗网”医疗体制：见图5-1。

Adopt the “three-class medical network” medical system: see Figure 5-1.

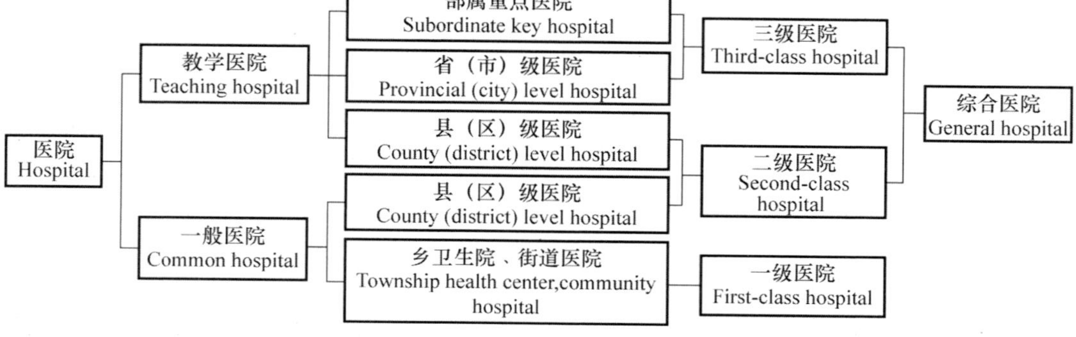

图5-1 “三级医疗网”医疗体制

Figure 5-1 The “three-class medical network” medical system

5.3.3　综合医院建设用地指标

5.3.3　Construction land index of general hospital

综合医院的建设用地包括急诊部、门诊部、住院部、医技科室、保障系统、行政管理和院内生活用房等七项设施的建设用地、道路用地、绿化用地、晾晒用地（用于洗涤物品的晾晒）和医疗废物与日产垃圾的堆放、处置用地，按下表计算。

The construction land of the general hospital includes the construction land, road land, greening land, drying land (used for the drying of washing articles), and the stacking and disposal land of medical waste and daily garbage of the emergency department, outpatient department, inpatient department, medical technology department, security system, administrative management and hospital living room, which are calculated according to the following table.

综合医院建设用地指标见表 5-1。

Construction land index of general hospital (m^2/bed), see Table 5-1.

表 5-1　综合医院建设用地指标

Table 5-1　Construction land index of general hospital (m^2/bed)

建设规模（床） Construction scale(beds)	200 ~ 300	40 ~ 500	600 ~ 700	800 ~ 900	1000
用地指标（m^2/床） Land index(m^2/bed)	117	115	113	111	109

5.3.4　综合医院功能分区

5.3.4　Functional division of general hospital

5.3.4.1　综合医院按功能划分三大区：医疗区、医技区、后勤供应区。

5.3.4.1　The general hospital is divided into three areas by function: medical area, medical technology area and logistics supply area.

1. 医疗区

1. Medical area

（1）门诊部：急诊部或急救中心。

(1) Outpatient department: emergency department or first-aid center.

（2）住院部：各科病区、手术、重症监护病房、产房等。

(2) Inpatient department: inpatient ward, operating room, intensive care unit, maternity ward, etc.

2. 医技区

2. Medical technology area

药剂、影像诊断、检验、病理、放疗、功能检查、高压氧、核医学、输血、内窥镜、理疗等。

Pharmacy, imaging diagnosis, examination, pathology, radiotherapy, functional examination, hyperbaric oxygen, nuclear medicine, blood transfusion, endoscope, physiotherapy, etc.

3. 技术供应服务区：制剂室、中心消毒供应室、营养厨房、各类设备用房、实验动物房、洗衣房、太平间、危险品库、垃圾房等。

3. Technical supply service area: manufacturing laboratory, central disinfection supply room, nutritious kitchen, various device rooms, laboratory animal room, laundry room, morgue, dangerous goods warehouse, garbage room, etc.

4. 行政管理区：办公室、病案库、信息中心、维修等。

4. Administrative area: office, medical record storeroom, information center, maintenance, etc.

5. 科研、教学区：教室、实验室、会议中心等。

5. Research and teaching area: classroom, laboratory, conference center, etc.

6. 生活区：职工厨房、食堂、值班室、住院医生和实习生宿舍、进修医生宿舍等。

6. Living quarter: staff kitchen, canteen, duty room, dormitory for resident doctor and intern, dormitory for trainee doctor, etc.

5.3.4.2 功能分区设置出入口：

5.3.4.2 Entrance and exit set by functional division:

出入口分类见表 5-2。

Entrance and exit classification table, see Table 5-2.

表 5-2 出入口分类

Table 5-2 Entrance and exit classification

<table>
<tr><td rowspan="5">主出入口
Main entrance and exit</td><td rowspan="5">人员出入口
Entrance and exit for staff</td><td>门诊病人出入口
Entrance and exit for outpatients</td><td rowspan="2">门急诊病人出入口
Entrance and exit for outpatients and emergency patients</td></tr>
<tr><td>急诊病人出入口
Entrance and exit for emergency patients</td></tr>
<tr><td colspan="2">住院病人和探视人员出入口
Entrance and exit for inpatients and visitors</td></tr>
<tr><td colspan="2">医务人员出入口
Entrance and exit for medical care staff</td></tr>
<tr><td colspan="2">传染科病人出入口
Entrance and exit for infectious diseases patients</td></tr>
<tr><td rowspan="4">次出入口
Secondary entrance and exit</td><td rowspan="2">供应入口
Supply entrance</td><td colspan="2">医疗物品入口
Entrance for medical articles</td></tr>
<tr><td colspan="2">生活物品入口
Entrance for articles for daily use</td></tr>
<tr><td rowspan="2">污物出口
Waste exit</td><td colspan="2">尸体出口
Corpse exit</td></tr>
<tr><td colspan="2">废弃物出口
Waste exit</td></tr>
</table>

5.3.5 洁污路线设置要点

5.3.5 Key points of cleaning route setting

5.3.5.1 出入口设置和道路布置要求功能明确、使用方便、洁污分离，力求高效率，并应避免和减少交叉感染。

5.3.5.1 The entrance and exit settings and road layout require clear functions, convenient use, separation of clean and dirty articles, and strive to achieve high efficiency, and shall avoid and reduce cross infection.

5.3.5.2 尸体运送路线应避免与出入院交叉，并避免经门诊诊察室附近和病房楼前面。

5.3.5.2 The corpse transportation route shall avoid crossing with the hospital entrance and exit, and avoid passing near the outpatient consulting room and in front of the ward building.

5.3.5.3 职工住宅与医院基地毗邻时，应予以分隔。

5.3.5.3 When the staff residence is adjacent to the hospital base, they shall be separated.

5.3.5.4 医院至少有两处出入口，一为人员出入口，二为供应入口；最好分为三处，将供应入口和污物出口分开；设有传染病科者，必须另设专用出入口。

5.3.5.4 The hospital shall have at least two entrances and exits: entrance and exit for personnel and the supply entrance; it is better to divide them into three parts, separating the supply entrance and the waste exit; if there is a department of infectious diseases, a special entrance and exit must be set up.

5.4 中医医院

5.4 Traditional Chinese medicine hospital

5.4.1 中医医院等级标准

5.4.1 Class standard of traditional Chinese medicine hospital

5.4.1.1 有关规定：《国家中医药管理局关于积极发展中医预防保健服务的实施意见》《中医预防保健服务提供平台建设基本规范》和《中医医院“治未病”科建设与管理指南》。

5.4.1.1 Relevant provisions: Opinions on NATCM on Actively Developing TCM Preventive Health Care Service, Basic Standards for Construction of TCM Preventive Health Care Service Delivery Platform and Guidelines for Construction and Management of “Preventive Treatment of Disease” Department in TCM Hospital.

5.4.1.2 基本要求：中医医院的门诊中医药治疗率不低于85%，病房中医药治疗率不低于70%。

5.4.1.2 Basic requirements: the outpatient treatment rate of traditional Chinese medicine in traditional Chinese medicine hospital is not less than 85%, and the ward treatment rate of traditional Chinese medicine is not less than 70%.

5.4.1.3 一级中医医院：住院床位总数20～79张。

5.4.1.3 First-class traditional Chinese medicine hospital: the total number of hospital beds is 20～79.

科室设置：

Department setting:

1. 临床科室：至少设有三个中医一级临床科室和药房、化验室、X光室。

1. Clinical departments: at least three Level Ⅰ clinical departments of traditional Chinese medicine, pharmacies, laboratories, and X-ray rooms.

2. 基本设备：心电图机、洗胃机、呼吸球囊、吸引器、心备手术刀包、显微镜、离心机、分光光度计、中药煎药设备、各类针具、紫外线杀菌灯、妇科检查台、给氧装置、X光机、针麻仪、高压灭菌设备、冰箱、蒸馏水装置。

2. Basic device: ECG machine, gastric lavage machine, breathing balloon, aspirator, necessary surgical kit, microscope, centrifuge, spectrophotometer, traditional Chinese medicine decocting device, various needling instruments, UV germicidal lamp, gynecological examining table, oxygen supply apparatus, X-ray machine, needle anesthesia instrument, autoclaving device, refrigerator and distilled water apparatus.

5.4.1.4 二级中医医院：住院床位总数80～299张。

5.4.1.4 Second-class traditional Chinese medicine hospital: the total number of hospital beds is 80～299.

科室设置：

Department setting:

1. 临床科室至少设中医内科、外科等五个以上中医一级临床科室。

1. Clinical departments: at least more than five Level Ⅰ clinical departments of traditional Chinese medicine, such as internal medical department and surgical department of traditional Chinese medicine.

2. 医技科室：至少设有药剂科、检验科、放射科等医技科室。

2. Medical technology departments: at least pharmacy department, laboratory department and radiology department.

3. 基本设备：心电图机、自动洗胃机、给氧装置、呼吸机、麻醉机、电针仪、手术器械、手术床、酸度计、分析天平、钾钠分析仪、培养箱、电冰箱、干燥箱、分光光度计、X光机、纤维胃镜、结肠镜、妇科检查台、蒸馏水器、高压灭菌设备、中药煎药设备、电动吸引器、显微镜、心脏除颤器、离心机、各类针具、B超、无影灯、骨科牵引床、尿分析仪、紫外线杀菌灯、洗衣机。

3. Basic device: ECG machine, automatic gastric lavage machine, oxygen supply apparatus, breathing machine, anaesthesia machine, electroacupuncture apparatus, surgical instrument, operating bed, acidometer, analytical balance, potassium and sodium analyzer, incubator, refrigera-

tor, drying oven, spectrophotometer, X-ray machine, fiber gastroscope and fiber colonoscope, gynecological examining table, distilled water apparatus, autoclaving device, traditional Chinese medicine decocting device, electric aspirator, microscope, cardiac defibrillator, centrifuge, various needling instruments, B-Ultrasonic, shadowless lamp, orthopedic traction table, urine analyzer, UV germicidal lamp and washing machine.

5.4.1.5　三级中医医院：住院床位总数 300 张以上。

5.4.1.5　Third-class traditional Chinese medicine hospital: the total number of hospital beds is over 300.

科室设置：

Department setting:

1. 临床科室：至少设有急诊科、内科、外科、妇产科、儿科、针灸科、骨伤科、肛肠科、皮肤科、眼科、推拿科、耳鼻喉科。

1. Clinical departments: at least emergency, internal medical, surgical, obstetrics and gynecology, pediatric, acupuncture, bone injury, anorectal, dermatology, ophthalmology, massage and otolaryngology departments.

2. 医技科室：至少设有药剂科、检验科、放射科、病理科、消毒供应室、营养部和相应的临床功能检查室。

2. Medical technology departments: at least pharmacy department, laboratory department, radiology department, pathology department, disinfection supply room, nutrition department and corresponding clinical functional examination room.

3. 基本设备：心电图机、自动洗胃机、给氧装置、电动呼吸机、多功能抢救床、心电监护仪、无影灯、麻醉机、麻醉监护仪、手术器械、荧光显微镜、尿分析仪、气血分析仪、自动生化分析仪、酶标仪、电冰箱、离心机、分光光度计、超净工作台、肺功能仪、X 光机、移动式 X 光机、膀胱镜、纤维胃镜、电检眼镜、裂隙灯、直接喉镜、动态心电图机、妇科检查台、骨科牵引床、石蜡切片机、冷冻切片机、高压灭菌设备、各类针具、药器柜、人流吸引器、电动吸引器、B 超、心脏除颤器、纤维结肠镜、万能手术床、乙状结肠镜、针麻仪、鼻咽镜、血细胞计数器、多普勒成像仪、钾钠分析仪、牙科综合治疗台、恒温箱、紫外线杀菌灯、干燥箱、电针仪、分析天平、中药煎药设备、洗衣机。

3. Basic device: ECG machine, automatic gastric lavage machine, oxygen supply apparatus, electric breathing machine, multifunctional rescue bed, ECG monitor, shadowless lamp, anaesthesia machine, anesthesia monitor, surgical instrument, fluorescence microscope, urine analyzer, blood gas analyzer, automatic biochemical analyzer, microplate reader, refrigerator, centrifuge, spectrophotometer, clean bench, spirometer, X-ray machine, mobile X-ray machine, cystoscope, fiber gastroscope, electric ophthalmoscope, slit lamp, direct laryngoscope, dynamic ECG machine, gynecological examining table, orthopedic traction table, microtome, freezing microtome, autoclaving device, various needling instruments, medicine cabinet, abortion aspirator, electric aspirator, B-ultrasound, cardiac defibrillator, fiber colonoscope, universal operating bed, sigmoid-

oscope, needle anesthesia instrument, nasopharyngoscope, hemocytometer, Doppler imager, potassium and sodium analyzer, dental comprehensive treatment table, thermostat, UV germicidal lamp, drying oven, electroacupuncture apparatus, analytical balance, traditional Chinese medicine decocting device and washing machine.

5.4.1.6 治未病中心：独立或设在中医医院内。

5.4.1.6 Center for preventive treatment of disease: independent or located in the traditional Chinese medicine hospital.

服务项目：

Service items:

1. 健康状态辨识及评估项目：中医体质辨识，中医经络、脏腑功能、血气状态评估等。

1. Health status identification and assessment items: traditional Chinese medicine constitution identification, traditional Chinese medicine meridian, viscera function, blood gas status assessment, etc.

2. 健康调养咨询服务：开具健康处方、养生功法示范指导、中药调养咨询指导等。

2. Health care consultation service: issuing health prescription, demonstration guidance on health care techniques, consultation guidance on traditional Chinese medicine care, etc.

3. 中医特色干预技术：包括针刺、灸法、拔罐、推拿、穴位贴敷、埋线、药浴、熏洗（蒸）、刮痧、砭石、音疗及热疗、电疗等其他理疗技术。

3. Characteristic intervention techniques of traditional Chinese medicine: including acupuncture, moxibustion, cupping, massage, acupoint application, catgut embedding, medicated bath, steaming and washing (fumigation), scraping therapy, needle stone, sound therapy, and other physiotherapy techniques such as thermal therapy and electrotherapy.

4. 产品类：如膏方、养生调养茶饮等。

4. Products: such as herbal paste, health care tea, etc.

此外，健康档案建立、慢性病健康管理、健康信息管理，以及管理效果评价等也可纳入治未病服务项目。

In addition, the establishment of health records, chronic disease health management, health information management, and management effect evaluation, etc. can also be included in the preventive treatment of disease service items.

设备配置：

Device configuration:

1. 健康状态信息管理设备。如办公桌、办公椅、计算机、打印机、电话、专用文件柜等。

1. Health status information management device. Such as office desks and chairs, computers, printers, telephones and special filing cabinets.

2. 健康状态辨识及其风险评估设备。如中医体质辨识系统、舌象仪、脉象仪、经络检测设备，体重仪、身高仪、血压计、心血管检测仪、肺功能仪、骨密度检测仪、心电图机、血糖监测仪等常规体检、理化、影像设备。

2. Health status identification and risk assessment device. Such as traditional Chinese medicine constitution identification system, tongue imaging instrument, pulse condition instrument, meridian detection device, weight meter, height meter, sphygmomanometer, cardiovascular detector, spirometer, bone density detector, ECG machine, blood glucose monitor and other conventional physical examination, physical and chemical, imaging device.

3. 健康咨询与指导设备。如健康教育宣传栏、影像等演示设备、多媒体教学设备及信息网络系统设备等。

3. Health consultation and guidance device. Such as health education bulletin board, video and other demonstration device, multimedia teaching device and information network system device.

4. 健康干预设备及器具。如针具、灸具、罐具、刮痧板、砭石，及中医电疗、磁疗、热疗设备等。

4. Health intervention devices and appliances. Such as needling instruments, acupuncture appliances, pots, scraping plates, needle stone, and traditional Chinese medicine electrotherapy, magnetic therapy and thermal therapy devices.

5.4.2 中医医院设计

5.4.2 Design of traditional Chinese medicine hospital

5.4.2.1 前期策划与可行性研究

5.4.2.1 Preliminary planning and feasibility study

1. 选址：土地用途应为医疗卫生用地，应符合中医医院建设标准。

1. Site selection: the land shall be used for medical and health purposes and shall conform to the construction standards of traditional Chinese medicine hospital.

2. 项目定位：确定医院级别，如市级、县级、地区级、省级、国家级，传承、发扬、光大中医文化和中医医疗技术。

2. Project orientation: determine the hospital level, such as municipal, county, regional, provincial and national level, to inherit, carry forward and promote the traditional Chinese medicine culture and traditional Chinese medicine medical technology.

3. 可行性研究：包括项目所在区域新建或改扩建中医医院的背景与必要性、项目建筑工程以及配套设施规模、建设方案、项目投资估算和社会效益等。

3. Feasibility study: including the background and necessity of new construction or renovation and expansion of traditional Chinese medicine hospital in the project area, the scale, construction proposal, project investment estimation and social benefits of the project construction engineering and auxiliary facilities, etc.

4. 确定运营模式，如：以急性病诊断治疗平台为基础，多个专病中心为特色的综合医疗体系。施行以中医诊断与治疗技术为特色，具有中医馆、国医大师的独特功能，融合现代医疗技术，形成诊断、治疗、康复为一体的全新医疗模式，并结合以治未病、健康体检、慢病治疗、医养结合、特色治疗的中医特色慢病管理中心。

4. Determination of operation mode, such as a comprehensive medical system based on the acute disease diagnosis and treatment platform and featuring multiple specialized disease centers. With the characteristics of diagnosis and treatment technology of traditional Chinese medicine, a new medical model integrating diagnosis, treatment and rehabilitation with the unique functions of traditional Chinese medicine hall and national medical masters and modern medical technology, is implemented in combination with a chronic disease management center with the characteristics of traditional Chinese medicine, including the preventive treatment of disease, physical examination, chronic disease treatment, combination of medical care and special treatment.

5. 其他建设内容：中医药科学研究、中医药教学与中医人才培养、中药制剂。

5. Other construction contents: scientific research of traditional Chinese medicine, teaching of traditional Chinese medicine and training of talents in traditional Chinese medicine, and preparation of traditional Chinese medicine.

6. 确定建设内容：例如“集医疗、教学、科研、预防、保健、康复为一体的国际化大型综合三级甲等研究型中医医院，包含医学院临床实习基地、中医学科研究基地、临床药学研究基地。”

6. Determination of construction content: for example, “an international large-scale comprehensive Third-class Level A research-type traditional Chinese medicine hospital integrating medical treatment, teaching, scientific research, prevention, health care and rehabilitation, including a clinical practice base of medical college, a research base of traditional Chinese medicine discipline and a research base of clinical pharmacy.”

7. 确定规模与分期建设：一次规划，分期建设或一次建成；项目总体和每一期技术经济指标。

7. Determination of scale and phased construction: one-time planning, phased construction or one-time construction; overall technical and economic indicators of the project and each phase.

5.4.2.2 建筑设计

5.4.2.2 Architectural design

除了做到医疗建筑通用性设计，还要体现中医特色。

In addition to the general design of healthcare buildings, it shall also embody the characteristics of traditional Chinese medicine.

1. 以人为本：医院的总平面布置及门诊、急诊、医技、住院和生活各功能分区的平面布置应充分考虑到患者就医的舒适、方便、安全、卫生的要求，某些医院可提供康复、保健、教学、科研设施。

1. People-oriented: the requirements of comfort, convenience, safety and hygiene of patients seeking medical treatment shall be fully considered in terms of the general layout of hospitals and the layout of outpatient department, emergency department, medical technology department, inpatient department and functional areas of life. Some hospitals can provide rehabilitation,

health care, teaching and scientific research facilities.

2. 技术科学合理先进：根据国内先进中医药发展理念和创新思维，在大力发扬中医传统的同时采用高新技术设备，使院区具有相应的科学性和先进性。在总体规划中科学合理布局，充分高效地利用土地资源，在具体科室的布置中，注重科学合理，流程短捷。

2. Scientific, reasonable and advanced technology: according to the domestic advanced development concept and innovative thinking of traditional Chinese medicine, vigorously develop the traditional Chinese medicine and adopt high-tech device, so that the hospital area will be scientific and advanced. In the overall planning, make scientific and reasonable layout, make full and efficient use of land resources, pay attention to scientific and reasonable layout of specific departments, and ensure short process.

3. 卫生安全经济高效：满足现代化中医医院对卫生安全的严格要求，防控院内交叉感染；做到功能分区明确、流程合理、人流物流路线清晰；合理安排建筑物的朝向、间距及院内园林绿化等措施，尽可能自然采光、通风，体现人与自然和谐共生，降低能耗和日常运行费用。

3. Health, safety, economy and efficiency: meet the strict health and safety requirements of modern traditional Chinese medicine hospitals, and prevent and control cross infection in hospitals; achieve clear functional area, reasonable process, and clear people flow and logistics routes; reasonably arrange the building orientation, spacing and landscaping in hospitals, and take natural lighting and ventilation as much as possible to reflect the harmonious coexistence of human and nature, and reduce energy consumption and daily operation cost.

4. 室内外环境有中医中药特色：与时俱进，体现中医药诊疗现代化，将传统医学文化、传统建筑文化和现代医疗建筑相互融合；根据中医中药特点，在景观绿化布置上配以具有中式园林风格的园景、小品建筑、国学体训、运动、康复设施、安全的中草药类植物。

4. The indoor and outdoor environment has the characteristics of traditional Chinese medicine: keep pace with the times, reflect the modernization of diagnosis and treatment of traditional Chinese medicine, and integrate traditional medical culture, traditional architectural culture and modern healthcare buildings; according to the characteristics of traditional Chinese medicine, the landscape greening layout is matched with garden scenery with Chinese garden style, pavilion, traditional Chinese training, sports, rehabilitation facilities, and safe Chinese herbal plants.

5. 总体规划应符合国家发展改革委、住房城乡建设部、各省市有关部门相关规定、指标、规范和标准，充分、高效地利用好土地资源，保护好原有的自然环境，并预留未来可持续发展的空间。

5. The overall planning shall conform to the relevant regulations, indicators, specifications and standards of the National Development and Reform Commission, the Ministry of Housing and Urban-Rural Development of the People's Republic of China and relevant departments of all provinces and cities, fully and efficiently utilize land resources, protect the original natural envi-

ronment, and reserve space for sustainable development in the future.

5.4.2.3 中医医院下设重点专科医院

5.4.2.3 Key specialized hospitals under traditional Chinese medicine hospitals

在进一步细化专业分科基础之上，中医医院可以通过基层布点，体现更多特色专科利民服务，这类专科网点的功能设置应有特色；建筑可以是新建，也可以是租赁，还可以是既有建筑改造。包括：

On the basis of further detailed division of specialties, traditional Chinese medicine hospitals can display more characteristic specialty services for the benefit of the people through basic-level distribution. The functional settings of such specialized branches should have their own characteristics; buildings can be newly built or leased, and existing buildings can be renovated. These include:

1. 针灸推拿分院；

1. Acupuncture and Massage Branch;

2. 中医理疗馆；

2. Traditional Chinese Medicine Physiotherapy Center;

3. 中医诊所；

3. Traditional Chinese Medicine Clinic;

4. "治未病" 中心。

4. Center for "Preventive Treatment of Disease".

5.5 中西医结合医院

5.5 Hospital of integrated traditional Chinese and Western medicine

中西医结合医院是将中医与西医两种医疗进行互补、融合的综合性医院。

The hospital of integrated traditional Chinese and Western medicine is a comprehensive hospital that integrates traditional Chinese medicine and Western medicine.

5.5.1 中西医结合医院等级标准

5.5.1 Class standard of hospitals of integrated traditional Chinese and Western medicine

5.5.1.1 一级中西医结合医院：住院床位总数 20～99 张。

5.5.1.1 First class hospital of integrated traditional Chinese and Western medicine: the total number of hospital beds is 20～99.

科室设置：

Department setting:

1. 临床科室：至少设有中西医结合内科、外科与预防保健科。

1. Clinical departments: at least internal medical department, surgical department and prevention and health care department of integrated traditional Chinese and Western medicine.

2. 医技科室：至少设有中药房、西药房、化验室、X光室、消毒供应室。

2. Medical technology departments: At least traditional Chinese medicine pharmacy, western medicine pharmacy, laboratory, X-ray room and disinfection supply room are arranged.

3. 基本设备：心电图机、（自动）洗胃机、X光机、给氧装置、呼吸球囊、呼吸机、电针仪、妇科检查台、高压灭菌设备、显微镜、离心机、紫外线杀菌灯、器械柜、抢救车、蒸馏水装置、各类针具、中药煎药设备、电冰箱、人工洗片装置、药品柜、必备手术刀包、吸引器。

3. Basic devices: ECG machine, (automatic) gastric lavage machine, X-ray machine, oxygen supply apparatus, breathing balloon, breathing machine, electroacupuncture apparatus, gynecological examining table, autoclaving device, microscope, centrifuge, UV germicidal lamp, instrument cabinet, rescue vehicle, distilled water apparatus, various needling instruments, traditional Chinese medicine decocting device, refrigerator, manual film washing apparatus, medicine cabinet, necessary surgical kit and aspirator.

5.5.1.2 二级中西医结合医院：住院床位总数100～349张。

5.5.1.2 Second-class hospital of integrated traditional Chinese and Western medicine: the total number of hospital beds is 100～349.

科室设置：

Department setting:

1. 临床科室：设有6个以上中西医结合一级临床科室，设中西医结合专科或专病研究室（组）。

1. Clinical departments: there are more than 6 first-class clinical departments of integrated traditional Chinese and Western medicine, and specialized departments or special disease study rooms (groups) of integrated traditional Chinese and Western medicine.

2. 医技科室：至少设有药剂科、检验科、放射科、病理科、消毒供应室。

2. Medical technology departments: at least pharmacy department, laboratory department, radiology department, pathology department and disinfection supply room.

3. 基本设备：心电图机、自动洗胃机、呼吸机、心脏除颤器、万能手术床、无影灯、胃肠减压器、万能产床、手术器械、各类针具、妇科检查台、干燥箱、电针仪、涡轮机、高压灭菌设备、紫外线杀菌灯、电冰箱、离心机、显微镜、分光光度计、分析天平、尿分析仪、恒温箱、酸度计、器械柜、中药煎药设备、冷热水净化系统、培养箱、冰冻切片机、石蜡切片机、电动吸引器、钾钠分析仪、心电监护仪、超声心电图机、麻醉机、给氧装置、产程监护仪、药品柜、器械柜、骨科牵引床、蒸馏水器、鼻咽镜、B超、牙钻机、牙科治疗椅、纤维胃镜、X光机、洗衣机。

3. Basic devices: ECG machine, automatic gastric lavage machine, breathing machine, cardiac defibrillator, universal operating bed, shadowless lamp, gastrointestinal decompressor, universal obstetric table, surgical instruments, various needling instruments, gynecological exami-

ning table, drying oven, electroacupuncture apparatus, turbine, autoclaving device, UV germicidal lamp, refrigerator, centrifuge, microscope, spectrophotometer, analytical balance, urine analyzer, thermostat, acidometer, instrument cabinet, traditional Chinese medicine decocting device, cold and hot water purification system, incubator, freezing microtome, microtome, electric aspirator, potassium and sodium analyzer, ECG monitor, ultrasonic ECG machine, anaesthesia machine, oxygen supply apparatus, delivery monitor, medicine cabinet, instrument cabinet, orthopedic traction table, distilled water apparatus, nasopharyngoscope, B-ultrasound, dental engine, dental chair, fiber gastroscope, X-ray machine and washing machine.

5.5.1.3 三级中西医结合医院：住院床位总数350张以上。

5.5.1.3 Third-class hospital of integrated traditional Chinese and Western medicine: the total number of hospital beds is more than 350.

科室设置：

Department setting:

1. 临床科室：至少设有急诊科、内科、外科、妇产科、儿科、耳鼻喉科、口腔科、眼科、皮肤科、针灸科、麻醉科、预防保健科。设立中西医结合专科或专病研究所（室）。

1. Clinical departments: at least emergency department, internal medical department, surgical department, obstetrics and gynecology department, pediatric department, otolaryngology department, stomatology department, ophthalmology department, dermatology department, acupuncture department, anesthesiology department, and prevention and health care department. There are specialized departments or special disease institutes (study rooms) of integrated traditional Chinese and Western medicine.

2. 医技科室：至少设有药剂科、放射科、检验科、病理科、血库、消毒供应室、病案室、营养部和相应的临床功能检查室。

2. Medical technology departments: at least pharmacy department, radiology department, laboratory department, pathology department, blood bank, disinfection supply room, medical record room, nutrition department and corresponding clinical functional examination room.

3. 基本设备：心电图机、自动洗胃机、呼吸机、心脏除颤器、肺功能仪、万能手术床、麻醉机、麻醉监护仪、高频电刀、胃肠减压器、产程监护仪、手术器械、骨科牵引床、妇科检查台、引产吸引器、裂隙灯、直接喉镜、电针仪、牙钻机、高压灭菌设备、X光机、电冰箱、钾钠分析仪、荧光显微镜、显微镜、分光光度计、分析天平、尿分析仪、恒温箱、酸度计、药品柜、器械柜、膀胱镜、电栓眼镜、移动式X光机、多功能抢救床、乙状结肠镜、中药煎药设备、冷热水净化系统、培养箱、多普勒成像仪、纤维结肠镜、石蜡切片机、纤维胃镜、电动吸引器、酶标分析仪、心电监护仪、超声心电图机、无影灯、给氧装置、手术显微镜、支气管镜、万能产床、动态心电图机、各类针具、牙科综合治疗台、干燥箱、自动生化分析仪、鼻咽镜、蒸馏水器、涡轮机、B超、紫外线杀菌灯、冰冻切片机、离心机。

3. Basic devices: ECG machine, automatic gastric lavage machine, breathing machine, cardiac defibrillator, spirometer, universal operating bed, anaesthesia machine, anesthesia monitor,

high-frequency electrotome, gastrointestinal decompressor, delivery monitor, surgical instruments, orthopedic traction table, gynecological examining table, induced labor aspirator, slit lamp, direct laryngoscope, electroacupuncture apparatus, dental engine, autoclaving device, X-ray machine, refrigerator, potassium and sodium analyzer, fluorescence microscope, microscope, spectrophotometer, analytical balance, urine analyzer, thermostat, acidometer, medicine cabinet, instrument cabinet, cystoscope, mobile X-ray machine, multifunctional rescue bed, sigmoidoscope, traditional Chinese medicine decocting device, cold and hot water purification system, incubator, Doppler imager, fiber colonoscope, microtome, fiber gastroscope, electric aspirator, enzyme-labeled analyzer, ECG monitor, ultrasonic ECG machine, shadowless lamp, oxygen supply apparatus, operation microscope, bronchoscope, universal obstetric table, dynamic ECG machine, various needling instruments, dental comprehensive treatment table, drying oven, automatic biochemical analyzer, nasopharyngoscope, distilled water apparatus, turbine, B-ultrasound, UV germicidal lamp, freezing microtome, and centrifuge.

5.5.2 中西医结合医院建筑设计

5.5.2 Architectural design of hospitals of integrated traditional Chinese and Western medicine

5.5.2.1 前期策划与可行性研究

5.5.2.1 Preliminary planning and feasibility study

1. 选址：土地用途应为医疗卫生用地，应符合中西医结合医院建设标准。

1. Site selection: the land shall be used for medical and health purposes and shall conform to the construction standards of hospital of integrated traditional Chinese and Western medicine.

2. 项目定位：确定医院级别，如市级、县级、地区级、省级、国家级，传承、发扬、光大中医文化和中医医疗技术。

2. Project orientation: determine the hospital level, such as municipal, county, regional, provincial and national level, to inherit, carry forward and promote the traditional Chinese medicine culture and traditional Chinese medicine medical technology.

3. 可行性研究：包括项目所在区域新建或改扩建中医医院的背景与必要性、项目建筑工程以及配套设施规模、建设方案、项目投资估算和社会效益等。

3. Feasibility study: including the background and necessity of new construction or renovation and expansion of traditional Chinese medicine hospital in the project area, the scale, construction proposal, project investment estimation and social benefits of the project construction engineering and auxiliary facilities, etc.

4. 确定运营模式，如：以急性病诊断治疗平台为基础，多个专病中心为特色的综合医疗体系。施行以中医诊断与治疗技术为特色，具有中医馆、国医大师的独特功能，融合现代医疗技术，形成诊断、治疗、康复为一体的全新医疗模式，并结合以治未病、健康体检、慢病治疗、医养结合、特色治疗的中医特色慢病管理中心。

4. Determination of operation mode, such as a comprehensive medical system based on the acute disease diagnosis and treatment platform and featuring multiple specialized disease cen-

ters. With the characteristics of diagnosis and treatment technology of traditional Chinese medicine, a new medical model integrating diagnosis, treatment and rehabilitation with the unique functions of traditional Chinese medicine hall and national medical masters and modern medical technology, is implemented in combination with a chronic disease management center with the characteristics of traditional Chinese medicine, including the preventive treatment of disease, physical examination, chronic disease treatment, combination of medical care and special treatment.

5. 其他建设内容：中医药科学研究、中医药教学与中医人才培养、中药制剂。

5. Other construction contents: scientific research of traditional Chinese medicine, teaching of traditional Chinese medicine and training of talents in traditional Chinese medicine, and preparation of traditional Chinese medicine.

6. 确定建设内容：例如“集医疗、教学、科研、预防、保健、康复为一体的国际化大型综合三级甲等研究型中医医院，包含医学院临床实习基地、中医学科研究基地、临床药学研究基地。”

6. Determination of construction content: for example, “an international large-scale comprehensive Third-class Level A research-type traditional Chinese medicine hospital integrating medical treatment, teaching, scientific research, prevention, health care and rehabilitation, including a clinical practice base of medical college, a research base of traditional Chinese medicine discipline and a research base of clinical pharmacy. ”

7. 确定规模与分期建设：一次规划，分期建设或一次建成；项目总体和每一期技术经济指标。

7. Determination of scale and phased construction: one-time planning, phased construction or one-time construction; overall technical and economic indicators of the project and each phase.

5.5.2.2 建筑设计

5.5.2.2 Architectural design

5.5.3 中西医结合医院下设重点专科医院

5.5.3 Key specialized hospitals under hospitals of integrated traditional Chinese and Western medicine

在进一步细化专业分科基础之上，中西医结合医院可以通过基层布点，体现更多专科利民服务：

On the basis of further detailed division of specialties, hospitals of integrated traditional Chinese and Western medicine can display more specialty services for the benefit of the people through basic-level distribution:

5.5.3.1 针灸推拿分院。

5.5.3.1 Acupuncture and Massage Branch.

5.5.3.2 中医理疗馆。

5.5.3.2 Traditional Chinese Medicine Physiotherapy Center.

5.5.3.3　中医诊所或设在药店里的中医坐诊。

5.5.3.3　Traditional Chinese medicine clinics or doctors of traditional Chinese medicine in pharmacies for treating patients illness

5.5.3.4　同时具有西医和中医的社区医疗中心。

5.5.3.4　Community medical centers with both Western medicine and traditional Chinese medicine.

这类小型专科医院的功能设置应有特色；建筑可以是新建，也可以是租赁，既有建筑改造。

The functional settings of such small specialized hospitals should have their own characteristics; buildings can be newly built or leased, and existing buildings can be renovated.

案例：见表 5-3。

Case: see Table 5-3.

表 5-3　案例

Table 5-3　Case

楼层 Floor	功能科室设置 Functional department setting
一层 1F	后勤物资仓库、物资配送中心（包括消毒医疗物资配送、被服配送、后勤物资配送）、大厅、挂号收费处、出入院登记处、住院收费处 Logistics material warehouse, material distribution center (including disinfection and medical material distribution, bedding and clothing distribution, logistics material distribution), hall, registration and charging office, admission and discharge office and inpatient charging office
二层 2F	挂号收费处、预防保健科、社康中心、治未病中心（健康咨询指导门诊、体检科、康复中心） Registration and charging office, prevention and health care department, community health service center, center for preventive treatment of disease (health consultation and guidance outpatient department, physical examination department and rehabilitation center)
三层 3F	营养咨询门诊、营养餐厅、厨房 Nutrition counseling outpatient department, nutrition dining room and kitchen
四层 4F	康复中心、宣教中心 Rehabilitation center and propaganda and education center
五层 5F	值班宿舍、营养食堂 Dormitories for on-duty personnel and nutrition canteen
六层 6F	针灸治疗室、推拿治疗室、艾灸室、药浴室、足疗室、熏蒸室、候诊室、办公室等 Acupuncture treatment room, massage treatment room, moxibustion room, medicated bathroom, foot massage room, fumigation room, waiting area, office, etc.
七层 7F	供制作片剂、注射剂、气雾剂、丸剂、散剂、膏剂等 Used for preparation of tablet, injection, aerosol, pill, powder, paste, etc.
八层 8F	健康档案管理中心、办公室等 Health record management center, office, etc.
九层 9F	针灸病区 Acupuncture ward

续表

楼层 Floor	功能科室设置 Functional department setting
十层 10F	康复病区 Rehabilitation ward
十一层 11F	推拿病区 Massage ward
十二层 12F	综合病区 Comprehensive ward
地下一层 B1	变配电房、空调机房、水泵房、废弃物中转站、污水处理站控制室 Power transformation and distribution room, air conditioning room, water pump house, waste transfer station and control room of sewage treatment station
地下二层 B2	污水处理池、设备仓库 Sewage treatment tank and device warehouse

5.6 妇幼保健院

5.6 Maternity and childcare hospital

妇幼保健医院是公共卫生服务体系的重要组成部分；妇幼保健医院的功能是通过优生优育、妇幼保健和计划生育技术服务，降低妇女、儿童的患疾率及死亡率，提高妇女、儿童的身心健康水平，保证我国“控制人口数量，提高人口素质”基本国策的实现。

Maternity and childcare hospital is an important part of the public health service system. The function of maternity and childcare hospital is to reduce the morbidity and mortality rate of women and children, improve the physical and mental health level of women and children, and ensure the realization of China’s basic national policy of “controlling population quantity and improving population quality” through eugenics, maternal and child health care and family planning technical services.

5.6.1 妇幼保健院等级标准

5.6.1 Class standard of maternity and childcare hospitals

妇幼保健院的基本要求：

Basic requirements of maternity and childcare hospitals:

5.6.1.1 一级妇幼保健院：住院床位总数 5 ~ 19 张。

5.6.1.1 First-class maternity and childcare hospital: the total number of hospital beds is 5 ~ 19.

科室设置：

Department setting:

1. 临床科室：妇女保健科、婚姻保健科、儿童保健科、计划生育科、妇产科、儿科、健康教育科、信息资料科。

1. Clinical departments: women's health care department, marital health care department, child healthcare department, family planning department, obstetrics and gynecology department, pediatric department, health education department and information and data department.

2. 医技科室：药房、化验室。

2. Medical technology departments: pharmacy and laboratory.

3. 基本设备：妇科检查床、产床、妇科治疗仪、电动吸引器、节育手术器械、新生儿复苏囊、儿童体格测量用具、超声雾化器、紫外线灯、氧气瓶、显微镜、离心机、血红蛋白测定仪、高压灭菌设备、健康教育基本设备、电冰箱、洗衣机。

3. Basic devices: gynecological examining table, obstetric table, gynecological treatment instrument, electric aspirator, birth control surgical instrument, newborn resuscitation kit, children's physique measuring appliance, ultrasonic nebulizer, ultraviolet lamp, oxygen cylinder, microscope, centrifuge, hemoglobin analyzer, autoclaving device, basic health education device, refrigerator and washing machine.

5.6.1.2 二级妇幼保健院：住院床位总数20～49张。

5.6.1.2 Second-class maternity and childcare hospital: the total number of hospital beds is 20～49.

科室设置：

Department setting:

1. 临床科室：妇女保健科、婚姻保健科、围产保健科、优生咨询科、乳腺保健科、儿童保健科、儿童生长发育科、妇儿营养科、儿童五官保健科、生殖健康科、计划生育科、妇产科、儿科、健康教育科、培训指导科、信息资料科。

1. Clinical departments: women's health care department, marital health care department, perinatal health care department, aristogenesis counselling department, breast health care department, child healthcare department, child growth & development department, gynecology and pediatrics nutritional department, child facial feature health care department, reproductive health department, family planning department, obstetrics and gynecology department, pediatric department, health education department, training guidance department and information and data department.

2. 医技科室：药剂科、检验科、影像诊断科、功能检查科、手术室、消毒供应室。

2. Medical technology departments: pharmacy department, laboratory department, imaging diagnosis department, functional examination department, operating room and disinfection supply room.

3. 基本设备：妇科检查床、产床、妇科治疗仪、电动吸引器、手术器械、综合手术台、乳腺透照仪、B超、心电图机、双目显微镜、多普勒胎心诊断仪、新生儿抢救台、儿童体格测量用具、200mA X光机、同视机、新生儿保温箱、儿童口腔保健椅、高压灭菌设备、儿童智力测查工具、洗衣机、电冰箱、血红蛋白测定仪、分光光度计、离心机、水浴箱、电视机、录放像机、救护车。

3. Basic devices: gynecological examining table, obstetric table, gynecological treatment instrument, electric aspirator, surgical instrument, multipurpose operating table, breast transilluminator, B-ultrasound, ECG machine, binocular microscope, Doppler fetal heart diagnostic instrument, neonatal rescue table, children's physique measuring appliance, 200nA X-ray machine, synoptophore, neonatal incubator, children's oral health chair, autoclaving device, children's intelligence testing tool, washing machine, refrigerator, hemoglobin analyzer, spectrophotometer, centrifuge, water-bath, television, video recorder and ambulance.

5.6.1.3 三级妇幼保健院：住院床位总数50张以上。

5.6.1.3 Third-class maternity and childcare hospital: the total number of hospital beds is more than 50.

科室设置：

Department setting:

1. 业务科室：妇女保健科、婚姻保健科、围产保健科、优生咨询科、女职工保健科、更年期保健科、妇儿心理卫生科、儿童生长发育科、儿童口腔保健科、儿童眼保健科、生殖健康科、计划生育科、妇产科、儿科、培训指导科、健康教育科、信息资料科。

1. Business departments: women' s health care department, marital health care department, perinatal health care department, aristogenesis counselling department, women workers' health care department, menopause health care department, women's and children's mental health department, child growth & development department, children's oral health department, children's eye health department, reproductive health department, family planning department, obstetrics and gynecology department, pediatric department, training guidance department, health education department and information and data department.

2. 医技科室：药剂科、检验科、影像诊断科、功能检查科、遗传实验室、手术室、消毒供应室、病案图书室。

2. Medical technology departments: pharmacy department, laboratory department, imaging diagnosis department, functional examination department, genetics laboratory, operating room, disinfection supply room and medical record room and library.

3. 基本设备：妇科检查床、产术、综合手术台、电动吸引器、腹部手术器械、高压灭菌设备、多普勒胎心诊断仪、新生儿保温箱、B超（线、扇）200mA以上X光机、心电图机、宫腔镜、新生儿抢救台、麻醉机、妇科治疗仪、乳腺透照仪、儿童体格测量用具、儿童智力测查工具、同视机、儿童口腔保健椅、裂隙灯、节育手术器械、超净工作台、半自动生化分析仪、分光光度计、尿液分析仪、血细胞计数仪、酶标仪、双目显微镜、恒温培养箱、万能显微镜、γ-计数仪、离心机、分析天平、洗衣机、文字处理机（打字机）、电视机、幻灯机、录放像机、投影仪。

3. Basic devices: gynecological examining table, obstetric operating table, multipurpose operating table, electric aspirator, abdominal surgical instruments, autoclaving device, Doppler fetal

heart diagnostic instrument, neonatal incubator, B-ultrasound, X-ray machine over 200mA, ECG machine, hysteroscope, neonatal rescue table, anaesthesia machine, gynecological treatment instrument, breast transilluminator, children's physique measuring appliance, children's intelligence testing tool, synoptophore, children's oral health chair, slit lamp, birth control surgical instrument, clean bench, semi-automatic biochemical analyzer, spectrophotometer, urine analyzer, blood cell counter, microplate reader, binocular microscope, constant temperature incubator, universal microscope, γ counter, centrifuge, analytical balance, washing machine, word processor (typewriter), television, slide projector, video recorder and projector.

5.6.2 妇幼保健院建筑设计

5.6.2 Architectural design of maternity and childcare hospital

5.6.2.1 前期策划与可行性研究

5.6.2.1 Preliminary planning and feasibility study

1. 选址：土地用途应为医疗卫生用地，应符合妇幼保健院建设标准。地理位置适中，交通方便；节约用地，不占或少占农田；便于利用城市基础设施；环境安静，不宜与市场、学校、幼儿园、公共娱乐场所、交通干线毗邻，且不宜远离居民区；地形力求规整，场地干燥，并有必要的防洪排涝设施；远离易燃、易爆物品的生产和储存区，并远离污染源和高压线路。

1. Site selection: the land shall be used for medical and health purposes and shall conform to the construction standards of maternity and childcare hospital. The geographical location should be moderate and the transportation should be convenient. The land utilization should be optimal, and no farmland will be occupied or less farmland will be occupied. The urban infrastructure should be easy to use. The environment should be quiet, and hospitals should be far from markets, schools, kindergartens, public places of entertainment and traffic arteries, and should not be far from residential areas. The topography should be regular, the site should be dry, and necessary flood control and drainage facilities should be provided. Hospitals should be far from the production and storage areas of flammable and explosive materials, as well as pollution sources and high-voltage lines.

2. 项目定位：确定医院级别，如市级、县级、地区级、省级、国家级；例如："三级甲等妇保院包括围产保健中心、妇女保健中心、儿童保健中心""集保健、医疗、教学、科研、计划生育、康复为一体的三级甲等妇幼保健院"。

2. Project orientation: determine hospital class, such as municipal, county, regional, provincial and national level; for example, "the Third-class Level A maternity and childcare hospital includes perinatal health care center, women's health care center, children's health care center, etc." "the Third-class Level A maternity and childcare hospital integrates health care, medical treatment, teaching, scientific research, family planning and rehabilitation."

3. 可行性研究：包括项目所在区域新建或改扩建妇幼保健院的背景与必要性、项目建筑工程以及配套设施规模、建设方案、项目投资估算和社会效益等。妇幼保健院应根据所在地区的人口数量及密度、经济、地理、交通和服务半径等因素，合理确定建设规模。

3. Feasibility study: including the background and necessity of new construction or renovation and expansion of maternity and childcare hospital in the project area, the scale, construction proposal, project investment estimation and social benefits of the project construction engineering and auxiliary facilities, etc. The construction scale of maternity and childcare hospital should be reasonably determined according to the population, density, economy, geography, transportation and service radius of the area.

4. 政策落实：各地市的《妇幼保健院、所建设标准》《综合医院建设标准》《妇幼保健机构评审标准》《市“十二五”区域卫生规划实施细则》《市“十二五”医疗卫生发展规划及远景目标》《妇幼保健机构管理办法》《中国妇女、儿童发展纲要》《妇幼卫生工作条例》《市妇女和儿童发展规划》《省医院分级管理办法实施细则》。

4. Implementation of policies: the Construction Standards of Maternity and Childcare Hospitals and Centers, the Construction Standard for General Hospitals, the Evaluation Standards of Maternal and Child Health Care Institutions, the Detailed Rules for Implementation of Municipal "12th Five-year" Regional Health Plan, the Municipal "12th Five-year" Medical and Health Development Plan and its Long-term Objectives, the Management Measures of Maternal and Child Health Care Institutions, the Outline for Development of Chinese Women and Children, the Regulations on Maternal and Child Health, the Municipal Development Plan for Women and Children, and the Detailed Rules for Implementation of Management Measures of Provincial Hospital Classification of all prefecture-level cities.

5. 确定运营服务目标，如：“使某市妇幼保健院实现以病人为中心的服务理念和科技兴院的战略方针，突出以保健为中心、以保障生殖健康为目的、保健与临床相结合，面向群体、面向基层和预防为主的妇幼卫生指导方针，使广大患者能及时康复。”

5. Determination of operational service objectives: for example, "to enable a municipal maternity and childcare hospital to realize the patient-centered service concept and the strategic policy of developing the hospital through science and technology, and to highlight the maternal and child health guidelines that focus on health care, aim at ensuring reproductive health, combine health care with clinic, focus on groups and grassroots and give priority to prevention, so that the majority of patients can recover in time."

6. 确定建设内容、规模与分期建设：一次规划，分期建设或一次建成；项目总体和每一期技术经济指标。

6. Determination of construction content, scale and phased construction: one-time planning, phased construction or one-time construction; overall technical and economic indicators of the project and each phase.

7. 进行医疗服务设计和医疗工艺设计：例如某市妇幼保健院诊疗科目：“儿科：呼吸道感染、腹泻、营养缺乏、缺钙等疾病的诊治；妇科：计划生育技术服务、盆腔炎、生殖道感染、附件炎等疾病的诊治；妊娠高血压疾病、子痫等高危妊娠的筛查与管理、正常分娩的处理等。”门诊保健综合楼功能用房需求：“产前诊断、生殖健康、医技功能检查室、诊室、康复宣教中心、中心药房、输液大厅、住院病房、设备机房、办公及档案管理等。”以此为依据展开设计，并做出建筑设计任务书。

7. Design of medical services and processes: for example, the diagnosis and treatment subject of a municipal maternity and childcare hospital: “pediatric: diagnosis and treatment of respiratory tract infection, diarrhea, nutritional deficiency, calcium deficiency and other diseases; gynecology: family planning technical services, and diagnosis and treatment of pelvic inflammatory disease, reproductive tract infection, adnexitis and other diseases; screening and management of pregnancy-induced hypertension, eclampsia and other high-risk pregnancies, and management of normal labor.” Requirements for functional rooms of multi-functional outpatient and health care building: “prenatal diagnosis, reproductive health, medical technology functional examination room, consulting room, rehabilitation propaganda and education center, central pharmacy, infusion hall, inpatient ward, device room, office and file management.” Based on this, the design should be carried out and the architectural requirements specifications should be prepared.

5.6.2.2 建筑设计

5.6.2.2 Architectural design

市级三级甲等妇幼保健院各类保健用房总建筑面积的比率：地市级孕产保健用房23%，儿童保健门诊用房32%，妇女保健部用房15%，医技科室用房18%，行政后勤用房13%。提供住院服务的妇幼保健机构宜按照88m^2/床的建筑面积指标增加相应的住院及与之配套的医技、后勤、保健等用房。

Proportion of the total building area of all kinds of health care rooms in the municipal Third-class Level A maternity and childcare hospitals: 23% for the prefecture-level maternity health care rooms, 32% for the children's health care outpatient rooms, 15% for the women's health care department rooms, 18% for the medical technology department rooms, and 13% for the administrative logistics rooms. Maternal and child health care institutions that provide inpatient services should increase the corresponding inpatient and supporting medical technology, logistics, health care and other rooms according to the building area index of 88m^2/bed.

妇幼保健院建筑设计除了做到医疗建筑通用性设计，还要体现自身特色：

The architectural design of maternity and childcare hospital should not only achieve the general design of healthcare buildings, but also embody its own characteristics:

1. 以人为本：妇幼保健院应既满足院各类功能的合理性，以保障各部分在相互协作的合理管理体系的正常运行，又要营造愉快舒适的就医环境，并通过简洁顺畅的流线组织、清晰合理的功能布局、简单易懂的标志系统引导患者及家属顺利完成就医和探访。

1. People-oriented: maternity and childcare hospital should not only satisfy the rationality of various functions of the hospital, so as to ensure the normal operation of all parts in the reasonable management system of mutual cooperation, but also create a pleasant and comfortable medical treatment environment, and guide patients and their families to successfully complete medical treatment and visits through simple and smooth streamline organization, clear and reasonable functional layout, and simple and understandable marking system.

2. 技术科学合理先进：建立具有鲜明个性和先进保健医学建筑设计理念的现代化

综合性妇幼保健院；提高医院的工作效率、突出疾病防治和为病人提供良好的服务；在具体科室的布置中，注重科学合理，流程短捷。

2. Scientific, reasonable and advanced technology: establish a modern comprehensive maternity and childcare hospital with distinct personality and advanced health care medical architectural design concept; improve the working efficiency of the hospital, highlight the disease prevention and treatment and provide good service for patients; pay attention to scientific and reasonable layout of specific departments, and ensure short process.

3. 卫生安全经济高效：体现妇女、儿童、家属、医护保健工作人员生理和心理特点；满足现代化医院对卫生安全的严格要求，防控院内交叉感染；做到功能分区明确、流程合理、人流物流路线清晰；合理安排建筑物的朝向、间距及院内园林绿化等措施，尽可能自然采光、通风，降低能耗和日常运行费用。

3. Health, safety, economy and efficiency: reflect the physiological and psychological characteristics of women, children, family members and health care staff; meet the strict health and safety requirements of modern hospitals, and prevent and control cross infection in hospitals; achieve clear functional area, reasonable process, and clear people flow and logistics routes; reasonably arrange the building orientation, spacing and landscaping in hospitals, and take natural lighting and ventilation as much as possible to reduce energy consumption and daily operation cost.

4. 关注人口发展与计划生育政策：立足现在、结合长远，合理安排，控制投资，并适度超前，在满足基本功能需要的同时，适当考虑未来发展。

4. Pay attention to the population development and family planning policy: based on the present and combined with the long term, make reasonable arrangements, control investment, and lead moderately; while meeting the basic functional needs, give due consideration to future development.

5. 总体规划应符合国家发展改革委、住房城乡建设部、各省市有关部门相关规定、指标、规范和标准，充分、高效地利用好土地资源，保护好原有的自然环境，并预留未来可持续发展的空间。

5. The overall planning shall conform to the relevant regulations, indicators, specifications and standards of the National Development and Reform Commission, the Ministry of Housing and Urban-Rural Development of the People's Republic of China and relevant departments of all provinces and cities, fully and efficiently utilize land resources, protect the original natural environment, and reserve space for sustainable development in the future.

5.6.3 妇幼保健院功能特点

5.6.3 Functional characteristics of maternity and childcare hospital

5.6.3.1 分区鲜明：病患和健康人群出入口相互分离，互不干扰。病患人员服务包括：急诊急救、综合门诊、妇科门诊、儿科门诊、综合病房、妇科病房、儿科病房。健康人群服务包括：产科门诊、体检保健、保健病房以及行政区域。两个分区均要与医技保持紧密联系，并通过交通枢纽相互联系；分区的设计应避免患病人群与健康人群交叉与干扰，缩短门诊、病房、医技三者之间的距离，方便医生和患者。

5.6.3.1 Distinct area: the entrances and exits of patients and healthy people are separated from each other without interference. Services for patients include emergency treatment, first-aid, comprehensive outpatient, gynecology outpatient, pediatric outpatient, comprehensive ward, gynecology ward, and pediatric ward. Services for healthy people include obstetric outpatient, physical examination and health care, health care ward and administrative areas. The two areas should keep close contact with medical technology and be connected with each other through transportation hubs. The area design should avoid the cross and interference between sick and healthy people, shorten the distance among outpatient, ward and medical technology, and facilitate doctors and patients.

5.6.3.2 产科分诊单元的布置较为特殊，为了照顾孕妇行动不便，尽量将大部分产科诊室以及孕妇学校布置在建筑底层，将产房、新生儿病房等布置在其相邻楼层，并通过绿色通道与产科病房直接相连。

5.6.3.2 The layout of obstetric triage units is relatively special. In order to take care of the inconvenient movement of pregnant women, most of the obstetric consulting rooms and schools for pregnant women should be arranged on the ground floor of the building as far as possible, and the maternity ward, neonatal ward, etc. should be arranged on the adjacent floors and directly connected with the obstetrics ward through the green channel.

5.6.3.3 急诊、急救中心规划设计要考虑产妇紧急分娩、儿童急诊隔离、观察与治疗，设热伤处置室。

5.6.3.3 The planning and design of emergency and first-aid centers should take into account the emergency delivery of lying-in women, emergency isolation of children, observation and treatment, and thermal injury treatment rooms should be arranged.

5.6.3.4 考虑家庭化病房、产房设计，注重未成年人隐私保护，满足特需服务要求。

5.6.3.4 The design of family wards and maternity wards should be considered, attention should be paid to the protection of minors' privacy, and VIP should be met.

5.6.3.5 分娩室、免疫缺损新生儿室、血液病房、烧伤病房等洁净病房应自成一区。

5.6.3.5 The clean wards such as delivery room, immunodeficiency newborn room, hematology ward and burn ward should be set in a separate area.

5.6.3.6 重症护理单元（ICU）、早产儿护理单元（NICU）应自成一区。

5.6.3.6 Intensive care unit (ICU) and neonatal intensive care unit (NICU) should be set in a separate area.

5.6.3.7 要充分考虑日照因素，保证病房、门诊等主要功能区域获得充足的日照时间。

5.6.3.7 The sunlight should be fully taken into account to ensure that the ward, outpatient and other major function areas can enjoy sufficient sunlight.

5.6.3.8 除了通用的无障碍设计之外，还应设置儿童无性别厕所、孕产妇宽大、坐式洁具设计的厕所，以及哺乳室。

5. 6. 3. 8 In addition to the universal barrier-free design, gender-free toilets for children and large seat-type sanitary ware design for pregnant should also be provided, and nursing rooms should be arranged.

5. 6. 3. 9 室内设计：应专门为妇女、儿童进行配色、材质、软装、照明及配电末端安全等专项设计。

5. 6. 3. 9 Interior design: special design should be specially made for women and children in color matching, material, soft decoration, lighting and safety of power distribution terminal.

5.7 专科医院

5.7 Specialized hospital

5. 7. 1 传染病医院

5. 7. 1 Infectious disease hospital

5. 7. 1. 1 概述

5. 7. 1. 1 Overview

传染病医院为收治法定传染病专科诊疗机构，包括消化道传染病、呼吸道传染病以及肝炎、艾滋病等其他传染病，医院设置门诊部、医技部、24 小时服务的急诊部和住院部。其设计应符合《传染病医院建筑设计规范》。

Infectious disease hospital is a specialist diagnosis and treatment institution to receive and cure patients with legal infectious diseases, including digestive tract infectious disease, respiratory infectious disease, hepatitis, AIDS and other infectious diseases. The hospital has the outpatient department, medical technology department, 24-hour emergency department and the inpatient department. The design shall conform to the Architectural Design Code for Infectious Diseases Hospital.

5. 7. 1. 2 总体布置要求

5. 7. 1. 2 General layout requirements

1. 应根据医院规模设置 2 ~ 3 个院区出入口，院区的主要出入口应避免安排在交通主干道上以免造成交通车流阻断。大型传染病院可分设门急诊出入门、住院探视出入口及后勤保障、污物出入口。较小规模医院可将前者合设。

1. 2 ~ 3 entrances and exits shall be set up in the hospital according to the size of the hospital. The main entrances and exits of the hospital shall not be arranged on the main road to avoid blocking the traffic flow. For large infectious disease hospitals, the entrances and exits for emergency treatment, visitation of in patients, logistics support, and waste can be set separately. For smaller hospitals, the former can be set in combination.

2. 应科学合理安排院区内各个功能科室与部门，使各个功能部门配置合理得当，人流物流顺畅便捷；洁污分区、洁污分流明确，互不穿插交叉，以利于控制与防止院内交叉感染。

2. Arrange the functional sub departments and departments in a scientific and reasonable way, so as to make the allocation of the functional departments reasonable and proper, and make the flow of people and logistics smooth and convenient. Clearly divide and distribute the clean and pollution zones, avoiding mutual cross, in order to control and prevent cross infection in the hospital.

3. 对涉及污染环境的污物（含医疗废弃物、污废水等）进行环境规划。

3. Carry out environmental planning for the pollutants (including medical waste, sewage, and others) that pollute the environment.

4. 传染病医院出入口附近布置急救车冲洗消毒场地。

4. Arrange the ambulance flushing and disinfection site near the entrance and exit of infectious disease hospitals.

5. 在门诊、急诊、急救和住院主要出入口处，必须有带雨篷的机动车停靠处。如设坡道时，应按无障碍要求另设计为残疾人专用的坡道。

5. Motor vehicle park places with canopies must be set at the main entrances and exits of outpatient, emergency, first aid and inpatient zones. If ramps are designed, the disabled only ramps shall be designed according to the accessibility requirements.

6. 门诊、急诊和病房应充分利用自然通风和天然采光。

6. Outpatient rooms, emergency rooms and wards shall make full use of natural ventilation and natural lighting.

5.7.1.3 用地与建筑面积指标

5.7.1.3 Land area and building area indicators

1. 传染病医院建筑用地面积指标表：见表 5-4。

1. Building Land Area Indicators of Hospitals for Infectious Diseases: see Table 5-4.

表 5-4 传染病医院建筑用地面积指标表

Table 5-4 Building Land Area Indicators of Hospitals for Infectious Diseases

建设规模（床位数） Construction scale (number of beds)	150	250	400	400 以上 Above 400
建筑用地面积指标（m^2/床） Building land area indicator (m^2/bed)	130	125	120	120

2. 传染病医院建设面积指标表：见表 5-5。

2. Construction Area Indicators of Hospitals for Infectious Diseases: see Table 5-5.

表 5-5 传染病医院建设面积指标表

Table 5-5 Construction Area Indicators of Hospitals for Infectious Diseases

建设规模（床位数） Construction scale (number of beds)	150	250	400	400 以上 Above 400
建设面积指标（m^2/床） Construction area indicator (m^2/bed)	80 ~ 82	~ 80	78 ~ 80	78

3. 传染病医院各类房间占建筑面积的比率（%）：见表5-6。

3. Proportion of All Kinds of Rooms in the Building Area of Hospitals for Infectious Diseases (%): see Table 5-6.

表5-6 传染病医院各类房间占建筑面积的比率（%）

Table 5-6 Proportion of All Kinds of Rooms in the Building Area of Hospitals for Infectious Diseases (%)

门、急诊 Outpatient and emergency	11
住院部 Inpatient department	45
医技科室 Medical technology department	17
保障系统 Support system	15
行政管理 Administrative management	6
院内生活 Hospital life	6

5.7.1.4 建筑设计

5.7.1.4 Architectural design

1. 一般规定

1. General provisions

（1）在门诊、急诊、急救和住院主要出入口处，必须有带雨篷的机动车停靠处；如设坡道时，应按无障碍要求另设计为残疾人专用的坡道。

（1）Motor vehicle park places with canopies must be set at the main entrances and exits of outpatient, emergency, first aid and inpatient zones; If ramps are designed, the disabled only ramps shall be designed according to the accessibility requirements.

（2）医院的各个分区和医疗用房应在交通本枢纽处，如大厅、电梯门、各科室入口处等设置明显的导向图标。

（2）Each partition and medical room of the hospital shall be set with obvious guiding icons at the transportation hub, such as hall, elevator door, and entrance of each department.

（3）电梯

（3）Elevator

① 二层及二层以上的门诊楼宜设电梯，四层及四层以上的门诊楼或病房楼应设电梯，且不得少于两部；当病房楼高度超过24m时，应设污物梯，有条件时可增设自动扶梯；

① The outpatient building in two or more floors should be equipped with elevators, and the outpatient building or ward building in four or more floors shall be equipped with elevators which are not less than two; when the height of the ward building exceeds 24m, it shall be equipped with waste elevators, escalators shall be added when possible;

② 供病人使用的电梯和污物梯，应采用病床电梯；

② The bed elevators shall be used for elevators and waste elevators for use of patients;

③ 电梯井道不得与主要用房贴邻。

③ The elevator shaft way shall not be adjacent to the main room.

(4) 楼梯

(4) Stair

① 楼梯的位置应同时符合防火疏散和功能分区的要求；

① The location of stairs shall meet the requirements of both fire evacuation and functional division;

② 主楼梯宽度不得小于1.65m, 踏步宽度不得小于0.28m, 高度不得小于0.16m;

② The width of the main stair, the step width and the height shall not be less than 1.65m, 0.28m, and 0.16m respectively;

③ 主楼梯和平台深度不宜小于2m。

③ The depth of the main stair and platform should not be less than 2m.

(5) 三层及三层以下无电梯的病房楼，以及观察室与抢救室不在同一层又无电梯的急诊部，均应设置坡道，坡道坡度按无障碍坡道设计，并应有防滑措施。

(5) Ramps shall be set for the ward building in three or less floors without elevator, and the emergency department of which the observation room and emergency room are not on the same floor, and has no elevator; the slope of ramps shall be designed as that of wheelchair accessible ramps, with anti-skid measures.

(6) 通行推床的室内走道净宽不应小于2.4m: 有高差者必须用坡道相接，坡道坡度按无障碍坡道设计。

(6) The clear width of the indoor passageway going through rollaway bed shall not be less than 2.4m: Those with height difference must be connected with ramps, and the slop of ramps shall be designed as that of wheelchair accessible ramps.

(7) 半数以上的病房应获得良好日照及景观。

(7) More than half of the wards shall have good sunlight and views.

(8) 门诊、急诊和病房应充分利用自然通风和天然采光。

(8) Outpatient rooms, emergency rooms and wards shall make full use of natural ventilation and natural lighting.

(9) 室内净高在自然通风条件下不应低于下列规定：

(9) The indoor clear height of buildings under natural ventilation shall not be lower than the following:

① 诊查室2.60m, 病房2.80m;

① Consulting room: 2.60m, ward: 2.80m;

② 医技科室2.80m, 或根据需要而定。

② Medical technology department: 2.80m, or determined as required.

（10）病房的允许噪声级（A 声级）昼间应小于或等于 50dB，夜间应小于或等于 40dB，隔墙与楼板的空气声的计权隔声量应大于或等于 40dB，楼板的计权标准化撞击声压级宜小于或等于 75dB。

(10) The allowable noise level (sound level A) of the wards shall be less than or equal to 50dB during the day, less than or equal to 40dB at night, the weighted transmission loss of the air sound of the partition wall and the floor shall be greater than or equal to 40dB, and the weighted impact sound pressure level of the floor should be less than or equal to 75dB.

（11）所有医疗用房必须满足医疗使用和医疗设备对环境的特殊要求。

(11) All medical rooms must meet the special requirements of medical use and medical devices for the environment.

（12）厕所

(12) Toilet

① 室内厕所设前室，门急诊、医技科室病人公用厕所应将前室设成宽式布置，不设门扇，并应设非手动开关龙头的洗手盆；

① The front chamber shall be set for the indoor toilets, and the front chambers in public toilets for patients of outpatient and emergency departments and medical technology department shall be arranged in a wide type, without door leaves, and the wash basins with non-manual switch faucets shall be set;

② 如采用室外厕所，宜用连廊与门诊、病房楼相接；

② If the outdoor toilets are used, the corridor should be used to connect to the outpatient and ward buildings;

③ 病人使用的厕所隔间的平面尺寸，不应小于 1. 10m × 1. 40m，门朝外开，门闩应能里外开启；

③ The plan view size of the toilet cubicle for use of patients shall not be less than 1. 10m × 1. 40m, the doors shall be opened outwards, and the latches shall be able to open inside and outside;

④ 病人使用的坐式大便器的坐圈宜采用“马蹄式”，蹲式大便器宜采用“下卧式”，大便器旁应装置“助力拉手”；

④ The seat ring of the seat-type closet pans for use of patients should be in “horseshoe type”, and the squat closet pans should be in down-horizontal type”, with the “assisted handle” equipped beside a closet pan;

⑤ 男、女公共厕所应各设一个无障碍隔间厕所；

⑤ Male and female public toilets shall each have an accessible cubicle toilet;

⑥ 新建无障碍厕位面积不应小于 1. 80m × 1. 40m；

⑥ The area of the new accessible toilet cubicle shall not be less than 1. 80m × 1. 40m;

⑦ 改建无障碍厕位面积不应小于 2. 00m × 1. 00m；

⑦ The area of the reconstructed accessible toilet cubicle shall not be less than 2. 00m × 1. 00m;

⑧ 厕位门扇向外开启后，入口净宽不应小于 0. 80m，门扇内侧应设关门拉手；

⑧ After the door leaf of toilet cubicle opens outwards, the entrance clear width shall not be less than 0.80m, and the inside of the door shall be equipped with closing handle;

⑨ 每个护理单元至少设一间带无障碍厕位的病房。

⑨ Each nursing unit shall have at least one ward with accessible toilet cubicle.

2. 门诊部与接诊部

2. Outpatient department and reception department

（1）与一般综合性医院相比，传染病医院门诊量相对比较小，日门诊人次与住院床位比例为1：1～0.6：1范围。

（1）Compared with general hospitals, the outpatient load of infectious disease hospital is relatively small, and the ratio of outpatient person-time per day to inpatient beds ranges from 1：1 to 0.6：1.

（2）门诊部与接诊部的位置应靠近院区的主要出入口以方便病患者急找；为接受其他医院转诊病人以及应对公共卫生突发事件，应根据医院规模，在医院门诊部入口附近设置接诊部或筛查部，在其毗邻设置隔离观察区，在入口部对不同传染病患者进行筛查分流，在突发公共卫生事件时可以及时排查并有效控制带有扩散传染病风险的病患者的活动范围，减少其扩散范围。

（2）The outpatient department and the reception department shall be located close to the main entrance and exit of the hospital area to facilitate patients to find them in emergency. In order to receive transfer treatment patients from other hospitals and respond to public health emergencies, a reception department or screening department shall be set near the entrance of the outpatient department according to the scale of the hospital. An isolated observation area shall be set adjacent to it. Screening and triage of patients with different infectious diseases shall be conducted at the entrance. In the case of public health emergencies, patients with risk of spreading infectious diseases and their range of activities can be timely checked and effectively controlled, to reduce the scope of spread.

（3）大型传染病医院门诊应考虑不同传染病种的门诊区域，包括肝炎门诊、消化道传染病门诊、呼吸道传染病门诊、艾滋病门诊、发热门诊、综合门诊等；各类传染病门诊科室包括候诊、诊室等，均应分别自成一区，相对独立。

（3）The outpatient department in large infectious disease hospitals shall consider to set outpatient service areas of different infectious diseases, including hepatitis outpatient area, digestive tract infectious disease outpatient area, respiratory infectious disease outpatient area, AIDS outpatient area, fever outpatient area, comprehensive outpatient area, etc. Outpatient departments of various infectious diseases, including waiting area and consulting room, shall be set in their respective separate area and relatively independent.

（4）门诊部一般包括入口厅、问讯处、挂号、划价交费、中西药房、各科诊室、医务人员值班更衣、医办、病人厕所；为减少与控制传染病患者就医环境中交叉感染的风险，应有足够的室内候诊就医空间。

（4）The outpatient department generally includes the outpatient hall, inquiry desk, rooms for registration and charging, traditional Chinese medicine pharmacy, western medicine pharmacy, consulting rooms of various departments, duty room and locker room for medical care staff, doctor office and patient toilet. In order to reduce and control the risk of cross infection in the medical environment for patients with infectious diseases, the infectious disease hospital shall have sufficient spaces for indoor waiting and medical treatment.

（5）传染病医疗建筑平面布局可采用双通道布置方式，即病患者与医务人员分别使用不同通道。医务人员进出工作区的口部应设置更衣室与卫生间，发热门诊医护人员可按卫生通过室要求设置。

（5）The healthcare buildings for infectious diseases can use two channel layout, that is, patients and medical care staff use different channels respectively. The entrance for the medical care staff getting in and out of the work area shall be equipped with dressing room and toilet, and the entrance for medical care staff of fever outpatient department can be set according to the requirements of the uncontaminated transition room.

（6）候诊厅及候诊廊可根据传染病医院的规模以及预计接纳的日平均门诊人次设置，大中型医院采用厅式候诊，小型医院采用走道式候诊。

（6）The waiting hall and waiting passageway will be set based on the scale of infectious disease hospital and the predicted daily average number of outpatient visits. Large and medium-sized hospitals adopt the waiting hall and small hospitals adopt the waiting passageway.

3. 住院部（包含若干病区或护理单元）

3. Inpatient department (including several wards or nursing units)

（1）应把不同类型的传染病依据床位规模大小用不同方式分开，独立成栋、成组或竖向组合布置，每类病种出入口独立。

（1）Different types of infectious diseases should be separated in different ways according to the size of beds. They should be arranged in separate buildings, groups or vertical combinations. The entrances and exits of each type of disease should be independent.

（2）不同传染病种患者应分别安排在不同病区构成不同护理单元，不得混合安排，如需要合并时应予以分隔。

（2）Patients with different infectious diseases should be arranged in different wards to form different nursing units, and should not be arranged in a mixed way, and should be separated if combination is required.

（3）在特大型传染病医院可设置1～2个负压病房的护理单元，可考虑正负压转换，平时与应急时期相结合，在中型医院可在其ICU病区内设置2～4间负压监护室。

（3）Nursing units of 1～2 negative pressure wards can be set up in super-large infectious disease hospitals, and positive-negative pressure conversion can be considered. Peacetime should be combined with emergency period. In medium-sized hospitals, 2～4 negative pressure intensive care units can be set up in ICU wards.

（4）传染病的护理单元应按传染病隔离要求分设病患者使用通道，包括垂直交通楼电梯，病患者由专用通道进入病室，医务工作人员使用专用垂直交通楼电梯及走廊进入工作区，医务工作人员进出工作区口部应设置卫生通过室。

(4) Nursing units for infectious diseases shall be provided with separate passages for patients according to the requirements for isolation of infectious diseases, including vertical transportation stairs and elevators. Patients shall enter the ward through special passages. Medical care staff shall use special vertical transportation stairs and elevators and corridors to enter the work area. Uncontaminated transition rooms shall be provided at the entrance of medical care staff entering and leaving the work area.

(5) 根据需要可设计安排疑似筛查病区，该病区应全部按单人病室布置，应考虑病患者确诊后的分流通道。

(5) According to the needs, suspected screening wards can be designed and arranged. The wards should all be arranged according to single patient rooms, and distribution passages for patients after diagnosis should be considered.

(6) 传染病区备餐间应分成相邻洁污独立小间，相互间设传递窗，患者餐食由工作人员在清洁备餐间分取，由传递窗送至各病房，患者用餐后残留食品餐具另由患者通道回收，送至污染备餐小间，倒弃废残食，回收餐具经蒸煮消毒后，由传递窗传向清洁小间。

(6) The pantries in the infectious diseases wards shall be divided into adjacent small rooms that clean and contaminated areas are separated, and transfer windows shall be set between them. The patient's food shall be divided by the staff in the clean pantry and delivered to each ward through the transfer window. The residual food and tableware after meals shall be recycled through the patient channel and delivered to the contaminated pantry. The waste and residual food shall be discarded. After being steamed and boiled for sterilization, the recycled tableware shall be transferred to the clean pantry through the transfer window.

(7) 烈性传染病区或有条件的医院可采用一次性餐具，病患者用餐后、残食、一次性餐具统一回收集中灭菌、销毁。

(7) Disposable tableware can be used in severe infectious disease wards or hospitals where conditions permit. After meals, the residual food and disposable tableware can be recycled, sterilized and destroyed.

(8) 烈性传染病室与走廊间应设有双门密闭式传递窗，病室与走廊间设缓冲过渡小间，将门错开布置避免气流倒灌，过渡小间内应设非手动式龙头刷手池，供医务人员进出病室刷手用。

(8) A double-door closed transfer window shall be provided between the severe infectious disease ward and the corridor. A buffer transition cubicle shall be provided between the ward and the corridor. The doors shall be staggered to avoid air flow backward. A non-manual faucet hand washing sink shall be provided in the transition cubicle for medical care staff to wash their hands when they enter and leave the ward.

4. 传染病医院的手术部应根据医院规模及其承担的任务，设计1～3间负压手术室，负压手术室宜安排布置在手术部靠近换床间的前端，应另设专用通道与一般换床通道分开。

4. 1 ~3 negative pressure operating rooms should be designed in the operating department of infectious disease hospital according to the size of the hospital and the tasks it undertakes. The negative pressure operating rooms should be arranged at the front end of the operating department close to the bed changing room. A special passage should be set up separately from the general bed changing channel.

5. 传染病医院的放射检查室应采用负压通风，每个放射检查室前设 2 ~3 间更衣小间。

5. Negative pressure ventilation should be adopted in radioscopy examination rooms of infectious disease hospitals. 2 ~3 small locker rooms should be set up in front of each radioscopy examination room.

6. 有条件的传染病医院应设负压病理解剖室并和太平间合建。

6. If possible, infectious disease hospital should set up negative pressure pathological anatomy room, which should be built together with the morgue.

7. 烈性传染病区的衣服、被单应由专门容器或包裹先送至大型环氧乙烷灭菌器熏蒸后，再由洗衣房接受洗涤。

7. Clothes and sheets from severe infectious disease wards should be sent to large ethylene oxide sterilizer for fumigation by special containers or packages before being washed in laundry.

8. 传染病医院或传染病区必须具备自然通风条件，各处门口不应设置空气幕；应设置机械通风系统，清洁区、半污染区、污染区的机械送、排风系统应按区域设置；医院门、急诊口部的筛查，其通风系统应独立设置；病房卫生间排风不应通过共用竖井排风，应结合病房排风统一考虑。

8. Infectious disease hospitals or infectious disease wards must have natural ventilation conditions, and air curtains should not be set up at the entrances of all places. Mechanical ventilation system shall be set up, and the mechanical air supply and exhaust systems in clean area, semi-contaminated area and contaminated area shall be set up according to the area. The ventilation system should be set up independently for the screening work area at the entrance of the outpatient department and emergency department of the hospital. Exhaust of the ward toilet should not be realized through a common shaft, but should be considered in combination with ward exhaust.

9. 负压隔离病房的外围护结构应采用接缝严密的建筑材料建造。

9. The peripheral protection structure of the negative pressure isolation ward shall be constructed of closely jointed building materials.

5. 7. 2 精神病医院

5. 7. 2 Psychiatric hospital

5. 7. 2. 1 一般要求

5. 7. 2. 1 General requirements

精神病医院是指主要提供综合性精神卫生服务的医疗机构。本条适用于精神病医院以及综合医院的精神病区，包括不需要内科/外科急性护理服务的精神病院住院病人、门诊病人的护理与处置。医疗保健设施通用设计适用于本设施。

Psychiatric hospitals refer to medical institutions that mainly provide comprehensive mental health services. This article is applicable to psychiatric wards of psychiatric hospitals and general hospitals, including the nursing and treatment of inpatients and outpatients in psychiatric hospitals that do not need acute internal medical/surgical nursing services. The general design of health care facilities is applicable to this facility.

共享服务：

Sharing service:

当精神病设施是另一个设施的一部分时，如餐饮、储藏、药房、洗衣房等这些服务应切合实际的予以共享。在某些情况下，所有辅助性服务可由主要设施满足；另一方面，有些计划内容与需求可能要求单独服务。

When the psychiatric facility is the part of another facility, services such as catering, storage, pharmacy, laundry should be shared practically. In some cases, all ancillary services can be met by the main facilities; on the other hand, some plan contents and requirements may require separate services.

机动床位：

Mobile beds:

在设施内应允许有一组房间的使用情况做出变动，依据病人年龄、性别、安全程度或治疗内容的不同，适应不同病人群体的需要。

The use for a group of rooms shall be allowed to change in the facility, and adapted to the needs of different patient groups according to the differences in patient age, gender, safety degree or treatment content.

5.7.2.2 精神专科医院建设标准与等级

5.7.2.2 Construction standard and level of psychiatric specialized hospital

我国公立精神专科医院建设项目的审批、核准、设计和建设必须符合《精神专科医院建设标准》(建标 176—2016)。

The examination, approval, design and construction of public psychiatric specialized hospital construction project in China must conform to the Construction Standard of Psychiatric Specialized Hospital(JB 176—2016).

1. 精神区域医疗中心

1. Psychosis regional medical center

(1) 应当为三级甲等精神专科医院或者具备相应精神专科能力的三级甲等综合医院（专科医院优先），所处地理位置交通便利，方便所在区域患者就医；能够提供服务的精神科床位数≥150 张；急性（短期）住院床位数占医院床位总数≥15%，符合精神科疑难急重患者收治标准患者≥50%；获得精神专业国家临床重点专科建设项目≥1 个，或拥有省部级重点实验室或国家重点学科等平台≥2 个。

(1) It should be a Third-class Level A psychiatric specialized hospital or a Third-class Level A general hospital with corresponding psychiatric specialized capacity (specialized hospital is

preferred), with convenient geographical location and transportation, so as to facilitate the medical treatment of patients in the area; the number of beds available in psychiatry department≥150; the number of acute (short-term) in hospital beds accounting for the total number of hospital beds≥15%, the patients who meet the receiving and curing standard of difficult, urgent and severe patients in psychiatry department≥50%; the projects obtaining the national clinical key specialty development in psychiatric discipline≥1, or the number of platforms such as provincial and ministerial-level key laboratories or national key disciplines≥2.

（2）临床/医技科室设置

（2）Clinical/medical technology department setting

独立设置精神病专业、精神医学专业、药物依赖专业、精神康复专业、社区防治专业、临床心理专业、司法精神专业的科室，营养科、医学检验科、医学影像科（含心电图、脑电图、神经肌肉电图等）、药剂科、中西医结合科等；设置儿童、老年、重症医学等亚专业，除普通门诊、专家门诊、联合会诊外，应当设置相应亚专业门诊，应有教学、科研设施；规划与建筑设计应遵循《精神专科医院建筑设计规范》（建标 176—2016），医院运营应符合《三级综合医院评审标准》（卫医管发〔2019〕33 号）。

Independent settings for departments of psychosis discipline, psychiatry discipline, drug dependence discipline, psychiatric rehabilitation discipline, community prevention discipline, clinical psychology discipline, forensic psychiatry discipline, nutritional department, medical laboratory department, medical imaging department (including ECG, electroencephalogram, neuromuscular electrograph), pharmacy department, integrated traditional Chinese and western medicine department, etc; settings of sub disciplines such as children, elderly, and critical care medical, in addition to general clinic, expert clinic and joint consultation, corresponding sub discipline clinic should be set up, with teaching and scientific research facilities; the planning and architectural design should follow the Code for Architectural Design of Psychiatric Specialized Hospital(JB 176—2016), and the hospital operation should conform to the Review Criteria for Third-class General Hospital(Health and medical management[2019]No. 30).

2. 一级精神病医院：精神科住院床总数 20～69 张。

2. First-class psychiatric hospital: the total number of hospital beds in psychiatry department is 20～69.

科室设置：

Department setting:

（1）临床科室：至少设有精神科门诊、精神科房（男、女病区分设）、预防保健室。

（1）Clinical departments: at least psychiatry department clinic, psychiatry department room (male and female wards are divided) and prevention and health care room.

（2）医技科室：至少设有药房、化验室、X 光室、消毒供应室。

（2）Medical technology departments: at least pharmacy, laboratory, X-ray room and disinfection supply room.

（3）基本设备：供氧装置、呼吸机、洗胃机、电动吸引器、心电图机、气管切开包、静脉切开包、导尿包、灌肠器、显微镜、火焰光度计、pH计、血细胞计数仪、离心机、自动稀释器、电冰箱、干燥箱、X光机、B超、脑电图仪、眼底镜、五官检查器、常用处置器械、药用天平、储存柜、器械柜、电休克治疗仪、体疗设备、电视机、录音机、紫外线灯、蒸馏装置、高压灭菌设备、洗衣机。

(3) Basic devices: oxygen supply device, breathing machine, gastric lavage machine, electric aspirator, ECG machine, tracheotomy kit, phlebotomy kit, urethral catheterization kit, enemator, microscope, flame photometer, pH meter, blood cell counter, centrifuge, autodilutor, refrigerator, drying oven, X-ray machine, B-ultrasound, electroencephalograph, ophthalmoscope, facial feature detector, common disposal instrument, pharmaceutical scales, storage cabinet, instrumentcabinet, electric shock therapeutic instrument, physiotherapy device, television, recorder, ultraviolet lamp, distillation unit, autoclaving device, washing machine.

3. 二级精神病医院：精神科住院床位总数70～299张。

3. Second-class psychiatric hospital: the total number of hospital beds in psychiatry department is 70～299.

科室设置：

Department setting:

（1）临床科室：至少设有精神科（内含急诊室、心理咨询室）、精神科男病区、精神科女病区、工娱疗室、预防保健室。

(1) Clinical departments: at least psychiatry department (including emergency room, psychological counseling room), male ward of psychiatry department, female ward of psychiatry department, occupational and recreational therapy room and prevention and health care room.

（2）医技科室：至少设有药房、化验室、X光室、心电图室、脑电图室、消毒供应室、情报资料室、病案室。

(2) Medical technology departments: at least pharmacy, laboratory, X-ray room, ECG and electroencephalogram room, disinfection supply room, intelligence information room and medical record room.

（3）基本设备：供氧装置、呼吸机、电动吸引器、洗胃机、心电图机、心电监护仪、气管切开包、显微镜、火焰光度计、血细胞计数仪、分光光度计、自动生化分析仪、血气分析仪、荧光光度计、血小板计数仪、PG计、自动稀释器、恒温箱、干燥箱、分析天平、离心机、超净操作台、电动振荡器、电冰箱、X光机、脑电图仪、脑电地形图仪、脑血流图仪、B超、眼底镜、五官检查器、常用处置器械、体疗设备、电休克治疗仪、超声治疗仪、音频电疗机、音乐治疗仪、生物反馈治疗机、电视机、录音机、扩音机、储存柜、紫外线灯、蒸馏装置、高压灭菌设备、洗衣机。

(3) Basic devices: oxygen supply device, breathing machine, electric aspirator, gastric lavage machine, ECG machine, ECG monitor, tracheotomy kit, microscope, flame photometer, blood cell counter, spectrophotometer, automatic biochemical analyzer, blood-gas analyzer, fluorophotometer, blood platelet counter, PG meter, autodilutor, thermostat, drying oven, analytical balance, cen-

trifuge, ultra-clean console, electric agitator, refrigerator, X-ray machine, electroencephalograph, electroencephalogram-topography instrument, cerebral blood flow chart, B-ultrasound, ophthalmoscope, facial feature detector, common disposal instrument, physiotherapy device, electric shock therapeutic instrument, ultrasonic treatment device, audiofrequency electrotherapy unit, music therapeutic instrument, biological feedback therapeutic machine, television, recorder, amplifier, storage cabinet, ultraviolet lamp, distillation unit, autoclaving device, washing machine.

4. 三级精神病医院：精神科住院床位总数300张以上。

4. Third-class psychiatric hospital: the total number of hospital beds in psychiatry department is more than 300.

科室设置：

Department setting:

（1）临床科室：至少设有精神科门诊（含急诊、心理咨询），4个以上精神科病区，男女病区分开，心理测定室、精神医学鉴定室、工娱疗室、康复科。

（1）Clinical departments: at least psychiatry department clinic (including emergency, psychological counseling), more than 4 wards of psychiatry department, with male and female ward divided, psychological testing room, psychiatry identification room, occupational and recreational therapy room and rehabilitation department.

（2）医技科室：至少设有药剂科、检验科、放射科、心电图室、脑电图室、超声波室、消毒供应室、情报资料室、病案室和3个以上的研究室。

（2）Medical technology departments: at least pharmacy department, laboratory department, radiology department, ECG room, electroencephalogram room, ultrasonic room, disinfection supply room, intelligence information room, medical record room and more than 3 laboratories.

（3）基本设备：供氧装置、呼吸机、洗胃机、气管插管、电动吸引器、心电图机、心电监护仪、按摩机、气管切开包、显微镜、火焰光度计、血细胞计数仪、血小板计数仪、自动生化分析仪、血气分析仪、血氨测定计、尿分析仪、酶自动分析仪、分光光度计、荧光光度计、pH计、分析天平、离心机、干燥箱、恒温箱、霉菌培养箱、电动振荡器、自动稀释器、净化操作台、电冰箱、X光机、B超、脑电地形图仪、脑血流图仪、五官检查器、常用处置器械、诱发电位仪、音乐治疗仪、超声治疗仪，音频电疗仪、生物反馈治疗机、电休克治疗仪、体疗设备、储存柜、电视机、录音机、扩音机、紫外线灯、蒸馏装置、高压灭菌设备、洗衣机。

（3）Basic devices: oxygen supply device, breathing machine, gastric lavage machine, tracheal intubation, electric aspirator, ECG machine, ECG monitor, massage machine, tracheotomy kit, microscope, flame photometer, blood cell counter, blood platelet counter, automatic biochemical analyzer, blood-gas analyzer, blood ammonia meter, urine analyzer, automatic enzyme analyzer, spectrophotometer, fluorophotometer, pH meter, analytical balance, centrifuge, drying oven, thermostat, mold incubator, electric agitator, autodilutor, purification console, refrigerator, X-ray machine, B-ultrasound, electroencephalogram-topography instrument, cerebral blood flow chart, facial feature detector, common disposal instrument, evoked potential device, music therapeutic instrument, ultrasonic treatment device, audiofrequency electrotherapy unit, biological feedback therapeutic machine, electric shock therapeutic instrument, physiotherapy device, storage cabinet, television, recorder, amplifier, ultraviolet lamp, distillation unit, autoclaving device, washing machine.

5.7.2.3 护理环境

5.7.2.3 Nursing environment

1. 精神病医疗设施的安全环境非常重要，但其内部没有一个环境是绝对安全而无风险的。多数企图自杀的人都在经受能治疗的精神失常或物质滥用失常或两者都有的痛苦。在精神病治疗设施中的病人被认为是自杀高危人群，因此在保持治疗环境的同时，应避免存在实际的危险性环境。一个已经建造好的环境，无论设计或建设得多好，都不能完全信赖其有绝对安全的防护措施，所以可能还需要增加防护栅栏、坚固的门锁、24 小时监护设备。

1. The safety environment of psychosis medical facilities is very important, however, none of its internal environment is absolutely safe and risk-free. Most people who attempt suicide are suffering from treatable psychiatric disorder, substance abuse disorder or both. Patients in psychosis treatment facilities are considered to be high risk group of suicide. Therefore, while maintaining the therapeutic environment, actual hazardous environment should be avoided. In the environment, no matter how well it is designed or built, cannot be fully trusted to have absolutely safe protective measures, therefore, it may be necessary to add protective fences, strong door locks and 24 – hour monitoring equipment.

2. 精神病医院建筑要为计划治疗内容提供合适的环境单元，应具有开放的感觉，要强调自然光。在建筑设计和维护的每一个方面，最重要的是应对所服务特殊病人群体的潜在危险性做出一些决定。在护理单元走道、用餐区和社交区，例如娱乐室和活动区均应考虑可视控制（包括电子监控），应避免可供藏匿的库房、盲角或盲区。护士站的开放性由治疗规划确定，应考虑到病人隐私和员工安全。

2. Psychiatric hospital buildings should provide appropriate environmental units for the planned treatment contents, and should have an open feeling. The natural light should be emphasized. In every aspect of architectural design and maintenance, it is important to make some decisions to deal with the potential dangers of the special patient groups served. Visual control (including electronic monitoring) shall be considered in the passageway, dining area and social area of nursing unit, such as entertainment room and activity area, the storehouse, blind corner or blind area for hiding should be avoided. The openness of nurse station should be determined according to the treatment planning, which should take into account patient privacy and staff safety.

5.7.2.4 治疗环境

5.7.2.4 Therapeutic environment

精神病院建筑应为计划治疗内容提供适宜的环境，包括：空间分区、隔离监护、户外活动、家庭化、工娱室等。

Psychiatric hospital buildings should provide suitable environment for the planned treatment content, including: spatial division, isolation and monitoring, outdoor activities, domestication, occupational and recreational rooms, etc.

5.7.2.5 安全设计

5.7.2.5 Safety design

1. 设计应为计划处置内容提供一定的安保级别。

1. The design shall provide a certain level of security for the planned disposal content.

2. 为防止伤害或自杀，在设计时要对建筑和室内的细节装饰面与设备做特殊考虑。

2. In order to prevent injury or suicide, special consideration should be given to the detail decoration surface and equipment of buildings and interior during design.

5.7.2.6 精神病院护理场所

5.7.2.6 Psychiatric facility nursing place

1. 综合性精神病护理单元

1. Comprehensive psychosis nursing unit

（1）典型精神病人病房每个房间应满足以下标准：房间的最大容量为两位病人；病房应设置窗户，应有桌子或书写台面供病人使用；每个病人应有到厕所的通道，而不必进入公用走道区；每间厕所服务的床位不要多于4床或2床，厕所内应有马桶、洗手处，厕所门应向外开或双向开；每位病人在自己的病房内应有独立的衣柜、锁柜，用于挂长大衣或存放个人财物，存放量应能满足每天更换衣服，共七天之需。

（1）Each room in a typical psychiatric ward should meet the following standards: the maximum accommodation of each room is two patients. The ward should be provided with windows, tables or writing tables for patients. Each patient should have access to the toilet without entering to the public passageway area; the bed served for each toilet should be no more than 4 or 2. The closestool, hand washing place should be provided in the toilet. The toilet door shall be opened outwards or opened in both directions; each patient should have an independent closet and locker in his or her own ward, which is used to hang long overcoats or store personal belongings. The storage capacity should be able to meet the needs of changing clothes every day for a total of seven days.

（2）应设安保室以减少患者脱逃、藏匿、自残或伤害他人或自杀的潜在可能性，从安保室到厕所、淋浴间与衣柜处应受到限制；安保室可集中设置在一个单元内，或各单元分散设置。

（2）Security room should be set up to reduce the potential possibilities of escape, hiding, self-harm, injury to others or suicide of patients. The places from the security room to the toilet, shower cubicle or closet should be restricted; the security room can be centralized in one unit or scattered in each unit.

（3）不需要设置供患者活动的区域，以防止神志模糊的患者在指定的住院区外闲逛。

（3）There is no need to set up an area for patients' activities to prevent patients with mental confusion from wandering outside the designated hospitalization area.

（4）若采取封闭的四合院形式，可提供病人的露天活动场所，该场所的一切设施应牢固、光滑、无伤害性、不能够被破坏。

(4) If a closed quadrangle is adopted, an open-air activity place for patients can be provided, and all facilities in the place shall be firm, smooth and harmless, and shall not be destroyed.

2. 特殊病人护理区域

2. Nursing area for special patient

（1）隔离室：为病人短时期使用，设在精神护理单元内，此房间提供给需要安全与防护的病人。每间病房只能有一位病人。每个精神护理单元内，每 24 张病床至少设 1 间隔离室，如医院有一个以上的精神护理单元，则隔离室的数量应随着医院内精神病病床总数而定。隔离室应位于护理人员能直接监视之处，几个隔离室可组合在一起，隔离处置室面积应至少 $6m^2$。根据任务书的需要设置约束性床位，约束性病房的面积应大于 $8m^2$。

(1) Isolation room: it is used by patients for a short period of time, and set up in the psychiatric nursing unit. The room is provided for patients who need safety and protection There can only be one patient for each ward. In each psychiatric nursing unit, at least 1 isolation room should be set up for every 24 hospital beds. If the hospital has more than one psychiatric nursing unit, the number of isolation rooms should be determined according to the total number of psychosis beds in the hospital. The isolation room should be located where the nursing personnel can directly monitor, several isolation rooms can be combined, and the area of the isolation and disposal room should be at least $6m^2$. Constraint beds shall be set up according to the requirements of the design specification, and the area of the constraint ward should be more than $8m^2$.

（2）隔离处置室：应有能直接进出厕所的前室，厕所门宽为 1. 2m; 应能防止病人藏匿、逃脱、伤害或自杀；室内不应有突出的尖角和棱缘，门向外开，门宽 1. 2m, 既能让员工通过观察窗看到病人，又能保证病人的私密性。

(2) Isolation and disposal room: there should be a front room with direct access to the toilet. The width of the toilet door is 1. 2m; it should be able to prevent the patient from hiding, escape, injury or suicide; there should be no protruding sharp corner and edge. The door should be opened outwards, with a width of 1. 2m. It not only enables employees to see patients through the observation window, but also ensures the privacy of patients.

（3）在精神病院应考虑设置负压隔离病房，其要求、数量依据感染控制要求确定。

(3) Negative pressure isolation ward should be set up in the psychiatric hospital. Its requirements and quantity should be determined according to the requirements of infection control.

（4）儿童精神病单元应与成人精神病单元分开，病房的容量最大为 4 名儿童；房间面积要求单床面积为 $10m^2$, 多床病房为每床 $8m^2$, 每个护栏床 $6m^2$; 应配备储藏室，存放玩具、设备、多余的护栏床、床以及父母夜间陪伴用的折叠床或躺椅。

(4) The child psychiatric unit should be separated from the adult psychiatric unit, and the maximum capacity of the ward should be 4 children; the room area shall require a single bed ar-

ea of $10m^2$, multi-bed wards of $8m^2$ per bed and guarded beds of $6m^2$ per bed; the storage room should be provided to store toys, equipment, extra guarded beds, beds and folding bed, deck chair for parents to accompany at night.

（5）活动区域：房间要求综合面积每个病人 $4m^2$，就餐和活动的总面积应至少为每个病人 $5m^2$，如果设置独立就餐区，每位病人至少 $2m^2$。

（5）Activity area: the room requires a comprehensive area of $4m^2$ for each patient, and the total area for dining and activities should be at least $5m^2$ for each patient. If there is an independent dining area, each patient should have at least $2m^2$.

（6）户外区：在需要治疗两周或更长时间的精神病院内，室外应有保护措施，使儿童能方便地到达安全的室外去游玩与治疗。

（6）Outdoor area: in the psychiatric hospitals requiring treatment for two weeks or more, the outdoor should have protective measures so that children can easily reach safe outdoor places for playing and treatment.

（7）老年病、阿尔茨海默症与其他痴呆病护理单元：除了满足标准的精神病单元的常规要求外，老年病单元还需要考虑以下因素：单床病房的面积至少为 $12m^2$，多床病房应至少为 $20m^2$，每间病房应备有毯子、枕垫和内衣裤存放处，门宽度最少为1.2m；其社交区域除社交活动综合区面积按每个病人 $3m^2$ 计算以外，其他按标准单元的规定。

（7）Nursing units for senile diseases, Alzheimer's disease and other dementia diseases: in addition to meeting the general requirements of the standard psychiatric unit, the senile disease unit also needs to consider the following factors: the area of single bed ward shall be at least $12m^2$, and that of multi-bed ward shall be at least $20m^2$. Each ward shall be equipped with blanket, pillow and underwear storage place, and the door width shall be at least 1.2m; in addition to the area of the social activity comprehensive area, which is calculated according to the $3m^2$ of each patient, other social activity area is determined according to the provisions of the standard unit.

（8）护士呼叫系统：应按标准设置护士呼叫系统，呼叫按钮插座应能方便地拆下或盖上，呼叫用绳索或拉线不允许超过15cm。

（8）Nurse calling system: the nurse calling system should be set up according to the standard. The call button socket should be easily removed or covered. The rope or stay wire for calling should not exceed 15cm.

（9）洗浴设施：在每一护理单元内，病人应至少有一个去浴缸的通道。

（9）Bathing facilities: patients should have access to at least one bathtub within each nursing unit.

（10）在每个护理单元内应提供轮椅储藏室。

（10）Wheelchair storage room should be provided in each nursing unit.

3. 涉法精神病单元

3. Psychiatric unit involving laws

（1）精神病标准单元适用于涉法单元；在单元入口处应设安全的前室或出入口，

需设处置区、警察、法庭以及其他安全设施。

(1) Psychosis standard unit is applicable to the unit involving laws; a safe front room or entrance and exit should be set up at the entrance of the unit, disposal area, police, court and other safety facilities should be required.

(2) 儿童、少年、青年病区应与老年病区分开。

(2) Child, juvenile and youth ward should be separated from the elderly ward.

5.7.2.7 精神病护理单元支持区域

5.7.2.7 Support area for psychosis nursing unit

1. 管理中心或护士站文件制作室：应提供单独的图表制作室，并有隔声和保护病人档案隐私的措施。员工办公室、多功能室可设置在精神病护理单元内或紧邻护理单元。还应有会客室、安静室、咨询室、讨论室、组团诊疗室、洗手处、药物配送站、营养站、清洁工作室或清洁物供应室、服务工作室、设备及供应物、轮椅储藏室、清洁布件存放间、应急设备库、供应物管理仓库、房间整理服务用房。

1. Documentation room for management center or nurse station: a separate illustration room should be provided with the measures for sound insulation and protection of the privacy of patient files. The room such as the staff office, multi-function room, should be set up in the psychosis nursing unit or adjacent to the nursing unit. It should also be needed for reception room, quiet room, consulting room, discussion room, group diagnosis room, hand washing place, drug supply station, nutrition station, cleaning room or cleaning supply room, service work room, equipment and supplies, wheelchair storage room, cleaning cloth storage room, emergency equipment warehouse, supply management warehouse and make-up service room.

2. 员工支持区域：包括员工休息设施和员工厕所、员工物品存放处，要足够宽敞。

2. Staff support area: it includes staff rest facilities, staff toilets and staff articles storage place, which should be spacious enough.

3. 病人支持区域：

3. Patient support area:

(1) 应至少设置两个社交室，一个适用于嘈杂的活动，另一个适用于安静活动房间，这两间活动室中每间的综合面积应至少每位病人 $3m^2$，总面积不少于 $12m^2$。如为每个病人增加 $2m^2$，此活动室可与餐饮活动共享，否则餐饮面积按每位病人 $2m^2$ 计。餐饮设施可在护理单元之外的一个集中区内。

(1) It shall be set up at least two social rooms, one is suitable for noisy activities and the other is suitable for quiet activities. The comprehensive area of each of these two activity rooms shall be at least $3m^2$ for each patient, and the total area shall not be less than $12m^2$. If $2m^2$ is added for each patient, this activity room can be shared with catering activities, otherwise the catering area should be calculated as $2m^2$ for each patient. The catering facilities can be located in a centralized area outside the nursing unit.

(2) 病人洗浴设施：除病房内的洗浴设施外，每 6 张床位应增设 1 支浴缸或淋浴器，洗浴设施的设计与位置应考虑病人方便；应为病人提供有自动洗衣机与烘干器的洗衣设施；应提供受员工控制的安全存放库，以存放被确认为有潜在危害的病人财物，如剃须刀、带丁爪的文件夹、香烟打火机等。

（2） Patient bathing facilities: in addition to the bathing facilities in the ward, 1 bathtub or shower should be added for every 6 beds. The design and location of bathing facilities should consider patient convenience; laundry facilities with automatic washing machine and dryer should be provided for the patients; a safe storage warehouse controlled by the staff should be provided to store the belongings of patients identified as potential hazards, such as razors, folders with tack claws, and cigarette lighters.

5.7.2.8 诊断与处置场所

5.7.2.8 Diagnosis and disposal place

1. 检查与处置室：位置要能够服务多个护理单元，房间内要有洗手处，设书桌和搁架，最小面积 $12m^2$；精神病院不需要提供影像科和核医学科。

1. Inspection and medical waste storage room: the location should be able to serve multiple nursing units, and the room should be equipped with hand washing, desk and shelf, with a minimum area of $12m^2$; Psychiatric hospital is not required to provide imaging department and nuclear medicine department.

2. 康复治疗部：精神病院的康复治疗，主要是为了心理功能的诊断与处置，也可能寻求体力功能不同程度的康复，它包含了一种或数种类别的服务。如果医院要求设正式的康复治疗服务，则应配以必要的设施与设备，以满足功能要求，当有两项或更多项康复服务时，这些设施与设备可以共享。

2. Rehabilitation department: the rehabilitation of psychiatric hospital is mainly for the diagnosis and disposal of psychological function, and may also seek rehabilitation of physical function in different degrees, which includes one or several types of services. If the hospital requires formal rehabilitation services, necessary facilities and equipment should be provided to meet the functional requirements. When there are two or more rehabilitation services, these facilities and equipment can be shared.

5.7.2.9 理疗

5.7.2.9 Physiotherapy

理疗对病人的精神康复有直接影响，在精神病院长期护理的病人与老年病人需要理疗，包含个人处置区、体育锻炼区与设施；空间设计有理疗支持区、病人支持区。

Physiotherapy has a direct impact on the psychiatric rehabilitation of patients. Long-term care of patients and elderly patients in psychiatric hospitals need physiotherapy, including personal disposal area, physical exercise area and facilities; the space is designed with physiotherapy support area and patient support area.

1. 作业疗法：

1. Occupational therapy:

工作室与工作台、展示室、洗手处、设备与供应物储藏室。

Work room and workbench, exhibition room, hand washing place, equipment and supply storage room.

2. 职业工作治疗：

2. Professional work therapy:

工作区应适合轮椅进出，组团工作室的大小，应适用于一次一个治疗组的需要，应设洗手处、设备与供应库房，对有危害的供应物、设备应提供安全存放室，对潜在有害的设备，应提供远程电器开关。

The working area should be suitable for wheelchair access. The size of the group work room should be suitable for the needs of one treatment group at a time. The hand washing place, equipment and supply warehouse should be set up. Safe storage room shall be provided for hazardous supplies and equipment, and remote electric switch shall be provided for potentially harmful equipment.

3. 休闲治疗：

3. Recreational therapy:

是通过空闲时的活动，帮助病人拓展和保持在社区中的起居能力。这些活动可在休闲治疗部门中，在特殊设施如体育馆中，其他区域如护理单元内的多功能房间中或户外进行，包括活动室、洗手处、设备供应物库房，电器开关应为远程开关。

It helps patients to expand and maintain their daily life ability in the community through leisure activities. These activities can be carried out in the recreational therapy department, special facilities such as gymnasium, other areas such as multi-functional room in nursing unit or outdoor, including activity room, hand washing place, equipment supply warehouse, and the electric switch should be remote switch.

4. 教育治疗：

4. Educational therapy:

对于儿童与青年进行教育治疗时，必须设置教师、学生课桌。教室的面积至少 $24m^2$，应设库房。

For child and youth educational therapy, it is necessary to set up teachers and students' desks. The area of the classroom should be at least $24m^2$, and a warehouse should be set up.

5.7.2.10 康复治疗支持区域

5.7.2.10 Rehabilitation support area

1. 可以共享或每项服务单独设置，包括接待与控制站、病人等候区、办公室、多功能室、房间整理服务用房。

1. It can be shared or set up separately for each service, including reception and control station, patient waiting area, office, multi-function room and make-up service room.

2. 员工支持区应提供方便的通道至厕所与更衣室，每个工作区应设安全室和橱柜；病人支持区应提供无障碍厕所。

2. Convenient access to toilets and dressing rooms should be provided in the staff support area, and each working area should be equipped with safety room and cabinet; accessible toilets should be provided in the patient support area.

3. 检验科

3. Laboratory department

需要的检测可以在现场进行，或由其他检验机构提供服务，应提供标本采集与处理条件，最少应具备实验台、洗涤盆、冷柜、设备供应物存放处、书写处与档案存放处。

The required tests can be carried out on site or provided by other inspection agencies. It shall provide specimen collection and treatment condition, and shall provide at least laboratory table, washing basin, refrigerator, equipment supply storage place, writing place and archive storage place.

4. 公共服务区

4. Public service area

包括药房、餐饮设施及物资管理、环境卫生、服务、保洁、工程服务与维修、医疗档案、员工办公及生活设施。

It includes pharmacy, catering facilities and material management, environmental health, service, cleaning, engineering service and maintenance, medical records, staff office and living facilities.

5. 家具

5. Furniture

要减少患者受伤的机会，家具的结构应承受直接性的催损，抽屉拉手应为凹型，以减少用作打结点的可能性；厕所洗浴器具要防止患者受伤和自杀，不允许设置毛巾杆、浴帘杆等形成杆件的把手，完全可以行走的病人厕所内不得安装扶手或扶手应可拆卸；各种设施与墙之间的空间应填塞，以防绳索系在上面用于上吊。

In order to reduce the chance of injury to patients, the structure of furniture should be subject to direct damage. Drawer handles should be concave to reduce the possibility of being used as tie points. Toilet bathing appliances should prevent patients from injury and suicide. It is not allowed to set handles that form rods such as towel bar and shower curtain rods. Handrails shall not be installed in toilets of patients that are fully walkable or should be detachable. The space between various facilities and the wall should be filled to prevent ropes from being tied to it for hanging.

6. 吊顶

6. Ceiling

（1）在患者厕所里的吊顶，要防止胡乱摆弄型的或有足够高度的病人触及。不允许设置非安全型、非紧扣的吊顶结构，病房内吊顶高度应足够高，以防触及。任何卫生间的管道、其他管道会有潜在危险性的部件，应设置在吊顶之上。

(1) The ceiling in the patient's toilet should be prevented from being touched by patients who like tampering with things or have sufficient height. Unsafe and unfastened ceiling structures are not allowed. The ceiling height in the ward should be high enough to prevent touching. Any toilet pipes and other pipes will be hidden from view and should be installed on the ceiling.

（2）吊顶板材料的穿孔应很小，以避免被作为系吊点，并有足够的高度，防止病人接触到。通风格栅应有安全性。在非监控区的病人淋浴喷头，应为内置型或可减少病人触及的可能。喷淋头和其他突出物都不应设置。

(2) The perforation of ceiling board material should be small to avoid being used as a hanging point and have sufficient height to prevent patients from touching it. Ventilation grille should be safe. Patient shower heads in non-monitoring areas should be built-in or can reduce the possibility of patient touching. Spray head and other protrusions should not be set.

（3）在病人厕所里的空调风口、灯具、喷淋头和其他附属物应是防胡乱摆弄型。

(3) Air conditioning vents, lamps, spray head and other accessories in patient toilets should be tamper-proof.

7. 机电末端

7. Electromechanical end

精神病院的所有机电设备设施的末端都应考虑是不能被精神病人胡乱摆弄的类型，也必须不能够有被病人用来自我伤害、上吊、自杀等使用的构造。

The end of all electromechanical equipment and facilities in psychiatric hospitals should be considered to be of a type that cannot be tampered with by mental patients, and must not have structures that can be used by patients for self-injury, hanging, suicide, etc.

5.7.3　肿瘤医院

5.7.3　Tumor hospital

5.7.3.1　适用性

5.7.3.1　Applicability

本条适用于肿瘤专科医院、综合医院肿瘤科、其他肿瘤疾病诊疗设施，医疗保健设施通用设计适用于本类设施。

This article is applicable to oncology department of oncology specialized hospital, general hospital and other facilities for diagnosis and treatment of oncological diseases. The general design of medical and health care facilities is applicable to such facilities.

共享服务：

Sharing service:

当肿瘤设施是另一个设施的一部分时，如餐饮、储藏、药房、洗衣房等这些服务应切合实际的予以共享。在某些情况下，所有辅助性服务可由主要设施满足；另一方面有些计划内容与需求可能要求单独服务。

When the tumor facility is the part of another facility, services such as catering, storage, pharmacy, laundry should be shared practically. In some cases, all ancillary services can be met by the main facilities; on the other hand, some plan contents and requirements may require separate services.

机动床位：

Mobile beds:

在设施内应依据病人年龄、性别、安全程度或治疗内容的不同，允许有一组房间的使用情况做出变动，以适应不同病人群体的需要。

In the facility, a group of rooms should be allowed to change according to the age, gender, safety level or treatment content of the patients to meet the needs of different patients.

5.7.3.2 医院等级

5.7.3.2 Hospital level

我国不设一级肿瘤医院，肿瘤医院等级分为：

There is no primary tumor hospital in China. The tumor hospital is classified into:

1. 二级肿瘤医院：住院床位总数 100 ~ 399 张。

1. Second-class tumor hospital: the total number of hospital beds is 100 ~ 399.

科室设置：

Department setting:

（1）临床科室：至少设有肿瘤外科、肿瘤内科、放射治疗科、中医（中西医结合）科、急诊室。

(1) Clinical departments: at least cancer surgery, oncology department, radiotherapy department, traditional Chinese medicine (integrated traditional Chinese and western medicine) department and emergency room.

（2）医技科室：至少设有药剂科、检验科、放射科、B 超室、手术室、病理（包括细胞）科、血库、消毒供应室、病案室、营养室。

(2) Medical technology departments: at least pharmacy department, laboratory department, radiology department, B-ultrasound room, operating room, pathology (including cytolog ical diagnosis) department, blood bank, disinfection supply room, medical record room and nutrition room.

（3）基本设备：心电图机、B 超、麻醉机、电止血器、显微镜、胃镜、支气管镜、生化分析仪、肺功能测定仪、病理切片机及染色设备、200mA 以上 X 光机、钴 60 机加速器、高压灭菌设备、洗衣机、电冰箱。

(3) Basic devices: ECG machine, B-ultrasound, anaesthesia machine, electric hemostat, microscope, gastroscope, bronchoscope, biochemical analyzer, lung function tester, pathological slicer and dyeing device, X-ray machine over 200mA, cobalt 60 accelerator, autoclaving device, washing machine, refrigerator.

2. 三级肿瘤医院：住院床位总数 400 张以上。

2. Third-class tumor hospital: the total number of hospital beds is more than 400.

科室设置：

Department setting:

（1）临床科室：至少设有肿瘤外科、肿瘤内科、肿瘤妇科、放射治疗科、中医（中西医结合）科、麻醉科急诊室、预防保健科。

(1) Clinical departments: at least cancer surgery, oncology department, tumor gynecology, radiotherapy department, traditional Chinese medicine (integrated traditional Chinese and western medicine) department, anesthesiology department, emergency room and prevention and

health care department.

（2）医技科室：至少设有药剂科、检验科、影像诊断科、内窥镜室、手术室、病理（包括细胞学诊断）科、输血科、核医学科、消毒供应室、病案室、营养部和相应的临床功能检查室。

(2) Medical technology departments: at least pharmacy department, laboratory department, imaging diagnosis department, endoscope room, operating room, pathology (including cytological diagnosis) department, blood transfusion department, nuclear medicine department, disinfection supply room, medical record room, nutrition department and corresponding clinical functional examination room.

（3）基本设备：心电图机、B 超、电手术刀、麻醉机、电止血器、显微镜、自动生化分析仪、自动血细胞计数仪、500mA 以上 X 光机、模拟定位机、照相机（同位素检查）、钴 60 治疗机、直线加速器、肺功能测定仪、病理切片机、支气管镜、胃镜、结肠镜、膀胱镜、高压灭菌设备、洗衣机、电冰箱。

(3) Basic devices: ECG machine, B-ultrasound, electric scalpel, anaesthesia machine, electric hemostat, microscope, automatic biochemical analyzer, automatic blood cell counter, X-ray machine over 500mA, analog positioning machine, camera (isotope examination), cobalt 60 therapy machine, linear accelerator, lung function tester, pathological slicer, bronchoscope, gastroscope, colonoscope, cystoscope, autoclaving device, washing machine, refrigerator.

5.7.3.3　护理环境

5.7.3.3　Nursing environment

1. 肿瘤医院医疗设施的安全环境非常重要，肿瘤治疗住院周期较长，设计对舒适性和安全性有更高的要求。其内部没有一个环境是绝对安全而无风险的。多数企图自杀的人和家属都在经受痛苦。在肿瘤治疗设施中的病人被认为是自杀高危人群，因此在保持治疗环境的同时，应避免存在实际的危险性环境。一个已经建造好的环境，无论设计或建设得多好，都不能完全信赖其有绝对安全的防护措施。所以可能还需要增加防护栅栏、坚固的门锁、24 小时监护设备。

1. The safety environment of medical facilities in tumor hospitals is very important, the hospitalization period of tumor treatment is long, and the design requires the improvement of comfort and safety. None of its internal environment is absolutely safe and risk-free. Most people who attempt suicide and their families are suffering. Patients in tumor treatment facilities are considered to be high risk group of suicide. Therefore, while maintaining the therapeutic environment, actual hazardous environment should be avoided. No matter how well designed or constructed, an already built environment cannot be fully trusted to have absolutely safe protective measures. Therefore, it may be necessary to add protective fences, strong door locks and 24-hour monitoring equipment.

2. 肿瘤医院建筑要为计划治疗内容提供合适的环境单元，应具有开放的感觉，要强调自然光在建筑设计和维护的每一个方面是最重要的，应对所服务的特殊病人的心理辅导做出一些安排，例如用餐区、社交区、娱乐室和活动区，可视控制包括电子监控，应考虑到病人隐私和员工安全。

2. Tumor hospital buildings should provide appropriate environmental units for the planned treatment contents, and should have an open feeling. It should be emphasized that natural light is the most important in every aspect of architectural design and maintenance. It is needed to make some arrangements for psychological counseling of special patients served, such as dining area, social area, entertainment room and activity area. Visual control includes electronic monitoring, and should take into account patient privacy and employee safety.

5.7.3.4 治疗环境

5.7.3.4 Therapeutic environment

肿瘤医院建筑应为计划治疗内容提供适宜的环境。化疗室应舒适宽敞，有良好的采光。

Tumor hospital buildings should provide suitable environment for planned treatment content. Chemotherapy room should be comfortable and spacious, with good lighting.

5.7.3.5 安全设计

5.7.3.5 Safety design

设计应为计划处置内容提供一定的安保级别。为防止自杀，对建筑和室内的细节装饰面与设备，要在设计时做特殊考虑。

The design shall provide a certain level of security for the planned disposal content. In order to prevent suicide, special consideration should be given to the detail decoration surface and equipment of buildings and interior.

5.7.3.6 肿瘤设施护理场所

5.7.3.6 Tumor facility nursing place

1. 综合性肿瘤病护理单元应包含以下内容：每个房间的最大容量为两位病人；病房应设置窗户，应有桌子或书写台面供病人使用；每个病人应有到厕所的通道，而不必进入公用走道区；每间厕所服务的床位不要多于4床或2床，厕所内应有马桶、洗手处，厕所门应向外开或双向开；每位病人在自己的病房内应有独立的衣柜、锁柜，用于挂长大衣或存放个人财物，存放量应能满足每天更换衣服，共七天之需。

1. The comprehensive oncology nursing unit should include the following contents: the maximum accommodation of each room is 2 patients. Windows should be set up in the ward, and tables or writing desks should be provided for patients. Each patient should have access to the toilet instead of entering the common walkway area. No more than 4 beds or 2 beds shall be served in each toilet. Toilets shall be equipped with closestools and hand washing places. Toilet doors shall be opened outwards or in both directions. Each patient should have independent wardrobes and lockers in his ward for hanging long coats or storing personal belongings. The storage amount should meet the need of changing clothes every day for a total of seven days.

2. 洗浴设施：在每一护理单元内，病人应至少有一个去浴缸的通道。

2. Bathing facilities: patients should have access to at least one bathtub within each nursing unit.

3. 在每个护理单元内应提供轮椅储藏室。

3. Wheelchair storage room should be provided in each nursing unit.

5.7.3.7　质子肿瘤治疗中心

5.7.3.7　Proton tumor treatment center

1. 肿瘤质子治疗区应设在地下，治疗舱的固定安装空间在其底部；其门诊和住院部分可设在地上，宜设置地面天窗给地下室提供良好的自然采光，质子治疗的物理师办公室、IBA 工程师办公室，以及治疗计划室，都临近天窗设置。

1. The tumor proton therapy area should be located underground, and the fixed installation space of the therapy cabin should be at the bottom. The outpatient and inpatient departments can be located on the ground. It is appropriate to set up ground skylights to provide good natural lighting for the basement. The proton therapy physicists' offices, IBA engineers' offices and treatment plans rooms are all located near skylights.

2. 质子治疗区包括质子回旋加速器旋转治疗舱和固定束治疗舱，同时还应就近设置直线加速器、模拟 CT 和模拟 MRI, 作为多种放疗的支持，应方便其与物理师之间协同工作。质子治疗、直加治疗、影像模拟定位，应分别有各自的等候区，同时特设儿童专区，包括儿童等候、游戏和预麻室。模具加工间也在同层，以提高治疗效率。

2. Proton treatment area includes proton cyclotron rotating treatment chamber and fixed beam treatment chamber, and linear accelerator, simulated CT and simulated MRI should also be set nearby as support for various kinds of radiotherapy, which should facilitate the cooperative work between them and convenient physicists. Proton therapy, direct therapy and image simulation positioning should have their own waiting areas respectively, while special children's areas should be set up, including children's waiting rooms, games and pre-anesthesia rooms. The mold processing rooms are also on the same floor to improve the treatment efficiency.

3. 放疗直线加速器和质子设备，都设置在靠地下室外壁，以降低辐射冲击并且降低造价。

3. The radiotherapy linear accelerator and proton equipment are both arranged on the outer wall of the basement to reduce radiation impact and cost.

4. 质子楼的物资供应、所有洁净物品、食物、药品和医护人员出入采用垂直运送。

4. The proton building supplies, all clean items, food, medicine and medical care staff are transported vertically.

5. 特需服务或 VIP 宜通过专门的电梯到达门诊、治疗、住院区。

5. Special services or VIP should reach the outpatient, treatment and hospitalization areas through special elevators.

6. 地下有防辐射要求的后砌隔墙（如迷道、治疗室隔墙等）采用实心混凝土砖，平面交叉错缝砌筑；防辐射隔墙混凝土砖的表观密度、混凝土强度等级等技术参数需要满足防辐射设计的要求，砌筑时尽量少用砂浆，保证可以人工拆除。质子区域建筑主体、建筑设备都需达到 IBA 国际交付标准后，才可进行质子设备的安装工作。

6. Solid concrete bricks shall be used for rear partition walls (e. g. maze roads, therapy

room partition walls, etc.) with radiation protection requirements under the ground, and shall be laid with plane crossing and staggered joints; technical parameters such as bulk density, concrete strength grade and the like of the radiation-proof partition wall concrete brick need to meet the requirements of radiation-proof design, mortar is used as little as possible during masonry, and manual removal is ensured. The installation of proton equipment can only be carried out after the main building body and building equipment in proton area meet IBA international delivery standards.

5.7.4 儿童医院

5.7.4 Children's hospital

5.7.4.1 适用性

5.7.4.1 Applicability

本条适用于儿童专科医院、综合医院儿科中心、儿科诊所设施，医疗保健设施通用设计适用于本类设施。

This article is applicable to the facilities of children's specialized hospitals, general hospitals' pediatric centers and pediatric clinics, and the general design of medical and health care facilities is applicable to such facilities.

共享服务：

Sharing service:

当儿科设施是另一个设施的一部分时，如餐饮、储藏、药房、洗衣房等这些服务应切合实际的予以共享。在某些情况下，所有辅助性服务可由主要设施满足；另一方面，有些计划内容与需求可能要求单独服务。

When pediatric facility is part of another facility, such services as catering, storage, pharmacy, laundry, etc. should be shared realistically. In some cases, all ancillary services can be met by the main facilities; on the other hand, some plan contents and requirements may require separate services.

机动床位：

Mobile beds:

在设施内应依据患儿年龄、性别、安全程度或治疗内容的不同，允许有一组房间的使用情况做出变动，以适应不同年龄段儿童群体的需要。

In the facility, a group of rooms should be allowed to change according to the age, gender, safety level or treatment content of the children to meet the needs of different age groups of children.

5.7.4.2 医院建设标准与等级

5.7.4.2 Construction standard and level of hospital

我国公立儿童医院建设项目的审批、核准、设计和建设必须符合《儿童医院建设标准》（建标 174—2016）。

The examination, approval, design and construction of public children's hospital construction project in China must conform to the Construction Standard of Children's Hospital (JB

174—2016).

1. 一级儿童医院：住院床位总数 20 ~ 49 张。

1. Primary children's hospital: the total number of hospital beds is 20 ~ 49.

科室设置：

Department setting:

（1）临床科室：至少设有急诊室、内科、预防保健科。

（1）Clinical departments: at least emergency room, internal medical department and prevention and health care department.

（2）医技科室：至少设有药房、化验室、X 光室、消毒供应室。

（2）Medical technology departments: at least pharmacy, laboratory, X-ray room and disinfection supply room.

（3）基本设备：氧气瓶、呼吸球囊、电动吸引器、心电图机、抢救车、必备手术器械、显微镜、离心机、电冰箱、X 光机、人工洗片装置、器械柜、药器柜、紫外线灯、高压灭菌设备、洗衣机，以及常水、热水供应。

（3）Basic devices: oxygen cylinder, breathing balloon, electric aspirator, ECG machine, rescue vehicle, necessary surgical instruments, microscope, centrifuge, refrigerator, X-ray machine, manual film washing apparatus, instrument cabinet, medicine cabinet, ultraviolet lamp, autoclaving device, washing machine, and normal water and hot water supply.

2. 二级儿童医院：住院床位总数 50 ~ 199 张。

2. Second-class children's hospital: the total number of hospital beds is 50 ~ 199.

科室设置：

Department setting:

（1）临床科室：至少设有急诊室、内科、外科、五官科、口腔科、预防保健科。

（1）Clinical departments: at least emergency room, internal medical department, surgical department, otolaryngology department, stomatology department, and prevention and health care department.

（2）医技科室：至少设有药剂科、检验科、放射科、手术室、病理科、消毒供应室、病案统计室。

（2）Medical technology departments: at least pharmacy department, laboratcry department, radiology department, operating room, pathology department, disinfection supply room and medical record statistics room.

（3）基本设备：给氧装置、呼吸机、电动吸引器、心电图机、心电监护仪、手术床、无影灯、麻醉机、相应的手术器械、显微镜、恒温培养箱、分析天平、自动生化分析仪、尿分析仪、离心机、电冰箱、X 光机、B 超、裂隙灯、直接喉镜、牙科综合治疗台、雾化吸入设备、婴儿保温箱、器械柜、敷料柜、蒸馏器、紫外线灯、高压灭菌设备、洗衣机，以及通风、降温烘干设备、常温水、热水、净化过滤系统。

（3）Basic devices: oxygen supply device, breathing machine, electric aspiratcr, ECG machine, ECG monitor, operating bed, shadowless lamp, anaesthesia machine, corresponding surgical instruments, microscope, thermostatic incubator, analytical balance, automatic biochemical analyzer, urine analyzer, centrifuge, refrigerator, X-ray machine, B-ultrasound, slit lamp, direct laryngoscope, dental comprehensive treatment table, atomizing inhalation device, infant incubator, instrument cabinet, surgical dressing cabinet, distiller, ultraviolet lamp, autoclaving device, washing machine, ventilation, cooling and drying device, room temperature water, hot water, purification and filtration system.

3. 三级儿童医院：床位：住院床位总数200张以上。

3. Third-class children's hospital: the total number of hospital beds is more than 200.

科室设置：

Department setting:

（1）临床科室：至少设有急诊科、内科、外科、耳鼻喉科、眼科、皮肤科、传染科、麻醉科、中医科、预防保健科。

（1） Clinical departments: at least emergency department, internal medical department, surgical department, otolaryngology department, ophthalmology department, dermatology department, infectious department, anesthesiology department, traditional Chinese medicine department and prevention and health care department.

（2）医技科室：至少设有药剂科、检验科、放射科、功能检查科、手术室、病理科、血库、消毒供应室、病案室、营养部。

（2） Medical technology departments: at least pharmacy department, laboratory department, radiology department, functional examination department, operating room, pathology department, blood bank, disinfection supply room, medical record room and nutrition department.

（3）基本设备：给氧装置、呼吸机、心电图机、心脏除颤器、电动吸引器、自动洗胃机、心电监护仪、万能手术床、无影灯、麻醉机、麻醉监护仪、牙科综合治疗台、涡轮机、显微镜、自动生化分析仪、血液气体分析仪、尿分析仪、电子血细胞计数仪、离心机、分析天平、恒温箱、X光机、移动式X光机、B超、脑电图机、裂隙灯、肺功能仪、婴儿保温箱、食道镜、支气管镜、结肠镜、膀胱镜、石蜡切片机、冷冻切片机、电冰箱、器械柜、敷料柜、洗衣机、紫外线灯、蒸馏器、高压灭菌设备，以及通风、降温烘干设备、常温水、热水、净化过滤系统、器械消毒设备（冲洗工具、去污、去热源）、热源监测设备（恒温箱、净化台、干燥箱）、净物存放、消毒灭菌密闭设备。

（3） Basic devices: oxygen supply device, breathing machine, ECG machine, cardiac defibrillator, electric aspirator, automatic gastric lavage machine, ECG monitor, universal operating bed, shadowless lamp, anaesthesia machine, anesthesia monitor, dental comprehensive treatment table, turbine, microscope, automatic biochemical analyzer, blood-gas analyzer, urine analyzer, electronic blood cell counter, centrifuge, analytical balance, thermostat, X-ray machine, mobile X-ray machine, B-ultrasound, electroencephalogram machine, slit lamp, spirometer, infant incubator, esophagoscope, bronchoscope, colonoscope, cystoscope, microtomes, freezing microtome, refrigerator, instrument cabinet, surgical dressing cabinet, washing machine, ultraviolet lamp, distiller, autoclaving device, ventilation, cooling and drying device, room temperature water, hot water, purification and filtration system, instrument disinfection device (flushing tools, decontamination, heating source removal), heating source monitoring device (thermostatic box, purification table, drying oven), clean item storage, disinfection and sterilization airtight device.

5.7.4.3 护理环境

5.7.4.3 Nursing environment

1. 儿童医院医疗设施的安全环境非常重要，其内部没有一个环境是绝对安全而无风险的。在保持治疗环境的同时，应避免存在实际的危险性环境。一个已经建造好的环境，无论设计或建设得多好，都不能完全信赖其有绝对安全的防护措施，所以可能还需要增加防护栅栏、门锁、24小时监护设备。儿童护理单元应根据年龄分段。

1. The safety environment of children's hospital medical facilities is very important, and none of its internal environment is absolutely safe and risk-free. While maintaining the therapeutic environment, actual hazardous environment should be avoided. In the environment, no matter how well it is designed or built, cannot be fully trusted to have absolutely safe protective measures, therefore, it may be necessary to add protective fences, door locks and 24-hour monitoring equipment. Child nursing units should be segmented according to age.

2. 儿童医院建筑要为儿童提供合适的环境单元，应具有开放的感觉，要强调自然光在建筑设计和维护的每一个方面，最重要的是应结合所服务的不同年龄段儿童、幼儿的心理做出一些安排，例如儿童娱乐室、亲子活动区、母婴室、用餐区，可视控制包括电子监控，应考虑到儿童的安全、学习。

2. Children's hospital buildings should provide suitable environmental units for children, should have an open feeling, should emphasize natural light in every aspect of architectural design and maintenance. The most important thing is to make some arrangements according to children of different ages and the psychology of infants served, such as children's entertainment room, parent-child activity area, mother-child room, dining area, visual control including electronic monitoring, should consider the safety and learning of children.

5.7.4.4 治疗环境

5.7.4.4 Therapeutic environment

儿童医院建筑应为儿童及其亲属提供适宜的诊疗环境，应安全、舒适、宽敞，有良好的采光，色彩符合儿童心理需求。

Children's hospital buildings should provide a suitable diagnosis and treatment environment for children and their relatives. They should be safe, comfortable and spacious, with good lighting and colors that meet children's psychological needs.

5.7.4.5 儿童医院建筑设计

5.7.4.5 Architectural design of children's hospital

1. 满足儿童医疗保健的专业流程，突出为儿童服务的特点；具有较大的适应性、灵活性、可发展性，既满足作为儿童综合医院和保健院的功能需要（医疗、教学、保健、康复、科研），又满足作为儿童专科医院的需要；考虑我国儿科医疗就诊陪诊家属达到1：2.5以上的特殊性，适应未来人口生育规模特点。

1. They meet the professional process of children's medical care and highlight the characteristics of serving children. They have great adaptability and flexibility, as well as enforceability and developability. They not only meet the functional needs (medical treatment, teaching, health care, rehabilitation and scientific research) of children's general hospitals and health care hospitals, but also meet the needs of children's specialized hospitals. They take into account the particularity that the ratio for family members attending pediatric medical consultations in China reaches more than 1∶2.5 and adapt to the characteristics of future population fertility scale.

2. 儿童医疗、保健、康复、教学、科研各项功能既相对独立，又有机融合；医技和后勤保障系统等方面资源可以共享，便于资源优化配置，统一管理。

2. The functions of children's medical treatment, teaching, health care, rehabilitation and scientific research are relatively independent and organically integrated. Resources such as medical technology and logistics support system can be shared to facilitate optimal allocation and unified management of resources.

3. 休息室、卫生间、哺乳区、儿童活动区应充分考虑儿童的方便和卫生管理。

3. Lounge, toilet, nursing area and children's activity area should give full consideration to children's convenience and hygiene management.

4. 靠近急诊科应设置热伤处理室、发热门诊、隔离观察病房。

4. Near the emergency department, thermal injury therapy room, fever clinic and isolation observation ward should be set up.

5.7.5 其他专科医院

5.7.5 Other specialized hospitals

5.7.5.1 适用性

5.7.5.1 Applicability

1. 医疗保健设施通用设计适用于其他各类专科医院；专科医院可有独立用地、独栋建筑、独立经营，也可位于综合医院内，以院中院形式与综合部分资源共享、共同经营。

1. The universal design of medical and health care facilities is applicable to various specialized hospitals. Specialized hospitals can have independent land use, single buildings and independent operation. They can also be located in a general hospital, sharing resources and being in joint management.

2. 用房包括门诊、医技、住院、行政、后勤保障、院内生活、科研、教学、值班宿舍。

2. The rooms include outpatient, medical technology, inpatient, administration, logistics support, hospital life, scientific research, teaching, dormitories for on-duty personnel.

3. 共享服务：当本设施是另一个设施的一部分时，如餐饮、储藏、药房、洗衣房、停车等这些服务应切合实际的予以共享；在某些情况下，所有辅助性服务可由主要设施满足；另一方面，有些计划内容与需求可能要求单独服务。

3. Sharing service: when this facility is part of another facility, such services as catering,

storage, pharmacy, laundry, parking, etc. should be shared realistically; in some cases, all ancillary services can be met by major facilities; on the other hand, some planning contents and requirements may require separate services.

5.7.5.2 口腔医院

5.7.5.2 Oral hospital

1. 二级口腔医院：牙科治疗椅 20 ~ 59 台，住院床位总数 15 ~ 49 张。

1. Second-class oral hospital: 20 ~ 59 dental treatment chairs, with a total of 15 ~ 49 hospital beds.

科室设置：

Department setting:

（1） 临床科室：至少设有口腔内科、口腔颌面外科和口腔预防保健科、口腔急诊室。

（1） Clinical departments: at least oral internal medical department, oral and maxillofacial surgical department, oral prevention and health care department and oral emergency room.

（2） 医技科室：至少设有药剂科、检验科、放射科、消毒供应室、病案室。

（2） Medical technology departments: at least pharmacy department, laboratory department, radiology department, disinfection supply room and medical record room.

2. 三级口腔医院：牙科治疗椅 60 台以上，住院床位总数 50 张以上。

2. Third-class oral hospital: more than 60 dental treatment chairs, with a total of more than 50 hospital beds.

科室设置：

Department setting:

（1） 临床科室：至少设有口腔内科、口腔颌面外科、口腔修复科、口腔正畸科、口腔预防保健科、口腔急诊室。

（1） Clinical departments: at least oral internal medical department, oral and maxillofacial surgical department, prosthodontics department, orthodontic department, oral prevention and health care department and oral emergency room.

（2） 医技科室：至少设有药剂科、检验科、放射科、病理科、消毒供应室、病案室、营养室。

（2） Medical technology departments: at least pharmacy department, laboratory department, radiology department, pathology department and disinfection supply room, medical record room and nutrition room.

5.7.5.3 眼科医院

5.7.5.3 Ophthalmic hospital

1. 眼科医院基本标准

1. Basic standards for ophthalmic hospitals

（1） 床位：总数 20 ~ 79 张（含日间观察床）。

（1） Beds: a total of 20 ~ 79 (including daytime observation beds).

（2）科室设置：

（2）Department setting:

临床科室：至少设有眼科、急救室。

Clinical departments: at least ophthalmology department and first-aid room.

医技科室：至少设有治疗室、手术室、检查室。

Medical technology departments: at least therapy rooms, operating rooms and examination rooms.

职能科室：至少设医疗质量与安全、医院感染管理、病案（统计）等部门。

Functional departments: at least medical quality and safety, hospital infection management, medical records (statistics) and other departments.

（3）房屋：

（3）Houses:

每床建筑面积≥40m^2(不含日间观察床)；病房每床净使用面积≥5m^2(不含日间观察床)；设置日间观察床的，床间距不少于0.8m;治疗室、手术室、检查室等每个房间需独立；总建筑面积应当符合医院业务需要。

The construction area of each bed is ≥40m^2 (excluding daytime observation beds); The net usable area of each bed in the ward is ≥5m^2 (excluding daytime observation beds); if the daytime observation bed is set, the bed spacing shall not be less than 0.8m; each room such as therapy room, operating room and examination room shall be independent. The total construction area shall meet the needs of the hospital business.

（4）目前我国不设一、二级眼科医院。

（4）At present, there are no first-class or second-class ophthalmic hospitals in China.

2. 三级眼科医院基本标准

2. Basic standards for third-class ophthalmic hospitals

（1）床位：80 张以上。

（1）Beds: 80 or more.

（2）科室设置：20 个科室，其中临床科室 9 个。

（2）Department setting: 20 departments, including 9 clinical departments.

临床科室：至少设有白内障、青光眼、角膜病、眼底病、眼外伤、屈光眼肌和肿瘤整形专科、麻醉科、眼预防保健科。

Clinical departments: at least cataract, glaucoma, corneal disease, ocular fundus disease, ocular trauma, refractive ocular muscle and tumor plastic specialty, anesthesiology department and eye prevention and health care department.

（3）医技科室：化验室、放射科、病理科、药剂科、供应室、手术室、验光视野室、超声波室、眼底荧光血管造影室、激光室、病案室。

（3）Medical technology departments: laboratory, radiology department, pathology department, pharmacy department, supply room, operating room, optometry visual field room, ultrasonic room, fundus fluorescein angiography room, laser room and medical record room.

（4）基本设备：给氧装置、呼吸机、心电图机、电动吸引器、心电监测仪、心脏除颤器、麻醉机、手术床、专科检查台、手术显微镜、麻醉监测仪、高频电刀、双极电凝器、检眼镜、视力表、裂隙灯、眼压计、验光仪及镜片箱、视野计、角膜曲率计、眼肌力仪、眼底照相机、超声波仪、手术显微镜、冷冻仪、玻璃体切割仪、激光治疗仪、火焰光度计、血球计数仪、分析天平、离心机、电冰箱、X光机、敷料柜、器械柜、高压灭菌设备、煮沸消毒锅、电烤箱、紫外线灯、洗衣机。

(4) Basic devices: Oxygen supply device, breathing machine, ECG machine, electric aspirator, ECG monitor, cardiac defibrillator, anaesthesia machine, operating bed, specialized examination table, operation microscope, anesthesia monitor, high-frequency electrotome, bipolar electrocoagulation device, ophthalmoscope, visual acuity chart, slit lamp, tenonometer, optometry and lens box, perimeter, corneal curvature meter, ocular muscle strength meter, fundus camera, ultrasonic instrument, operation microscope, cryoscope, vitrectomy instrument, laser therapeutic instrument, flame photometer, blood cell counter, analytical balance, centrifuge, refrigerator, X-ray machine, surgical dressing cabinet, instrument cabinet, autoclaving device, boiling and disinfecting pot, electric oven, ultraviolet lamp, washing machine.

（5）房屋：每床建筑面积≥$40m^2$，病房每床净使用面积≥$6m^2$，总面积≥$3200m^2$。

(5) Housing: the construction area of each bed is ≥$40m^2$, the net usable area of each bed in the ward is ≥$6m^2$, and the total area is ≥$3200m^2$.

3. 门诊各科设计

3. Design of outpatient departments

（1）眼科：等候厅、标准暗诊室（每间诊室配一台裂隙灯）、更衣区（拉帘）、处置室。

(1) Ophthalmology department: waiting room, standard dark consulting room (each room is equipped with a slit lamp), dressing area (curtain), and medical waste storage room.

（2）儿科：诊室、儿童游戏场，位置靠近视光学中心。

(2) Pediatric department: consulting room, children's playground, close to optometry center.

（3）内科：1间诊室，1间心电室，设在功能科附近。

(3) Internal medical department: 1 consulting room and 1 electrocardiogram room, located near the functional department.

（4）视光中心：防控中心、角膜塑形镜中心（检查、谈话、试戴），验光区、配镜区、加工区、弱视训练室、等候区，办公室、库房。

(4) Optical centers: prevention and control center, orthokeratology center (examination, conversation and trial wear), optometry area, lens matching area, processing area, amblyopia training room, waiting area, office and warehouse.

（5）准分子中心：独立一区、等候厅、接待室、谈话室、检查室，不含手术室（设在中心手术部）。

(5) Excimer center: independent area 1, waiting room, reception room, conversation room and examination room, excluding operating room (located in the central operating department).

（6）干眼中心：等候厅、诊室、检查室、处置室，独立一区。

（6）Dry eye center: waiting room, consulting room, examination room, medical waste storage room, independent area 1.

（7）门诊集中处置室：独立一区，各科共享。

（7）Outpatient centralized medical waste storage room: independent area, shared by all departments.

（8）设 1 间急诊室。

（8）Set up an emergency room.

4. 药剂科：设药品库、器械库、毒麻药品库、西药房、中药房、煎药室。

4. Pharmacy department: drug storehouse, instrument storehouse, narcotic drug storehouse, western medicine pharmacy, traditional Chinese medicine pharmacy and decocting room.

5. 医技部：设功能科（特检室）、检验科、1 间放射 DR 机房；设消毒供应中心、靠近手术室；可设眼库、眼病研究所、动物眼实验室，三者相邻布置。

5. Medical technology department: functional department (special examination room), a laboratory department and a radiation DR machine room; disinfection supply center, close to the operating room; eye bank, eye disease research institute and animal laboratory, which are arranged next to each other.

6. 中心手术部：包括Ⅲ级准分子激光手术室、Ⅰ级净化手术室、Ⅲ级普通手术室、医护人员生活区、手术等候区、净化设备、麻醉气体站房。

6. Central operating department: including third level excimer laser operating room, third level purification operating room, third level general operating room, medical care staff living area, family waiting area, purification equipment, anesthesia gas station room.

7. 住院病区：除了设处置室、功能检查室、暗室之外，病房其他部分按护理单元通用设计。

7. Inpatient ward: Except for the medical waste storage room, functional examination room and darkroom, other parts of the ward are designed according to the general purpose of the nursing unit.

5.7.5.4 心血管病医院

5.7.5.4 Cardiovascular hospital

1. 我国不设一、二级心血管病医院。

1. At present, there is no first-class or second-class cardiovascular hospital in China.

2. 三级心血管病医院：住院床位总数 150 张以上。

2. Third-class cardiovascular hospital: the total number of hospital beds is more than 150.

3. 科室设置：

3. Department setting:

（1）临床科室：至少设有急诊科、心内科（并设重症监护室）、心外科（并设重症监护室）、麻醉科。

(1) Clinical departments: at least emergency department, cardiology department (with ICU), cardiology surgical department (with ICU) and anesthesiology department.

(2) 医技科室：至少设有药剂科、检验科、放射科、输血科、手术室、核医学科、消毒供应室、病案室。

(2) Medical technology departments: at least pharmacy department, laboratory department, radiology department, blood transfusion department, operating room, nuclear medicine department, disinfection supply room and medical record room.

(3) 基本设备：呼吸机、除颤器、麻醉机、心电监护仪、临时心内起搏器、体外循环机、体内（外）除颤器、血气分析仪、并型计数器、免疫分析仪、全自动生化分析仪、1/10000 分析天平、恒温箱、X 光机、床旁 X 光机、心血管造影机、伽马相机、彩色血流显像仪、超声图像分析仪、床旁超声心动图机、心电图运动试验仪、放射性活度测量仪、数据处理系统、电影放映机、负荷运动试验设备、消毒灭菌密闭柜、电冰箱、高压灭菌设备、洗衣机。

(3) Basic devices: breathing machine, defibrillator, anaesthesia machine, ECG monitor, temporary intracardiac pacemaker, extracorporeal circulation machine, intracorporeal (external) defibrillator, blood-gas analyzer, parallel counter, immune analyzer, full-automatic biochemical analyzer, 1/10000 analytical balance, thermostat, X-ray machine, bedside X-ray machine, cardiovascular imaging machine, gamma camera, color blood flow imaging machine, ultrasonic image analyzer, bedside ultrasonic cardiogram machine, ECG exercise tester, radioactivity measuring instrument, data processing system, movie projector, load exercise test device, disinfection and sterilization airtight cabinet, refrigerator, autoclaving device and washing machine.

5.7.5.5 血液病医院

5.7.5.5 Hematology hospital

1. 目前我国不设一、二级血液病医院。

1. At present, there are no first-class or second-class hematology hospitals in China.

2. 三级血液病医院：住院床位总数 200 张以上，其中专科床位不少于 120 张。

2. Third-class hematology hospital: the total number of inpatient beds is more than 200, including not less than 120 specialized beds.

3. 科室设置：

3. Department setting:

(1) 临床科室：至少设有急诊室、血液内科（含三级科室）：血液一科（各类贫血）、血液二科（白血病及各类恶性血液病患）、血液三科（出凝血疾病）、血液四科（骨髓移植科）、预防保健科。

(1) Clinical departments: at least emergency rooms and hematology departments (including three-level departments): hematology department Ⅰ (various anemia), hematology department Ⅱ (leukemia and various malignant hematological patients), hematology department Ⅲ (coagulation diseases), hematology department Ⅳ (bone marrow transplantation) and prevention and health care department.

（2）医技科室：至少设有药剂科、检验科（包括细胞形态室）、放射科、功能检查室、手术室、输血科、病理科、消毒供应室、病案室。

（2） Medical technology departments: at least pharmacy department, laboratory department (including cell morphology room), radiology department, functional examination room, operating room, blood transfusion department, pathology department, disinfection supply room and medical record room.

（3）基本设备：显微镜、自动生化分析仪、血液成分分离机、血液细胞计数仪、全自动凝血测定仪、全自动微生物检测仪、血气分析仪、紫外线分光光度计、血液黏度计、超低温冰柜、低速冷冻离心机、恒温培养箱、超净工作台、X 光机、体外生理监护仪、动态心电监测仪、彩色超声多普勒诊断仪、胃镜、结肠镜、血液辐射治疗仪、自动呼吸机、全功能麻醉机、冷冻切片机、消毒灭菌密闭柜、电冰箱、高压灭菌设备、洗衣机。

（3） Basic devices: microscope, automatic biochemical analyzer, blood component separator, blood cell counter, full-automatic coagulation analyzer, full-automatic microorganism detector, blood-gas analyzer, ultraviolet spectrophotometer, blood viscosity meter, ultra-low temperature freezer, low-speed freezing centrifuge, constant temperature incubator, clean bench, X-ray machine, external physiological monitor, dynamic ECG monitor, color ultrasonic Doppler diagnostic instrument, gastroscope, colonoscope, blood radiation therapeutic instrument, automatic breathing machine, full-function anaesthesia machine, freezing microtome, disinfection and sterilization airtight cabinet, refrigerator, autoclaving device and washing machine.

5.7.5.6 皮肤病医院

5.7.5.6 Dermatological hospital

1. 目前我国不设一、二级皮肤病医院。

1. At present, there are no first-class or second-class dermatological hospitals in China.

2. 三级皮肤病医院：住院床位总数 100 张以上。

2. Third-class dermatological hospital: the total number of hospital beds is more than 100.

3. 科室设置：

3. Department setting:

（1）临床科室：至少设有皮肤内科、皮肤外科、真菌病科、康复理疗科、中西医结合科、性病科、预防保健科。

（1） Clinical departments: at least skin internal medical department, skin surgical department, mycosis pathology, rehabilitation physiotherapy, integrated traditional Chinese and western medicine, venereal disease pathology, prevention and health care department.

（2）医技科室：至少设有药剂科（含制剂室）、检验科（含真菌检验）、放射科、手术室、病理科、治疗室、消毒供应室、病案室。

（2） Medical technology department: at least pharmacy department (including preparation room), laboratory department (including fungus inspection), radiology department, operating room, pathology department, therapy room, disinfection supply room and medical record room.

（3）基本设备：外呼吸机、心电图机、心电监护仪、显微镜、荧光显微镜、自动生化分析仪、自动免疫分析仪、血细胞计数仪、尿液分析仪、X光机、B超、肌电图机、八导生理仪、X光治疗机、激光治疗机、冷冻治疗设备、光治疗设备、水治疗设备、电治疗设备、冷冻切片机、超薄切片机、消毒灭菌密团柜、电冰箱、高压灭菌设备、洗衣机。

(3) Basic devices: external breathing machine, ECG machine, ECG monitor, microscope, fluorescence microscope, automatic biochemical analyzer, active immunity analyzer, blood cell counter, urine analyzer, X-ray machine, B-ultrasound, electromyography machine, eight-channel physiological instrument, X-ray therapy machine, laser therapy machine, cryotherapy device, phototherapy device, water therapy device, electrotherapy device, freezing microtome, ultrathin microtome, disinfection and sterilization airtight cabinet, refrigerator, autoclaving device and washing machine.

5.7.5.7 整形外科医院

5.7.5.7 Plastic surgery hospital

1. 目前我国不设一、二级整形外科医院。

1. At present, there are no first-class or second-class plastic surgery hospitals in China.

2. 三级整形外科病医院：住院床位总数120张以上。

2. Third-class plastic surgery hospital: the total number of hospital beds is more than 120.

3. 科室设备：

3. Department equipment:

（1）临床科科室：至少设有整形外科、麻醉科。

(1) Clinical departments: at least plastic surgery and anesthesiology departments.

（2）医技科室：至少设有药剂科、检验科、放射科、手术室、病理科、消毒供应室。

(2) Medical technology departments: at least pharmacy department, laboratory department, radiology department, operating room, pathology department and disinfection supply room.

（3）基本设备：呼吸机、麻醉机、心电监护仪、体外除颤器、自动血压监测仪、吸入麻醉药浓度测定仪、整形外科手术相应的各种手术器械、显微镜、1/10000分析天平、血气分析仪、自动生化分析仪、尿分析仪、血细胞计数仪、免疫酶标仪、离子分析仪、酸度仪、恒温培养箱、超净工作台、X光机及暗室成套设备、脉搏氧饱和度监测仪、呼气末二氧化碳浓度测定仪、冰冻切片机、洗衣机、消毒灭菌设备、紫外线灯、高压灭菌设备、电冰箱。

(3) Basic devices: breathing machine, anaesthesia machine, ECG monitor, external defibrillator, automatic blood pressure monitor, inhalation anesthetic concentration meter, various surgical instruments corresponding to plastic surgery, microscope, 1/10000 analytical balance, blood-gas analyzer, automatic biochemical analyzer, urine analyzer, blood cell counter, immunoenzyme reader, ion analyzer, acidity meter, thermostatic incubator, clean bench, X-ray machine and darkroom complete device, pulse oxygen saturation monitor, end-expiratory carbon dioxide

concentration meter, freezing microtome, washing machine, disinfection and sterilization device, ultraviolet lamp, autoclaving device and refrigerator.

5.7.5.8 美容医院

5.7.5.8 Beauty hospital

1. 床位牙椅：住院床位总数50张以上，美容床20张以上，牙科治疗椅10张以上。

1. Beds and dental chairs: the total number of hospital beds is more than 50, beauty beds are more than 20, and dental treatment chairs are more than 10.

2. 科室设置：

2. Department setting:

（1）临床科室：至少设有美容外科、口腔科、皮肤科、理疗科、中医科、设计科、麻醉科。

（1）Clinical departments: at least cosmetic surgical department, stomatology department, dermatology department, physiotherapy, traditional Chinese medicine, design, anesthesiology department.

（2）医技科室：至少设有药剂科、检验科、放射科、手术室、病理科、技工室、影像室、消毒供应室、病案室。

（2）Medical technology departments: at least pharmacy department, laboratory department, radiology department, operating room, pathology department, mechanic room, imaging room, disinfection supply room and medical record room.

（3）基本设备：呼吸机、电动吸引器、心电监护仪、体外除颤器、自动血压监测仪、口腔综合治疗台、超声洁治器、涡轮机、光敏固化灯、银汞搅拌机、正颌外科器械、光固化烤塑机铸造与烤瓷设备、X光牙片机、口腔全景X光机、麻醉机、二氧化碳激光机、高频电治疗机、皮肤磨削机、离子喷雾器、文眉机、皮肤测量仪、1/10000分析天平、自动生化分析仪、尿分析仪、酶标仪、恒温培养箱、酸度仪、电冰箱、超净工作台、石蜡切片机、器械柜、高压灭菌设备、紫外线灯、洗衣机、X光机及暗室成套设备、血气分析仪、超声波美容治疗机、多功能健胸治疗机，以及美容外科手术相应的各种手术器械。

（3）Basic devices: breathing machine, electric aspirator, ECG monitor, external defibrillator, automatic blood pressure monitor, oral comprehensive treatment table, ultrasonic cleaner, turbine, photosensitive curing lamp, silver mercury mixer, orthognathic surgical instrument, photocuring plastic baking machine casting and porcelain device, X-ray dental film machine, oral panoramic X-ray machine, anaesthesia machine, carbon dioxide laser machine, high-frequency electric therapy machine, skin dermabrasion machine, ion sprayer, eyebrow tattoo machine, skin measuring instrument, 1/10000 analytical balance, automatic biochemical analyzer, urine analyzer, microplate reader, constant temperature incubator, acidity meter, refrigerator, clean bench, microtome, instrument cabinet, high-pressure sterilization device, ultraviolet lamp, washing machine, X-ray machine and darkroom complete set of device, blood-gas analyzer, ultrasonic beauty therapy machine, multifunctional breast enhancement therapy machine, and various surgical instruments corresponding to beauty surgery.

5.8　其他卫生保健设施

5.8　Other health care facilities

5.8.1　康复设施

5.8.1　Rehabilitation facilities

5.8.1.1　组成

5.8.1.1　Composition

1. 康复设施可以由医院或者有组织的康复部门、门诊诊所、康复中心以及其他服务于单一性残疾或多种残疾的设施组合而成。残疾的类型包括但不限于：脑血管病、头部损伤、脊髓损伤、被截肢者、并发性骨折、关节炎、神经性退化、遗传性疾病和心脏病患者。

1. Rehabilitation facilities may be composed of hospitals or organized rehabilitation departments, outpatient clinics, rehabilitation centers and other facilities serving single or multiple disabilities. Types of disability include but are not limited to: Cerebrovascular diseases, head injuries, spinal cord injuries, amputees, concurrent fractures, arthritis, neurodegeneration, hereditary diseases and heart diseases.

2. 当住院病人为康复设施中的一部分病人时，每一护理单元应提供以下内容：每间病房最多四个病人，多人病房内每张病床床脚处的间距为 1.2m, 作为设备与床间通道。病房应能够通风采光、能开窗。多床病房内应为每位病人提供拉帘，保障私密性，每间病房应有残疾人厕所，保证轮椅进出，有病人使用的衣柜。护士站应提供护士呼叫系统，设置检查/治疗室。

2. If some patients served by the rehabilitation facilities are in-patients, each nursing unit shall meet the following requirement: Each ward can hold 4 patients at most, and a spacing of 1.2m shall be reserved at the foot of each bed in the ward with multiple beds to serve as the passage between the equipment and the bed. The window of the ward shall be openable for ventilation and daylighting. A curtain shall be provided for each patient in the ward with multiple beds to ensure the privacy. Each ward shall be provided with an accessible toilet convenient for the wheelchair access, and a wardrobe for patients. The nurse station shall be provided with a nurse call system, and an examination/therapy room shall be set.

3. 医疗保健设施通用设计适用于康复设施。

3. The general design of health care facilities is applicable to the rehabilitation facility.

5.8.1.2　护理单元支持区域

5.8.1.2　Support area for psychosis nursing unit:

1. 包括管理中心或护士站，位于每个护理区内或者在护理区能方便利用的地方，每一个护理楼层至少有一个支持区，每一个支持区的安排和位置，可以服务一个以上的护理单元。

1. On each nursing floor in each nursing area or in the accessible place within the nursing area, including management center or nurse station, one support area shall be set at least, and the arrangement and the location of each support area shall ensure the service to more than one nursing unit.

2. 管理中心或护士站：应有护士与医生用的图表制作设施，护士办公室、洗手盆，洗手盆应靠近护士站和药物配送站，一个洗手盆可同时服务护士站与药物配送站。药物配送站应有能24小时方便迅速地发药给病人的措施，药品可来自药品准备室、整体式药品配送装置。应设配餐间，并靠近病人，有冰箱、储存柜、洗手池。设清洁工作室或清洁库房，污染库房、设备和供应物品库房。

2. Management center or nurse station: there shall be charting facilities for nurses and doctors, nurse office room and wash basin. The wash basin shall be set close to the nurse station and drug supply station, and one wash basin can serve the nurse station and the drug supply station. The drug supply station shall have the measures to deliver the drugs to patients timely within 24 hours. Drugs may be provided by the drug preparation rooms or by the automatic drug supply device. A diet preparation room shall be set close to patients, and be provided with a refrigerator, a storage cabinet and a wash basin. A cleaning room or a cleaning supplies warehouse, a contaminant warehouse, an equipment room and a supplies warehouse shall be set.

3. 员工支持区域：应配备员工休息室和厕所、员工库房，护理人员个人壁柜应位于值班站之处或在中心位置。

3. Staff support area: staff rooms, toilets and staff' s storage room shall be set, and a wall locker for storing the nursing workers' personal items shall be set at the duty station or at the central position.

5.8.1.3 病人支持区域

5.8.1.3 Patient support area

1. 不需经过公共走道的厕所，应靠近每个集中洗浴区，厕所的门宽可让轮椅通过，在应急时能够让人从外面进入。公共厕所里每一个马桶配一个洗手盆。

1. The toilet that can be accessed without passing through the public passageway shall be set close to each centralized bath area, and the door width of the toilet shall ensure that the wheelchair can pass through, and can be accessed by the people from the outside in case of emergency. Each toilet in the public toilet shall be provided with a wash basin.

2. 病人洗浴设施：除在病房里的洗浴设施外，还应按每八张床位配置一个浴缸或一支淋浴器的洗浴设施，每个浴缸和淋浴器应配有独立的房间或私密性围护，为个人使用洗浴装置、擦身、梳理、停放轮椅以及为助浴人员提供空间。集中洗浴设施中的淋浴器占用面积至少为1.22m^2，地上无凸缘，为轮椅病人使用。

2. Bath facilities for patients: besides the bath facilities in the ward, one bathtub or one shower shall be provided for every eight beds. Each bathtub and shower shall be equipped with

an independent room or a privacy shelter to provide the space for the patient to use the bath apparatus, dry themselves, dress and park the wheelchair, as well as the bath-assistant service personnel. Showers for the centralized bath facilities shall at least occupy an area of 1.22m^2, and the floor shall have no flange and be accessible by the wheelchair patients.

3. 特殊病人护理区域：每个护理单元可以设少量空气感染隔离病房，用作应急护理房间，也可以将若干个负压病房组合成一个独立的隔离单元。

3. Special patient nursing area: in each nursing unit, a small number of air-borne infection isolation wards can be set as the emergency nursing room, or an independent isolation unit composed of several negative pressure wards can be set.

4. 病人生活区：就餐、娱乐与日间病房：住院部每床面积 6m^2。

4. Patient living area: dining, recreation and daytime ward: Each bed in the inpatient department shall have an area of 6m^2.

门诊部如就餐是日间护理计划中的一部分，则门诊每人 6m^2，否则只提供娱乐与日间用房。就餐室内应设置洗手盆。

For the outpatient department, if dining forms a part of the day care plan, a space of 6m^2 shall be provided for each person; otherwise, only the recreation room and the daytime room will be provided. In addition, a wash basin shall be set in the dining room.

5.8.1.4 日常起居活动单元

5.8.1.4 Daily living activity unit

是提供指导日常起居活动的单元，该单元应包括病房、浴室、厨房、走楼梯操练房。个人服务区包括理发、美容，日常起居单元的各种活动可服务于此目的。

As a unit providing the guidance in daily living activities, the unit shall be composed of ward, bathroom, kitchen and staircase exercise room. The personal service area shall include the area for hairdressing and beauty service, and all activities of daily living units can serve this purpose.

5.8.1.5 诊断与治疗场所

5.8.1.5 Diagnosis and therapy area

包括医疗评估单元，每个康复设施应有一个医疗评估单元、检查室、评估室、办公室。

It shall include the medical evaluation unit, and each rehabilitation facility shall be provided with a medical evaluation unit, an examination room, an evaluation room and offices.

5.8.1.6 其他所需单元

5.8.1.6 Other required units

每个康复设施除了有医疗评估单元以外，还应有以下单元中的一个或多个单元：

In addition to the medical evaluation unit, each rehabilitation facility shall be provided with one or more of the following units:

心理服务单元、社交性服务单元、职业服务单元、理疗单元、作业治疗单元、修复与校正单元、言语与听觉单元、牙科单元、影像科（包括肌电图、DR、CT、MRI、核医学)、检验设施、家庭保健服务，以及药房、餐饮单元、消毒设施、布件服务、物资管理、环境卫生服务、工程服务与维护用房。

Psychological service unit, social service unit, occupational service unit, physiotherapy unit, operation therapy unit, recovery and correction unit, speech and hearing unit, dental unit, imaging department (including PET, DR, CT, MRI and nuclear medicine) inspection facilities, family health service, pharmacy, catering unit, disinfection facilities, linen service, material management, environmental health service, engineering service and maintenance room.

5.8.1.7 公共区域

5.8.1.7 Public area

入口应无障碍，满足轮椅出入，门厅应设接待、等候处、公共厕所、公用电话、饮用水装置、轮椅存放处、便利店。

The entrance shall be wheelchair accessible, and the reception center and waiting areas, public toilets, public telephones, water dispensers, wheelchair storage area and convenience stores shall be set in the hall.

管理与相关支持区域：设会谈室、综合与个人办公室、多功能室、病人储藏室、综合储藏室、员工及护工设施（如：柜子）。

Management and related support areas: Meeting rooms, comprehensive and personal offices, multi-purpose rooms, storage rooms for patients, comprehensive storage rooms and the facilities for staffs and nursing workers (such as cabinets) shall be set.

5.8.1.8 细节与装饰面的一般标准

5.8.1.8 General standards for details and decorative surfaces

1. 康复设施内的病人是残疾程度不同的病人，所以应为这些使用者建立很高的安全标准，以减少事故发生；走道里若摆放饮水机、电话亭、自动售货机及移动设备装置时，不应妨碍走道交通宽度，或使走道宽度减少到低于最低要求；吊顶的高度最低为2.4m。

1. Since patients in rehabilitation facilities are patients with different degrees of disability, high safety standards shall be established for these users to reduce accidents; if water dispensers, telephone booths, vending machines and mobile apparatus are arranged in the passageway, the arrangement shall ensure that width of the passageway is not affected, or the width of the passageway shall be reduced to be less than the minimum width; the ceiling height shall be 2.4m at least.

2. 门的型式：除电梯门外，走道与房间之间的门或通向使用房间的门应是摆动型的；门宽度要符合残疾人使用标准；除了个人用的小柜的门以外，其他的门不应开向走道，以免妨碍交通或减小走道所需的宽度。

2. Door type: except for the elevator door, all doors from the passageway to rooms or doors access to rooms shall be the use room shall be swing door; the width of the door shall conform to the accessibility standard for the disabled; except for doors of the lockers for personal use, other doors shall not be opened towards to the passageway so as not to block the traffic or reduce the

width required for the passageway.

3. 病人洗浴和厕所的门应该能从外面让人进入；门应向外开启，使门开启时不会挤压着可能在房间内已经虚脱的病人；门框与伸缩缝的盖子应与地板面平接，以方便轮椅和推车。

3. Doors of the patient bath rooms and toilets shall be accessible from the outside; the door shall be opened outward so as not to squeeze the patient who may have collapsed in the room; the cover plate between the door frame and the expansion joint shall be flush with the flcor surface to ensure the accessibility both for wheelchairs and carts.

4. 病房内的窗扇应设计成开启时，为防止意外跌落事故，应提供安全屏障，经常开启的窗和外门应设防虫纱窗纱门；门、侧窗、借光窗、离地 50cm 以内的窗的玻璃应采用安全玻璃、夹胶玻璃或能防碎或碎时无危险切割边缘的塑料透光材料，类似的材料也应用在游戏室与体操室墙上的洞口处。

4. The window sash in the ward shall be designed to prevent accidental falling when it is opened, and safety barriers shall be provided. Insect-proof screen windows and doors shall be provided for opened windows and exterior doors that may be opened frequently; glasses used for doors, side windows, daylighting windows and those windows with height less than 50cm above the ground shall be safety glass, laminated glass or be made of plastic light-transparent materials that are scatter-proof or have no sharp cutting edge in case of being broken. Moreover, the glass made of such materials shall also be applied to the openings on the walls of the play room and the gymnastics room.

5. 淋浴间的门与浴盆围挡上应采用安全玻璃或塑料透光材料，残疾人使用的洗手盆，应符合残疾人设计标准，镜子的布置应方便轮椅病人与站着的病人使用，在所有的洗手处应有干手装置，洗手盆与洗手处应牢固地固定，应能承受不小于 114kg 的垂直荷载。

5. The door of the shower cubicle and the bathtub enclosure shall be made of safety glass or plastic light-transparent material. The wash basin for the disabled shall conform to the design standards, and the mirror shall be accessible for both wheelchair patients and other patients. A hand dryer shall be arranged at all hand washers. The washbasin and the hand washer shall be firmly fixed and be capable to bear vertical load no less than 114kg.

6. 在所有病人用的厕所、浴缸、淋浴器与坐浴处应设置扶手；扶手端头的构造应能防止勾住病人的衣服；淋浴的连杆应可以考虑做短暂支撑，在浴室内应设嵌入式肥皂盒。

6. All toilets, bathtubs, showers and sitz bath stalls for patients shall be provided with handrails; the end of the handrail shall be designed to prevent the patient's clothes from being caught; the shower bar shall be designed to be used as a temporary support, and an embedded soap box shall be set in the bathroom.

7. 病人经过的走道两侧应有扶手，扶手端头的构造应防止勾住病人的衣服。

7. The passageway that the patients may pass shall be provided with handrails on both

sides, and the end of the handrail shall be designed to prevent the patient's clothes from being caught.

8. 娱乐室、健身房以及会产生冲击噪声的房间，除非采取了特殊的降噪措施，否则不应位于病房区的上面。

8. Recreation rooms, rehabilitation training rooms and rooms that may produce impact noise shall not be set above the ward area, unless special noise reduction measures are taken.

9. 地面应平整无不规则表面，以防使用矫形器材的病人绊倒；有水的房间，地面应防滑，墙面应可清洗，病人的走道、护士站、休息室、娱乐室、餐厅、等候室应设吸声吊顶。

9. The floor shall be flat and free from irregularities to prevent the patients using the orthopedic equipment from tripping; for rooms provided with water, the floor shall be skid resistant, the wall shall be washable, and the patient channels, nurse stations, rest rooms, recreation rooms, dining rooms and waiting rooms shall be provided with acoustical ceilings.

10. 被管道、风管、导管穿过的楼板和墙面要密封，以防害虫侵入，构件的结合处也应密封。

10. Floors and walls through which pipes, air ducts and ducts pass shall be sealed to prevent pests from invading, and joints of components shall also be sealed.

5.8.2 康复治疗设施

5.8.2 Rehabilitation facilities

5.8.2.1 特点

5.8.2.1 Characteristics

为长期进行康复护理的住员提供体能与/或职业病治疗。若康复护理设施属于综合医院或其他设施中的局部时，有些服务可适当共享。

It is designed to provide physical training and/or occupational disease therapy for patients undergoing long-term rehabilitation nursing. If the rehabilitation and nursing facilities form a part of a general hospital or other facilities, some services can be shared as appropriate.

5.8.2.2 诊断与治疗场所

5.8.2.2 Diagnosis and therapy area

1. 有住员体能与职业病治疗条件，包括实施指定的每种治疗所需要的房间与设备，洗手处应位于治疗单元内，提供住员私密性要求的条件，住员使用区域应符合室内外无障碍设计规范。

1. It shall be provided with the physical training and occupational diseases therapy conditions for inpatients, including rooms and devices required for implementing various designated therapies. The hand washer shall be set in the therapy unit. The conditions meeting the patient privacy requirements shall be provided, and the area for inpatients shall meet the design specification for indoor and outdoor accessibility.

2. 康复治疗的支持区有文档室、唱片室、活动管理室、供应物设备仓库、房间整理服务室，可以在单元内或临近单元设置。

2. The rehabilitation therapy support area, including document room, record room, activity management room, supplies and equipment warehouse and housekeeping service room, can be set in or near the unit.

3. 公众等候区及公共厕所应与门诊病人的支持区、活动区、厕所、淋浴间分开。

3. Public waiting area and the public toilet shall be separated from the support area, activity area, toilets and shower cubicle for outpatients.

4. 应提供餐饮设施、洗衣房、布件服务、物资管理、保洁、工程服务与维修用房。

4. In addition, catering facilities, laundry, linen service, material management, cleaning, engineering service and maintenance rooms shall be also provided.

5.8.2.3　公共区域：按通用设计。

5.8.2.3　Public area: According to the general design.

5.8.3　成人日间保健护理服务设施

5.8.3　Adult day-care service facilities

5.8.3.1　特点

5.8.3.1　Characteristics

1. 成人日间保健护理服务具有集合性的内容，设计要满足功能和认知能力受损成人的需要，是为不可独自在家，以求安全和安康的成人提供护理的非公共机构性场所，提供受保护的场所、保健和支持性服务。

1. Adult day-care service area is an integrated non-public institutional place designed to provide the nursing service to adults who have suffered from function and cognitive ability impairment and cannot be at home alone, to ensure their safety and health.

2. 有些设施还为特殊人群提供特殊服务，如阿尔茨海默氏病人、病情在进展的残疾人、脑伤者、精神疾病人、人体免疫缺损病人、艾滋病人和视觉及听觉损伤病人。成人日间护理设施是为老年人及残疾人提供护理的统一体系的组成部分。

2. Some facilities also provide special services for special groups, such as Alzheimer' s patients, patient with disabilities, brain injury, mental disease, human immunodeficiency, AIDS and visual and hearing impairment. Adult day-care facilities are the part of a unified system of care for the elderly and the disabled.

5.8.3.2　成人日间护理形式

5.8.3.2　Form of adult day-care

1. 多功能设施是一个有组织、涉及面广、非住宅型、在一个受保护的场所内提供保健、社交和支持性等各种服务的设施，这些形式的设施规模较大，以适应服务内容改变的需要。

1. Multi-functional facilities are the organized and non-residential facilities with wide coverage that provide various services, such as health care, social and supporting services, in a protected place. These forms of facilities have a large scale to meet the needs of changing service content.

2. 特殊设施是一个有组织、涉及面广、非住宅型的设施，它提供保健、社交等各种支持性服务，以及为一个对象明确的人群提供特殊服务，这类设施有其独特的需求，影响着活动房间的适用要求。

2. Special facilities are the organized and non-residential facilities with wide coverage that provide health care, social services and other supporting services, as well as special services for a clear object population. These facilities are suitable for unique needs and have applicable requirements in the activity rooms.

3. 当成人日间保健护理中心位于一个医院或其他服务机构内时，它自己应该有能识别的房间，其支持性房间可以被共享。

3. When an adult day-care center is located in a hospital or other service institutions, it should have its own room that can be identified and its supporting room can be shared.

4. 病人活动室和休息室应靠近护士站、洗手处、厕所，这些患者的活动空间应该比其他病人的活动空间要宽敞，应设置处置室、护士站、药品和设备库房，这些区域要有洗手盆。

4. Patient activity room and rest room should be close to nurse station, lavatory and toilet. The activity space of these patients should be more spacious than that of other patients. Medical waste storage room, nurse station, medicine and equipment warehouse should be set up. These areas should be equipped with wash basin.

5. 设施内的家具应坚实、安全，参与者在以它做支持行走或坐下时不易被倾倒；家具应有合适比例，可以让敏捷度受限的人方便使用，且可让脚搁地休息。

5. The furniture in the facility should be solid and safe, and the users are not easy to be dumped when walking or sitting with it as support; the furniture should have an appropriate proportion, which can make it convenient for people with certain disabilities to use, and can make their feet rest on the ground.

5.8.4 康复医院

5.8.4 Rehabilitation hospitals

5.8.4.1 我国不设一级康复医院，《康复医院基本标准》（2012 年版）为新建二、三级康复医院的验收标准，是提供综合性康复医疗服务的科室。选址上考虑到外来患者接待，服务周边人口的康复治疗大厅宜设在门诊部，并方便住院患者抵达；相对专业的疾病康复，如儿童康复、心理康复、精神康复等，则需要设专区。

5.8.4.1 There is no first-class rehabilitation hospital in China. The Basic Standard for Rehabilitation Hospital (2012) is the acceptance standard of the new second-class and third-class rehabilitation hospitals, which are the hospitals providing comprehensive rehabilitation medical services. For the location, the reception of non-local patients should be considered, the rehabilitation hall serving the surrounding population should be set up in the outpatient department, which is convenient for the arrival of inpatients; special areas should be set up for relatively professional disease rehabilitation, such as pediatric rehabilitation, psychological rehabilitation, psychiatric rehabilitation, etc.

5.8.4.2 康复门诊包括：综合康复诊室、功能测评室、咨询室等。其中听力语音康复门诊和低视力康复门诊比较特殊，如声控室、听力测定实验室、检查暗室、验光暗室、视野检查室等可借助耳科、眼科的配套设施来完善。

5.8.4.2 Rehabilitation clinic includes: comprehensive rehabilitation consulting room, function evaluation room, consulting room, etc. Among them, the hearing and speech rehabilitation clinic and the low vision rehabilitation clinic, such as sound control room, audiometry room, examination darkroom, optometry darkroom and the perimetry examination room can be improved by the auxiliary facilities of the otology and ophthalmology department.

5.8.4.3 康复治疗包括：运动治疗室（PT）、作业治疗室（OT）、言语治疗室（ST）、物理治疗室、传统康复治疗室、职业康复治疗室、心理治疗室等。其中 PT 室通常采用大空间，便于布置康复设备，并方便治疗师同时照护多个患者。

5.8.4.3 Rehabilitation therapy room includes: physical therapy room, operational therapy room, speech therapy room, physiotherapy room, traditional rehabilitation therapy room, operational rehabilitation therapy room, psychological therapy room, etc. The PT room usually has a large space, which is convenient for the arrangement of rehabilitation equipment, and enables the therapist to care for multiple patients at the same time.

5.8.4.4 三级康复医院

5.8.4.4 Third-class rehabilitation hospitals

1. 床位

1. Bed

住院床位总数 300 张以上，其中康复专业床位 75% 以上。

The total number of hospital beds is more than 300, of which over 75% are the beds for rehabilitation discipline.

2. 科室设置

2. Department setting

（1）临床科室：至少设骨关节康复科、神经康复科、脊髓损伤康复科、儿童康复科、老年康复科、心肺康复科、疼痛康复科、听力视力康复科、烧伤康复科中的 6 个科室，以及内科、外科和重症监护室。

（1）Clinical departments: at least 6 departments among bone and joint rehabilitation department, neurological rehabilitation department, spinal cord injury rehabilitation department, pediatric rehabilitation department, geriatric rehabilitation department, cardiopulmonary rehabilitation department, pain rehabilitation department, hearing and vision rehabilitation department, burn rehabilitation department, and internal medical department, surgical department and ICU.

（2）治疗科室：至少设物理治疗室、作业治疗室、言语治疗室、传统康复治疗室、康复工程室、心理康复室和水疗室。

（2）Treatment departments: at least physiotherapy room, operational therapy room, speech therapy room, traditional rehabilitation therapy room, rehabilitation engineering room, psychological rehabilitation therapy room and hydrotherapy room.

（3）评定科室：至少设运动平衡功能评定室、认知功能评定室、言语吞咽功能评定室、作业日常活动能力评定室、心理评定室、神经电生理检查室、心肺功能检查室、听力视力检查室、职业能力评定室中的7个。

（3） Evaluation departments: at least 7 of the movement balance function evaluation room, cognitive function evaluation room, swallowing and speech function evaluation room, operational and daily activity ability evaluation room, psychological assessment room, neurotic electrophysiology examination room, cardio-pulmonary function examination room, hearing and vision examination room and operational ability evaluation room.

（4） 医技科室：至少设医学影像科、检验科、药剂科、营养科、门诊手术室、消毒供应室。

（4） Medical technology departments: at least medical imaging department, laboratory department, pharmacy department, nutritional department, outpatient operating room and disinfection supply room.

（5） 职能科室（部门）：至少设医疗质量管理部门、护理部、医院感染管理科、器械科、病案（统计）室、信息科、社区康复服务部门等科室（部门）。

（5） Functional departments: at least medical quality management department, nursing department, hospital infection management department, equipment department, record (statistics) room, information department and community rehabilitation service department.

3. 场地

3. Site

（1） 每床建筑面积≥95m^2。病房每床净使用面积≥6m^2，床间距≥1.2m。

（1） Construction area of each bed ≥95m^2. Net usable area of each bed in the ward is ≥6m^2, and the bed spacing is≥1.2m.

（2） 康复治疗区域总面积≥3000m^2。

（2） Total area of rehabilitation treatment area ≥3000m^2.

（3） 医院建筑设施执行国家无障碍设计相关标准。

（3） The hospital construction facilities shall implement the relevant standards of national barrier-free design.

4. 设备

4. Devices

（1） 基本设备：参照同级综合医院基本设备并结合本专业实际需要配置。

（1） Basic devices: refer to the basic devices of the general hospital of the same level and the configuration combined with the actual needs of the discipline.

（2） 专科设备：康复评定、运动治疗、物理因子治疗、作业治疗、认知、言语、吞咽治疗、传统康复治疗、康复工程、水疗。

（2） Specialized equipment: equipment for rehabilitation evaluation, exercise therapy, physical factor treatment, operation therapy, cognitive, speech and swallowing treatment, traditional rehabilitation treatment, rehabilitation engineering and hydrotherapy.

（3） 信息化设备：在住院部、信息科等部门配置自动化办公设备，保证医院信息化建设符合国家相关要求。

（3） Information equipment: automatic office equipment should be equipped in the inpatient department, information department and other departments to ensure that the hospital information construction meets the relevant national requirements.

（4） 病房床单元基本装备同三级综合医院。

（4） Basic devices of ward bed unit are the same as that of third-class general hospital.

（5） 有能满足日常诊疗业务需要的其他设备。

（5） Other devices meet the needs of daily diagnosis and treatment business.

5. 8. 4. 5 二级康复医院

5. 8. 4. 5 Second-class rehabilitation hospitals

1. 床位

1. Bed

住院床位总数 100 张以上，其中康复专业床位占 75% 以上。

The total number of hospital beds is more than 100, of which over 75% are the beds for rehabilitation discipline.

2. 科室设置

2. Department setting

（1） 临床科室：至少设置骨关节康复科、神经康复科、儿童康复科、老年康复科、听力视力康复科、疼痛康复科中的 3 个科室以及内科、外科、重症监护室。

（1） Clinical departments: at least 3 departments, including bone and joint rehabilitation department, neurological rehabilitation department, pediatric rehabilitation department, geriatric rehabilitation department, hearing and vision rehabilitation department, pain rehabilitation department, and internal medical department, surgical department and ICU.

（2） 治疗科室：至少具备物理治疗、作业治疗、言语治疗、传统康复治疗功能。

（2） Treatment departments: at least the functions of physical treatment, operation therapy, speech treatment and traditional rehabilitation treatment.

（3） 评定科室：至少具备运动平衡功能评定、认知功能评定、言语吞咽功能评定、作业日常生活活动能力评定、神经电生理检查、听力视力检查中的 5 项功能。

（3） Evaluation departments: at least 5 functions of exercise and balance functions evaluation, cognitive function evaluation, speech and swallowing functions evaluation, operational daily activity evaluation, neuro electrical physiological examination, hearing and vision examination.

（4） 医技科室：至少设置超声科、检验科、放射科、药剂科和消毒供应室。

（4） Medical technology departments: at least ultrasonic department, laboratory department, radiology department, pharmacy department and disinfection supply room.

（5） 职能科室（部门）：至少设医疗质量管理部门、护理部、医院感染管理科、信息科、器械科、病案（统计）室、社区康复服务科室（部门）。

（5） Functional departments: at least medical quality management department, nursing department, hospital infection management department, information department, equipment department, record (statistics) room and community rehabilitation service department.

3. 场地

3. Site

(1) 每床建筑面积≥85m^2。病房每床净使用面积≥6m^2，床间距≥1.2m。

(1) Construction area of each bed ≥85m^2. Net usable area of each bed in the ward is ≥6m^2, and the bed spacing is ≥1.2m.

(2) 康复治疗区域总面积≥800m^2。

(2) Total area of rehabilitation treatment area ≥800m^2.

(3) 医院建筑设施执行国家无障碍设计相关标准。

(3) The hospital construction facilities shall implement the relevant standards of national barrier-free design.

4. 设备

4. Devices

(1) 基本设备：参照同级综合医院设备并结合本专业实际需要配置。

(1) Basic devices: refer to the devices of the general hospital of the same level and the configuration combined with the actual needs of the discipline.

(2) 专科设备：康复评定、运动治疗、物理因子治疗、作业治疗、认知言语治疗、传统康复治疗。

(2) Specialized equipment: rehabilitation evaluation, exercise therapy, physical factor treatment, operation therapy, cognitive, speech treatment, traditional rehabilitation treatment.

(3) 信息化设备

(3) Information device

在住院部、信息科等部门配置自动化办公设备，保证医院信息化建设符合国家相关要求。

Automatic office equipment should be equipped in the inpatient department, information department and other departments to ensure that the hospital information construction meets the relevant national requirements.

(4) 病房床单元基本装备同二级综合医院。

(4) Basic devices of ward bed unit are the same as that of second-class general hospital.

(5) 有能满足诊疗业务需要的其他设备。

(5) Other devices meeting the needs of diagnosis and treatment business.

5.8.5 健康管理

5.8.5 Health management

5.8.5.1 健康管理工作包括：健康信息采集、健康检测、健康评估、个性化健康管理方案、健康干预。从事健康管理的机构可以是独立运营的商业公司，也可以是位于综合医院和专科医院内的体检中心或体检科。

5.8.5.1 Health management includes: health information collection, health testing, health assessment, personalized health management proposal, and health intervention. Institutions engaged in health management may be independent commercial companies, or physical examination centers or departments in general hospitals and specialized hospitals.

5.8.5.2 指标

5.8.5.2 Indicators

医院内设置体检功能可以为医院带来更多的患者，有利于早期发现疾病、早期干预和治疗，既是保健也是一种营销方式。《综合医院建设标准》（建标 110—2008）未给出体检设施建设指标，并指出："健康体检设施及其所需的面积指标，应根据实际需要报批。"实际项目设计中宜采取"量体裁衣"法，如：取用一个门诊单元、一个防火分区或一栋单体建筑的一层乃至两层楼，以实现独立分区管理，且能够独立核算建筑经济指标或运维财务指标。

Setting up the physical examination centers or departments in the hospital can bring more patients, which is conducive to the early detection of diseases, early intervention and treatment, and the health care is also a marketing method. Construction Standard for General Hospitals(JB 110—2008) does not give the constructive physical examination indicators, and states that "The health examination facilities and the required area indicator shall be submitted for approval according to actual needs. " The actual project design shall act according to actual circumstances, for example: taking one outpatient unit, one fire compartment or one or even two floor(s) of an individual building to realize independent zone management, and to be able to independently calculate economic indicator for the building or financial indicator for operation and maintenance.

5.8.5.3 策划

5.8.5.3 Planning

1. 国家投资的公立医院若未申请到健康体检设施面积指标而要建设体检科，则需挤占其他功能的面积指标。非国家投资的医院围绕市场和商业服务目标设立科室，通过强化体检中心建设，加大体检专用设备的投入，可以吸引更多人前往，故需要较大的建筑面积。

1. If a public hospital invested by the state wants to build a physical examination department without applying for the area indicator of health examination facilities, it will have to occupy the area indicator of other functions. Non-state-invested hospitals set up departments around the market and commercial service objectives. By strengthening the construction of medical examination centers and increasing the investment in specialized medical examination device, more people can be attracted to the hospitals, thus requiring a larger construction area.

2. 我国省立和市立医院中需设干部保健，主要为干部和公务员提供健康管理，包括体检；可以将干部保健与体检中心、特需服务合并设在一起；某些医院还结合健康管理设立老年病中心或干部医疗保健中心，形成规模化的独立院区和品牌效应，通过这些功能策划，可以实现医护人员、医疗设备、机电系统、建筑乃至土地的资源共享。

2. Health care center for cadres is required in provincial and municipal hospitals in China, which mainly provides health management, including physical examination, for cadres and civil servants; the health care center for cadres can be combined with physical examination centers

and VIP; some hospitals also set up geriatric centers or health care center for cadres in combination with health management to form large-scale and independent hospital areas and brand effects. Through these functional planning, the sharing of resources among medical care staff, medical device, mechanical and electrical systems, buildings and even land can be realized.

3. 有场地条件的健康管理机构可以设置住院体检病房乃至病区/护理单元，供体弱或特需人士入住，进行若干天的体检、保健、检后专家会诊、诊断治疗。

3. Health management organizations with sufficient land can set up inpatient physical examination wards and even ward/nursing units for weak or special person to check in for several days of physical examination, health care, expert consultation after examination, diagnosis and treatment.

5.8.5.4 设计

5.8.5.4 Design

1. 设备

1. Devices

健康管理机构和医院体检中心应设置必要的检查设备，应由医疗工艺设计做出设备规划；体检设备和保障设备有条件时可以与挂钩医院或所在医院的医技设备资源共享。

Health management organizations and hospital physical examination centers should set up necessary examination device, and the device planning should be made by design for medical process; physical examination device and support equipment, as well as the medical device resources of the designated hospital or the hospital where they are available, can be shared when conditions permit.

2. 工艺

2. Process

体检中心是为健康人群服务的，应与医疗部门分开；医院内设立体检中心的流程应符合医院医疗工艺一级流程设计，应纳入信息系统规划。根据体检规模的大小，可以考虑开设独立的特需体检及 VIP 区，独立设立女性体检区、男性体检区。批量招工的工人体检、军队入伍体检不需要设独立的分区，但应在管理时考虑分流；体检中心可配套设置移动体检车，提供上门体检服务，包括采样、拍片，检查报告在体检中心出具。

Physical examination centers serve healthy people and should be separated from medical departments; the process of setting up a medical examination center in a hospital should conform to the first-level process design of hospital medical process and should be incorporated into the information system planning. According to the size of the physical examination, it may be considered to set up an independent physical examination for special need and VIP area, and an independent female examination area and a male examination area. There is no need to set up separate examination zones for mass recruitment of workers and army recruits, but triage should be considered in management; the physical examination center can be equipped with a mobile physical examination vehicle to provide on-site physical examination services, including sampling and X-raying, and the examination report will be issued by the physical examination center.

3. 功能

3. Function

健康体检要考虑体检人员的隐私，体检区、办公区和辅助用房应分开，其功能分三个部分：

Health examination should consider the privacy of personnel examination. The examination area, office area and auxiliary room should be separated, and their functions are divided into three parts:

（1）检前检后区：包括接待服务台、健康问卷评估、健康咨询室、心理评估室、更衣室、信息采集台、健康体检资料汇总分析室、健康体检报告发放室、检查后咨询室、健康教育室、受检者及家属等候区、工作人员办公区、餐厅、卫生间、污物间。

（1）Pre-examination and post-examination areas: including reception desk, health evaluation room, health consulting room, psychological assessment room, locker room, information collection desk, health examination data collection and analysis room, health examination report distribution room, post-examination consultation room, health education room, waiting area for examinees and their relations, staff office area, dining room, toilet and waste room.

（2）体检区：包括引导台、静脉采血、一般检查、内科、外科、妇科、耳鼻喉科、眼科、口腔科、心电图、放射、超声、检验、骨密度、肺功能、动脉粥样硬化检测等特殊检查室。

（2）Physical examination area: including guide desk, venous blood sampling room, general examination room, internal medical department, surgical department, gynecology department, otolaryngology department, ophthalmology department, stomatology department, ECG room, radiology department, ultrasonic room, examination room, bone density room, lung function test room, atherosclerosis detection room, etc.

（3）工作人员办公区：包括更衣室、营销部、主任办公室、行政办公室、医生办公室、资料室、主检医生办公室、随访室、教室、茶水间、卫生间、库房。

（3）Staff office area: including locker room, marketing department, director's office, administrative office, doctor office, reference room, responsible physician office, follow-up room, classroom, tea room, toilet and warehouse.

健康管理机构除了体检功能之外还需要设置营销展示区、接洽室、档案室、办公区。

Health management organizations need to set up marketing display, contact, files and office areas in addition to physical examination function area.

5.8.6　预防保健

5.8.6　Prevention and health care

5.8.6.1　预防保健是进行卫生防疫的部门，包括传染病疫情管理及防治预防保健，包括计划免疫、预防接种、慢性病、地方病、寄生虫病的防治工作。预防保健机构可以设在医院里，作为预防保健科，但当条件允许的时候，预防保健的执行措施应被分散放在社区服务中心，医院内的预防保健科只做行政指导工作，不接待外来保健人群，保健工作分散在各个社区医疗中心。

5.8.6.1 Prevention and health care department is the department for hygiene and epidemic prevention, including epidemic management of infectious diseases, prevention and health care department, as well as planning immunization, providing preventive vaccination, prevention and treatment of chronic diseases, endemic diseases and parasitic diseases. Prevention and health care organizations can be set up in hospitals as prevention and health care departments, but when conditions permit, implementation measures of preventive health care should be taken in community service centers. The prevention and health care departments in hospitals only do administrative guidance and do not receive health care groups. Health care is carried out in community medical centers.

5.8.6.2 当预防保健作为综合医院内的一个科设置时，医院内预防保健用房的建筑面积，应按编制内每位预防保健工作人员 $20m^2$ 增加建筑面积。应设独立的出入口，远离患者，避免交叉感染，应考虑儿童活动空间和家属陪护空间。

5.8.6.2 When a prevention and health care department is set up in a general hospital, the construction area of preventive health care in the hospital should be increased by $20m^2$ per preventive health care worker in the establishment. Independent entrances and exits should be set up to keep away from patients and avoid cross infection. Children's activity space and family accompanying space should be considered.

5.8.6.3 预防保健科应设有儿童保健门诊、妇女保健门诊、相关管理科室和宣传教育会议室，应设候诊区、分诊台、儿童厕所和儿童检查室。儿童保健也可以放在医院儿科中心的一个独立区域，以实现医生护士的共享，节约场地资源，这种情况下应注意保健的儿童不能跟患者儿童在一个娱乐场所游乐，以避免交叉感染。

5.8.6.3 The prevention and health care department shall have children's health care clinic, women's health care clinic, relevant management departments and publicity and education meeting rooms, and shall set up waiting areas, triage desk, children's toilets and children's examination rooms. Children's health care can also be located in an independent area of the pediatric center of the hospital, so as to realize the sharing of doctors and nurses and save site resources. In this case, attention should be paid to the fact that health-care children cannot play with patient children in the same entertainment place to avoid cross infection.

5.8.6.4 预防保健科分为候诊区、保健管理区、儿童保健区、妇女保健区和医护工作区。有条件的城市和区域的儿童和妇女保健，首先设置在市区两级妇幼保健院里，或由妇幼保健院下辖的社区医疗中心安排监督执行保健措施。这样有利于使保健措施得到更高级别的指导，及时反馈疾患信息到医院，有利于实现分级诊疗服务，并使妇女儿童享受到更好的保健服务。一些民营医疗机构设置了预防保健科，这有利于从公立医疗机构中分流出一部分人，使他们享受到较好的服务。

5.8.6.4 The prevention and health care department is divided into waiting areas, health management areas, children's health care areas, women's health care areas and health care work areas. Children's and women's health care in cities and regions where conditions permit should be first set up in municipal and regional two-level maternity and childcare hospitals, or the community medical centers under the jurisdiction of maternity and childcare hospitals to arrange supervision and implementation of health care measures. This is conducive to higher-level guidance of health care measures, timely feedback of disease information to hospitals, realization of graded diagnosis and treatment services, and better health care services for women and children. Some private medical institutions have set up prevention and health care departments, which are conducive to the outflow of some people from public medical institutions and enable them to enjoy better services.

5.8.6.5　在一些专科医院或者大型综合医院的大专科里，如口腔科、眼科，也可以设置防保科目/部，这些是针对口腔、眼睛的专项检查，在口腔科有如小儿龋齿、换牙保健；在眼科有少儿近视、小儿弱视、斜视的检测。这些专科的检查和预防保健都非常有利于少儿的身体健康，是国家规定的预防保健项目之外的更优越的预防保健项目。

5.8.6.5　In some specialized hospitals or special departments of large general hospitals, such as stomatology department and ophthalmology department, prevention and protection subjects/departments can also be set up. These are special examinations for oral cavity and eyes, in stomatology department, like children's dental caries and dental replacement health care, and in ophthalmology, like tests of myopia, amblyopia and strabismus in children. These specialized examinations as well as prevention and health care are very beneficial to children's physical health, and are superior prevention and health care items besides those stipulated by the state.

5.8.6.6　预防保健科的设计应兼顾不同年龄段儿童、妇女等的舒适性、体验感，要有良好的通风采光，环境宜人，装饰优美；卫生间区域应设置哺乳、婴儿尿布换理区，应设置无障碍设施；应设置良好的信息通信设备，确保预防保健的信息数据传输到卫生疾控部门。

5.8.6.6　The design of prevention and health care department should take into account the comfortableness and experience of children and women of different ages, and should have good ventilation and lighting, pleasant environment and beautiful decoration. Toilet area should be equipped with lactation, diaper changing spaces and barrier-free facilities. Good information and communication device should be set up to ensure that information and data on prevention and health care are transmitted to health and disease control departments.

5.8.7　临终关怀

5.8.7　Hospice care

5.8.7.1　临终关爱护理是一项有医学指导、各学科相互配合，为临终病人和家属以及其他重要人员减轻痛苦的护理与服务，包括病人死亡后对家属的心理辅导。临终关爱护理着重于帮助临终病人在死亡过程中保持尊严和生活质量，并尽可能缓和或控制不愉快的症状。

5.8.7.1　Hospice care and nursing is a kind of nursing and service with medical guidance and cooperation of various disciplines to relieve the pain of dying patients, their families and other important personnel, including psychological counseling for families of patients after their death. Hospice care focuses on helping dying patients maintain dignity and quality of life in the process of death, and alleviating or controlling unpleasant symptoms as much as possible.

5. 8. 7. 2 临终关怀设施包括从临终状态到死后运至殡仪馆前若干环节的功能设施，包括配套建设不同宗教的礼拜、祈祷室、安慰室、临终关怀家庭病房、临终护理单元。临终关怀家庭病房宜位于较为隐蔽的区域，有自然采光、独立通风排风，室内装饰面应考虑吸声，减少对周边的干扰，并靠近尸体转运电梯。

5. 8. 7. 2 Hospice care facilities include functional facilities in several links from terminal stage to transportation to funeral parlor after death, including setting up chapels, prayer rooms, comfort rooms of different religions, hospice care homely wards and hospice nursing units. Hospice care homely wards should be located in a relatively hidden area with natural lighting, independent ventilation and exhaust. Sound absorption should be considered on the interior decoration surface to reduce interference to the surroundings and be close to the body transfer elevator.

5. 8. 7. 3 每个临终关爱护理单元规模限制宜为25 床以内；晚期绝症病人或垂死的人往往极度疲劳，使用的维持生命装置令人蒙羞，应分散、分组安排在非集中的服务区，应全部为单人间，应配备空气感染隔离病房。

5. 8. 7. 3 The size limit of each hospice nursing unit should be within 25 beds. Patients with terminal diseases or dying people are often extremely tired. The life-sustaining apparatus used are humiliating. They should be scattered and grouped in non-centralized service areas. All should be equipped with single rooms having air-borne infection isolation wards.

5. 8. 7. 4 临终关爱房间大小应基于护理内容、特殊的室内家具、衣物储藏柜的要求，有访客座椅、陪伴床位，有可以看见室外自然光的窗户，有条件的可以设家庭式厨房和就餐设施，供陪人和访客使用。

5. 8. 7. 4 The size of the hospice care room should be based on the nursing content, special indoor furniture and clothing storage cabinet requirements, with guest seats, accompanying beds, windows that can see outdoor natural light, and family kitchens and dining rooms for accompanying people and visitors if conditions permit.

5. 8. 7. 5 提供临终关爱服务的医院、养老机构内，可设教堂、佛堂、礼拜堂、安宁室等。

5. 8. 7. 5 Hospitals and pension institutions that provide hospice care services can have churches, Buddha halls, chapels, hospice rooms, etc.

5. 8. 7. 6 应考虑提供通向室外或露台、阳台的通道，户外空间提供平和的环境变化，适宜焦虑与不安的病人、家属散步，花园象征着生与死的循环，是从容和精神镇定的源泉。

5. 8. 7. 6 Consideration should be given to providing access to outdoor or terrace, balcony. Outdoor space needs to provide placid environmental changes, suitable for anxious and restless patients and family members to walk. Garden symbolizes life and death cycle and is the source of unhurried state and mental calm.

5.9 疗养院

5.9 Sanatorium

5.9.1 床位：住院床位总数 100 张以上。

5.9.1 Beds: the total number of hospital beds is more than 100.

5.9.2 科室设置

5.9.2 Department setting

5.9.2.1 临床科室：至少设有两个疗养区，至少设有传统康复医学室、体疗室。

5.9.2.1 Clinical departments: at least two convalescent areas, traditional rehabilitation medicine rooms and body therapy rooms.

5.9.2.2 医技科室：至少设有药房、化验室、X 光室、心电图室、超声波室、理疗室、消毒室。

5.9.2.2 Medical technology departments: at least pharmacy, laboratory, X-ray room, electrocardiogram room, ultrasonic room, physiotherapy room and disinfection room.

5.9.2.3 基本设备：呼吸机、吸痰器、心电图机、除颤机、显微镜、血细胞计数仪、生化分析仪、分光光度计、自动稀释器、电泳仪、离心机、电冰箱、干燥箱、水浴箱、X 光机、A 超或 B 超、姿势矫正镜、墙拉力器、前臂旋转训练器、滑轮吊环、划船器、手指肌训练器、各种助行器、中频治疗仪、超短波治疗仪、红外线治疗机、磁疗机、针灸用具、按摩用具、颈椎牵引设备、腰椎牵引设备、关节功能评定装置、肌力计、高压灭菌设备、密闭灭菌柜、洗衣机，以及常温水、热水、蒸馏水、净化过滤系统。

5.9.2.3 Basic devices: breathing machine, sputum aspirator, ECG machine, defibrillation machine, microscope, blood cell counter, biochemical analyzer, spectrophotometer, autodilutor, electrophoresis instrument, centrifuge, refrigerator, drying oven, water bath box, X-ray machine, A-mode or B-mode ultrasound, posture correcting mirror, wall tensioner, forearm rotation trainer, pulley lifting ring, rowing machine, finger muscle trainer, various walking aids, intermediate frequency therapeutic instrument, ultrashort wave therapeutic instrument, infrared therapeutic instrument, magnetotherapy machine, acupuncture appliances, massage appliances, cervical vertebra traction device, lumbar vertebra traction device, joint function evaluation apparatus, muscle strength meter, autoclaving device, closed sterilization cabinet, washing machine, as well as room temperaturel water, hot water, distilled water, purification and filtration system.

5.9.3 规划设计

5.9.3 Planning and design

5.9.3.1 行业标准：《疗养院建筑设计标准》（JGJ/T 40—2019）。

5.9.3.1 Industrial standard: Standard for Architectural Design of Sanatorium (JGJ/T

40—2019).

5.9.3.2 选址：疗养院与敬老院联合建设，土地资源共享。

5.9.3.2 Location: sanatorium and nursing home will be jointly built and land resources will be shared.

5.9.3.3 功能设计

5.9.3.3 Function design

1. 设疗养/住宅楼、休闲活动场、影剧院、养殖场、种植场、绿地等。

1. Set up convalescence/residential buildings, leisure activity fields, theaters, farms, planting fields, green spaces, etc.

2. 疗养院/楼的一楼设入口区域、公共空间和托儿所，可居住的房间在二层及以上，每个单元设 12 到 14 个床位。这种设计不仅可以有效地利用空间，便于护理人员的工作，还方便居民的日常生活。每个单元内都备有一个开放式厨房、餐厅和起居室，起居室外有一个带防护栏的阳台。

2. The first floor of the sanatorium/building is equipped with entrance area, public space and nurseries. The habitable rooms are on the second floor or above, and each unit is equipped with 12 to 14 beds. This design can not only effectively use the space, facilitate the work of nursing staff, but also facilitate the daily life of residents. Each unit is equipped with an open kitchen, dining room and living room. There is a guard rail equipped balcony outside the living room.

3. 疗养院可以采用宾馆客房式布局，房间带卫生间。

3. Sanatorium can use hotel room style layout and rooms are equipped with toilet.

5.10 应急医疗设施

5.10 Emergency medical facilities

5.10.1 人民防空医疗工程

5.10.1 Medical works for civil air defense

5.10.1.1 应根据战时功能和防护要求划分染毒区、第一密闭区和第二密闭区（即清洁区），相邻两区之间应设置密闭隔墙；当电梯通至清洁区或第一密闭区时，其间应设置密闭通道或应有可靠的封堵措施；当电梯通至染毒区时，其间应设置一道防护密闭门或应有封堵措施。

5.10.1.1 The contaminated area, the first enclosed area and the second enclosed area (i. e. clean area) shall be divided according to the wartime function and protection requirements, and a sealed partition shall be set between the two adjacent areas. When the elevator passes to the clean area or the first enclosed area, a sealed passage should be set up or reliable blocking measures should be taken. When the elevator passes to the contaminated area, a protective sealed door should be set up or blocking measures should be taken.

5.10.1.2 人防医疗工程的室内净高不宜小于2.60m，通行担架的内部通道净宽不宜小于1.80m。

5.10.1.2 The indoor clear height of medical works for civil air defense should not be less than 2.60m, and the clear width of the internal passage for passing stretchers should not be less than 1.80m.

5.10.1.3 出入口和染毒区

5.10.1.3 Entrance and exit and contaminated area

1. 中心医院不应少于3个出入口，且其中至少有2个直通室外地面的出入口，并应分别作为战时的第一、第二主要出入口；急救医院、救护站不应少于2个出入口，且其中至少有1个直通室外地面的出入口，并应作为战时的主要出入口。

1. The central hospitals should not have less than 3 entrances and exits, of which at least two are directly connected to the entrances and exits on the outdoor ground and should be used as the first and second main entrances and exits in wartime respectively. Emergency hospitals and first-aid stations shall not have less than 2 entrances and exits, of which at least one is directly connected to the entrances and exits on the outdoor ground and should be used as the main entrances and exits in wartime.

2. 中心医院的第一主要出入口和急救医院、救护站的主要出入口应设置一道防毒通道，并应与分类急救部相连；中心医院的第一主要出入口应采用坡道式，且宜按通行救护车设计。

2. The first main entrance and exit of the central hospital and the main entrance and exit of the emergency hospital and the first-aid station shall be provided with a gas protection passage, which shall be connected with the classified first-aid department. The first main entrance and exit of the central hospital should be of ramp type and should be designed according to the passing ambulance.

3. 人防医疗工程的各次要出入口、备用出入口以及清洁区与普通地下建筑之间的连通口均应设置一道密闭通道；中心医院第二主要出入口应设置两道防毒通道和洗消间，洗消间各房间的使用面积不宜小于9.0m^2。

3. A sealed passage shall be set up for the secondary entrance and exit, standby entrance and exit and the connection between the clean area and ordinary underground building in the medical works for civil air defense. The second main entrance and exit of the central hospital should be equipped with two gas protection passages and decontamination rooms. The use area of each room in the decontamination room should not be less than 9.0m^2.

(1) 第一密闭区

(1) The first enclosed area

从主要出入口经防护密闭门到第一防毒通道 > 经密闭门到分类厅 > 经密闭门到第二消毒通道 > 经密闭门到清洁区。主要出入口设排风扩散室、防爆波活门、排风竖井。第一防毒通道设污物间、盥洗室、厕所。分类厅设急救观察室、诊疗室、更衣室、淋浴房。

From the main entrance and exit to the first gas protection passage through a protective sealed door > to the classification hal through a sealed door > to the second disinfection passage through a sealed door > to the clean area through a sealed door. The exhaust diffusion room, blast valve, and exhaust shaft are set at the main entrance and exit. The waste room, wash room, and toilet are set at the first gas protection passage. The first-aid observation room, diagnosis and treatment room, undressing room, and shower room are set at the classification hall.

(2) 第二密闭区

(2) The second enclosed area

第二密闭区应位于第一密闭区之后，用第二防护通道和洗消间与第一密闭区相连接，设有手术部、医技部、护理单元和公共设施等。

The second enclosed area shall be located behind the first enclosed area. It is connected with the first enclosed area through the second protective passage and decontamination room, and is provided with operating department, medical technology department, nursing unit, public facilities, etc.

4. 手术部设计

4. Design of operating department

手术部应邻近护理单元设置，并应自成一区；由手术室、洗手间、无菌辅料器械室、麻醉药械室、医护办公室、换鞋处、男女更衣浴厕室、污物室、石膏室等组成；平面布置应符合功能流程和洁污分区要求，入口处应设卫生通过区，换鞋处应有防止洁污交叉的措施和推床的洁污转换措施，通往外部的门应采用弹簧门或自动启闭门；大手术室平面净尺寸不应小于5. 40m ×5. 10m，小手术室平面净尺寸不应小于3. 30m ×4. 80m。

The operating department should be set up adjacent to the nursing unit and should be set in a separate area. It consists of operating room, toilet, sterile dressing and device room, anaesthetic and device room, office for medical care staff, shoes changing place, locker room, bathroom and toilet for men and women, waste room and plaster room. The layout shall conform to the functional process and the requirements for division of clean and contaminated areas. A uncontaminated transition room shall be set up at the entrance. Measures to prevent the cross between clean and contaminated areas and measures of conversion of clean and contaminated areas for pushing the bed shall be taken at the shoes changing place. The door leading to the outside shall be a spring door or an automatic opening and closing door. The net plane size of large operating rooms shall not be less than 5. 40m ×5. 10m, and that of small operating rooms shall not be less than 3. 30m × 4. 80m.

5. 人防急救功能空间关系

5. Spatial relationship of first-aid functions for civil air defense

(1) 空间分为：急救单元、手术单元、门诊单元、住院单元。

(1) Spaces: first-aid unit, operation unit, outpatient unit, and inpatient unit.

(2) 功能包括：急救处理、急救手术、早期治疗、影像检查、住院护理、饮食、住宿。

(2) Functions: first-aid treatment, first-aid surgery, early treatment, imaging, inpatient, food and drink, and accomodation.

(3) 主要用房：心肺复苏室、感染休克抢救室、胸科手术室、骨科手术室、外科处理室、X光透视室、药品发放室、B超室、护士站、住院病房、医护休息室、餐厅。

(3) Main rooms: CPR room, infection shock resuscitation room, operating room for thoracic surgery, orthopedic operating room, surgical treatment room, X-ray room, drug distribution room, B-ultrasound room, nurse station, inpatient ward, rest room for medical care staff, and canteen.

(4) 功能关系：急救、门诊、检查、手术、住院、后勤保障按照战时医疗工艺流程布局。

(4) Function relationship: First-aid, outpatient, diagnosis, surgery, inpatient, and logistics support are arranged based on the wartime medical process flow layout.

5.10.2 突发性公共卫生事件应急预案

5.10.2 Emergency plan for public health emergencies

5.10.2.1 突发性公共事件分四级：Ⅰ级-特别重大、Ⅱ级-重大、Ⅲ级-较大、Ⅳ级一般。

5.10.2.1 Public health emergencies are divided into four levels: Level Ⅰ-extraordinarily serious, Level Ⅱ-serious, Level Ⅲ-relatively serious and Level Ⅳ-general.

5.10.2.2 公共卫生事件：包括瘟疫/传染病、群体不明原因疾病、食品安全与职业危害事故、动物瘟疫以及其他严重影响公众健康和生命安全的事件。

5.10.2.2 Public health emergencies: including plague/infectious diseases, mass diseases of unknown origin, food safety and occupational hazard accidents, animal plague and other events that seriously affect public health and life safety.

5.10.2.3 应急预案：是面对突发性公共卫生事件如自然灾害、重特大事故、环境公害及人为破坏的应急管理、指挥、救援计划等。

5.10.2.3 Emergency plan: it is an emergency management, command and rescue plan for public health emergencies such as natural disasters, extraordinarily serious and serious accidents, environmental hazards and man-made damages.

5.10.3 平疫结合医院

5.10.3 Hospital in combination of peacetime and epidemic time

5.10.3.1 平时作为医院正常业务用房，公共突发事件爆发时，可立即转换成后备传染病楼。既能应对突发事件接诊疑似及传染病人，又能在非疫情时发挥楼宇经济价值，而非闲置，达到平疫结合、平战结合。

5.10.3.1 It can be used as a normal business building of the hospital in peacetime, when a public health emergency breaks out, it can be immediately converted into a standby infectious disease building. It can not only treat suspected and infectious patients in case of emergency, but also give full play to the economic value of buildings when no epidemic occurs, rather than idle, so as to achieve the combination of peacetime and epidemic time as well as peacetime and wartime.

5.10.3.2 对于高层平疫结合医疗建筑，要把后备传染楼、普通病房楼的病人主电梯厅与医护、探访人员专用电梯厅、清洁物品运输电梯厅、手术部专用电梯分开来，要把后备传染楼的患者电梯做成可以双向出入口、双电梯厅的形式，以兼顾平时和疫情期间的使用要求。传染楼设专用出入口大厅，医、患分开，在医护人员专用电梯厅前端和传染病人电梯厅前端各设一组入楼洗消区，经换鞋、一次更衣、淋浴洗消、二次更衣四道程序后方可进入清洁的电梯厅，以保障电梯空间的洁净。

5.10.3.2 For high-rise healthcare buildings in combination of peacetime and epidemic time, the main elevator hall for patients in the standby infectious building and ordinary ward building should be separated from the special elevator hall for medical care staff and visiting personnel, cleaning article transportation and operating department. The elevator for patients in the standby infectious building should be made into the form of two-way entrance and exit and double elevator hall, so as to meet the use requirements in peacetime and epidemic time. The infectious building should be equipped with special entrance and exit hall to separate doctors and patients. A group of entrance decontamination areas should be respectively arranged at the front end of the special elevator hall for medical care staff and the front end of the elevator hall for infectious patients. The clean elevator hall can only be entered after four procedures of changing shoes, changing clothes once, showering and decontamination and changing clothes twice, so as to ensure the cleanliness of the elevator space.

5.10.4 应急医院

5.10.4 Emergency hospital

5.10.4.1 临时医院：搭建临时板房、临时改造非医疗设施。

5.10.4.1 Temporary hospitals: build temporary portable houses and temporarily transform non-medical facilities.

5.10.4.2 移动方舱医院：是解放军野战机动医疗系统的一种，在各种应急救治中也有广泛使用。“方舱医院”一般由医疗功能单元、病房单元、技术保障单元等部分构成，是一种模块化卫生装备，具有紧急救治、外科处置、临床检验等多方面功能。

5.10.4.2 Mobile cabin hospital: it is one of the field mobile medical systems of the PLA and is also widely used in various emergency treatments. “Mobile cabin hospital” is generally composed of medical functional units, ward units, technical support units and other parts. It is a modular health device and has many functions such as emergency treatment, surgical treatment and clinical examination.

5.10.4.3 方舱医院

5.10.4.3 Mobile cabin hospital

1. 指地方上借鉴军方战地救护设施，利用政府公共建筑临时改造建立的临时医院，用以最大限度地扩充收治患者的容量，避免确诊的患者在家庭或者在社会上得不到救治和隔离，而造成对其他人员的交叉感染，让医务人员能够集中为大批患者提供治疗和护理。

1. It refers to temporary hospitals built by drawing lessons from military field rescue facili-

ties and temporarily transforming government public buildings to maximize the capacity of patients to be received and cured, so as to avoid cross infection of other personnel caused by the lack of treatment and isolation of diagnosed patients at home or in society, and to enable medical care staff to provide centralized treatment and care for a large number of patients.

2. 方舱设置有洗消方舱、储物方舱、药品方舱、灭菌方舱、检验方舱、手术方舱、急救方舱、电站方舱等。

2. Cabins such as decontamination cabin, storage cabin, medicine cabin, sterilization cabin, examination cabin, operation cabin, first-aid cabin and power station cabin are arranged.

3. 方舱的组成：急救、手术、检验、X 光等多个模块化系统及辅助设备。

3. The cabin has first-aid, operation, examination, X-ray and other modular systems and auxiliary device.

4. 各个模块化系统既可单独使用，也可无缝连接建成大规模移动医院，既可以共享方舱内的设备资源，又便于对伤病员进行快速、高效处置。

4. Each modular system can be used independently or can be seamlessly connected to build a large-scale mobile hospital, which can not only share the devices and resources in the cabin, but also facilitate rapid and efficient treatment of the sick and wounded.

5. 舱内无菌程度需要时应可以达到三级甲等医院的标准。

5. The degree of sterility in the cabin should be up to the standard of Third-class Level A hospitals when necessary.

6. 方舱内应具备流动水洗手设备、药品及无菌物品贮存、器械消毒灭菌、持续的电源供应等条件，并可以开展手术，进行检验、彩超、X 光射线检查。

6. The cabin shall be equipped with hand washing device with running water, storage of drugs and sterile articles, disinfection and sterilization of instruments, continuous power supply and other conditions, and operation, examination, color ultrasound and X-ray examination can be performed in it.

7. 感染控制

7. Infection control

（1）对既有建筑做洁净和感染分区和流线的临时性分离，在感染区内植入不到顶隔断（1.5～2.5m）把数百上千位患者划分为 20～30 床一组的护理单元，在大空间内附着一两个小房间作为隔离或休息空间。

（1） A temporary separation of clean and infected areas and flow lines should be made for the existing buildings. Unreached partitions (1.5～2.5m) should be installed in the infected areas to divide hundreds of thousands of patients into a group of 20～30 beds of nursing units, and one or two small rooms should be arranged in large spaces as isolation or rest spaces.

（2）患者入院流程方面：医务人员在诊疗过程中，要先排除流感等一些呼吸系统疾病，把重症病人集中到综合实力最强的医疗机构，进入方舱医院的都是确诊的患有相同疾病的患者，不会造成交叉感染。方舱医院设有专门的感染控制团队加以指导，因此不会造成患者之间的交叉感染。

(2) Patient admission process: during diagnosis and treatment, medical care staff should first eliminate some respiratory system diseases such as influenza and concentrate the severe patients in the medical institutions with the strongest comprehensive strength. Patients entering the mobile cabin hospital are patients diagnosed with the same disease and will not cause cross infection. The mobile cabin hospital has a special infection control team to guide, so it will not cause cross infection between patients.

(3) 降低感染率方面：加强疫情的监测，落实各项防控措施，特别是严格隔离疑似患者和密切接触者。

(3) Reduction of infection rate: strengthen the monitoring of the epidemic and implement various prevention and control measures, especially strictly isolate suspected patients and close contacts.

(4) 提高收治率方面：加强预检分诊和发热门诊的管理，提高早期识别和鉴别的能力。

(4) Improvement of admission rate: strengthen the management of pre-examination and triage and fever outpatient, and improve the ability of early identification and discrimination.

(5) 提高治愈率方面：不断优化技术方案、改进诊疗方案。

(5) Improvement of cure rate: continuously optimize the technical proposal and improve the diagnosis and treatment proposal.

(6) 降低病死率方面：进一步加强重症患者的管理，把重症病人集中到综合实力最强的医疗机构，组成国家级的专家组，加强对重症患者医疗救治的指导。

(6) Reduction of fatality rate, further strengthen the management of severe patients, concentrate severe patients in the medical institutions with the strongest comprehensive strength, form a national expert group, and strengthen the guidance for medical treatment of severe patients.

8. 方舱医院医疗单元设置

8. Establishment of medical units in mobile cabin hospitals

(1) 方舱医院按 50 床为一个医疗单元，每单元配 4 医 12 护，护理团队采用“国家支援队 + 地方医疗队”模式，主要专业包括：重症医学科、感染科、呼吸科、循环内科和麻醉科等专业。

(1) In the mobile cabin hospital, every 50 beds are divided into a medical unit and each unit is equipped with 4 doctors and 12 nurses. The nursing team adopts the mode of “national support team + local medical team”. The main specialties include intensive care unit, infection department, pneumology department, circulation internal medical department and anesthesiology department.

(2) 在诊疗流程上，方舱医院和正规医院存在着很大的差别，正规医院的诊疗模式主要采取诊疗小组，包括护理组和医生组，医生组包括主任医师（或者副主任医师）、主治医师和住院医师，一个梯队管理 8 ~ 10 个病人。

(2) In the process of diagnosis and treatment, there are great differences between the mo-

bile cabin hospital and the standard hospital. The diagnosis and treatment mode of the standard hospital mainly adopts the diagnosis and treatment group, including the nursing group and the doctor group. The doctor group includes the chief physician (or deputy chief physician), the attending physician and the resident physician, and an echelon manages 8 ~ 10 patients.

（3）舱内配置观看电视娱乐区、开水供应区、被服衣物食品领取区、医护人员办公区、公共厕所。

（3）The cabin is equipped with television viewing and entertainment, boiled water supply area, bedding, clothing and food collection area, medical care staff office area and public toilets.

（4）因每 200 ~ 300 名患者在同一个病房接受治疗，必须适当分区、相对隔离；由于管理的病人比较多，在医务工作者人数相对较少的情况下，以最大潜力来管理大规模的患者，解决患者多、床位少的难点，床位排布密集，一目了然，从而缩短护理半径，降低护理工作量，提高护理效率。

（4）Since every 200 ~ 300 patients are treated in the same ward, they must be properly partitioned and relatively isolated. Due to the large number of patients under management, when the number of medical care staff is relatively small, large-scale patients should be managed with the greatest potential to solve the difficulties of more patients and fewer beds. Beds should be densely arranged and clear, thus shortening the nursing radius, reducing the nursing workload and improving the nursing efficiency.

9. 方舱医院医疗模式

9. Medical mode of mobile cabin hospital

（1）采用值班模式，一般一天 24h 排四个班次；每个班次五个医生，其中一个医生负责收治患者，为有住院指征的患者开住院证，进入方舱医院治疗，其余医生分区进行患者的管理与治疗；每个医生值班 6h，考虑到穿、脱隔离衣的时间较长，每位医生每天在隔离舱内工作 7 ~ 8h，每个医生管理约 100 个病人。

（1）The duty mode is adopted, with four shifts arranged 24 hours a day. There are five doctors in each shift, one of which is responsible for receiving and curing patients, issuing hospitalization certificates for patients with hospitalization indications, and entering the mobile cabin hospital for treatment, and the other doctors are responsible for patient management and treatment in different areas. Each doctor is on duty for six hours. Considering the longer time to wear and remove isolation gown, each doctor works in the isolation cabin for 7 ~ 8 hours every day, and each doctor manages about 100 patients.

（2）一个护理小组有 9 名成员，负责 300 张床位，护理工作为“一站式服务”：测量生命体征、发放口服药品、采集标本、执行医嘱。

（2）A nursing group has 9 members and is responsible for 300 beds. The nursing work is a “one-stop service”: measuring vital signs, distributing oral drugs, collecting specimens and executing medical advices.

（3）方舱医院收治已确诊的轻型和普通型患者；所谓轻型，主要是症状比较轻，包括发热、咳嗽，少数人伴有腹泻、呕吐的症状；普通型指影像学可见肺炎表现，没有达到重型和危重型的标准。

(3) The mobile cabin hospital will receive and cure the diagnosed light and ordinary patients. The so-called light patients are mainly characterized by mild symptoms, including fever and cough, and a few people accompanied by diarrhea and emesis; ordinary patients refer to the patients with manifestation of pneumonia visible in imaging, without meeting the criteria of severe patients and critical patients.

第6章

医用专项工程设计

Chapter Ⅵ

Medical Special Engineering Design

袁清 骆艳 朱希
Qing Yuan Yan Luo Xi Zhu

6.1 空气净化工程

6.1 Air purification engineering

6.1.1 空气净化工程概要

6.1.1 Summary of air purification engineering

6.1.1.1 空气净化的意义

6.1.1.1 Significance of air purification

为防止空气生物性、物理性和化学性污染可能造成的医院内部感染和健康问题，空气净化技术越来越多地被应用到医院环境的空气卫生控制，区别于工业领域的空气净化概念和室内空气品质的通用概念，医院空气净化的核心概念是消毒杀菌，严格控制微生物污染。医疗净化工程，是指通过空气净化技术提供无菌无尘的医疗环境。医院需要净化处理的场所较多，从专业角度分为两大类，普通空气净化类和专业空气净化类。

In order to prevent infection and health problems in the hospital caused by biological, physical and chemical air pollution, air purification technology is increasingly applied to air hygiene control of hospital environment. Different from the air purification concept and general indoor air quality concept in the industrial field, the core concepts of hospital air purification are disinfection and sterilization, and strict control of microbial pollution. Medical purification engineering refers to providing a sterile and dust-free medical environment through air purification technology. There are many places that need purification treatment in hospitals, which can be divided into two categories from a professional point of view, i. e. ordinary air purification category and professional air purification category.

6.1.1.2 普通空气净化

6.1.1.2 Ordinary air purification

普通空气净化类是指进行一般性的空气净化处理，达到一定的净化及灭菌要求，如：医院的较大聚集性场所、有一定污染源的场所，这类系统往往是通过空调系统结合适当空调处理系统来完成十万级以上的净化空气处理。一般空气处理分为两种模式，一种为中央处理方式，另一种为末端处理方式。中央处理方式成本较低，如在空气处理机上加装某种灭菌及净化装置，优势是整体性处理，但由于空调管路的影响，会影响最终出风口的空气净化结果；末端式空气处理方式较为灵活，如果末端灭菌装置处理效果好、适应性强及相对成本低，可普遍采用。

Ordinary air purification refers to general air purification treatment to meet certain purification and sterilization requirements, such as large gathering places in hospitals and places with certain pollution sources. Such systems often complete purification air treatment of over 100000 level through air conditioning systems combined with appropriate air conditioning treatment systems. General air treatment is divided into two modes, i. e. central treatment mode and terminal treatment mode. The cost of the central treatment mode is relatively low. If a certain sterilization and purification apparatus is additionally installed on the air treatment machine, the advantage is integrated treatment. However, due to the influence of air conditioning pipelines, the air purification result of the final air outlet will be affected. The terminal air treatment mode is relatively flexible. If the terminal sterilization apparatus has good treatment effect, strong adaptability and low relative cost, it can be generally adopted.

6.1.1.3 专业空气净化

6.1.1.3 Professional air purification

专业空气净化是指要达到严格的空气净化及隔离要求，达到严格的净化等级要求及灭菌要求，如：医院净化手术室、ICU、NICU、中心供应室、白血病净化病房、烧伤病房及隔离病房等，包括有生物安全等级要求的实验室及检验室等。

Professional air purification refers to meeting strict requirements for air purification and isolation, as well as strict requirements for purification grade and sterilization, such as hospital purification operating rooms, ICU, NICU, central supply rooms, leukemia purification wards, burn wards and isolation wards, including laboratories and examination rooms with biological safety level requirements.

专业净化工程是以高效空气过滤技术为主，结合工艺流程及压差控制，控制精准的温湿度及严格标准洁净度的专业工程。建造前期要科学规划，专业设计，还要专业施工，专业验收。

The professional purification engineering is a professional engineering mainly based on high-efficiency air filtration technology and combined with process flow and differential pressure control, so as to control accurate temperature and humidity and strictly standard cleanliness. Scientific planning and professional design are required in the early stage of construction, professional construction and professional acceptance are also required.

6.1.2 手术室空气净化

6.1.2 Air purification in operating room

手术室是为病人提供手术及抢救的场所，是医院的重要技术部门。手术室已成为反映一个医院实力和技术水平的重要标志。

Operating room is a place to provide operation and rescue for patients, and is an important technical department of the hospital. Operating room has become an important symbol reflecting the strength and technical level of a hospital.

高效安全的手术室空气净化系统，可保证手术室的无菌环境，从而可以满足器官移植、心脏、血管、人工关节置换等手术所需的高度无菌环境。洁净手术室分级：洁净手术室以空气洁净度与细菌浓度两个主要指标分为四级，并以空气洁净度级别作为必要保障条件。Ⅰ级：特别洁净手术室；Ⅱ级：标准洁净手术室；Ⅲ级：一般洁净手术室；Ⅳ级：准洁净手术室。

An efficient and safe air purification system in the operating room can ensure a sterile environment in the operating room, which can meet the needs of highly sterile environment for organ transplantation, heart, blood vessels, artificial joint replacement and other operations. Classification of clean operating room: the clean operating room is classified into four levels according to the two main indexes of air cleanliness and bacterial concentration, and the air cleanliness level is taken as the necessary guarantee condition. LevelⅠ: specially clean operating room; LevelⅡ: standard clean operating room; Level Ⅲ: generally clean operating room; Level Ⅳ: quasi-clean operating room.

6.1.2.1 标准化模块化装配式手术室相关标准

6.1.2.1 Relevant standards for standardized modular fabricated operating rooms

1. 面积标准：普通手术室标准面积30m^2，大手术室标准面积45m^2，超大手术室标准面积70m^2。标准化手术室的意义在于满足医疗严格要求与建筑条件的最佳配置。

1. Area standard: standard area of general operating room is 30m^2, that of large operating room is 45m^2, and that of super-large operating room is 70m^2. The significance of standardized operating room is to meet the strict medical requirements and the optimal configuration of building conditions.

2. 单间手术室标准化模块化设计。以30m^2标准手术室为例：手术室的长短轴、开门方向，各面墙壁的装置配置，医疗设备装置配置，都以标准化模块化设计。

2. Standardized and modular design of single room operating room. Take the standardized operating room of 30m^2 as an example: the major and minor axes and the direction of opening the door for the operating room. The allocation of all devices on each wall and the allocation of all medical devices and apparatuses are subject to standardized and modular design.

3. 装配式手术室规划：在规划设计阶段必须考虑装配式手术室建设需求，只有完成装配式手术室的精准设计，才会产生装配式手术室的高质高效建设结果。

3. Planning of fabricated OR: The construction requirements of the fabricated operating room must be considered in the planning stage. Only when the precise design of fabricated operating room is completed can the high-quality and efficient construction results of the fabricated operating room be achieved.

4. 手术室数字化运维管理系统建设：基于手术室控制的复杂要求及运维的高可靠性的要求，手术室应建立全数字化、网络化、全信息化的运维保障平台；方便现场的实时便捷的运行管理及远程维护管理；保证手术室系统安全可靠运行，节能环保。

4. Construction of digital O&M management system of the operating room: Based on the complex requirements of operating room control and the requirements of O&M high reliability, the operating room shall be provided with a fully-digital, networked and fully-informationalized O&M guarantee platform; shall facilitate the real-time and convenient operation management and remote maintenance management on the site; and, ensure the safe and reliable operation of the operating room system, and achieve energy conservation and environment protection.

5. 手术室的区域划分

5. Division of operating room

手术室须严格划分为限制区（无菌手术间）、半限制区（污染手术间）和非限制区。

The operating room must be strictly divided into restricted area (sterile operating room), semi-restricted area (contaminated operating room) and unrestricted area.

限制区包括无菌手术间、洗手间、无菌室、储药室等。半限制区包括急诊手术间或污染手术间、器械敷料准备室、麻醉准备室、消毒室。非限制区设更衣室、石膏室、标本间、污物处理间、麻醉复苏室和护士办公室、医护人员休息室、餐厅、手术病人家属休息室等。值班室和护士办公室应设在入口近处。

The restricted area consists of sterile operation room, toilet, sterilizing room, drug storage room, etc. The semi-restricted area consists of emergency operating room or contaminated operation room, instrument and dressing preparation room, anesthesia preparation room and disinfection room. The non-restricted area consists of dressing room, plaster room, specimen room, waste disposal room, post anesthesia care unit and nurse office room, medical care staff rest room, restaurant, rest room for surgical patients' families, etc. The duty room and nurse office room shall be located near the entrance.

6. 手术室的布局

6. Layout of operating room

手术室常规采用双通道方案，无菌手术通道和非洁净处置通道。其中无菌手术通道包括医护人员通道、患者通道、洁净物品供应通道。

The operating room shall be designed into a two-channel structure. It includes sterile operating channel and non-clean disposal channel. The sterile operating channel includes medical care staff channel, patient channel and clean article supply channel.

7. 手术室内装备配置清单

7. Configuration list of equipment in operating room

器械柜 1 台、麻醉柜 1 台、药品柜 1 台、观片灯 1 台、电源插座 3 台（其中 1 台带 380V 插座）、书写台 1 台、多功能控制面盘 1 台、气体终端控制箱 1 台、送风天花板 1 台、自动门 1 樘。部分百级手术室需配保温柜、保冷柜各一台。

1 instrument cabinet, 1 anesthesia cabinet, 1 medicine cabinet, 1 film viewer, 3 power sockets (including one 380V socket) , 1 writing desk, 1 multi-functional control panel, 1 gas terminal control box, 1 air supply ceiling and 1 automatic door. Some hundred-level operating rooms shall be equipped with one thermotank and one refrigerator.

传统建设模式已经远远不能适应快速发展的需求。高质、高效及高标准的手术室建设模式成为快速发展医疗建设的模式。标准化、模块化、工业化手术室预制拼装建造模式是唯一解决途径。

The traditional construction mode is far from meeting the needs of rapid development. The construction mode of high-quality, efficient and high-standard operating room has become a mode for rapid development of medical construction. The only solution is the prefabrication and assembly construction mode of standardized, modularized and industrialized operation room.

6. 1. 2. 2　手术室设计的关键点

6. 1. 2. 2　Key points of operating room design

标准化、模块化、装配式设计是建设高质高效手术室的重要前提条件，一对一的空调系统是防止交叉感染的重要保障条件。气流组织的控制：医院手术室净化工程设计中，空调对手术室空气途径的感染控制有效且不可替代。手术室空调系统服务面积小但能耗大，且使用时间长，因此节能须着重考虑。

Standardization, modularization and fabricated design is an important prerequisite for the construction of high-quality and efficient operating room. One-to-one air conditioning system is an important guarantee condition to prevent cross infection. Control of air distribution: Air conditioning is effective and irreplaceable for infection control of air routes in operating room in hospital operating room purification engineering design. The air conditioning system in operating room has a small service area and large energy consumption. As it will be used for a long term, energy conservation must be considered in particular.

6.1.2.3 手术室空气净化标准化设计

6.1.2.3 Standardized design of air purification in operating room

由于医疗任务的标准化及相关建筑条件的约束，考虑到洁净度的控制及节能的诸多因素，为获得高性能优质性价比的手术室建设效果，按标准化手术室设计是必须的。

Combining the standardization of medical tasks and the constraints of related construction conditions, considering many factors such as cleanliness control and energy conservation of the operating room, the standardized design of operating room must be provided in order to obtain a high performance, high quality and cost-effective operating room construction effect.

1. 手术室面积标准化：$30m^2$ 普通标准手术室、$45m^2$ 标准大手术室、$70m^2$ 复合手术室。

1. Standardization of operating room area: $30m^2$ general standard operating room, $45m^2$ standard large-sized operating room and $70m^2$ compound operating room.

2. 装备标准化：三柜体器械柜、麻醉柜、药品柜，以及手术室内各种装备全部标准固定的安装位置、高度、尺寸标准化配置。

2. Equipment standardization: The three-cubicle instrument cabinet, anesthesia cabinet, medicine cabinet, and all kinds of equipment in the operating room shall be subject to standardized configuration based on the standardized and fixed position, height, and dimensions.

3. 空调设备标准化：手术室一对一空调，既可以满足洁净的需求又可以满足院感高品质高性能的要求。

3. Standardization of air conditioning device: The one-to-one air conditioning device in operating room can meet the requirements of cleanliness and hospital comfort, with high quality and high performance.

4. 控制系统标准化：全数字化、全信息化及网络化控制系统，集控制、管理、远程及实时维修于一体。

4. Standardization of control system: It is a fully-digital, fully-informationalized and networked control system, which integrates control, management, and remote & real-time maintenance.

5. 医疗设备配套标准化：标准化的安装位置及固定点的灯、床、塔的集合。标准化手术室在于满足医疗严格要求与建筑条件的最佳配置，最大限度满足医疗空间及使用的要求。

5. Standardization of medical device supporting components: Set of lamps, beds and towers, with standardized installation positions and fixing points. The significance of standardized operating room is to meet the strict medical requirements and the optimal configuration of building conditions, and to meet the requirements for medical space and its purpose to the greatest extent.

6.1.2.4 手术室模块化设计

6.1.2.4 Modular design of operating room

手术室结构模块化设计：为保证手术室长期的质量稳定性，手术室墙体表面装饰材料宜采用不锈钢，结构采用重型龙骨；预制手术室结构：0.8mm 不锈钢 + 18mm 石膏板 + 60mm × 30mm × 2mm 方通；板间接缝采用医用耐高温 10mm 硅胶多层密封条。工厂一次成型，预制拼装，符合医用消毒和清洁的要求。

Modular design of operating room structure: In order to ensure the long-term quality stability of the operating room, the stainless steel shall be used as the wall surface decoration materials of the operating room, and the heavy keel shall be used as the structure; structure of prefabricated operating room: 0.8mm stainless steel + 18mm gypsum board + 60mm × 30mm × 2mm square tube; the medical high-temperature-resistant 10mm silicone multilayer sealing strips shall be used as the joints between the plates. It shall be formed at the factory at one time, prefabricated and assembled, which meets the requirements of medical disinfection and cleaning.

6.1.3 空气净化设计咨询

6.1.3 Design consultation on air purification

手术室净化工程具有很强的综合性，涉及多个学科及多项专业技术，需要协调各方意见，手术室按标准化设计是统一各方意见的唯一标准。

The operating room purification engineering is highly integrated, involving many disciplines and many professional technologies. It is necessary to coordinate the comments of all parties. The standardized design of the operating room is the only standard to unify the comments of all parties.

6.1.4 ICU 空气净化工程

6.1.4 ICU purification engineering

ICU 重症监护病房，收治内科、外科等各科病人中患有呼吸、循环、代谢及其他全身功能衰竭的病人，并对他们集中进行强有力的呼吸、循环、代谢及其他功能的全医疗任务管理。

ICU (intensive care unit) is designed to receive and cure patients suffering from functional failure in respiratory, circulatory, metabolic and other system from internal medical department, surgical department and other departments, and to intensively carry out strong comprehensive medical task management in respiratory, circulatory, metabolic and other functions.

由于 ICU 重症监护的需求，必须对 ICU 的空气净化处理、空气流向及压差进行控制。这对患者的交叉感染控制有重要意义，也对医护人员有着重要保护作用。

Considering the demand of intensive care in ICU, it is necessary to control air purification treatment, air flow direction and pressure difference in ICU. Such option is of great significance to the control of cross infection of patients and also has an important protective effect on medical care staff.

首先，ICU 应该有特殊的地理位置，设置于方便患者转运、检查和治疗的区域，并考虑以下因素，如接近主要服务对象病区、手术室、影像学科、化验室和血库等，在横向无法实现“接近”时，应该考虑楼上楼下的纵向“接近”；其次，ICU 设有中心监护站及 ICU 开放式病床每床的占地面积为 15～18m^2；ICU 配备必要的正/负压隔离病房，面积为 18～25m^2。ICU 病房的洁净度宜不低于十万级，患者区域与医护人员的办公区域应相对独立，空调须独立设置系统，医患人员公共区域应考虑适当的空气流向处理，设计必须考虑职业保护。

Firstly, ICU shall be located at a special place geographically and set up in an area which is convenient for patient transfer, examination and treatment. The following factors shall be taken into account, for example close to the main targeted inpatient ward, operating rooms, imaging disciplines, laboratories and blood banks. If “close to” cannot be realized horizontally, vertically “close to” shall be considered upstairs and downstairs. Then, ICU shall be equipped with a central monitoring station and ICU open beds, with an area for each bed 15～18m^2. ICU shall be equipped with necessary positive/negative pressure isolation wards, with an area of 18～25m^2. The cleanliness of ICU ward should be no less than 100000 level. The patient area shall be independent from the medical care staff office area. An independent air conditioning system shall be provided. Appropriate air flow processing shall be considered in the common area of patients and medical care staff. Occupational protection shall be considered in the design.

ICU 的基本辅助用房包括医师办公室、主任办公室、工作人员休息室、中央工作站、治疗室、配药室、仪器室、更衣室、清洁室、污废物处理室、值班室、盥洗室等。有条件的 ICU 可配置其他辅助用房，包括示教室、家属接待室、实验室、营养准备室等。辅助用房面积与房面积之比应达到 1.5：1 以上。

The basic auxiliary rooms in ICU include physician’s office, director’s office, staff lounge, central workstation, treatment room, dispensing room, instrument room, dressing room, cleaning room, waste disposal room, duty room, wash room, etc. If possible, other auxiliary rooms may be designed in the ICU, including demonstration classroom, family reception room, laboratory, nutrition preparation room, etc. The area ratio of auxiliary room to ward shall be larger than1.5：1.

ICU 材料及装备设计选择：结构系统宜采用装配式专业净化结构系统，具有表面抗菌性强、易于清洁、大面积无缝等特点，满足净化及快速安装需求，同时保证适当的扩展及可调整性。

Design and selection of ICU materials and equipment: The fabricated professional purification structure system shall be selected, which is easy to clean and seamless in large area, with strong surface antibacterial property, etc. to meet the requirements of purification and rapid installation, and ensure appropriate expansion and adjustability.

ICU 设计关键步骤如下：

Key steps in ICU design is as follows:

第一步，确定医院总体的医疗功能，根据医院的总床位数确定 ICU 的床位数。特别是 ICU 净化工程医疗模式的确定，要明确医院的定位、特长，并考虑未来发展的趋势。

Step 1: Establish the overall medical function of the hospital, and according to the total number of beds in the hospital, determine the number of beds in ICU. In particular, while determining the medical mode of ICU purification engineering, the hospital's positioning and specialty shall be clearly defined and the future development trend shall be considered.

第二步，规划医疗流程、平面交通和功能分布方面的问题，根据场地的特性做相应的平面规划。这个时候往往会受到各种各样的建筑条件限制，比如场地面积非常小，容积率高。核心通道可保证医疗流程的顺畅。

Step 2: Plan the medical flows, plane transportation and function distribution, and make corresponding planning according to the characteristics of the site. At this time, it is often restricted by various construction conditions, such as extremely small site area and high plot ratio. Core channels ensure smooth medical flows.

第三步，规划空气净化的模式。在 ICU 定位的大框架下，开始设计各个相应建设版块的水、气、电等，特别是净化空调的模式，病区一套空调系统，辅助区域一套空调系统，既保证 ICU 净化工程有效满足规范要求，又满足院感的要求避免了交叉感染。两个区的空调系统相互独立。

Step 3: Plan the air purification mode. Under the general framework of ICU positioning, it is necessary to start the design of water, gas and electricity for each corresponding construction area, especially the mode of purifying air conditioning. That is, one air conditioning system shall be provided in the ward and the auxiliary area respectively so as to ensure that the ICU purification engineering can effectively meet the specifications and meet the requirements of hospital comfort, thus avoiding cross infection. The air conditioning systems in both areas shall be independent from each other.

第四步，对结构系统的安装实施第三方检测。

Step 4: Carry out third-party inspection on installation of structure system.

6.1.5 消毒供应室空气净化

6.1.5 Air purification of disinfection supply room

中心供应室即消毒供应室，医院中各医疗部门所使用的清洁物品需在这里经严密消毒，国家有标准的供应室建设验收规范。

The central supply room is the disinfection supply room. Cleaning articles used by various medical departments in the hospital shall be strictly disinfected here. There is standard acceptance criteria for the construction of supply rooms.

消毒供应室设计要点如下：

Key points of disinfection supply room design are as follows:

1. 环境设置要求

1. Environmental setting requirements

中心供应室的位置要布置在一个相对独立、清洁、无污染源、无异味、无粉尘的区域。不可将供应室位置设在穿越交通线的走廊上，供应室的门不应开向公共交通走道。

The central supply room shall be located in a relatively independent and clean environment, free from pollution sources, odor and dust. The supply room shall not be located in the corridor crossing the traffic line, and the door of the supply room shall not be opened to the public traffic passageway.

2. 建筑设计要求

2. Requirements for building design

平面设计应利于消毒供应中心实现“物品由污到洁”的工作流程，不得出现污物交叉或物品逆流；天花板、墙壁应光滑无缝隙，不应有颗粒和纤维脱落，便于清洗和消毒；墙角宜采用弧形设计以减少死角。地面应防滑、易清洗、耐腐蚀。滴漏应采用防反溢式，下水道出口应采取防鼠措施，污水应集中排入医院污水处理系统。灭菌物品存放区不宜设洗手设施。洗手设施建议采用非接触式。电源插座应采用嵌墙式防水安全型，照明设施应满足器械检查等功能需要。

Graphic design shall be conducive for the disinfection supply center to realize the work flow of “articles from dirty to clean”, without dirt mixing or reverse flow of articles; the ceiling and walls shall be smooth and seamless without falling of particles and fibers, and shall be easy to clean and disinfect; the corner of the wall shall be designed into arc structure to reduce dead angle. The ground shall be skid-proof, easy to clean and corrosion-resistant. The floor drain shall be of anti-overflow type, the sewer outlet shall be provided with rat prevention measures, and the sewage shall be intensively discharged into hospital sewage treatment system. Handwashing facilities shall not be set up in the sterilization articles storage area. Non-contact handwashing facilities are recommended. The power socket shall be of recessed waterproof and safe type, and the lighting facilities shall meet the functional requirements such as instrument inspection.

3. 中心供应组成与特点

3. Composition and characteristics of central supply room

医院中心消毒供应室分为污染区、清洁区、无菌区以及工作人员更衣、淋浴卫厕、休息办公等用房。

The central disinfection supply room of the hospital shall be divided into contaminated area, clean area, sterile area, as well as staff dressing room, bathroom, toilet, rest room and office room.

（1）污染区：回收室、洗涤室。

（1）Contaminated area: recovery room and washing room.

（2）清洁区：包装室、敷料室、储藏室。

（2）Clean area: packing room, surgical dressing room and storage room.

（3）无菌区：高压蒸汽灭菌室、发放室。

（3）Sterile area: autoclave steam sterilization room and distribution room.

（4）工作人员单独的更衣间及淋浴卫生间等设施。

（4）Separate dressing room for staff, and facilities such as bathroom and toilet.

4. 中心供应布局形式

4. Layout form of central supply room

（1）周围环境清洁、无污染源，为一个相对独立的区域。

（1）The surrounding environment shall be clean, and free of pollution sources. The central supply area shall be a relatively independent area.

（2）设在全院医疗部门的适中位置，以便节约人力，缩短运输距离。

（2）It shall be located in the middle of medical department of the hospital to save manpower and shorten transportation distance.

（3）小型医院中，适宜将中心消毒供应室设在手术室及产房附近的同一层。

（3）In small-sized hospitals, it is suitable to locate the central disinfection supply room on the same floor near the operating room and maternity ward.

（4）在大型综合医院中，由于手术室对清洁器械和敷料要求供应量大，而且对消毒的质量要求严格。为了减轻供应室的负担，并保证手术室供应物品的无菌质量，多由手术室独自设立辅助中心供应室来单独消毒供应。病房、门诊与医技的消毒供应，则由全院中心消毒供应室负责。

（4）In large-sized general hospitals, considering a large supply demand of cleaning equipment and surgical dressings and strict disinfection quality requirements, in order to reduce the burden of the supply room and ensure the sterile quality of the supplies in the operating room, the operating room shall be provided with an auxiliary central supply room for separate disinfection supply. The disinfection supply of wards, outpatient services and medical technologies shall be the responsibility of the central disinfection supply room of the hospital.

（5）严格按“三区制——污染区、清洁区、无菌区”设置，并应按单向流程布置。工作人员辅助用房自成一区。

（5）It shall be set up in strict accordance with the “three-area system: contaminated area, clean area, sterile area”, and arranged according to one-way flow. Auxiliary rooms for staff shall be separately designed into an area.

（6）进入污染区、清洁区和无菌区的人员应设立卫生通道。

（6）Uncontaminated transition shall be provided for personnel entering the contaminated areas, clean areas and sterile areas.

（7）能保障消毒灭菌所需的供水供气便利。

（7）Guarantee the convenience of water and gas supply necessary for disinfection and sterilization.

6.1.6 设计与管理

6.1.6 Design and management

设计与管理是做好医院净化工程的关键，设计的基点是满足医疗需求，树立以病人为中心的人性化设计理念。应用现代化的科技、信息理念设计实施。须重视前期规划，为业主制订详细的任务书及远近期总体规划，使之有计划、有步骤地逐步过渡，实现规划、方案设计、初步设计、施工图设计的有序递进。

Design and management is the key to do a good job in hospital purification engineering. The basic requirement of design is to meet medical demands and establish a patient-centered people-oriented design concept. The modern science & technology and information concepts shall be used for design implementation. Attention must be paid to the preliminary planning, including preparation of detailed specifications and long-term and short-term overall plans for the owners so as to make a planned and step-by-step transition and realize the orderly progression of planning, SD, PD and CD.

6.1.6.1 清晰医院的净化工程定位管理。

6.1.6.1 Know well about positioning management of hospital purification engineering.

6.1.6.2 难点是工程与医疗装备的配套管理。`

6.1.6.2 The difficult point lies in the supporting management for both engineering and medical device.

6.1.6.3 重点是净化工程的洁净度及压差控制的专项设计管理。

6.1.6.3 The key point lies in the special design management with respect to cleanliness and differential pressure control for purification engineering.

6.1.7 净化工程项目管理

6.1.7 Purification engineering project management

树立设计为项目服务的宗旨，设计应成为项目管理的有效工具。

Establish an objective that the design serves the project. The design shall become an effective tool for project management.

6.1.7.1 设计应满足医疗需求。

6.1.7.1 The design shall meet medical demands.

设计应满足净化工程的一、二、三级的工艺流程。这些对洁污分区以及医患分流有比较高要求的医疗需要深入设计。通过设计的过程再调整医院整体的交通组织和功能规划，使方案趋向完美。

The design shall meet the requirements of the primary, secondary and tertiary technological process for the purification engineering. In-depth design is required for the medical services which have relatively high requirements for division of clean and contamination areas and diversion of doctors and patients. Through the design process, the overall traffic organization and functional planning of the hospital shall be readjusted to improve the proposal.

设计不仅要对净化工程的设计专业到位，而且要精通医疗设备的安装使用条件，实现医疗工程环境与医疗设备融合的一体化设计。

The design shall be not only specialized in the design of purification engineering, but also proficient in the installation and service conditions of medical device, thus realizing the integrated design of medical engineering environment and medical device.

6.1.7.2 设计应满足建筑要求。

6.1.7.2 The design shall meet the building requirements.

确定医院总体的医疗功能，其中包括整体规模、面积指标以及医院的医疗模式。特别是医疗模式的确定，要明确医院的定位、特长，并考虑未来发展的趋势。确定主要科室的核心：如床位数、房间数以及大型设备的数量。医学建造要求转化为各种参数进行设计，材料选择、建造结构及力学和流体力学、电路优化等研究，多元化比较，选择最优模式。

Determine the overall medical function of the hospital, including overall scale, area index and medical mode of the hospital. In particular, while determining the medical mode, the hospital's positioning and specialty shall be clearly defined and the future development trend shall be considered. Determine the core elements of major departments: such as the number of beds, rooms and large-sized devices. The medical construction requirements shall be converted into various parameters for design. The optimal mode shall be selected through diversified comparison after study on material selection, construction structure, mechanics, fluid mechanics and circuit optimization.

6.1.7.3 造价问题

6.1.7.3 Cost

造价的设计也是设计的重要内容之一，应根据医疗任务及投资规模做设计，不仅要体现建筑功能设计还应体现造价设计功能。

Cost design is also one of the important contents of design, which shall be performed according to medical tasks and investment scale. It shall not only reflect the function of building design, but also the function of cost design.

6.2 电离辐射防护与电磁屏蔽

6.2 Ionizing radiation protection and electromagnetic shielding

6.2.1 工程技术概要

6.2.1 Summary of engineering technology

6.2.1.1 X 射线（DR、MRI、DSA、CT 等医疗放射设备）诊断等电离辐射技术是现代医学的重要组成部分，电离辐射医学应用目前已成为最大的人工生成辐射来源。

6.2.1.1 X-ray (DR \ MRI \ DSA \ CT and other medical radiation equipment) diagnosis and other ionizing radiation technologies play a significant role in modern medical sciences. At present, the medical application of ionizing radiation is the largest source of artificially generated radiation.

6.2.1.2 X 射线检查既为疾病的筛查和明确诊断服务，也能够穿透细胞，破坏 DNA 诱发病变，要掌握好适应症，做好防护工作，尽量减少不必要的照射。

6.2.1.2 X-ray examination not only serves for disease screening and definite diagnosis, but also penetrates cells and destroys DNA-induced lesions. In this view, it is necessary to master indications, ensure effective protection and minimize unexpected radiations.

6.2.1.3 医院设备电离辐射自然环境与检查室防护内环境相互构成防护工程，射线防护是一种高能电磁波技术，只要达到对人体可能产生伤害的剂量情况下都应采取必要的防护措施。

6.2.1.3 The natural environment of ionizing radiation of hospital equipment and the protective environment of examination room constitute a protective building. Radiation protection refers to a high-energy electromagnetic wave technology. Necessary protective measures shall be taken once a dose that may cause harm to human body is reached.

6.2.1.4 医疗防护工程建设程序

6.2.1.4 Construction procedures of medical protection engineering

1. 选址：首先是设备选址，考虑到设备的危害性、承重性、运输性，大多数医院放置在一层或地下一层，也可放置在独立的放疗中心。

1. Site selection: Equipment location is the top issue. In consideration of hazard, load capacity and transportation, related equipment will be arranged on the first floor or underground floor or in an independent radiotherapy center in most hospitals.

2. 环境卫生评价报告：防护工程建设前向卫生监督部门报告，医院委托有相应资质的评价单位编写《职业病危害评价报告》，组织专家论证会进行评价，卫生监督部门审批通过后方可建设。

2. Environmental and hygienic assessment report: Before the construction of medical protection engineering, report to the local health supervision authority. The hospital shall entrust an assessment organization with corresponding qualifications to prepare the Occupational Disease Hazard Assessment Report. Meanwhile, organize experts for assessment. The construction can be started only with the review and approval of health supervision authority.

3. 医疗设备选型及设计施工。

3. Medical equipment selection and design.

4. 工程竣工后逐步完成审批、工程验收，医院委托有资质的单位进行防护功能检测，达到《医用 X 射线治疗放射防护要求》（GBZ 131—2017）后并出具相应的检测报告。然后编制《职业病危害防治控制评价效果报告》，组织专家论证进行现场评价，向卫生监督部门提出申请验收。

4. Review, approval and acceptance shall be completed after the project is completed. Then the hospital shall entrust a competent organization for the inspection of protection function, which issues the corresponding inspection report after the project has complied with Require-

ments for Radiological Protection in Medical X-ray Therapy. Then prepare the Report on the Effect of Occupational Disease Hazard Prevention and Control Assessment. Organize related experts for field assessment and submit an application to the health supervision authority for acceptance.

5. 防护工程须拿到工程验收报告、办理大型医疗设备配置证后，方可投入使用。

5. Before the project is put into operation, it is necessary to obtain the acceptance report and the large-sized medical equipment configuration certificate.

6. 人员培训及医院管理。

6. Personnel training and hospital management.

6.2.2 CT机房设计

6.2.2 CT equipment room design

6.2.2.1 选址

6.2.2.1 Site selection

设备位置及平面布置应符合下列要求：宜在底层设置，并应自成一区，且应与门、急诊部和住院部邻近布置，并有便捷联系；有条件时，患者通道与医护工作人员通道应分开设置。

The equipment location and plane layout shall meet the following requirements. In detail, it is recommended that CT equipment should be deployed on the first floor and in an isolated zone. Moreover, the room shall be located close to the outpatient department, emergency department and in-patient department and must have an easy access with these departments. When actual conditions permit, separate the patient passage and medical care staff passage.

6.2.2.2 配置标准

6.2.2.2 Configuration criteria

检查室应配备CT机架、检查床、稳压装置及主机用UPS，控制室和设备间可合用，应配备控制台及办公装置、计算机柜（成像装置）、电源开关柜及相应控制柜。

The examination room shall be configured with CT gantry, patient table, voltage stabilizer and host UPS. One room can be shared as the control room and equipment room. In addition, arrange a console table, related office supplies, computer cabinet (imaging device), power supply cabinet and corresponding control cabinets.

6.2.2.3 建筑要求

6.2.2.3 Building criteria

1. CT室的门净宽不应小于1.20m，控制室门净宽宜为0.90m，应考虑设备运输及进场。

1. The net width of CT room shall not be less than 1.20m. The recommended net width of control room door is 0.90m. Take the equipment transportation and entry into account.

2. 透视室与CT室的观察窗净宽不应小于0.80m，净高不应小于0.60m。在机房内，CT装置宜斜向安放，以利于操作者观察受检者。机房出入门应处于散射辐射相对低的位置。

2. The recommended net width of observation window for the fluoroscopic room and CT equipment room shall not be less than 0. 80m and the net height shall not be less than 0. 60m. It is recommended to place the CT equipment obliquely in the equipment room, thus facilitating the operator to observe the patient. The door of the equipment room shall be set at the position with relatively low scattered radiation.

3. CT 机房应有足够的使用空间，面积应不小于 $50m^2$，单边长度不应小于 4m，机房内不应堆放无关杂物。

3. Ensure a sufficient usable area not smaller than $50m^2$ for the CT equipment room. A single side shall not be less than 4m. Any unnecessary item is not allowed inside the equipment room.

4. 地基基础用不低于 C25 的细石混凝土现场一次浇筑，表面原浆压光。如基础不能做到 500mm 厚，则可采用整间机房地面整体浇筑 300mm 厚 C25 以上细石混凝土。其中基础位置从表面到 150mm 厚处为素混凝土，150mm 以下为钢筋混凝土。设备固定用的膨胀螺栓的膨胀受力点须保证在原楼板中。

4. Use fine stone concrete not lower than C25 to cast the foundation in one process and polish the surface original mortar. If the foundation concrete thickness is less than 500mm, cast the whole equipment room with 300mm-thick fine stone concrete above C25. As for the foundation, cast with plain concrete from the surface to the level of 150mm thickness and with reinforced concrete below the level of 150mm. All expansion stress points of expansion bolts used for fixing the equipment must be located in the original floor slabs.

5. 检查室地面的整体水平度要求：机房一般要求下沉 20cm(包括梁)，设计时需要考虑回填材料的质量。

5. Requirements for the overall levelness of the examination room floor: It is generally required that the settlement of equipment room shall be 20cm (including beams). Take the weight of backfill into account during design.

6. 检查室内吊顶的净高度宜为 2800mm。如配有悬吊式高压注射器，建议此高度不低于 2900mm。所有电缆沟（槽）均需有可打开的外盖，电缆沟（槽）盖板必须有足够的承重能力，建议用 3mm 以上不锈钢盖板。

6. The recommended net height of suspended ceiling inside the examination room is 2800mm. If there is a high pressure injector, the recommended height of suspended ceiling shall not be less than 2900mm. All cable trenches (troughs) must be provided with openable covers. The cover plate of these cable trenches (troughs) must have a sufficient load capacity. Here stainless steel cover plate thicker than 3mm is recommended.

7. CT 室的设计与建设在给排水、暖通、电气、消防以及环境评估和院感等方面的相关要求。

7. Specific requirements of the CT room design and construction for the water supply & drainage, HVAC, electrical engineering, fire protection, environmental assessment and hospital infection control.

6.2.2.4 电气要求

6.2.2.4 Electrical requirements

1. CT 机电子加速器应不少于两个回路供电，其中主机部分应采用专用回路供电。单台 X 射线机供电线路导线截面应按满足 X 射线机电源内阻要求选用，并应对选用的导线截面进行电压损失校验。根据负荷用电性质，在配电设计上有条件时还宜设置备用电源回路，保证事故状态下供电。

1. Make sure that at least two power lines are available for the electronic accelerator of CT equipment. Provide a special power line for the main equipment. The cross section of power cables for a single set of X-ray equipment shall comply with the internal resistance of power supply for X-ray equipment. Meanwhile, perform voltage loss verification to the cross section of the chosen power cables. When actual conditions permit during power distribution design, standby power supply is recommended according to the load characteristics, thus ensuring normal power supply under emergency.

2. X 射线诊断机的线路保护电器，应按该机使用时的瞬时最大电流值进行选择；如果使用快速熔断器作线路保护，可直接以计算所得的瞬时最大电流值，选用快速熔断器。

2. Select the line protector of X-ray equipment according to the maximum transient current while the equipment is running. If a fast fuse is used for line protection, directly calculate the maximum transient current.

3. X 线诊断设备、CT 机、MRI 机、DSA 机、ECT 机等诊疗设备工作室的照明开关，宜设置在控制室内或在工作室及控制室内设双控开关；当为 X 射线机设置配套的电源开关箱时，电源开关箱应设在便于操作处，并不得设在射线防护墙上。

3. All lighting switches of operation rooms for X-ray diagnostic equipment, CT equipment, MRI equipment, DSA equipment, ECT equipment and other diagnosis and therapy equipment shall be set inside the corresponding control room. In detail, install double control switches in these operation rooms and control rooms. When configured for the X-ray equipment, the power switch cabinet shall be located for easier operations. Such cabinet must not be installed on the radiation protection wall.

4. 放射线设备的供电线路应采用铜芯绝缘电线或电缆。

4. Copper core insulated wires or cables shall be used for the power line of X-ray equipment.

5. 在 X 射线机房装设的与 X 射线诊断机配套使用的电源开关和保护装置，应按不小于 X 射线机瞬时负荷的 50% 和长期负荷 100% 中的较大值进行参数计算，并选择相应的电源开关和保护电器。

5. Calculate the parameters of power switch and protection device deployed inside the X-ray equipment room and used together with the X-ray equipment according to the greater one between the two values (not less than 50% of the transient load, and 100% of long-term load of X-ray equipment), and then select the corresponding power switch and protector.

6. 设有射线防护的房间，应采用地面非直通电缆沟槽的布线方式，并应避免直接通向射线防护房间；在1类及2类医疗场所的患者区域内，应做局部等电位联结，并应将下列设备及导体进行等电位联结。

6. For the rooms with radiation protection, power cables shall be laid into the non-straight cable trenches on the floor. Avoid directly laying power cables into the rooms under radiation protection. Set local equipotential bonding in the patient area of medical service centers of Class 1 and 2 and perform equipotential bonding to the following equipment and conductors.

7. 诊疗设备监控系统应对CT机进行监控。

7. The diagnosis and therapy equipment monitoring system shall monitor the CT equipment.

8. 1类和2类医疗场所内，宜提供2个不同电源的照明回路。在1类医疗场所，每个房间宜有1个由安全电源供电的灯具。

8. It is recommended to provide the medical service centers of Class 1 and 2 with two different lighting power lines. In the medical service centers of Class 1, every room shall be provided with one lamp powered by security power supply.

9. X射线诊断等诊疗设备主机室0. 75m水平面照度标准值为300lx; X射线诊室显色指数（Ra）不应小于80。

9. The standard illumination on 0. 75m horizontal plane in the main equipment room of X-ray diagnosis room and other main equipment rooms shall be 300lx. The color rendering index of X-ray diagnosis room shall not be lower than 80.

6. 2. 2. 5 消防设计

6. 2. 2. 5 Fire protection design

1. 建筑内部的配电箱、控制面板、接线盒、开关、插座等不应直接安装在低于B1级的装修材料上，用于顶棚和墙面装修的木质类板材，当内部含有电器、电线等物体时，应采用不低于B1级的材料。

1. It is not allowed to directly install distribution boxes, control panels, junction boxes, switches and sockets onto the decorative materials lower than B1 inside the hospital building. When electrical equipment and power cables are set inside the wooden boards for the decoration of ceiling and wall decoration, the grade of such wooden board shall not be lower than B1.

2. CT机房应采用耐火极限不低于2. 00h的不燃烧体与其他部分隔开。

2. The CT equipment room shall be isolated with other parts by nonflammable materials with the fire resistance rating not shorter than 2. 00h.

6. 2. 2. 6 环评防辐射要求

6. 2. 2. 6 Requirements for radiation protection in the environmental assessment

1. 机房外的人员可能受到照射的年有效剂量应小于0. 25mSv.

1. The annual effective dose that the personnel outside the equipment room may be exposed to shall be less than 0. 25mSv.

2. 在X射线机室、同位素治疗室、电子加速器治疗室、CT机扫描室的入口处，应设置红色工作标志灯。标志灯的开闭应受设备的操纵台控制。

2. Set red operation indicators at the entrance of X-ray equipment room, isotope treatment room, electronic accelerator treatment room and CT equipment room. The ON/OFF status of red operation indicators shall be controlled by the console of corresponding equipment.

3. CT机房的出入门和观察窗应与同侧墙具有同等的屏蔽防护。防护窗应略大于窗口，防止窗与墙接壤缝隙泄漏辐射。

3. The entrance/exit and observation window of CT equipment room must have the same strength of radiation protection with the same side wall. The observation window shall be slightly larger than the room window to prevent radiation leakage from the clearance between window and wall.

4. CT机房设有吊顶时，可在吊顶处采取附加屏蔽，使吊顶和机房顶的总屏蔽满足对屋顶上方驻留人员的防护要求。

4. When there is suspended ceiling in the CT equipment room, set an additional shield at the ceiling, thus ensuring the total shielding at the suspended ceiling and room roof can meet the requirement for protecting the personnel on the floor above from radiations.

5. 在距机房外表面0.3m处，空气比释动能率小于7.5uGy/h。

5. The specific kinetic energy released in atmosphere shall be less than 7.5uGy/h 0.3m away from the external surface of the equipment room.

6. 200mA及以上的摄影用X射线设备应有可安装附加滤过板的装置，并配备不同规格的附加滤过板。

6. X-ray equipment for tomography at 200mA and above shall be provided with the device to install additional filter plates. Moreover, prepare additional filter plates of different specifications.

7. CT机房有用/非有用线束方向铅当量为2mm(一般工作量)和2.5mm(较大工作量)。

7. The lead equivalent in useful/useless wiring harness directions of CT equipment room shall be 2mm (general workload) and 2.5mm (greater workload).

8. 机房门外应有电离辐射警告标志、放射防护注意事项、醒目的工作状态指示灯，灯箱处应设警示语句。

8. Set an ionizing radiation warning sign, radiation precautions and eye-catching operation status indicator outside the door of equipment room and set warning words on the light box.

9. 机房门应有闭门装置，且工作状态指示灯和与机房相通的门能有效联动。

9. The equipment room shall be provided with door closer. The operation status indicator shall be effectively interlocked with the equipment room door.

6.2.2.7 CT设备设计要求

6.2.2.7 CT equipment design requirements

1. 设备布局：根据建筑及设备选型布局。

1. Equipment layout: Deploy CT equipment according to the building structure and equipment selection.

2. 电源要求：CT 设备提供两套供电回路，三相四线制或三相五线制，50m 回路单根铜线经 $120mm^2$，90m 回路单根铜线经 $185mm^2$，地线阻值不大于 2Ω。

2. Power supply: Provide two three-phase four-wire or five-wire power lines for the CT equipment. For a single 50m power line, the cross section of its copper wire is $120mm^2$. For a single 90m power line, the cross section of its copper wire shall be $185mm^2$. The resistance of grounding cable shall not exceed 2Ω.

3. 环境要求：温度 18 ~26℃，湿度 30% ~60%，根据设备产热量建议设备间空调制冷量不小于 5 匹。

3. Environmental requirements: Temperature 18℃ to 26℃; Humidity 30% to 60%. The recommended air cooling capacity of air conditioning unit for the equipment room shall not be less than about 4000W.

4. 地面承重：每平方米承重不小于 1500kg。

4. Floor load capacity: The load capacity per square meter of floor shall not be less than 1500kg.

6.2.3 MRI 机房设计

6.2.3 MRI equipment room design

6.2.3.1 MRI 磁共振成像是利用生物磁自旋的成像技术及原子核自旋运动的特点，在外加磁场内，经射频脉冲冲激后产生信号，用探测器检测并输入计算机，经过计算机处理转换后在屏幕上显示图像。

6.2.3.1 MRI is an imaging technology using biological magnetism spin. Signals are generated after an external magnetic field is excited by RF pulses through the nuclear spin. Then detectors are used to detect and output the signals to a computer. After the signals are processed and converted by the computer, an image is displayed on the computer screen.

6.2.3.2 核磁共振 MRI 是大型精密医学影像诊断设备，应通过科学设计医用磁共振设备机房并合理布局安装 MRI 设备，保证 MRI 设备生成图像的质量，同时保障机房其他设备的正常运行及工作人员、患者和公众人员的身体健康。

6.2.3.2 MRI is a type of large-sized high-precision medical image diagnosis equipment. Implement scientific designs for the MRI equipment room for medical purpose and reasonably deploy and install MRI equipment to ensure the imaging quality and the normal operation of other equipment in the equipment room and protect the health of medical care staff, patients and the public.

6.2.3.3 MRI 机房设计要求

6.2.3.3 MRI equipment room design requirements

1. 选址：要考虑周围环境对磁共振设备的影响，首先是关注周围道路上移动车辆对 MRI 设备的影响，其次是附近楼内电梯、建筑设备的影响，宜距离机房 10m 以外。磁共振设备质量为 6 ~8t, 设备的安装运输通道也是提前考虑场地选址的条件之一，

1. Equipment room selection: Take into account the impact brought by the surrounding environment on MRI equipment. At first, focus on the impact of running vehicles on the roads around on the MRI equipment. Then consider the impact caused by elevators inside buildings

nearby and building equipment, which shall be 10m far away from the equipment room. A set of MRI equipment weighs 6 to 8 tons. The passage for the equipment transportation and installation is one of priorities during the site selection.

2. 各辅助检查室的位置：满足空间面积的需要，考虑空调及通风安全。

2. Location of all auxiliary examination rooms: Ensure a sufficient space area and take into account the safety of air conditioning and ventilation.

3. 平面布局：宜根据医院自己的具体需要、未来的发展空间、建筑物的整体布局及投资费用等方面综合考虑，选择适合本身的平面布局。

3. Plane layout: It is recommended to select an appropriate plane layout based on the actual need, future development space, overall building layout and specific investment of the hospital.

4. 普通 MRI 机房的面积，一般要求在 $45m^2$ 以上，柱子轴距宜 8500mm 以上。

4. In most cases, a common MRI equipment room shall be larger than $45m^2$and the spacing between columns shall be larger than 8500mm.

5. 结构设计：选择悬挂磁体时，还需对滑动轨道固定钢梁进行设计，钢梁同时需满足强度、挠度及磁体设备的功能要求；选择固定磁体时，应进行降板降梁设计，满足磁体下地面钢筋含量不超过设备厂家规定的要求。

5. Structural design: When there is suspended magnet, it needs to design the steam beam of sliding track. Make sure that the steel beam can meet the requirements for strength, deflection and magnet function. While fixing the magnet, reduce the elevation of floor slab and beam and make sure that the content of reinforcement for the floor under the magnet does not exceed the limit specified by the equipment manufacturer.

6. 电气设计：按照磁体设备要求，宜采用从配电室直接铺设电缆供应，保证独立供电，且应选择优质的多股铜芯电缆，而其他设备（包括空调机组、照明等）可由大楼配电箱供电；主机宜采用两路供电，冷水机组应采用两路供电。

6. Electrical design: According to the specific requirements from the magnet, it is recommended to directly lay power cables from the power distribution room to ensure an independent power supply. Meanwhile, Select high-quality multi-strand copper core power cables. Other equipment (including air conditioning unit and lighting) can be powered by the power distribution box of the building. Ensure two power lines for the main equipment and two power lines for the water chilling unit.

装修材料中的龙骨、装饰面板均应采用非磁材料，室内照明灯具需综合考虑磁体扫描对室内照明的不同需求。

All keels and decorative panels as decorative materials shall be non-magnetic materials. Interior lamps shall be set in response to actual needs of magnet scanning for interior lighting.

7. 电磁屏蔽处理：MRI 室的六面屏蔽壳体所采用的屏蔽板（包括壁板、顶板、底板）必须由具有良好导电导磁性能的金属网或金属复合材料构成，包括 MRI 的自动移动门和诊断室窗户，同样需按 MRI 特殊要求进行电磁屏蔽处理，为此 MRI 采用射频屏蔽门及屏蔽窗，所有进入室内的净化管道、电气管线、控制线、信号线、各种医疗气体管道都必须采用非磁性材料处理，且电线电缆需要经过电源滤波器后方能接入区域。空调净化送风口、回风口须装通风波导。MRI 屏蔽壳体应采用单点接地，其接地电阻≤2Ω，必须小于避雷接地的接地电阻。屏蔽壳体未与地连接时，其与地线间的绝缘电阻应不小于 10kΩ。

7. Electromagnetic shielding: All shielding slabs (including wall slab, roof slab and base slab) for the six shielding enclosures of MRI equipment room shall be made from metal meshes or metal composite materials with excellent electromagnetic conductivity. Other parts, including automatic sliding door of MRI equipment room and window of diagnosis room, shall be protected against electromagnetic radiation according to the special requirements of MRI. Therefore, install RF shielding door and shielding window for the MRI equipment room. All purification pipelines, electrical pipelines, control cables, signal cables and different types of medical gas pipelines into the equipment room must be made from non-magnetic materials. All power cables must be configured with a power filter before being connected into the equipment room. Ventilation waveguides must be installed at the air inlet and outlet of air conditioning unit. The shielding enclosure of MRI equipment shall be grounded through a single contact with the grounding resistance not more than 2Ω and less than the grounding resistance for lightning protection. If the shielding enclosure is not connected to ground, the insulation resistance between the enclosure and grounding cable shall not be less than 10kΩ.

6.2.4 专项设计咨询

6.2.4 Special design consultation

6.2.4.1 融合设备安装条件及建筑环境条件，深入了解具体的医疗流程、电离辐射防护与电磁屏蔽的医疗使用功能。熟悉设备的参数及安装条件，把握好建筑与设备安装条件相互融合的地方。

6.2.4.1 Integrate the equipment installation conditions into the building environment conditions and gain a deep understanding towards specific medical flows and medical functions of ionizing radiation protection and electromagnetic shielding. Get familiar with the equipment parameters and installation conditions and implement effective control over the integration of building conditions and equipment installation conditions.

6.2.4.2 医学实际应用及扩展

6.2.4.2 Practical medical applications and expansions

在医院建筑设计中，主要是流线组织、功能分区和建筑空间的合理设计。医学的实际应用需要提供专业的设计服务。在医院防护建筑设计中，首先，建筑设计师应重视并熟悉医疗 CT、MRI 的功能，使其与业主的需求相协调，而不是被动地受业主的影响。建筑设计需最大限度地满足医疗和防护的需求，适应医院的未来发展。

Streamline organization, function division and reasonable building space design are the main factors of hospital building design. With respect to practical applications, hospital buildings shall feature professional design services. During the hospital protective building design, the architectural designers shall firstly attach importance to and be familiar with the functions of CT and MRI to be coordinated with the Owner' s need rather passively affected by the Owner. The architectural design scheme shall meet the greatest requirements for medical and protective purpose and adapt to the hospital development.

6. 2. 4. 3 医疗建筑防护设计材料选择

6. 2. 4. 3 Selection of protective design materials for medical buildings

1. 射线防护混凝土相对密度较大，一次成型可以有效地节约空间，成本低但不宜更改，铅板防护材料的选择具有施工方便适合做机房的顶部及地面防护、稳定性好，节省机房空间，但成本较高，适宜改造工程。防护材料钡沙，主要成分硫酸钡，防护效果好，但施工难度大，适合做机房墙体及地面。钡沙成本低，适宜新建项目的设计使用。防护观察窗广泛用铅玻璃有较好的透光性，方便医护人员观察检查。

1. The concrete for radiation protection is characterized by a large bulk density. One-step casting can effectively save space and needs low cost. Concrete protective structure cannot be easily changed. However, lead protective plate can be easily installed and used for the roof and floor of equipment room for its excellent stability. Moreover, it saves equipment room space but requires a high cost. In this view, it is suitable for reconstruction projects. As a type of protective material, barium sulfate sand mainly consists of barium sulfate and shows a high efficiency of protection. But the construction difficulty is great. It can be used for the wall and floor of equipment room. It is cheap and suitable for new projects. Lead glass with high light transmittance is used for the radiation protection of observation windows, thus facilitating the medical care staff for observation and inspection.

2. 现代医院建筑防护设计应更加注重室内公共空间的处理，为患者创造舒适宜人的医疗环境，缓解患者入院前的焦虑，增强患者战胜疾病的信心。

2. During the protection design for modern hospital buildings, more attention shall be paid to handling the interior public space, creating a comfortable and pleasant medical environment for patients, easing the anxiety of patients before admission and boosting the confidence of patients in overcoming diseases.

6. 3 核辐射防护与衰变处理

6. 3 Nuclear radiation protection and decay treatment

6. 3. 1 概要

6. 3. 1 Summary

6. 3. 1. 1 核医学是利用核技术来诊断、治疗和研究疾病的一门学科，主要分为治疗及诊断两大类，均需要利用放射性元素。

6. 3. 1. 1 Nuclear medicine refers to the science of using nuclear technology for disease diagnosis, treatment and study. Its main functions include treatment and diagnosis, both of which need radioactive elements.

6.3.1.2 核医学分子影像技术主要包括 PET(正电子发射计算机断层扫描仪) 和 SPECT(单光子发射计算机断层扫描仪) 两大检查技术，在分子影像学研究中占据着极其重要的地位。在放射性粒子植入治疗的内照射治疗中，放射性核素所发射的射线和显像用的射线种类完全不同，其特点是杀伤力强，射程短，仅能杀死 2 ~ 3mm 内的病变组织，而对周围的正常组织没有影响，就像导弹定向摧毁目标一样。

6.3.1.2 Nuclear medicine molecular imaging technology mainly includes PET (positron emission computed tomography) and SPECT (single photon emission computed tomography), which play an extremely important role in the molecular imaging research. During internal radiation therapy, radioative implant is put inside the body. The radiation emitted by radioactive particles is completely different from that used for imaging. It has strong lethality and short range and can only kill lesion tissues within 2 ~ 3mm, but does not affect normal tissues around it, just like missiles destroying targets directionally.

6.3.2 核辐射防护工艺流程

6.3.2 Process flows of protection against nuclear radiation

核医学工作场所依据管理需要可分为三区，即控制区、监督区和非限制区。

Nuclear medicine workplace can be divided into three types of area according to actual management requirements, including controlled area, supervised are and unrestricted area.

6.3.2.1 控制区

6.3.2.1 Controlled area

在其中连续工作的人员一年内受到的照射剂量可能超过年限值 3/10 的区域，如制备、分装放射性药物的操作室、给药室、治疗病人的床位区等。

It includes the areas where the continuous working personnel may be exposed to more than 3/10 of the annual limit in one year, for example the operation room for the preparation and distribution of radioactive medicines, dispensing room and patient bed area.

6.3.2.2 监督区

6.3.2.2 Supervised area

在其中连续工作的人员一年内受到的照射剂量一般不超过年限值 3/10 的区域，如使用放射性核素的标记实验室、显像室、诊断病人的床位区、放射性核素或药物的储存区、放射性废物储存区等。

It includes the areas where the continuous working personnel may be exposed to not more than 3/10 of the annual limit in one year, for example labeling laboratory using radioactive nuclide, imaging room, patient diagnosis bed, storage area of radioactive nuclide or medicine and storage area of radioactive wastes.

6.3.2.3 非限制区

6.3.2.3 Unrestricted area

在其中连续工作的人员一年内受到的照射剂量一般不超过年限值 1/10 的区域，如工作人员办公室、电梯、走廊等。

It includes the areas where the continuous working personnel may be exposed to less than 1/10 of the annual limit in one year, for example medical care staff offices, elevators and corridors.

6.3.3 防护设计

6.3.3 Protection design

6.3.3.1 防护结构设计：按照工作场所分级对活性实验室、病房、洗涤室、显像室等场所室内表面及装备结构分等级进行防护设计，根据使用放射性核素的种类、特性和活度，确定防护等级。

6.3.3.1 Protective structure design: Perform classified protective designs to interior surfaces of radioactivity laboratories, wards, washing rooms and imaging rooms and to equipment structures according to the classification of workplaces. Meanwhile, determine the level of protection according to the type, feature and activity of radioactive nuclide in use.

6.3.3.2 空气防护设计：对控制区、监督区和非限制区的空气流向做压差控制，生产和操作放射性核素或药物的通风橱，在半开的条件下风速不应小于1m/s，污染的空气应集中高空排放，排气口应高于附近50m范围内建筑物的屋脊3m，并设有活性炭过滤装置或其他专用过滤装置，排出空气浓度不应超过有关限值。

6.3.3.2 Air protection design: Implement differential pressure control towards the air flow direction of controlled areas, supervised areas and unrestricted areas. Make sure that the air speed is not lower than 1m/s when the ventilating cabinet for preparing and operating radioactive nuclide or medicine is half open. The contaminated air shall be drained to a high place. The air outlet shall be 3m higher than the highest ridge of buildings 50m around and be configured with active carbon filter or other special filter. The concentration of drained air must not exceed specific limit.

6.3.3.3 污染物排放设计：接受治疗的病人应使用专用便器或设有的专用浴室和厕所。其排放管道应做防护处理，排污物应通过专门衰变池处理后方能正常排放。

6.3.3.3 Pollutant discharge design: All patients under treatment shall use special urinals or be provided with special bathroom and toilet. The discharge pipeline shall be protected and all pollutants shall be treated in a special decay tank and then normally discharged.

6.3.3.4 放射性物质的储存室应定期进行剂量监测及警示，无关人员不得入内。

6.3.3.4 Perform dose monitoring and warning on a regular basis inside the storage room of radioactive substances. Any irrelevant personnel shall enter the storage room.

6.3.4 衰变处理

6.3.4 Decay treatment

核医学科专用放射性污水池（衰变池）设计方法：根据核医学诊疗所使用的放射性核素种类及其废液和其他污水排放量，并按放射性废液排放限值要求，设计放射性污水池的结构和容积，同时避免使用繁复的数学计算过程。核医学科排放的放射性废液及污水，流经三级结构的衰变池后，必须满足国家标准排放要求。

Sewage tank (decay tank) shall be designed for nuclear medicines. As for the design method, the structure and volume or radioactive sewage tank shall be designed according to types of radioactive nuclide used in nuclear medical diagnosis and treatment and the emission volume of other types of sewage as well as the emission limit of radioactive waste liquid, while complicated mathematical calculation process shall be avoided. Radioactive waste liquid and sewage discharged from nuclear medicine must meet the national standard for discharge after having flowed through the three-level decay tank.

6.3.4.1 液体衰变池设计

6.3.4.1 Liquid decay tank design

1. 原则上要求衰变池的结构和容积必须保证核医学科所排放的放射性废液满足国家医院放射性废水的排放标准。衰变池的设计应满足以下要求：

1. In principle, the structure and volume of the decay tank must ensure the radioactive waste liquid from nuclear medicines meets the national emission standard for the radioactive wastewater from hospitals. The decay tank design shall meet the following requirements:

(1) 衰变池采用三级分隔连续式衰变池，池内设导流墙，推流式排放。衰变池的容积按医院放射性废水可排放标准浓度计算。

(1) A three-stage separation continuous decay tank is chosen with a diversion wall inside for thrust-flow discharge. The volume of the decay tank is calculated according to the standard concentration of radioactive wastewater discharged from the hospital.

(2) 根据国家环保总局 2003 年发布的《医院污水处理技术指南》，医院放射性废水可排放浓度范围为 $3.7\times10^2\sim3.7\times10^5$Bq/L。

(2) According to the Technical Guidelines for Hospital Wastewater Treatment issued by the State Environmental Protection Administration in 2003, the allowable discharge concentration of hospital radioactive wastewater is $3.7\times10^2\sim3.7\times10^5$Bq/L.

(3) 在衰变池前设置化粪池，用以沉淀消化固形物，其所含的放射性也得以衰减并防止固形物进入衰变池。

(3) Set a septic in front of the decay tank to precipitate and absorb solid matters, and the radioactivity contained in the septic tank can be attenuated and the solid matters can be prevented from entering the decay tank.

2. 衰变池的容积计算方法

2. Decay tank volume calculation method

衰变池可控容积范围为 $2\sim5m^3$(具体建设容积大小可根据医院的实际情况选定)。

The controllable volume of the decay tank is $2m^3$ to $5m^3$ (the specific volume can be determined in response to the actual need of the hospital).

3. 衰变池结构设计

3. Structural design of the decay tank

由于放射性废液排入到衰变池是随时间变化的动态过程，为了使衰变池的物理结构更加符合污水排入、流动、核素衰变的动态过程，将衰变池设计为三级连续式放射性衰变池。收集放射性废水的管道应采用耐腐蚀的特种管道，一般为不锈钢管道或塑料管；放射性污水池每池容积为总容积的 1/3（废水液面距离盖板的空间不算入其中）．考虑有生活粪便污水一并排出，在衰变池前设置专用化粪池，用以沉淀消化固形物，专用化粪池内沉渣如难以排出，可进行酸化，促进排入衰变池。其所含的放射性也得以衰减并防止固形物进入衰变池，衰变池必须恰当选址，池底和池壁应坚固、耐酸碱腐蚀和无渗透性，应有防止泄漏措施，化粪池进水口略高于出水口 5cm，池盖与废水液面应有 20cm 的距离，衰变池盖板的混凝土厚度不小于 30cm。如根据医院要求需加厚混凝土的屏蔽厚度，则根据屏蔽计算公式：$N = N_0 e^{-\mu d}$。式中，N_0，N 分别为屏蔽层前后探测器所测到的 γ 射线剂量率（或注量、注量率等），μ 代表能量为 Eγ 的光子在密度为 ρ 的物质中的线衰减系数，d 为屏蔽厚度。

Since the discharge of radioactive waste liquid into the decay tank is a dynamic process with changing time, the decay tank is designed as a three-stage continuous radioactive decay tank to make its physical structure better conform to the dynamic process of sewage discharge, flow and nuclide decay. The pipes for collecting radioactive wastewater shall be corrosion-resistant special pipes. In general, they are stainless steel pipes or plastic pipes. The volume of each radioactive sewage tank is one third of the total volume (the space between the wastewater level and the cover slab is not included). Considering that domestic excrement and sewage are discharged together, a special septic tank is arranged in front of the decay tank to precipitate and absorb solid matters. If the sediment in the special septic tank is difficult to discharge, acidification can be carried out to promote the discharge into the decay tank. The radioactivity contained therein can also be attenuated and solid matters can be prevented from entering the decay tank. The decay tank must be properly located. The bottom and wall of the tank must be firm, acid and alkali resistant and impermeable. Measures shall be taken against leakage. The water inlet of septic tank is slightly higher than the water outlet by 5cm. The distance between the tank cover and the wastewater level shall be 20cm. The concrete thickness of the decay tank cover slab shall not be less than 30cm. If the concrete shield needs to be thickened at the request of the hospital, please refer to the shielding calculation formula: $N = N_0 e^{-\mu d}$. In the formula, n is the γ-ray dose rate (fluence or fluence rate) measured by the detector before and after the shielding layer respectively, μ represents the linear attenuation coefficient of photons with energy Eγ in the substance with the density of ρ, and d is the shielding thickness.

4. 材料要求

4. Material requirements

（1）外围池壁及中间隔层墙体均采用 Mu10 砖，M10 水泥砂浆砌筑（有无地下水均同）。

（1）Mu10 bricks are used for the peripheral tank wall and laid one by one with M10 cement mortar (the same method for the cases with or without groundwater).

（2）预制混凝土强度为 45MPa，抗冻等级为 F200。

（2）Select the 45MPa precast concrete with the antifreeze grade of F200.

（3）底板：无地下水时采用 C10 混凝土，有地下水时用 C15 混凝土，下面均素土夯实。

（3）Base slab: Use C10 concrete when there is no groundwater, use C15 concrete when there is groundwater, and tamp the bottom with plain soil.

（4）抹面：内壁一律用 1∶2 水泥砂浆抹面 15mm 厚，无地下水时一律原浆勾缝。

（4）Plaster: The inner wall shall be plastered with 1∶2 cement mortar to a thickness of 15mm, and caulked with original mortar when there is no groundwater.

（5）根据医院可提供建设衰变池的位置，衰变池的长、宽及深度可略为调整，但总容积则仍需按要求设计。

（5）The length, width and depth of the decay tank can be slightly adjusted according to the location of the decay tank provided by the hospital, but the total volume still needs to be designed.

（6）收集放射性废水及粪便的管道应采用耐腐蚀的特种管道，一般为不锈钢管道或塑料管；建议采用 PVC 下水管，粪便器以后的管道尺寸为 $DN150$-$\phi168$mm（DN 的含义：代表管材的公称直径，符号 ϕ: 代表管材的外径表示）。

（6）The pipes for collecting radioactive wastewater and excrement shall be corrosion-resistant special pipes. In general, they are stainless steel pipes or plastic pipes. It is recommended to use PVC downcomer, and the pipe size after the fecal disposal device is $DN150$-$\phi168$mm（DN: the character ϕ represents the nominal diameter of the pipe and DN represents the outer diameter of the pipe）.

（7）如粪便立管不能用作粪便透气管或无粪便立管的厕所应加设透气管。

（7）Set a vent pipe if the excrement downcomer cannot be used as an excrement downcomer or there is no excrement downcomer.

6.3.4.2 固体衰变池设计方案

6.3.4.2 Solid decay tank design scheme

1. 污物桶收集

1. Collection by means of contaminant tub

（1）设置 1 个供收集废物的污物桶（建议设置 2 个污物桶，方便备用），且必须具有外防护层和电离辐射标志。

（1）Set a tub for the collection of wastes（but it is recommended to set two such tubs with one for standby）and make sure that the tub has outer protective layer and ionizing radiation warning sign.

（2）污物桶放置点应避开工作人员作业和经常走动的地方。

（2）The contaminant tub shall be set to avoid the locations where the staff work and move frequently.

（3）污物桶内应放置专用塑料袋直接收纳废物，装满后的废物袋应及时转送储存室。

（3）Special plastic bags shall be placed in the contaminant tub to directly store the waste, all full waste bags must be timely transferred to the storage room.

(4) 废物袋或废物包、废物桶及其他存放废物的容器必须在显著位置标有废物类型、核素种类、比活度范围和存放日期的说明，内装注射器及碎玻璃等物品的废物袋应附加外套。

(4) Waste bags, contaminant tubs and other containers for waste storage must be marked with the descriptions of waste type, nuclide type, specific activity range and storage date at prominent positions. Waste bags containing injectors, broken glass and other similar items shall be sheathed.

(5) 废物经过存放衰变，比活度降低到 7.4×10^4Bq/kg 以下后，即可作医用非放射性废物处理。

(5) After the decay through storage is realized and the specific activity is lower than 7.4×10^4Bq/kg, the waste can be treated as medical non-radioactive wastes.

(6) 衰变池收集。若医院节约放射性固体收集装置的开支，建议建造一个放射性固体废物收集和存放的衰变池，衰变池必须恰当选址（应避开工作人员作业和经常走动的地方）。

(6) Radioactive waste collection by means of decay tank. If the hospital plans to save expenses on radioactive solid collection devices, it is recommended to build a decay tank for solid radioactive waste collection and storage. The decay tank must be properly located (to avoid the places where the staff work and move frequently).

2. 材料要求

2. Material requirements

① 外围池壁均采用 Mu10 砖、M10 水泥砂浆砌筑（有无地下水均同）；

① Mu10 bricks are used for the peripheral wall and the intermediate partition wall, and laid one by one with M10 cement mortar(the same method for the cases with or without groundwater);

② 衰变池池盖采用预制混凝土，强度为 45MPa, 抗冻等级为 F200, 池盖上必须具有铅板外防护层和电离辐射标志；

② Precast concrete with the strength of 45MPa and the antifreeze level of F200 is used for the cover of the decay tank. The cover must be provided with an outer lead plate protective layer and an ionizing radiation warning sign;

③ 根据医院可提供建设衰变池的位置，衰变池的长、宽及深度可略微调整，但总容积仍需按要求设计。

③ The length, width and depth of the decay tank can be slightly adjusted according to the location of the decay tank provided by the hospital, but the total volume still needs to be designed as required.

6.3.5 专项设计咨询

6.3.5 Special design consultation

6.3.5.1 在建设设计中考虑辐射安全技术、医护人员防护、防护材料选择、合理优化防护施工，以及安全监测与事故预警应急机制。

6.3.5.1 Radiation safety technology, protection of medical personnel, selection of protective materials, reasonable optimization of protective engineering construction, and safety monitoring and early accident warning and emergency response mechanisms shall be considered in the construction design.

6.3.5.2 环境和卫生评价部门出具报告。

6.3.5.2 The environment and health assessment department shall issue a report.

6.3.5.3 选择防护学科的地理位置，医院内选址，在满足规范的同时达到便捷流程合理。

6.3.5.3 As for the geographical location of protection specialty, select a site inside the hospital to not only meet relevant codes but also ensure a convenient and reasonable workflow.

6.3.5.4 审核功能区域的划分。

6.3.5.4 The division of functional areas.

6.3.5.5 辐射防护工程需要其他建筑专业的协作。

6.3.5.5 Radiation protection needs the collaboration with other construction specialties.

6.3.5.6 防护工程管理：评价设计的合理性、节能及环境、标识、品质的保障等。

6.3.5.6 Protection engineering management: Evaluate the design rationality, energy conservation, environmental protection, labeling and quality assurance.

6.3.5.7 落实重点警示标识、医疗设备与防护门的有效联动。

6.3.5.7 Both warning sign and protective door interlocked with the equipment room door.

6.4 医用气体系统

6.4 Medical gas system

6.4.1 医用气体气源

6.4.1 Medical gas source

6.4.1.1 医用气体系统应根据医疗需求设置。

6.4.1.1 Medical gas system shall be deployed in response to actual medical needs.

6.4.1.2 气源站应根据医院总体规划确定，医用气体管道布置应合理。

6.4.1.2 Determine the layout of gas source station according to the overall planning of the hospital and make sure that all medical gas pipelines are reasonably laid.

6.4.1.3 医用废气的排放不应对医院及周边环境产生影响。

6.4.1.3 The emission of medical waste gas shall not affect the hospital and degrade the surrounding environment.

6.4.1.4 高压气瓶以及液态储罐供应的医用气体，应按日用量计算，并应贮备不少于3d的备用气量；采用制气机组供气时，应设置备用机组，采用分子筛制氧机组时，还应设高压氧气汇流排。当最大机组发生故障时，其他机组的供气能力应能满足系统设计最大负荷。

6.4.1.4 The medical gas supplied by high-pressure gas cylinders and liquid storage tanks shall be calculated according to the daily gas consumption, and the reserve shall guarantee at least three days of normal gas consumption. When the gas supply is guaranteed by gas generating unit, set another unit for standby. If molecular sieve oxygen generator is provided, set a high-pressure oxygen manifold. When the largest unit is fault, the gas supply capacity of other units shall meet the maximum design load of the system.

6.4.1.5 医院应设置氧气和负压吸引系统，可根据需要设置压缩空气、氧化亚氮、氮气、二氧化碳、氩气，以及麻醉废气排放等系统，气源应满足终端处气体参数要求。

6.4.1.5 Hospitals shall set the oxygen and negative pressure suction system, and set the compressed air, nitrous oxide, nitrogen, carbon dioxide, argon, and anesthesia exhaust emission systems as when necessary, the gas source shall meet the gas parameter requirements at the terminal.

6.4.1.6 手术部、监护病房、急救、抢救室供氧管道应单独从氧气站接出。

6.4.1.6 The oxygen supply pipes of the operating department, ICU, first aid and emergency room shall be connected separately from the oxygen station.

6.4.1.7 供气站应设供气异常报警装置，备用机组应设置自动投入使用装置。

6.4.1.7 The gas supply station shall be provided with a gas supply failure alarm device, the standby unit shall be configured with an automatic switching device.

6.4.1.8 医院宜采用无油空气压缩机，压缩空气应设过滤除菌设备。

6.4.1.8 Oil-free air compressors are recommended for hospitals. Filtration and sterilization devices shall be provided for the compressed air.

6.4.1.9 医用气体气源应设超压排放安全阀，气体应排至室外安全地点。

6.4.1.9 Install an overpressure relief valve for the medical gas source to discharge gas to an outdoor safe place.

6.4.1.10 设置分子筛制氧机组制氧站，应符合下列要求：

6.4.1.10 The setup of molecular sieve oxygen generator unit and station shall meet the following requirements:

1. 制氧站宜独立设置或设置在建筑物屋顶。

1. It is recommended to set an oxygen generator station independently or on the building roof.

2. 氧气汇流排间与机器间的隔墙耐火极限不应低于1.5h，氧气汇流排间与机器间之间的联络门应采用甲级防火门。

2. The fire resistance rating of partition wall between oxygen manifold room and generator room shall not be shorter than 1.5h. Use grade-A fire door as the door between the oxygen manifold room and equipment room.

3. 氧气储罐与机器间的隔墙耐火极限不应低于1.5h，氧气储罐与机器间之间的联络门应采用甲级防火门。

3. The fire resistance rating of partition wall between oxygen storage tank and generator room shall not be shorter than1.5h. Use grade-A fire door as the door between the oxygen stor-

age tank and equipment room.

6.4.1.11 采用液氧供氧方式时，大于500L的液氧罐应放在室外；室外液氧罐与办公室、病房、公共场所及繁华道路的距离应大于7.50m。

6.4.1.11 When liquid oxygen is used for the oxygen supply, the liquid oxygen tank larger than 500L shall be placed outdoors. The distance from the outdoor liquid oxygen tank to offices, wards, public areas and busy roads shall be more than 7.50m.

6.4.1.12 负压吸引机房应单独设置，其排放气体应经过处理后排入大气。

6.4.1.12 The negative pressure suction equipment room shall be independently set and its exhaust gas shall be discharged into atmosphere after treatment.

6.4.1.13 洁净手术部医用气体及相关装置

6.4.1.13 The medical gas and related devices available for the clean operating department

1. 有氧气、压缩空气、负压（真空）吸引、氧化亚氮、氮气、二氧化碳和氩气以及废气回收排放等，其中应配置的为氧气、压缩空气和负压吸引装置，氩气可随设备需要配置。

1. Involve oxygen, compressed air, negative (vacuum) suction, nitrous oxide, nitrogen, carbon dioxide and argon, and exhaust gas recovery and emission. Among them, oxygen, compressed air and negative suction devices shall be provided, and argon device can be configured as required by the main equipment.

2. 供给洁净手术部用的医用气源，不论气态或液态，都应按日用量要求储备足够的备用量，不宜少于3d。

2. Provide a sufficient reserve of medical gas regardless of gas or liquid for the clean operating department according to the daily gas demand and ensure at least three days of consumption.

3. 洁净手术部用气宜从中心供给站单独接入；当有专供手术部使用的中心站时，该站应设于临近洁净手术部的非洁净区域。中心站气源应设两组，应一用一备，并应具备人工和自动切换及报警功能。

3. It is recommended to obtain gas from the central gas supply station for the clean operating department. When there is a special central station for the operating department, the station shall be located in a non-clean area close to the clean operating department. Set two groups of gas source for the central gas supply station with one in operation and the other for standby. The functions of manual/automatic switch and alarm shall be available.

4. 供洁净手术部的气源为集中系统时，应设超压排放安全阀，开启压力应高于最高工作压力0.02MPa，关闭压力应低于最高工作压力0.05MPa，安全阀排放口必须设在室外安全地点。各种气体终端应设维修阀并应有调节装置和指示，检修门不应设在手术室内。终端面板根据气体种类应有明显标志。

4. When used as the gas source of clean operating department, the centralized gas supply system shall be configured with a pressure relief valve. Its opening pressure shall be 0.02MPa higher than the highest working pressure and its closing pressure shall be 0.05MPa lower than the highest working pressure. The discharge port of relief valve must be set in an outdoor safe

place. All gas terminals shall be provided with maintenance valves and regulators and indicators. The maintenance doors shall not be located in the operating room. Obvious signs shall be set on the terminal panel according to specific types.

5. 进入洁净手术部的各种医用气体应设气体压力显示及流量和超压欠压报警装置。氧气报警不应采用电接点压力表。

5. All types of medical gas into the clean operating department shall be provided with gas pressure indicator and flow and overpressure/underpressure alarm devices. Electrical contact pressure gauge shall not be used for oxygen alarm.

6. 洁净手术部医用气源还应符合现行国家标准《医用气体工程技术规范》（GB 50751—2012）的要求。

6. The medical gas source for clean operating department shall conform to the current national standard (GB 50751—2012) Technical code for medical gases engineering.

6.4.2 医用气体终端

6.4.2 Medical gas terminal

6.4.2.1 医用气体管终端应安全可靠，终端内部应清洁且密封良好。

6.4.2.1 The terminal of medical gas pipes shall be safe and reliable and must be cleaned and sealed inside.

6.4.2.2 医用气体的终端压力应符合表 6-1 的规定。

6.4.2.2 The terminal pressure of medical gases shall conform to the regulations in Table 6-1.

医用气体的终端压力：见表 6-1。

Terminal pressure of medical gases: see Table 6-1.

表 6-1　医用气体的终端压力

Table 6-1　Terminal pressure of medical gases

医用气体 Medical gases	供气压力（MPa) Gas supply pressure (MPa)
氧气 Oxygen	0.40 ~ 0.45
氧化亚氮 Nitrous oxide	0.35 ~ 0.40
医用真空 Medical vacuum	−0.03 ~ −0.07
压缩空气 Compressed air	0.45 ~ 0.95
氮气 Nitrogen	0.80 ~ 1.10
氩气 Argon	0.35 ~ 0.40
二氧化碳 Carbon dioxide	0.35 ~ 0.40

6.4.2.3 气体终端应采用国际单位制标准，接口制式应统一。

6.4.2.3 All gas terminals are subject to international standard units of system and unified interface system.

6.4.2.4 洁净手术部医用气体终端可选用悬吊式和暗装壁式各一套。

6.4.2.4 The medical gas terminals in the clean operating department can be either suspended type or concealed wall type.

6.4.2.5 不同种类气体终端接头不得有互换性。

6.4.2.5 Connectors for different types of gas terminals shall not be interchangeable.

6.4.2.6 气体终端接头应选用插拔式自封快速接头，接头应耐腐蚀、无毒、不燃、安全可靠、使用方便，寿命不宜少于20000次。

6.4.2.6 Plug type self-sealing quick connectors characterized by corrosion resistance, no toxicity, flammable resistance, safety, reliability, operation convenience and service life not less than 20000 times shall be selected for the gas terminals.

6.4.2.7 每类终端接头配置数量应按表6-2确定。终端压力、流量、日用时间应按表6-3确定。

6.4.2.7 The quantity of connectors for each type of terminal shall be determined according to Table 6-2. The terminal pressure, flow rate and daily operation duration shall be determined according to Table 6-3.

每床每套终端接头最少配置数量（个）：见表6-2。

Minimum number (Pcs.) of Connectors for Each Set of Terminal Per Bed: see Table 6-2.

表6-2 每床每套终端接头最少配置数量（个）

Table 6-2 Minimum number (Pcs.)

用房名称 Room name	氧气 Oxygen	压缩空气 Compressed air	负压（真空）吸引 Negative (vacuum) suction
手术室 Operating room	2	2	2
恢复室 Recovery room	1	1	2
预麻室 Narcotic drugs reserve	1	1	1

注：1. 预麻室如需要可增设氧化亚氮终端。

Note: 1. Nitrous oxide terminal can be added for the narcotic drugs reserve when necessary.

2. 腹腔手术和心外科手术除配置上表所列气体终端外，还应配置二氧化碳气体终端。

2. In addition to the gas terminals listed in the above table, carbon dioxide gas terminals shall be configured for abdominal surgery and cardiac surgery.

3. 神经外科、骨科和耳鼻喉科还应配置氮气终端。

3. Nitrogen terminals shall also be provided for neurosurgery, orthopedics and ENT department.

终端压力、流量、日用时间：见表6-3。

Terminal pressure, flow rate and daily operation duration: see Table 6-3.

表 6-3　终端压力、流量、日用时间

Table 6-3　Terminal pressure, flow rate and daily operation duration

气体种类 Type of gas	终端压力 Terminal pressure /MPa	终端流量 Terminal flow rate /L/min	平均日用时间 Average daily operation duration/min	同时使用率 Simultaneous utilization rate/%
氧气 Oxygen	0. 40 ~ 0. 45	10 ~ 80 （快速置换麻醉气体用） 10 ~ 80 (for the quick replacement of anesthetic gas)	120(恢复室 1440) 120 (1440 for recovery room)	50 ~ 100
负压（真空）吸引[1] Negative (vacuum) suction[1]	−0. 03 ~ −0. 07	15 ~ 80	120(恢复室 1440) 120 (1440 for recovery room)	100
压缩空气 Compressed air	0. 40 ~ 0. 45	20 ~ 60	60	80
压缩空气[2] Compressed air[2]	0. 90 ~ 0. 95	230 ~ 350	30	10 ~ 60
氮气 Nitrogen	0. 90 ~ 0. 95	230 ~ 350	30	10 ~ 60
氧化亚氮 Nitrous oxide	0. 40 ~ 0. 45	4 ~ 10	120	50 ~ 100
氩气 Argon	0. 35 ~ 0. 40	0. 5 ~ 15	120	80
二氧化碳 Carbon dioxide	0. 35 ~ 0. 40	6 ~ 10	60	30

注：1. 负压手术室负压（真空）吸引装置的排气应经过高效过滤器后排出。

Note: 1. The exhaust gas of negative (vacuum) suction device in negative pressure operating room can be discharged after being treated by HEPA filters.

2. 此项用于动力设备，如设计氮气系统，该项也可以不设。

2. This item is used for power equipment. This item could be ignored in the design of nitrogen system.

6. 4. 2. 8　洁净手术室壁上气体终端装置应与墙面平齐，缝隙密封，部位宜临近麻醉师工作位置。终端面板与墙面应齐平严密，装置底边距地 1. 0 ~ 1. 2m。

6. 4. 2. 8　The gas terminal on the wall of the clean operating room shall be flush with the wall surface with the gap sealed. Moreover, it shall be located nearby the working position of anesthetist. The terminal panel shall be flush and tight with the wall surface. The terminal bottom edge is 1. 0m to 1. 2m above the floor.

6. 4. 3　医用气体配管

6. 4. 3　Medical gas pipes

6. 4. 3. 1　医用气体管道应选用紫铜管或不锈钢管，负压吸引和手术室废气排放输送管可采用镀锌钢管。管道、阀门和仪表附件安装前应进行脱脂处理。

6. 4. 3. 1　Red copper or stainless steel pipes shall be selected as the medical gas pipes, whereas galvanized steel pipes are used as the exhaust gas discharge pipe of negative suction and operating room. Before installation, all the pipes, valves and instrument accessories shall be degreased.

6.4.3.2 供氧管道不应与电缆、腐蚀性气体和可燃气体管道敷设在同一管道井或地沟内。敷设有供氧管道的管道井，宜有良好通风。

6.4.3.2 Oxygen supply pipes shall not be laid in the same piping shaft or trench with cables, corrosive gas and combustible gas pipes. The piping shaft with oxygen supply pipes shall be effectively ventilated.

6.4.3.3 氧气管道架空时，可与各种气体、液体（包括燃气、燃油）管道共架敷设。共架时，氧气管道宜布置在其他管道外侧，并宜布置在燃油管道上面。供应洁净手术部的医用气体管道应单独设支吊架。

6.4.3.3 When laid overhead, the oxygen supply pipe can be installed onto the same support with various gas and liquid (including fuel gas and oil) pipes. Under such circumstance, it is recommended to lay oxygen supply pipes outside other pipes and above fuel pipes. Independent support and hanger shall be provided for the medical gas pipes of the clean operating department.

6.4.3.4 除氧气管道专用的导电线外，其他导电线不应与氧气管道敷设在同一支架上。

6.4.3.4 All conducting wires, excluding the special conducting wires for oxygen supply pipes, shall not be laid onto the same support with oxygen support pipes.

6.4.3.5 氧气管与其他管线之间距离应符合表 6-4 的规定，当间距无法满足时，应采取安全可靠的技术措施。

6.4.3.5 The distance from the oxygen supply pipe to other pipes shall conform to relevant regulations in Table 6-4. Otherwise, safe and reliable technical measures shall be taken.

氧气管与其他管线之间距离：见表 6-4。

List of distances from the oxygen supply pipe to other pipes: see Table 6-4.

表 6-4 氧气管与其他管线之间距离

Table 6-4 List of distances from the oxygen supply pipe to other pipes

名称 Name	平行净距/m Parallel clear distance/m	交叉净距/m Crossed clear distance/m
给排水管 Water supply and drainage pipe	0.25	0.10
热力管 Heating pipe	0.25	0.10
燃气管、燃油管 Gas pipe and fuel pipe	0.50	0.30
绝缘导线或电缆 Insulated conductor or cable	0.50	0.10

6.4.3.6 病区及洁净手术部内的氧气干管上，应设置手动紧急切断气源的装置。

6.4.3.6 Manual emergency cut-off devices shall be arranged on the main oxygen supply pipe in the ward and clean operating department.

6.4.3.7 穿过墙壁、楼板的氧气管道应敷设在套管内，并应用石棉或其他不燃材料将套管间隙填实。氧气管道不宜穿过不使用氧气的房间，必须通过时，在房间内的管道上不应有法兰或螺纹连接接口。

6.4.3.7 All oxygen supply pipes through walls and floor slabs shall be laid into bushing. It also needs to fill the space between the pipe and bushing with asbestos or other nonflammable materials. It is not recommended to lay the oxygen supply pipe through any room that does not use oxygen. If the pipe must be laid through such room, any flange or threaded connection is not allowed on the pipe in the room.

6.4.3.8 医用气体管道应做导静电接地装置。

6.4.3.8 All medical gas pipes shall be provided with electrostatic conduction grounding device.

6.4.3.9 医用气体管道与支吊架接触处，应做防静电腐蚀绝缘处理。

6.4.3.9 The treatment of electrostatic and corrosion resistance and insulation shall be carried out at the contact between any medical gas pipe and support & hanger.

6.4.3.10 含湿医用气体管道，应采取防冻措施。

6.4.3.10 Antifreeze measures shall be taken for medical gas pipes containing moisture.

6.4.3.11 医用真空管道，应坡向总管和缓冲罐，坡度不应小于 2‰。

6.4.3.11 Medical vacuum pipes shall be inclined to the main pipe and buffer tank with a gradient not less than 2‰.

6.4.3.12 管道材质内表面应光滑、耐腐蚀、耐磨损以及吸附和解析气体作用小。

6.4.3.12 The inner surface of these pipes shall be smooth and can resist corrosion and abrasion. Meanwhile, it shall have little effect on absorbing and analyzing gas.

6.4.3.13 气体在输送导管中的流速不应大于 10m/s。

6.4.3.13 The flow rate inside the gas discharge pipe shall not exceed 10m/s.

6.4.3.14 气体配管的连接方式应按国家标准《洁净室施工及验收规范》（GB 50591—2010）的要求执行。

6.4.3.14 The connection of gas pipes shall conform to the current national standard (GB 50591—2010) Code for construction and acceptance of clean room.

6.4.3.15 管道安装前应清洁内部，并应有防止二次污染措施；安装后应用无污染和无油空气或氮气以大于等于 20m/s 的速度吹除。

6.4.3.15 Before installation, clean the inner surface of pipes and take measures against secondary pollution. After installation, use clean and oil-free air or nitrogen to blow away dust at the speed higher than or equal to 20m/s.

6.4.3.16 洁净手术部的气体配管

6.4.3.16 Gas pipes for clean operating department

1. 应选用脱氧铜管或不锈钢管，负压吸引和废气排放输送导管也可采用镀锌钢管或 PVC 管。

1. Deoxidized copper pipes or stainless steel pipes should be used. For negative pressure suction and exhaust emission, galvanized steel pipes or PVC pipes can also be used.

2. 气体管道安装应单独做支吊架，不应与其他管道共架敷设；与燃气管、腐蚀性气体管的距离应大于 1.5m，且应有隔离措施；与电线管道平行距离应大于 0.5m，交叉距离应大于0.3m，当空间无法保证时，应做绝缘防护处理。

2. These pipes shall not be installed to the same support with other pipes. In addition, the distance to gas pipe and corrosive gas pipe shall be more than 1.5m and isolation measures shall be taken. The distance to cable pipes shall be more than 0.5m and the crossing distance shall be larger than 0.3m. When the space is not sufficient, insulation shall be carried out for protection.

3. 气体配管的安装支吊架间距应符合表6-5的规定；铜管、不锈钢管道与支吊架接触处，应做绝缘处理。

3. The spacing between support and hanger for the gas pipes shall conform to relevant regulations in Table 6-5. The contact of copper pipes and stainless steel pipes with the support and hanger shall be insulated.

支吊架间距：见表6-5。

List of spacing between support and hanger: see Table 6-5.

表6-5 支吊架间距

Table 6-5 List of spacing between support and hanger

管道公称直径/mm Nominal diameter of pipes/mm	4~8	9~12	13~20	21~25	≥25
支吊架间距/m Spacing between support and hanger/m	1	1.5	2	2.5	3

4. 凡进入洁净手术室的各种医用气体管道应做导静电接地，接地电阻不应大于10Ω，中心供给站的高压汇流管、切换装置、减压出口、低压输送管路和二次再减压出口处都应做导静电接地，接地电阻不应大于10Ω。

4. All types of medical gas pipes in the clean operating room must be grounded for electrostatic conduction with the grounding resistance not more than 10Ω. In addition, the high-pressure manifold, switching device, pressure relief outlet, low-pressure transfer pipe and secondary pressure relief outlet shall be grounded for electrostatic conductance with the grounding resistance not more than 10Ω.

6.4.3.17 医用气体配管、阀门和仪表安装前应清洗内部并应进行脱脂处理，用无油压缩空气或氮气吹除干净，封堵两端备用，不得存放在油污场所。

6.4.3.17 Before installation, all medical gas pipes, valves and instruments shall be cleaned and degreased with oil-free compressed air or nitrogen. Then two ends of the pipe shall be sealed for standby. These pipes shall not be stored in a greasy place.

6.4.3.18 暗装管道阀门的检查门应采取密封措施；管井上下隔层应封闭；医用气体配管不应与燃气、腐蚀性气体、蒸汽以及电气、空调等管线共用管井。

6.4.3.18 The inspection door of concealed pipes and valves shall be sealed. The partition of piping shaft shall be enclosed. Medical gas pipes shall not share the same piping shaft with gas pipes, corrosive gas pipes, steam pipes, electrical pipes and air conditioning pipes.

6.4.3.19 负压吸引气流入口处应有安全调压装置；手术过程中使用的负压吸引装置应有防止污液倒流装置。

6.4.3.19 A safety regulating device shall be set at the negative suction air flow inlet. The negative suction device used during operation shall be provided with the device against the backflow of contaminated liquid.

6.4.4 医用气体验收

6.4.4 Acceptance of medical gases

6.4.4.1 医用气体验收应符合设计要求。

6.4.4.1 The acceptance of medical gases shall meet specific design requirements.

6.4.4.2 医用气体系统强度试验及漏率试验应符合表6-6的规定。

6.4.4.2 The strength test and leak rate test for the medical gas system shall conform to relevant regulations on Table 6-6.

医用气体系统强度试验及漏率试验：见表6-6。

Strength test and leak rate test for the medical gas system: see Table 6-6.

表6-6 医用气体系统强度试验及漏率试验

Table 6-6 Strength test and leak rate test for the medical gas system

系统名称 System name	正压气体 Positive pressure gas	负压气体 Negative pressure gas	时间 Time
强度试验 Strength test	最高工作压力的1.15倍 115% of the highest working pressure	0.2MPa	10min无变化 No change within 10min
漏率试验 Leak rate test	最高工作压力下小于等于0.5%/h Less than or equal to 0.5%/h under the highest working pressure	最高工作压力下小于等于1.5%/h Less than or equal to 1.5%/h under the highest working pressure	保压24h Maintain pressure for 24h

注：1. 强度试验所用介质不得二次污染系统；

Note: 1. The medium for strength test should not produce secondary pollution to the system.

2. 强度试验要做好防护，保证安全；方法：逐步升压，观察其变化情况；

2. Ensure effective safety protection during the strength test. Increase pressure gradually and observe the change.

3. 漏率试验所用介质尽可能采用同等系统所输送的介质；

3. Try to use the same medium transported by the system for the leak rate test.

4. 使用不同气体终端的手术室，各抽测手术室总数的15%，应符合表6-3的要求；负压吸引终端的手术室抽测其手术室总数的20%，均应符合表6-3的要求。

4. For an operating room using different types of gas, 15% of total pipes for the operating room shall be sampled and tested according to Table 6-3. The percentage shall be 20% for the operating room with a negative suction terminal according to Table 6-3.

6.5 自动物流

6.5 Automatic logistics

6.5.1 适用性

6.5.1 Applicability

6.5.1.1 自动物流是指借助信息技术、光电技术、机械传动装置等一系列技术和设施，在设定的区域内运输物品。

6.5.1.1 Automatic logistics refer to a series of technologies and facilities such as information technology, photoelectric technology and mechanical transmission device to transport goods within a specified area.

6.5.1.2 大型医院运送物品种类多、运送量大、运送频繁、时间效率和传输精准性要求高，在人工成本较高的情况下，采用自动物流可以提高运送效率、节省人力。其他医疗保健设施应根据实际物流量、人工费综合考虑是否采用。

6.5.1.2 A large hospital has many types of materials to be transported. The transport volume is large and the transport frequency is high. There is a high requirement for the time efficiency and transport accuracy. Now labor cost is very high. Therefore, the application of automatic logistics helps promote transport efficiency and save labor force. Other health care facilities shall confirm whether an automatic logistics system is deployed according to actual logistics and labor costs.

6.5.2 医院物流输送系统类型

6.5.2 Types of hospital logistics transport system

6.5.2.1 自动物流系统分小型物流、中型物流、大型物流三类。小型物流分为气动物流和轨道小车物流，传输 5 ~ 15kg 物品，中型物流传输 30 ~ 50kg 物品，大型物流传输 100kg 以上的物品。

6.5.2.1 There three types of automatic logistics system, including small, medium and large logistics systems. The small logistics system is divided into pneumatic logistics and rail trolley logistics to transport the items of 5kg to 15kg. The medium transport system is used to transport the items of 30kg to 50kg and the large transport system is used to transport the items heavier than 100kg.

6.5.2.2 医院内可采用不同类型的自动物流输送系统，既可运送药品、小型医疗器械、单据、标本、血液、血样、X 光片、敷料、处方、办公用品等小型物品，也可运送输液、被服、手术器械包、餐饮、医疗废弃物等中等或者体积较大的物品。

6.5.2.2 A hospital can use different types of automatic logistics transport system to transport not only drugs, small medical devices, documents, specimens, blood, blood samples, X-ray films, dressings, prescriptions and office supplies but also infusions, bedding, surgical kits, food, medical wastes and other medium or large-sized items.

6.5.3 各类物流传输系统适用性

6.5.3 Applicability of different types of logistics transport system

6.5.3.1 气动物流系统：以专用传输瓶为载体、以压缩气体为动力，体积小、速度快，主要运输质量轻的标本、急用药品，适合小型物品、急诊用品的快速传送，解决小规模、高效率物品的运送。

6.5.3.1 Pneumatic logistics system: A special transport cylinder is used as the carrier and compressed air as the power. It features small size and high speed. It mainly transport lightweight specimens and emergency drugs, which is suitable for the rapid transport of small items and emergency items and ensures the efficient transport of small items.

6.5.3.2 轨道小车物流系统：是在计算机控制下利用载物小车在专用轨道上传输物品的系统，可以运输相对较重、体积较大的物品，包括输液药剂、批量的检验标本、供应室的小型物品。由收发工作站、轨道载物小车、物流轨道、轨道转轨器、自动隔离门、中心控制设备、控制网络等设备构成。

6.5.3.2 Rail trolley logistics system: It refers to a transport system using a trolley on rail under the control of computer. It can transport heavy and large items, including infusion solution, batch test specimens and small items of supply room. It consists of receiving and dispatching workstation, rail trolley, logistics rail, rail changer, automatic isolation door, central control equipment and control network.

6.5.3.3 中型物流输送系统：传输量大、传输效率高、维护成本低、占用建筑面积空间较大；依据医院建筑布局和物品取送方便来确定，不适合设置偏远、零星物流站点；通过类似电梯的设备在竖向楼层间实现物资输送，需要垂直井道，水平输送设备分为落地安装和吊装两种形式，吊装占用吊顶空间；若水平与竖向贯通构成系统，需设水平转换层，可位于地下室或设备层。

6.5.3.3 Medium-scale logistics transport system: It is characterized by a large transport capacity, high transport efficiency but low maintenance cost and occupies a small building area. It shall be determined according to the hospital building layout and material acquisition & distribution. It is not recommended to set remote and scattered logistics stations. Materials can be transported vertically between floors by using the equipment similar to an elevator. A vertical shaft is needed. The horizontal transport equipment can be installed to floor and suspended. The suspended ceiling will be occupied by the suspended horizontal transport equipment. If the horizontal passage and the vertical passage constitute a system, it needs to set a horizontal transfer floor, which can be located in the basement or on the equipment floor.

6.5.3.4 自动导引车传输系统（AGV）：是在计算机和无线局域网络控制下的无人驾驶自动导引运输车，经磁、激光等导向装置引导，沿程序设定路径运行并停靠到指定地点，完成物品移载、搬运等作业功能。主要用于取代劳动密集型的手推车，运送患者餐食、衣物、医院垃圾、批量的供应室消毒物品等，实现楼宇和楼层间的传送。

6.5.3.4 AGV (Automated Guided Vehicle) transport system: AGV refers to an unmanned automated guided vehicle under the control of computer and WLAN. It is guided by magnetic force, laser and other guiding devices to run along the programmed path and stop at the designated place to complete the transfer and transport of related items. It is mainly used to replace labor-intensive handcars to transport food and clothes of patients, medical wastes, and batch disinfection items of supply rooms, thus realizing the transport inside a building and between floors.

四类物流系统技术指标对比表：见表6-7。

Comparisons of technical indicators for the four types of logistics systems: see Table 6-7.

表 6-7 四类物流系统技术指标对比

Table 6-7 Comparisons of technical indicators for the four types of logistics systems

指标 Indicators	类型 Type			
	气动物流 Pneumatic logistics	轨道小车 Rail trolley	中型物流 Medium logistics	AGV
传输质量 Transport weight	5. 5kg	10 ~ 15kg	30 ~ 50kg	100 ~ 400kg
传输容量 Transport capacity	4. 5L	28L	89L	500L
可传物品比率 Proportions of transportable items	医用物品 30% 可以传输 30% of hospital items can be transported	医用物品 60% 可以传输 60% of hospital items can be transported	医用物品 95% 可以传输 95% of hospital items can be transported	医用物品 100% 可以传输 100% of hospital items can be transported
传输速度 Transport speed	5 ~ 8m/s	水平 0. 6m/s、 垂直 0. 4m/s Horizontal 0. 6m/s and vertical 0. 4m/s	水平 0. 4 ~ 1. 0m/s, 垂直 1. 0 ~ 3. 0 m/s Horizontal 0. 4 ~ 1. 0m/s and vertical 1. 0 ~ 3. 0m/s	水平 0. 6m/s Horizontal 0. 6m/s
井道尺寸 Shaft size	无 None	2150(长) mm × 2000(宽) mm 2150 (length) mm × 2000 (width) mm	1800(长) mm × 1800(宽) mm 1800 (length) mm × 1800 (width) mm	需专用电梯 Need special elevator
等候时间 Waiting time	高峰期排队等候 Queue up at peak hour	小车存放在车库里，科室不配备小车，需要叫车等候，其在高峰期等候时间较长 The trolley is stored in garage. Each department is not provided with a trolley. It needs to call a trolley and the wait time is long at peak hours	科室配备很多的传输箱，没有等候现象，随发随走 Each department must be provided with lots of transport boxes. It does not need to wait. Items can be transported immediately	没有等候现象，随发随走 It does not need to wait and items can be transported immediately
主传物品 Main transported items	血液、标本、药品、小型器材 Blood, specimen, drug and small equipment	血液、标本、药品、小型器材、文件、档案、部分输液 Blood, specimen, drug and small equipment, document, archive and part of infusions	血液、标本、药品、文件、器械、中心配液、中心供应物品、被服、配餐等几乎所有物品 Almost all items, including blood, specimens, drugs, documents, devices, central solution distribution, central supplies, bedding and clothing and food	血液、标本、药品、文件、器械、中心配液、中心供应物品、被服、配餐等几乎所有物品 Almost all items, including blood, specimens, drugs, documents, devices, central solution distribution, central supplies, bedding and clothing and food

续表

指标 Indicators	类型 Type			
	气动物流 Pneumatic logistics	轨道小车 Rail trolley	中型物流 Medium logistics	AGV
外表 Appearance	美观 Nice	小车外观漂亮，水平轨道安装在吊顶下方 The trolley has a nice appearance and the horizontal rail is installed below the suspended ceiling	水平轨道全部实施隐蔽施工；整体协调美观，不占用天花板以下空间 All horizontal rails are concealed The whole structure is coordinated and nice and will not occupy the space below the ceiling	外观漂亮，智能驱动，科技含量高 It has a nice appearance, smart drive and high technical content.
优点 Advantages	速度快、效率高 High speed and efficiency	比气动物流传输量大，一条轨上可以有多个小车同时发送 It has a larger transport capacity than pneumatic logistics. Multiple trolleys can run on the same rail	传输量很大，不受体积限制；传输箱使用存放方便，物品传输过程中始终水平放置 It has a transport capacity and the transport volume is not limited. Transport boxes can be easily used and stored and are horizontal during transportation	传输量很大，不受体积限制；水平行走无需固定通道，物品传输过程中始终水平放置；自动充电 It has a larger transport capacity and the transport volume is not limited. Its horizontal movement does not need a fixed path. During transportation, all the items are always kept horizontal. Automatic charging is available
缺点 Disadvantages	体积小、质量轻，只能传输一些急救物品 It has a small size and light weight and can only transport some emergency items	车离不开轨，物品翻转、物品形状受限 The trolley cannot work without rail and it is difficult to turn over materials. Limited types of materials can be transported	水平传输线在吊顶里，占用 0. 6m × 0. 6m 独立空间通道 0. 6m × 0. 6m independent passage will be occupied by the horizontal transport path above the suspended ceiling	水平行走容易受到附近人员干扰 The horizontal flow is easily disturbed by the personnel nearby
维护成本 Maintenance cost	维护成本低 Low maintenance cost	结构复杂，维护成本高，8% Complex structure, high maintenance cost, 8%	结构简单，维护成本适中，3% ~5% Simple structure, moderate maintenance cost, 3% ~5%	结构复杂，维护成本高 Complex structure and high maintenance cost
适合传输 Suitable for transport	随机性比较强的小规模物品 Small items with strong randomness	规格小于车体内部尺寸的物品；不易滴漏、破碎的物品 Items smaller than the internal space of trolley; items that will not drip or break easily	可以满足医院 95% 的物品传输 95% of the hospital items can be transported.	可以满足医院 100% 的物品传输 100% of the hospital items can be transported

6.5.4 自动物流传输站点设置

6.5.4 Settings of automatic logistics transport stations

医院需设置物流站点的科室与站点位置：见表6-8。

The hospital departments that need to set logistics stations and the location of each station are shown in Table 6-8.

表6-8 医院需设置物流站点的科室与站点位置

Table 6-8 The hospital departments that need to set logistics stations and the location of each station

设置科室 Department settings	传输物品 Transported items	工作站点位置 Workstation settings
检验科 Laboratory department	血液、体液、尿液、粪便等检验样本和检验报告 Blood, body fluid, urine, excrement and other test samples and test reports	位于物品收发处、物品接收台旁 It is located at the item receiving and dispatching areas and beside the item receiving desk
病理科 Pathology department	病理检验标本、检验报告 Pathological examination specimens and test reports	
静脉配置中心 Venous infusion solution preparation center	袋装、瓶装的静脉输液 Bagged or bottled venous infusion solution	
药房 Pharmacy	药品 Drugs	
中心供应室 Central supply room	医用材料及敷料、一次性无菌用品、小型手术包、小型治疗器械包等 Medical materials and dressings, disposable sterile items, small surgical kits and small therapeutic kits	
急诊科 Emergency department	药品、血液制品、小型手术包等，其中优先接收药品、麻醉剂和血液制品 Drugs, blood products, small surgical kits, etc. Drugs, anesthetics and blood products are preferentially transported	位于护士站旁 Located beside the nurse station
住院护理单元/ICU Inpatient care unit/ICU	检验样本、药品、静脉输液、一次性无菌用品等 Test samples, drugs, venous infusion solution, disposable sterile items, etc.	
门诊采血点 Outpatient blood connection point	血液标本 Blood specimen	
体检科 Physical examination department	血液、体液、尿液、粪便等检验样本和检验报告 Blood, body fluid, urine, excrement and other test samples and test reports	

续表

设置科室 Department settings	传输物品 Transported items	工作站点位置 Workstation settings
手术部 Operating department	血液、药物、器械、辅料、标本、耗材、餐食等 Blood, drugs, devices, auxiliary materials, specimens, consumables, food, etc.	位于物品收发处 It is located at the receiving and dispatching area
血库 Blood bank	血液制品 Blood products	
功能科室 Functional departments	药物、耗材等 Drugs, consumables, etc.	
影像科、内镜中心 Imaging department and endoscopy center	药品、一次性物品、X 光片、报告等 Drugs, disposable items, X-ray films and reports	
档案室 Archives	病历、病案等 Medical records, etc.	

6.5.5 物流中心控制室

6.5.5 Logistics central control room

6.5.5.1 物流中心控制系统必须安装在专用控制室内，控制室的面积大小取决于物流系统的规模，应位于采光较好、靠近医院物流量最大最频繁的核心区域。

6.5.5.1 The central control system of logistics must be installed inside a special control room. The size of this control room is subject to the scale of logistics system. Such control room shall be located nearby the core area with the largest logistics demand and highest transport frequency of the hospital with sufficient sunlight.

6.5.5.2 应给监控维保工作人员提供长期良好的工作环境，附近有厕所，便于就近查访检修。

6.5.5.2 The monitoring and operation & maintenance personnel shall be provided with a long-term good working environment with toilets nearby for the convenience of inspection and repair.

6.5.5.3 室内设中央监控设备、全院总体物流规划及运行状况自动显示电子屏。

6.5.5.3 Central monitoring equipment and automatic display electronic screen shall be deployed in the control room to show the general logistics planning and operation of the whole hospital.

6.5.6 自动物流防火设计

6.5.6 Fire protection design for the automatic logistics

6.5.6.1 自动物流系统贯穿医院建筑的水平及竖向防火分区，其井道、轨道或管线穿越防火分区时，须设置甲级电动防火窗/帘，窗/帘洞口处设置烟感报警装置，报警系统与整个医院集中报警系统联动，并接至医院的报警主机。

6.5.6.1 An automatic logistics system will pass through the horizontal and vertical fire zones of the hospital. While its shaft, rail or pipeline is set across fire zones, and grade-A electric fire windows/curtains must be set. In addition, it needs to set smoke-sensing alarm devices at the openings of window/curtain. The alarm system shall be linked with the centralized alarm system and connected to the alarm host of the hospital.

6.5.6.2 防火窗/帘洞口应能确保系统运行时物流管道/车/箱体正常通过，在火灾时能自动关闭，关闭时能够截断通道，确保各防火区域的密闭性，自动关闭反应时间需满足消防规范要求。

6.5.6.2 The openings of fire window/curtain shall ensure a normal passage of logistics pipelines/trolleys/boxes and can be closed automatically in case of fire. The passage can be cut off when the opening is closed to ensure the tightness of each fire zone. The response time of automatic cut-off shall conform to relevant codes on fire protection.

6.6 医院停车规划

6.6 Hospital parking planning

6.6.1 前期规划

6.6.1 Early planning

6.6.1.1 应进行医院新建、改扩建规划的交通影响评价。

6.6.1.1 Traffic impact assessment shall be carried out for the planning of construction, renovation and expansion of hospitals.

6.6.1.2 取得医院周边近期、远期公共交通设施规划条件、公交车站布点、定位尺寸，获取医院四周市政交通道路发展规划。

6.6.1.2 It is necessary to obtain the planning of short-term and long-term public transportation facilities, bus station layout and positioning dimensions around the hospital, and to obtain the development planning of municipal transportation roads around the hospital.

6.6.1.3 明确市政道路红线、辅道、相邻禁止开口过境路、高架路与医院用地红线的关系，规划以及市政部门对医院机动车出入口数量、开口宽度的许可、机动车出入口转弯半径规定。

6.6.1.3 It also needs to clarify the relation between the red line of municipal roads, auxiliary roads, adjacent crossing roads with no openings, elevated roads and the red line of hospital land, and the planning and the regulations of municipal engineering departments on the number of hospital motor vehicle entrances and exits, the regulations on the opening width, and the turning radius of motor vehicle entrances and exits.

6.6.1.4 确定地上、地下机动车和非机动车停车数量计算依据，地上、地下停车数量比例。

6.6.1.4 Moreover, determine the criteria of calculating the number of vehicles and non-motor vehicles parked on the ground and underground, and the proportion of vehicles parked on the ground and underground.

6.6.1.5　确定地下非静态停车空间，如下穿市政道路、地下出租车站、地下洗车场、生活垃圾转运站的建筑面积，此类空间需要车辆通行和车辆临时停靠。

6.6.1.5　Determine the construction area of underground non-static parking spaces that pass through municipal roads, underground taxi stations, underground car washing bays and domestic waste transfer stations. Such spaces require vehicles to pass through and temporarily stop.

6.6.2　停车类型与设计

6.6.2　Parking type and design

6.6.2.1　院内停车类型包括120救护车、医院货运车辆、传染病转运车辆停靠及洗消、无障碍停车、充电桩停车、大型车辆（体检车、救援车、大中型汽车）、普通停车（小型汽车、微型汽车）、垃圾转运车、殡仪馆运尸车。

6.6.2.1　The types of parking in the hospital include the parking of ambulances, hospital freight vehicles, the parking, washing and sterilization of infectious disease transfer vehicles, barrier-free parking, charging pile parking, large vehicles (physical examination vehicles, rescue vehicles, large and medium-sized vehicles), and ordinary parking (small cars, mini-cars), garbage transfer vehicles, and corpse transport vehicles of funeral home.

6.6.2.2　应结合场地和规划设计条件，确定采用地上停车、地下停车、地上停车楼、露天立体机械停车设备、室内仓储式全自动机械停车的停车方式。

6.6.2.2　The ground parking, underground parking, ground parking buildings, outdoor three-dimensional mechanical parking equipment, indoor warehouse type automatic mechanical parking and other parking methods shall be determined according to the site conditions and planning & design conditions.

6.6.2.3　地面和地下各类停车场库的机动车与行人道路、停车港湾、行人等候区、货物装卸区、洗车区应符合道路交通安全设计规范要求。

6.6.2.3　Motor vehicle and pedestrian roads, parking bays, pedestrian waiting areas, cargo handling areas and car washing areas of various parking lots on the ground and underground shall meet the requirements of road traffic safety design specifications.

6.6.2.4　医疗药品、医疗垃圾、餐食、生活垃圾等货品的装卸区设计，在地上应满足场地卫生、感控、安全要求，在地下应满足车辆净高、转弯掉头、通风、排风、消防要求，且不应妨碍正常动态交通通行动线，车辆卸货港湾区应配设场地冲洗、排水或消毒装置。

6.6.2.4　The design of ground loading and unloading areas for drugs, medical garbage, meals, household garbage and other items shall meet the requirements of site hygiene, sensory control and safety, and the requirements of vehicle clear height, turning around, ventilation, exhaust and fire protection. Moreover, they shall not disturb normal and dynamic traffic routes. The bending area of vehicle unloading bay shall be provided with site flushing, drainage or disinfection devices.

6.6.2.5 药品、食品等卸货区，应考虑室内或室外包装箱拆除的空间，避免影响医院场地和其他交通的日常使用。

6.6.2.5 For the unloading areas such as medicines and food, the indoor or outdoor space for removing packing boxes shall be considered to avoid affecting the daily use of hospital sites and other traffic.

6.6.2.6 室内外停车场库设计应符合停车场库设计规范和建筑防火设计规范，需配建标识系统、停车场自动化管理系统，超大型地下车库宜设置自动导航驻车系统、自动识别与收费系统。

6.6.2.6 The design of indoor and outdoor parking lots and garages shall conform to the design specifications and building fire prevention design specifications. Identification system and automatic management system of parking lots and garages shall be provided. The super-large underground garages shall be equipped with automatic navigation parking system, automatic identification and charging system.

6.7 直升机停机坪

6.7 Helipad

6.7.1 直升机停机坪是供直升机垂直起降的场地，要求配备相应的助航设备、航管、通信设备、气象设施、消防救援设备、机场标志标识等，使之能符合直升机安全起降的要求。

6.7.1 Helipad is a site for helicopters to take off and land vertically. It shall be configured with corresponding navigation aids, navigation control, communication equipment, meteorological facilities, fire fighting and rescue equipment and airport signs to meet the requirements for safe taking off and landing of helicopters.

6.7.2 国家级和区域级医疗中心、传染病医院、有建设条件的大型三级综合医院应建设直升机停机坪。

6.7.2 National and regional medical centers, infectious disease hospitals and large-scale third-class general hospitals with construction conditions shall build a helipad.

6.7.3 停机坪的选址

6.7.3 Helipad location

6.7.3.1 应在医院医疗工艺设计时统一规划，要满足直升机在医院内 24 小时随时起降的应急救援功能。

6.7.3.1 It shall be planned in hospital medical process design to meet the emergency rescue function of helicopters taking off and landing 24 hours a day in the hospital.

6.7.3.2　要保证医疗停机坪自身的可靠性、净空的可靠性、现场控制的可靠性、起降时周边障碍物快速清除的应急能力。

6.7.3.2　It is of great necessity to ensure the reliability of the helipad for medical purpose, the reliability of clearance, the reliability of field control, and the emergency preparedness of rapid removal of surrounding obstacles when taking off and landing.

6.7.3.3　停机坪与急诊科、手术部、应急传染隔离病区之间的流程为一级流程，应确定选址、最大机型、助航设施与医院急救系统的衔接、下机后抵达目标地的无障碍电梯、坡道、通道路径。

6.7.3.3　The process from the helipad to the emergency department, operating department and emergency infectious isolation ward is a first-class process. The site selection, the largest aircraft type, the connection between the navigation aid facilities and the hospital emergency system, and the barrier-free elevator, ramp and passage route to the destination after landing shall be determined at the same time.

6.7.3.4　停机坪的形式有屋顶和地面两类。

6.7.3.4　There are two types of helipad: roof and ground.

1. 地面停机坪应考虑场地尺寸、地质条件、起降的地效因素、周边障碍物在起降时能否快速清除、抵达目标地的专用通道以及地面人员对直升机降落时的干扰因素。

1. As for a ground helipad, it needs to consider the site size, geological conditions, ground effect factors of taking off and landing, whether the surrounding obstacles can be removed quickly during taking off and landing, the special passage to the destination, and the interference of ground personnel with helicopter landing.

2. 屋顶高架停机坪应避开高楼、高压线塔、通信塔、烟囱、高大树木。

2. The elevated helipad on a roof shall avoid tall buildings, high-voltage line towers, communication towers, chimneys and tall trees.

6.7.4　停机坪的设计

6.7.4　Helipad design

6.7.4.1　包括选型、结构、构造、疏散、无障碍通道、助航设施、标志、标识、配电照明、防雷、消防救援及保障系统等。

6.7.4.1　It involves the type selection, structure, configuration, evacuation, barrier-free access, navigation aids, signs, markings, power distribution & lighting, lightning protection, fire rescue and security systems.

6.7.4.2　应与医院建筑工程各专业同时设计、施工和验收。

6.7.4.2　Design, construction and acceptance shall be carried out simultaneously with various specialties of hospital construction engineering.

第7章

建筑信息模型BIM

Chapter Ⅶ

BIM (Building Information Modeling/Model)

汤 琼

Qiong Tang

7.1 工程设计 BIM

7.1 Engineering design BIM

7.1.1 BIM 的用途

7.1.1 Use of BIM

在建筑项目设计中实施 BIM 的目的是提高项目设计质量和效率，从而减少后续施工期间的摩擦和返工，保障施工周期，节约项目资金。它的主要特征体现在以下 5 个方面：

The ultimate goal of implementing BIM in construction project design is to improve the quality and efficiency of project design, thus reducing negotiation and reconstruction during subsequent construction, ensuring the construction cycle and saving project cost. Its value in the architectural design is mainly reflected in the following five aspects:

7.1.1.1 可视化：BIM 将专业、抽象的二维建筑描述通俗化、三维直观化，使设计师和业主等非专业人员对项目需求是否得到满足的判断更为明确、高效，决策更为准确。

7.1.1.1 Visualization: BIM popularizes professional and abstract two-dimensional building descriptions and visualizes them into three dimensions, making it clearer and efficient for designers and owners and other non-professional personnel to judge whether the project needs are met and make more accurate decisions.

7.1.1.2 协调：BIM 将专业内多成员间、多专业、多系统间原本各自独立的设计成果（包括中间结果与过程），置于统一、直观的三维协同设计环境中，避免因误解或沟通不及时造成不必要的设计错误，提高设计质量和效率。

7.1.1.2 Coordination: BIM integrates the original independent design results (including intermediate results and processes) among multiple members, disciplines and systems in a unified and intuitive three-dimensional collaborative design environment to avoid the design errors caused by misunderstanding or untimely communication and promote design quality and efficiency.

7.1.1.3 模拟：BIM 将原本需要在真实场景中实现的建造过程与结果，在数字虚拟世界中预先实现，可以最大限度减少未来真实世界的遗憾。

7.1.1.3 Simulation: BIM realizes the construction processes and results that must be previously realized in real scenes in the digital virtual world, which can minimize the regrets in the real world in the future.

7.1.1.4 优化：前面的三大特征，使得设计优化成为可能，进一步保障真实世界的完美。这点对目前越来越多的复杂造型建筑设计尤其重要。

7.1.1.4 Optimization: With the above three features, design optimization is possible to further guarantee the perfection in the real world. This is especially important for more and more complex architectural designs.

7.1.1.5　出图：基于 BIM 成果的工程施工图及统计表将最大限度保障工程设计企业最终产品的准确、高质量和富于创新。

7.1.1.5　Documentation: Engineering construction drawings and statistical tables based on BIM results will ensure the accuracy, quality and innovation of final products of engineering design enterprises to the greatest extent.

7.1.2　BIM 的介入点

7.1.2　Point of BIM application

根据具体项目设计需求、BIM 团队情况、设计周期等条件，可以选择在以下不同的设计阶段中实施 BIM。

According to specific project design requirements, BIM team profiles, design cycles and other conditions, BIM can be implemented in the following different design stages.

7.1.2.1　概念设计阶段：在前期概念设计中使用 BIM，在完美表现设计创意的同时，可以进行各种面积分析、体形系数分析、商业地产收益分析、可视度分析、日照轨迹分析等。

7.1.2.1　Conceptual design: BIM is used in the preliminary conceptual design, which can perform all types of area analysis, shape coefficient analysis, commercial real estate income analysis, visibility analysis and sunshine trajectory analysis while perfectly expressing the design creativity.

7.1.2.2　方案设计阶段：此阶段使用 BIM，特别是对复杂造型设计项目将起到重要的设计优化、方案对比（例如曲面有理化设计）和方案可行性分析作用。同时建筑性能分析、能耗分析、采光分析、日照分析、疏散分析等都将对建筑设计起到重要的设计优化作用。

7.1.2.2　SD: BIM is used in this stage, especially for complex modeling design projects, which will play an important role in the design optimization, scheme comparison (for example the rational curve design) and scheme feasibility analysis. At the same time, building performance analysis, energy consumption analysis, daylight analysis, sunlight analysis and evacuation analysis will play an important role in the optimization of architectural designs.

7.1.2.3　施工图设计阶段：对复杂造型设计等用二维设计手段施工图无法表达的项目，BIM 则是最佳的解决方案。在施工图设计任务重、时间紧的情况下，不妨采用 AutoCAD + BIM 的模式，前提是基于 AutoCAD 成果用 BIM 深化设计，以尽可能保证设计质量。

7.1.2.3　Construction drawing design: BIM is the best solution for complex modeling design and other projects that cannot be expressed in construction drawings by two-dimensional design methods. In case of heavy task and tight schedule during the construction drawing design, the combination of AutoCAD and BIM may be applied, provided that BIM is used to deepen design based on AutoCAD results to ensure the highest design quality.

7.1.2.4　专业管线综合：进行大型医疗项目复杂专业管线设计，BIM 是彻底、高效解决这一难题的唯一途径。

7.1.2.4 Special pipeline design: BIM is the only way to solve this problem thoroughly and efficiently for complex special pipeline design for large-scale medical center.

7.1.2.5 可视化设计：效果图、动画、实时漫游、虚拟现实系统等项目展示手段也是 BIM 应用的一部分。

7.1.2.5 Visual design: Project display approaches, such as rendering, animation, real-time roaming and virtual reality (VR) system, are also a part of BIM applications.

7.1.3 BIM 的实施模式

7.1.3 BIM implementation mode

7.1.3.1 BIM 外包服务：将特定阶段的 BIM 设计业务外包给有实力的 BIM 服务商独立完成，设计企业在服务商的成果上进一步深化设计。在此过程中 BIM 服务商扮演的是“辅助设计”的角色。

7.1.3.1 BIM outsourcing service: BIM designs at a specific stage can be outsourced to a competent BIM service provider and then the service provider completes the corresponding service independently. The design enterprise shall deepen the design based on the outcomes of the service provider. BIM service providers play the role of auxiliary designer in this process.

7.1.3.2 BIM 项目合作设计：在项目设计的特定阶段，由设计方和 BIM 服务商合作完成设计。在此过程中 BIM 服务商更多起着技术培训、设计难题技术支持、局部参与设计的“协助设计”的作用，项目设计依然主要由设计师完成。通过多个项目、不同类型项目的 BIM 实施，达到在设计师中逐步推广普及 BIM 应用的目的。

7.1.3.2 BIM cooperative project design: At a specific stage of project design, the designer and BIM service provider can cooperate to complete the design. In this process, BIM service providers play a more important role in technical training, technical support in solving design problems, and “assisting in design” with partial participation in design. The project design is still mainly completed by the designers. The BIM application can be gradually popularized among designers through BIM implementation in multiple projects and different types of projects.

7.1.3.3 自建 BIM 团队：成立专职的 BIM 团队或部门，使用 BIM 技术服务于设计院内部所有需要 BIM 的项目。该方式的重点是团队自身的建设与技术水平的提升，在运作前期，BIM 服务商的专业培训和技术支持至关重要。

7.1.3.3 Self-built BIM team: Establish a full-time BIM team or department to use BIM technology to serve all projects that need BIM within the design institute. In particular, this mode focuses on building a team and promoting its technical competency. At the early stage of operation, the professional training and technical support of BIM service providers are really important.

7.1.4 BIM 中的协同设计

7.1.4 Collaborative design in BIM

协同设计又细分为 2D 协同设计与 3D 协同设计，BIM 设计属于 3D 协同设计。

Collaborative design is further divided into 2D collaborative design and 3D collaborative design. BIM design is 3D collaborative design.

3D 协同设计在专业内和专业间的模式不同。

The modes of 3D collaborative design are different within and between disciplines.

7.1.4.1 专业内 3D 协同设计：是一种数据级的实时协同设计模式，即工作组成员在本地计算机上对同一个 3D 工程信息模型进行设计，每个人的设计内容都可以及时同步到文件服务器上的项目中心文件中，甚至成员间还可以互相借用属于对方的某些建筑图元进行交叉设计，从而实现成员间的实时数据共享。

7.1.4.1 Intra-discipline 3D collaborative design: It is a data-level real-time collaborative design mode. In detail, members of the working group use local computers for the same 3D engineering information model design. All design outcomes of each member can be synchronized to the project center file on the file server in real time. These members can even refer to some architectural graphic elements of other members for cross design, thus realizing real-time data sharing among all members.

7.1.4.2 专业间 3D 协同设计：当每个专业都有了 3D 工程信息模型文件时，即可通过外部链接的方式，在专业模型（或系统）间进行管线综合设计，这个工作可以在设计过程中的每个关键时间点进行。

7.1.4.2 Interdisciplinary 3D collaborative design: When each discipline has a 3D engineering information model file, comprehensive pipeline design can be carried out between professional models (or systems) through external links. This task can be performed at every key point in the design process.

7.1.5 BIM 中的协同作业

7.1.5 Collaboration in BIM

协同作业是设计之外的各种设计文件、办公文档管理、人员权限管理、设计校审流程、计划任务、项目状态查询统计等与设计相关的管理功能，是设计方与业主、施工方、监理方、材料供应商、运营商等与项目相关各方进行文件交接、沟通的管理系统，是提升全产业链各环节效率的手段。

Collaboration refers to the management functions beyond designs for various documents with office document management, personnel authorization management, design review process, planning task, project status query & statistics and other design-related management functions. It involves the collaboration management system for the designer with the owner, construction department, supervision department, material supplier, operator and other project stakeholders for document exchanges and communications. It is an important way to improve the efficiency of all steps in the whole industrial chain.

7.1.6 BIM 在医疗项目中的运用

7.1.6 Application of BIM in medical projects

7.1.6.1 场地与交通组织分析

7.1.6.1 Analysis of site and traffic organization

通过 BIM 结合地理信息系统（GIS），对场地及拟建的建筑物空间数据进行建模，通过 BIM 及 GIS 软件的强大功能，迅速得出令人信服的分析结果（如土方平衡量、排水泄洪方案等)，帮助项目在规划阶段评估指定场地的使用条件和特点，从而做出新建项目最理想的场地位置、交通流线组织关系、建筑主体布局等关键决策。

Based on the combination of BIM and GIS, the spatial data of the site and proposed building are modeled. With the powerful features of BIM and GIS software, convincing analysis results (such as earthwork balance, drainage and flood discharge scheme) can be quickly obtained to help assess the conditions and characteristics of the designated site during planning, thus determining the most ideal site location, traffic streamline organization relation, main building layout and other key decisions for a new project.

7.1.6.2 模拟空间发展

7.1.6.2 Simulate space development

BIM 能够帮助项目团队在建筑规划阶段，通过对空间进行分析来理解复杂空间的标准和法规，从而节省时间，为团队提供更多增值活动的可能。特别是在客户讨论需求、选择以及分析最佳方案时，借助 BIM 及相关分析数据，可以做出关键性的决定。

BIM can help the project team to understand the standards and regulations on complex space through spatial analysis during the construction planning, thus saving time and providing the team with the possibility of performing more value-added activities. In particular, BIM and related analysis data can be used to make critical decisions while customers are discussing related requirements and choosing and analyzing the best solutions.

在建筑策划阶段，BIM 还会帮助建筑师随时查看初步设计是否符合业主的要求，是否满足建筑策划阶段得到的设计依据，通过 BIM 连贯的信息传递或追溯，大大减少设计阶段因不合理设计造成修改的巨大浪费。

During construction planning, BIM will help architects to check whether the PD has met the requirements of the Owner and the design criteria obtained at this stage anytime. Through consecutive information transmission or tracing of BIM, the huge waste caused by caused by unreasonable designs at the design stage will be greatly reduced.

7.1.6.3 评估设计方案

7.1.6.3 Evaluate the design scheme

1. 在方案论证阶段用 BIM 评估设计方案的布局、照明、安全、声学、色彩及是否符合相关规范，利用建筑外部细节来迅速分析设计和施工中可能需要应对的问题。

1. During the scheme demonstration, use BIM to evaluate the layout, lighting, safety, acoustics, color and compliance with relevant codes in the design scheme. Analyze the problems that may occur during design and construction according to external details of buildings.

以某医院某科室门诊区域的设计为例，利用 BIM 模拟测算，判别门诊设计的合理性。该科室日常常规参数如下：

Here the design of outpatient service area in a department of a certain hospital is taken as an example. We can use BIM for analog calculation to judge the rationality of outpatient service designs. The daily conventional parameters for this department are listed as follows:

（1）常规门诊量：150 人（最高峰 250 人）；

(1) Conventional outpatient capacity: 150 patients (max. 250 patients) ;

（2）峰值门诊时段：9：00～11：00（平均 1 人/5 分钟）；

(2) Peak outpatient service period: 9:00～11:00 (1 patient/5 minutes on average);

（3）平均就诊时间：20min;

(3) Average diagnosis time: 20 minutes;

（4）患者可容忍等候时间：老人 45min，中青年 30min。

(4) Tolerable waiting time for patients: 45min for the old and 30min for the young.

通过对上述数据进行模拟动态测试，可以对设计方案进行论证，具体内容包括：

Design schemes can be analyzed by performing simulated dynamic tests to the above data. The following factors are analyzed.

（1）人群是否始终或长时间处于聚集状态，从而判断整个科室诊室区域面积是否足够；

(1) Analyze whether the crowd is in a gathering state all the time or for a long time so as to judge whether the area of the whole department is sufficient;

（2）什么时间就诊人群开始聚集，聚集在何处，以此判断整个诊室区域面积、诊室数量和候诊空间的比例是否合理；

(2) Analyze when and where the patients begin to gather so as to judge whether the area of the whole clinic, the number of clinics and the proportion of waiting space are reasonable;

（3）根据诊量高峰与低谷的比例，调整部分专科门诊的开放时间，如某些慢性专科门诊高峰时段不开门，而在低谷时段开放；

(3) Adjust the opening hours of some specialty clinics according to the ratio of peak outpatient capacity to valley outpatient capacity. For example, some chronic disease clinics are not open during peak hours but they are open during valley hours.

（4）根据对患者就诊路径、就诊时间、等候时间规律的判别，考虑在诊室区域植入相关医技、治疗功能。

(4) Consider integrating related medical techniques and treatment functions according to the rules of patients on diagnosis path, diagnosis time and waiting time.

2. 通过 BIM 优化诊室设计方案，使之更高效、舒适、方便，达到诊室设计效果最佳状态。

2. BIM helps optimize clinic design schemes to make all clinics more efficient, comfortable and convenient, thus achieving the best clinic design effect.

3. 方案论证阶段借助 BIM 提供不同的解决方案以供项目投资方进行选择，通过数据对比和模拟分析，找出不同解决方案的优缺点，帮助项目投资方迅速评估建筑投资方案的成本和时间。

3. BIM can be used during scheme demonstration to provide different optional solutions for the project investors. Through data comparison and simulation analysis, the advantages and disadvantages of different solutions can be found out, helping project investors to quickly evaluate the cost and time of construction investment schemes.

4. 通过 BIM 来评估所设计的空间，可以获得较高的互动效应，以便从使用者和业主那里获得积极的反馈。设计的实时修改往往基于最终用户的反馈，在 BIM 平台下，项目各方关注的焦点问题比较容易直观地展现并迅速达成共识，相应地，决策所需的时间会比以往减少。

4. Designers can through BIM to acquire a higher level of interaction, thus obtaining positive feedback from user sand the Owner. Real-time changes of design are often based on the feedback from end users. With the help of BIM platform, it is easy to intuitively show the core issues for all stakeholders of the project and quickly reach a consensus. Accordingly, the time spent on making a decision is reduced.

7.1.6.4 可视化设计

7.1.6.4 Visualization

使用 BIM 技术，让设计师真正回归到了三维的世界，使用三维的思考方式来完成建筑设计，同时也使业主真正摆脱了技术壁垒的限制，随时了解自己的投资与回报。

The application of BIM technology helps designers really return to the three-dimensional world and use three-dimensional thinking for architectural designs. Meanwhile, it releases owners from technical barriers to know their investment and return anytime.

7.1.6.5 多专业协同设计

7.1.6.5 Multi-disciplinary collaborative design

借助于 BIM 的技术优势，协同的范畴从单纯的设计阶段扩展到建筑全生命周期，需要规划、设计、施工、运营等各方的集体参与，因此具备了更广泛的意义，从而带来综合效益的大幅提升。

With the technical strength of BIM, collaborations are extended from design to the entire lifecycle of buildings. This requires the extensive participation of planning, design, construction and operation. In this more, collaboration has a greater significance, thus bringing about a substantial increase in comprehensive benefits.

7.1.6.6 建筑性能化分析

7.1.6.6 Building performance analysis

利用 BIM 技术，建筑师在设计过程中创建的虚拟建筑模型已经包含了大量的设计信息（包括几何信息、材料性能、构件属性等），只要将模型导入相关的性能化分析软件，就可以得到相应的分析结果，原本需要专业人士花费大量时间输入大量专业数据的过程，如今可以自动完成，这大大降低了性能化分析的周期，提高了设计质量，同时也使设计公司能够向业主提供更专业的技能和服务。

With the application of BIM technology, the virtual building model created by architects during design process has already contained a large amount of design information (including geometric information, material performance and component attributes). As long as the model is imported into relevant performance analysis software, the corresponding analysis results can be obtained. The process that originally required professionals to spend a lot of time on importing a large amount of professional data can now be completed automatically, which greatly shortens the cycle of performance analysis, improves the design quality, and also enables the design company to provide more professional skills and services to the owner.

7.2 项目管理与运维 BIM

7.2 Project management and operation & maintenance BIM

企业管理属于一个循环的过程，而项目管理指的是项目活动中循环运用专门的知识、技能、工具和方法。其核心任务是目标控制，通常能涉及三个主要的目标：质量、成本、进度。

Enterprise management is a cycle, while project management refers to the cycle of expertise, skills, tools and methods in project activities. Target control is the core task. In general, three major targets, including quality, cost and progress, are involved.

7.2.1 质量管理

7.2.1 Quality management

全面质量管理过程遵循 PDCA 循环［PDCA 即计划（Plan）、执行（Do）、检查（Check）、处理（Act）］. BIM 技术与 PDCA 结合，能更好地为工程项目的质量管理服务。

The whole process of total quality management follows PCDA cycle (PCDA including plan, do, check and act). In combination with PDCA, BIM can better serve the quality management of engineering projects.

7.2.1.1 P 阶段在 BIM 环境下，统一明确任务和目标，分别制订专业质量计划，同时保证计划的连续性、严谨性，进一步保证项目的经济合理性。

7.2.1.1 Under the BIM environment for P (Plan), specify tasks and objectives, respectively work out special quality plans, ensure the consistency and accuracy of quality assurance plans and further ensure the economic efficiency of projects.

7.2.1.2 D 阶段由于 BIM 的可视性，相互之间的沟通更加明确，一些问题更加明了，从而保证项目的正常进行。

7.2.1.2 At the stage of D (Do), the communication is clearer and some problems can be seen more easily due to the visibility of BIM, thus ensuring the normal project progress.

7.2.1.3 C 阶段由于 BIM 的模拟性，可以将二维设计图纸转化为三维模型，纠正偏差，使设计等达到质量要求。

7.2.1.3 At the stage of C (Check), the simulation capacity of BIM can convert 2D design drawings into 3D design drawings to correct deviations and make the design meet quality requirements.

7.2.1.4 A 阶段 BIM 技术在处置阶段，主要是通过协作平台，发布指令，随时监控处理情况，及时有效处理问题。

7.2.1.4 At the stage of A (Act), BIM can issue instructions through the collaboration platform, monitor the handling anytime and handle any problem timely and effectively.

7.2.2 成本管理

7.2.2 Cost management

成本管理共分为两个阶段：计划阶段和合同实施阶段。

Cost management is divided into two steps, including planning and contract implementation.

计划阶段主要是对成本进行估算，在 BIM 软件中，输入楼层信息，定义构件参数，然后进行三维模型的建立，可以进行准确的工程估算、概算，同时发现设计中的若干问题，再进行设计优化。BIM 模型能够将构件的各种信息分类统计并输出报告，同时也可以将信息文件云共享，提高了工作的准确性和效率。

Cost is mainly estimated during planning. Import floor information into BIM software, define component parameters and then perform three-dimensional modeling for the accurate engineering estimation and budget. Meanwhile, some design problems can be found and then corresponding designs can be optimized. BIM can classify and count different types of component information and export a report. Meanwhile, BIM can share the information file online, thus increasing the accuracy and efficiency.

在合同实施阶段，运用 BIM 软件导出的成本，当发生工程变更时，可及时变更信息修改模型，计算工程变更量再导出成本，可以动态地监测当前的成本，帮助工程师做出正确及时的决策。BIM 软件还可以进行成本分析，从而多角度地分析造价问题。

While performing a contract, use BIM software to export costs. When any change occurs to the project, BIM can timely update the information and model. Calculate the amount of engineering change and then export the cost. BIM can dynamically monitor the current cost and help engineers make correct and instant decisions. BIM software can also be used for cost analysis, thus analyzing cost in different views.

7.2.3 进度管理

7.2.3 Schedule control

7.2.3.1 制订进度计划

7.2.3.1 Construction schedule formulation

通过 BIM 软件导出工程量清单，套用工程量清单和计价定额，计算出各阶段的人、材、机的消耗量，然后与各方协调，进行流水计算。

BIM software is used to export the bill of quantities. Then calculate the consumption of laborers, materials and machines at different stages according to the bill of quantities and pricing quota and then coordinate with all stakeholders for the calculation of frequency of continuous construction.

7.2.3.2 控制进度计划

7.2.3.2 Schedule control

运用 BIM 软件导出进度计划，同时查看 4D 模拟施工，为下一工序做好充分准备。通过 GPS 定位和现场测量确定准确坐标，然后通过网络设备传回到基于 BIM 的 4D 施工进度跟踪与控制系统，从而准确及时地衡量工程进度。

Use the BIM software to export the schedule and view 4D simulated construction at the same time to make full preparations for the next process. Accurate coordinates are determined through GPS positioning and field measurement and then transmitted to the BIM-based 4D construction schedule tracking and control system through network equipment, thus controlling the project progress accurately and timely.

7.2.3.3 调整进度计划

7.2.3.3 Schedule adjustment

当发生工程变更或其他影响进度的情况时，及时将信息输入到 BIM 中，这样管理者能迅速得到变更的工程量或者资金，并马上做出安排。就在这样不断的对比过程中，各方能更好地把握项目的进展。

Timely import the corresponding information into BIM so that the management personnel can quickly acquire the changed quantities or funds and then make instant arrangements. After continuous comparisons, all the stakeholders can better control the project progress.

7.2.4 BIM 运维

7.2.4 BIM operation & maintenance

BIM 不是一个简单的医院建筑数字模式，它更是一个数字化的信息平台。

BIM is not a simple digital model for hospital buildings, but a digital information platform.

例如，在医院日常运营中，监控系统可以自动发现某个水泵控制阀门出现故障，查阅在库存记录中已无该阀门配件，于是提出采购申请—财务审核—主管领导审批—采购—安装（维修清单）—设备信息重新录入—最后重新进入设备运营监测。

For example, the monitoring system can automatically find the failure of a certain water pump control valve during the daily operation of hospital. If there is no spare part for this valve in the inventory records, follow the sequence of purchase application, financial audit, approval by competent leaders, purchase, installation (maintenance list), re-entry of equipment information and re-entry of equipment under operation monitoring.

整个过程涵盖了楼宇自动化系统、物业管理系统、财务系统、资源管理系统、ERP 系统等，这一切都建立在 BIM 的基础上，将原有离散的控制系统、执行系统和决策系统整合在 BIM 的平台上。

The entire process has covered building automation system, property management system, financial system, resource management system and ERP system, which are based on BIM. The original discrete control system, implementation system and decision system are integrated onto the BIM platform.

7.2.4.1 运营信息集成

7.2.4.1 Operation information integration

BIM 模型结合运维管理系统可以充分发挥空间定位和数据记录的优势，合理制订维护计划，分配专人专项维护工作，以降低建筑物使用过程中突发状况的维修风险的次数。对一些重要设备还可以跟踪维护工作的历史记录，以便对设备的适用状态提前做出判断，提高维护效率。

In combination with operation & maintenance management system, BIM gives spatial positioning and data recording into full play, reasonably works out maintenance plans and allocates special personnel for maintenance to reduce maintenance risks during building operation. For some important equipment, it can track the historical record of maintenance to judge the applicability of equipment in advance, promoting the maintenance efficiency.

7.2.4.2 设施及资产管理

7.2.4.2 Facility and asset management

BIM 中包含的大量建筑信息能够顺利导入现有的资产管理系统，这对于资产管理而言，大大减少了系统初始化在数据准备方面的时间及人力投入。此外由于传统的资产管理系统本身无法准确定位资产位置，通过 BIM 结合 RFID 的资产标签芯片还可以使资产在建筑物中的定位及相关参数信息一目了然，实现精确定位，快速查询。

A large amount of building information contained in BIM can be successfully imported into the current asset management system, which greatly reduces the time and labor cost in data preparation for the initialization for asset management system. The traditional asset management system cannot accurately locate assets. In combination with RFID asset label chipsets, the location and related parameter information of assets in the building can be clearly seen, thus achieving accurate positioning and quick search.

7.2.4.3 辅助能源管理

7.2.4.3 Auxiliary energy management

BIM 模型可以验证建筑物是否按照特定的设计规定和可持续标准建造，可以通过模拟更换整栋建筑所使用的材料设备，创建假设的解决方案，来显示建筑物性能更好或更差的状态。通过这些分析模拟，最终确定、修改系统参数甚至系统改造计划，以提高整个建筑的性能。

BIM will verify whether buildings are completed according to specific design regulations and sustainable standards, display better or worse performance of buildings by simulating the replacement of materials and equipment used in the whole building and creating hypothetical solutions. We can finalize and modify system parameters and even the system renovation plan to promote the performance of the whole building based on above analysis and simulation.

7.2.4.4 空间管理

7.2.4.4 Space management

通过 BIM 模型结合空间追踪系统可以合理分配建筑物空间，追踪当前空间的使用情况，确保设施空间资源最大利用率，还能根据统计数据协助日后空间改造时的空间使用需求。

The combination of BIM and space tracking system can reasonably allocate building space, track the current space utilization, ensure the maximum utilization rate of facility space resources, and also assist the space utilization requirements in future space renovation according to statistical data.

7.2.4.5 灾害应急模拟分析

7.2.4.5 Disaster emergency simulation analysis

1. 利用 BIM 模型及相应灾害分析模拟软件，可以在灾害发生前以模型和灾害预警信息为基础，模拟灾害发生的过程，分析灾害发生的原因，制定避免灾害发生的解决措施，以及发生灾害后人员疏散、救援支持的应急预案。

1. BIM and corresponding disaster analysis & simulation software can be used to simulate the disaster process based on the model and disaster early warning information before the disaster breaks out, analyze the causes of the disaster, work out solutions to avoid the disaster, and formulate emergency plans for evacuation and rescue support after the disaster occurs.

2. 当灾害发生后，BIM 模型可以提供救援人员紧急状况点的完整信息。此外楼宇自动化系统能及时获取建筑物及设备的状态信息，通过 BIM 和楼宇自动化系统的结合，使得 BIM 模型能清晰地呈现出建筑物内部紧急状况的位置，甚至到紧急状况点最合适的路线，救援人员可以由此做出正确的现场处置，提高应急行动的成效。

2. When a disaster breaks out, BIM can provide complete information on the emergency points of rescue personnel. In addition, the building automation system can timely acquire the status information of buildings and equipment. In collaboration with the building automation system, BIM can clearly show the location of emergency situations in buildings and even the most suitable route to emergency points, so that rescue personnel can correctly handle an emergency and improve the effect of emergency actions.

7.3 BIM 软件的选择

7.3 BIM software selection

BIM 软件工具从设计阶段、施工阶段到营运管理，不同阶段都有不同领域的需求，非单一软件所能负荷。主流 BIM 软件见表 7-1。

BIM software has requirements for the design time, constructure phase and operation management that involve different fields. These features cannot be implemented through a single software program. Mainstream BIM software programs are listed as shown in Table 7-1.

表 7-1 主流 BIM 软件

Table 7-1 Mainstream BIM software programs

BIM 软件名称 BIM Software Name	软件厂商 Software Manufacture	BIM 应用方向 BIM Application
Revit Architecture/CAD	Autodesk	概念设计，可行性研究，建筑设计 Concept design, feasibility research, architectural design

续表

BIM 软件名称 BIM Software Name	软件厂商 Software Manufacture	BIM 应用方向 BIM Application
Revit Structure	Autodesk	结构设计 Constructional design
AutoCad MEP/ Revit MEP	Autodesk	MEP 设计，加工图 MEP design, operation drawing
Navisworks Manage	Autodesk	碰撞检查，施工管理 Clash detection, construction managemen
QTO	Autodesk	工程数量 Project quantity
Bently BIM Suite	Bently	概念设计，多专业 Concept design, multi-disciplinary
Project Wise Navigator	Bently	碰撞检查，时程规划，施工管理 Clash detection, scheduling, construction management
Digital Project Designer	Gehry Technologies	多专业建模，结构分析，协调与施工管理 Multi-disciplinary modeling, structural analysis, coordination and construction management
Digital Project MEP, System Routing	Gehry Technologies	MEP 设计，模型协调，加工图 MEP design, model coordination, operation drawing
SketchUp Pro	Google	概念设计，多专业 Concept design, multi-disciplinary
ArchiCAD	Graphisoft	概念设计，建筑 + MEP Concept design, architecture + MEP
MEP Modeler	Graphisoft	MEP 设计 MEP design
Vectorworks	Nemetschek	建筑设计 Architectural design
Fastrak	CSC(UK)	结构设计 Constructional design
SDS/2	Design Data	结构设计，加工图 Constructional design, operation drawing
RISA	RISA Technologies	结构设计 Constructional design
Tekla Structures	Tekla	概念设计，结构设计，加工图，时程规划 Concept design, constructional design, operation drawing, scheduling
Cadpipe HVAC	AEC design Group	MEP 设计 MEP design
Fabrication for ACAD MEP	East Coast CAD/CAM	MEP 设计 MEP design
CAD-Duct	Micro Application Packages Ltd.	MEP 设计，加工图 MEP design, operation drawing

续表

BIM 软件名称 BIM Software Name	软件厂商 Software Manufacture	BIM 应用方向 BIM Application
DuctDesigner 3D PipeDesigner 3D	QuickPEn International	MEP 设计，预制加工 MEP design, pre-processing
HydraCAD	Hydratec	消防设计 Fire protection design
AutoSPRINK VR	M. E. P. CAD	消防设计 Fire protection design
FireCad	Mc4 Software	消防设计 Fire protection design

软件选用原则：优先考虑项目特性、建筑物用途、业主的需求，再考虑预算、软件特色以及兼容性，最后选择业界的领导品牌（市占有率高者）。

Principles of software selection: Priority should be given to the project characteristics, use of buildings and requirements of owners. Then, the budget, software features and compatibility should be considered. On that basis, industry-leading brands (with large market share) can be selected.

7.3.1 民用建筑（多专业）设计，可选用 AutodeskRevit;

7.3.1 For design of civil building (multi-disciplinary), AutodeskRevit can be selected;

7.3.2 工业工程、公共建设、医疗保健或基础设施设计，可选用 Bentley、Revit;

7.3.2 For design of industrial engineering, public construction, health care or infrastructure, Bentley and Revit can be selected;

7.3.3 建筑师事务所，可选择 ArchiCAD、Revit 或 Bentley;

7.3.3 For architects offices, ArchiCAD, Revit or Bentley can be selected;

7.3.4 专业度、困难度较高的工程且预算较充裕的项目，可选用较高阶的 DigitalProject 或 CATIA.

7.3.4 For professional projects that is complex and with sufficient budget, the advanced DigitalProject or CATIA can be selected.

参考文献

References

［1］马丽平. 现代医院管理实务［M］. 北京：中国商务出版社，2016.

［1］Ma Liping. Practice of Modern Hospital Management［M］. Beijing: China Commerce and Trade Press, 2016.

［2］沈崇德，朱希. 医院建筑医疗工艺设计［M］. 北京：研究出版社，2018.

［2］Shen Chongde, Zhu Xi. Design for Medical Technology of Hospital Buildings［M］. Beijing: Research Press, 2018.

［3］《中国医院建设指南》编撰委员会. 中国医院建设指南（上中下册）［M］.（4 版）. 北京：研究出版社，2019.

［3］Editorial committee of Guideline of Chinese Hospital Construction. Guideline of Chinese Hospital Construction (Volumes Ⅰ, Ⅱ and Ⅲ)［M］. (the fourth edition). Beijing: Research Press, 2019.

［4］江观全，王圣通. 如何开展医院工程项目管理［J］. 中国医院建筑与装备，2013，88-91.

［4］Jiang Guanquan, Wang Shengtong. How to Do the Hospital Project Management［J］. Chinese Hospital Architecture & Equipment, 2013, 88-91.

［5］中华人民共和国国家卫生和计划生育委员会. 综合医院建筑设计规范：GB 51039—2014［S］. 北京：中国计划出版社，2015.

［5］National Health and Family Planning Commission of PRC. Code for Design of General Hospital: GB 51039—2014［S］. Beijing: China Planning Press, 2015.

［6］中华人民共和国住房和城乡建设部，中华人民共和国国家质量监督检验检疫总局. 传染病医院建筑设计规范：GB 50849—2014［S］. 北京：中国计划出版社，2015.

［6］Ministry of Housing and Urban-Rural Development of PRC, General Administration of Quality Supervision, Inspection and Quarantine of PRC. Code for Design of Infectious Diseases Hospital: GB 50849—2014［S］. Beijing: China Planning Press, 2015.

［7］中华人民共和国住房和城乡建设部，中华人民共和国国家质量监督检验检疫总局. 医院洁净手术部建筑技术规范：GB 50333—2013［S］. 北京：中国建筑工业出版社，2014.

［7］Ministry of Housing and Urban-Rural Development of PRC, General Administration of Quality Supervision, Inspection and Quarantine of PRC. Architectural Technical Code for Hospital Clean Operating Department: GB 50333—2013［S］. Beijing: China Architecture & Building Press, 2014.

[8] 中华人民共和国住房和城乡建设部，中华人民共和国国家质量监督检验检疫总局. 急救中心建筑设计规范：GB 50939—2013 [S]. 北京：中国计划出版社，2014.

[8] Ministry of Housing and Urban-Rural Development of PRC, General Administration of Quality Supervision, Inspection and Quarantine of PRC. Code for Design of Emergency Center: GB 50939—2013 [S]. Beijing: China Planning Press, 2014.

[9] 中华人民共和国住房和城乡建设部，中华人民共和国国家质量监督检验检疫总局. 精神专科医院建筑设计规范：GB 51058—2014 [S]. 北京：中国计划出版社，2015.

[9] Ministry of Housing and Urban-Rural Development of PRC, General Administration of Quality Supervision, Inspection and Quarantine of PRC. Code for Design of Psychiatric Hospital: GB 51058—2014 [S]. Beijing: China Planning Press, 2015.

[10] 中华人民共和国国家质量监督检验检疫总局，国家标准化管理委员会. 医用空气加压氧舱：GB/T 12130—2005 [S]. 北京：中国标准出版社，2006.

[10] General Administration of Quality Supervision, Inspection and Quarantine of PRC, Standardization Administration. Code for Design of Hyperbaric Oxygen Chamber: GB12130—2005 [S]. Beijing: Standards Press of China, 2006.

[11] 中国中元国际工程有限公司. 新型冠状病毒感染的肺炎传染病应急医疗设施设计标准：T/CECS 661—2020 [S]. 北京：中国建筑工业出版社，2020.

[11] China Ippr International Engineering Co., Ltd. The Design Standard of Infectious Disease Emergency Medical Facilities for Novel Coronavirus Pneumonia: T/CECS 661—2020 [S]. Beijing: China Architecture & Building Press, 2020.

[12] 中华人民共和国住房和城乡建设部，中华人民共和国国家质量监督检验检疫总局. 建筑工程抗震设防分类标准：GB 50223—2008 [S]. 北京：中国建筑工业出版社，2008.

[12] Ministry of Housing and Urban-Rural Development of PRC, General Administration of Quality Supervision, Inspection and Quarantine of PRC. Standard for Classification of Seismic Protection of Building Constructions: GB 50223—2008 [S]. Beijing: China Architecture & Building Press, 2008.

[13] 广东省住房和城乡建设厅. 建筑结构荷载规范：DBJ 15—101—2014 [S]. 北京：中国建筑工业出版社，2015.

[13] Department of Housing and Urban-Rural Development of Guangdong Province. Load Code for the Design of Building Structures: DBJ 15—101—2014 [S]. Beijing: China Architecture & Building Press, 2015.

[14] 中华人民共和国住房和城乡建设部，中华人民共和国国家质量监督检验检疫总局. 建筑抗震设计规范：GB 50011—2010 [S]. 北京：中国建筑工业出版社，2016.

[14] Ministry of Housing and Urban-Rural Development of PRC, General Administration of Quality Supervision, Inspection and Quarantine of PRC. Code for Seismic Design of Buildings: GB 50011—2010 [S]. Beijing: China Architecture & Building Press, 2016.

[15] 中华人民共和国住房和城乡建设部，中华人民共和国国家质量监督检验检疫

总局. 供配电系统设计规范：GB 50052—2009 [S]. 北京：中国计划出版社，2010.

[15] Ministry of Housing and Urban-Rural Development of PRC, General Administration of Quality Supervision, Inspection and Quarantine of PRC. Code for Design Electric Power Supply Systems: GB 50052—2009 [S]. Beijing: China Planning Press, 2010.

[16] 中华人民共和国住房和城乡建设部，中华人民共和国国家质量监督检验检疫总局. 建筑照明设计标准：GB 50034—2013 [S]. 北京：中国建筑工业出版社，2014.

[16] Ministry of Housing and Urban-Rural Development of PRC, General Administration of Quality Supervision, Inspection and Quarantine of PRC. Standard for Lighting Design of Buildings: GB 50034—2013 [S]. Beijing: China Architecture & Building Press, 2014.

[17] 中华人民共和国住房和城乡建设部，中华人民共和国国家质量监督检验检疫总局. 低压配电设计规范：GB 50054—2011 [S]. 北京：中国计划出版社，2012.

[17] Ministry of Housing and Urban-Rural Development of PRC, General Administration of Quality Supervision, Inspection and Quarantine of PRC. Code for Design of Low Voltage Electrical Installations: GB 50054—2011 [S]. Beijing: China Planning Press, 2012.

[18] 中华人民共和国住房和城乡建设部，中华人民共和国国家质量监督检验检疫总局. 综合布线系统工程设计规范：GB 50311—2016 [S]. 北京：中国计划出版社，2017.

[18] Ministry of Housing and Urban-Rural Development of PRC, General Administration of Quality Supervision, Inspection and Quarantine of PRC. Code for Engineering Design of Generic Cabling System: GB 50311—2016 [S]. Beijing: China Planning Press, 2017.

[19] 中华人民共和国住房和城乡建设部，中华人民共和国国家质量监督检验检疫总局. 智能建筑设计标准：GB 50314—2015 [S]. 北京：中国计划出版社，2015.

[19] Ministry of Housing and Urban-Rural Development of PRC, General Administration of Quality Supervision, Inspection and Quarantine of PRC. Standard for Design of Intelligent Building: GB 50314—2015 [S]. Beijing: China Planning Press, 2015.

[20] 中华人民共和国住房和城乡建设部. 医疗建筑电气设计规范：JGJ 312—2013 [S]. 北京：中国建筑工业出版社，2014.

[20] Ministry of Housing and Urban-Rural Development of PRC. Code for Electrical Design of Medical Buildings: JGJ 312—2013 [S]. Beijing: China Architecture & Building Press, 2014.

[21] 中华人民共和国住房和城乡建设部，中华人民共和国国家质量监督检验检疫总局. 建筑物防雷设计规范：GB 50057—2010 [S]. 北京：中国计划出版社，2011.

[21] Ministry of Housing and Urban-Rural Development of PRC, General Administration of Quality Supervision, Inspection and Quarantine of PRC. Code for Design Protection of Structures against Lightning: GB 50057—2010 [S]. Beijing: China Planning Press, 2011.

[22] 中华人民共和国住房和城乡建设部，中华人民共和国国家质量监督检验检疫总局. 建筑机电工程抗震设计规范：GB 50981—2014 [S]. 北京：中国建筑工业出版社，2015.

[22] Ministry of Housing and Urban-Rural Development of PRC, General Administration of Quality Supervision, Inspection and Quarantine of PRC. Code for Seismic Design of Mechanical and Electrical Equipment: GB 50981—2014 [S]. Beijing: China Architecture & Building Press, 2015.

[23] 中华人民共和国住房和城乡建设部，中华人民共和国国家市场监督管理总局. 建筑给水排水设计标准：GB 50015—2019 [S]. 北京：中国计划出版社，2020.

[23] Ministry of Housing and Urban-Rural Development of PRC, State Administration for Market Regulation of PRC. Standard for Design of Building Water Supply and Drainage: GB 50015—2019 [S]. Beijing: China Planning Press, 2020.

[24] 北京市建筑设计研究院，等. 医院污水处理设计规范：CECS 07: 2004 [S]. 北京：中国计划出版社，2004.

[24] Beijing Institute of Architectural Design, et al. Code for Design of Hospital Sewage Treatment: CECS 07: 2004 [S]. Beijing: China Planning Press, 2004.

[25] 中华人民共和国建设部，中华人民共和国国家质量监督检验检疫总局. 人民防空地下室设计规范：GB 50038—2005 [S]. 北京：中国计划出版社，2006.

[25] Ministry of Development of PRC, General Administration of Quality Supervision, Inspection and Quarantine of PRC. Code for Design of Civil Air Defense Basement: GB 50038—2005 [S]. Beijing: China Planning Press, 2006.

[26] 中华人民共和国住房和城乡建设部. 民用建筑绿色设计规范：JGJ/T 229—2010 [S]. 北京：中国建筑工业出版社，2011.

[26] Ministry of Housing and Urban-Rural Development of PRC. Code for Green Design of Civil Buildings: JGJ/T 229—2010 [S]. Beijing: China Architecture & Building Press, 2011.

[27] 黄中. 医院通风空调设计指南 [M]. 北京：中国建筑工业出版社，2019.

[27] Huang Zhong. Guidelines for Design of Hospital Ventilation and Air Conditioning [M]. Beijing: China Architecture & Building Press, 2019.

[28] 中华人民共和国住房和城乡建设部，中华人民共和国国家质量监督检验检疫总局. 医用气体工程技术规范：GB 50751—2012 [S]. 北京：中国计划出版社，2012.

[28] Ministry of Housing and Urban-Rural Development of PRC, General Administration of Quality Supervision, Inspection and Quarantine of PRC. Technical Code for Medical Gases Engineering: GB 50751—2012 [S]. Beijing: China Planning Press, 2012.

[29] 中华人民共和国住房和城乡建设部，中华人民共和国国家质量监督检验检疫总局. 建筑信息模型应用统一标准：GB/T 51212—2016 [S]. 北京：中国建筑工业出版社出版，2017.

[29] Ministry of Housing and Urban-Rural Development of PRC, General Administration of Quality Supervision, Inspection and Quarantine of PRC. Unified Standard for Building Information Modeling: GB/T 51212—2016 [S]. Beijing: China Architecture & Building Press, 2017.

[30] 王青. BIM 在建筑设计中的实例应用 [J]. 中国医院建筑与装备，2014 (01).

[30] Wang Qing. Application of BIM Technology in Architectural Design [J]. Chinese Hospital Architecture & Equipment, 2014(01).

[31] 苏元颖. BIM 在医院建设和运营中的作用及实施 [J]. 中国医院建筑与装备, 2014 (01).

[31] Su Yuanying. Role and Implementation of BIM in Hospital Construction and Operation [J]. Chinese Hospital Architecture & Equipment, 2014(01).

[32] 苏元颖, 彭阳, 田佩龙. 基于 BIM 的运维管理研究与应用综述 [J]. 图学学报, 2015, 36 (5).

[32] Su Yuanying, Peng Yang, Tian Peilong. A Review for Researches and Applications of BIM-Based Operation and Maintenance Management [J]. Journal of Graphics, 2015, 36(5).